Functional Analysis

Functional Analysis

GEORGE BACHMAN and LAWRENCE NARICI

DEPARTMENT OF MATHEMATICS
POLYTECHNIC INSTITUTE OF BROOKLYN
BROOKLYN, NEW YORK

ACADEMIC PRESS   New York   San Francisco   London

A Subsidiary of Harcourt Brace Jovanovich, Publishers

ACADEMIC PRESS, INC.
111 Fifth Avenue, New York, New York 10003

United Kingdom Edition published by
ACADEMIC PRESS, INC. (LONDON) LTD.
24/28 Oval Road, London NW1

LIBRARY OF CONGRESS CATALOG CARD NUMBER: 65-26389

PRINTED IN THE UNITED STATES OF AMERICA

Preface

This book is meant to serve as an introduction to functional analysis and can be read by anyone with a background in undergraduate mathematics through the junior year. The most important courses in this undergraduate background are undoubtedly linear algebra and advanced calculus. For the convenience of the reader, however, a brief glossary of definitions and notations relevant to linear algebra is supplied in Sec. 1.1. The only exception to these prerequisites occurs in a few examples discussed in various places throughout the text. These examples are meant for the reader with a little more mathematical maturity — in particular, with some knowledge of real variables. The beginner, however, will lose nothing relevant to the main theme of development if he simply omits these examples on a first reading.

We have also attempted to make the proofs of all theorems as detailed as possible and in several places have given slightly longer proofs, mainly because we felt that they were clearer and demanded less mathematical background than certain shorter versions.

The exercises in most instances are meant to test the reader's understanding of the text material and should be considered as an important part of the book. In a few instances, the exercises are intended to push the reader's knowledge a bit further and appropriate references are given.

Thus it is hoped that the book will present a careful, detailed introductory treatment of functional analysis to the advanced undergraduate or beginning graduate mathematics major as well to those physics and engineering majors who are finding increasing applications of functional analysis in their disciplines. A second volume including such topics as topological vector spaces, Banach algebras and applications, and operator topologies is envisioned for those who might desire to pursue these matters further.

Last, it is our pleasant privilege to thank some of our many friends and colleagues who encouraged us in this undertaking. In particular, our thanks go to M. Maron and H. Allen who read the original lecture notes from which this book developed and who brought many corrections and improvements to our attention.

v

We would also like to single out Professor George Sell who read the entire manuscript with great care and made many helpful suggestions concerning content, presentation of certain proofs, and overall clarity. His perspicacious comments were most helpful and have, we feel, enhanced the development of certain sections of the book considerably.

Brooklyn, New York   G.B.

L.N.

Contents

Functional Analysis

CHAPTER **1**

Introduction to Inner Product Spaces

In this chapter, some of the conventions and terminology that will be adhered to throughout this book will be set down. They are listed only so that there can be no ambiguity when terms such as "vector space" or "linear transformation" are used later, and are not intended as a brief course in linear algebra.

After establishing our rules, certain basic notions pertaining to inner product spaces will be touched upon. In particular, certain inequalities and equalities will be stated. Then a rather natural extension of the notion of orthogonality of vectors from the euclidean plane is defined. Just as choosing a mutually orthogonal basis system of vectors was advantageous in two- or three-dimensional euclidean space, there are also certain advantages in dealing with orthogonal systems in higher dimensional spaces. A result proved in this chapter, known as the Gram–Schmidt process, shows that an orthogonal system of vectors can always be constructed from any denumerable set of linearly independent vectors.

Focusing then more on the finite-dimensional situation, the notion of the *adjoint* of a linear transformation is introduced and certain of its properties are demonstrated.

1.1 Some Prerequisite Material and Conventions

The most important structure in functional analysis is the *vector space (linear space)*—a collection of objects, X, and a field F, called *vectors* and *scalars*, respectively, such that, to each pair of vectors x and y, there corresponds a third vector $x + y$, the *sum* of x and y in such a way that X constitutes an abelian group with respect to this operation. Moreover, to each pair α and x, where α is a scalar and x is a vector, there corresponds a vector αx, called the *product* of α and x, in such a way that

$$\alpha(\beta x) = (\alpha\beta)x \qquad (\alpha, \beta \text{ scalars}, x \text{ a vector}), \qquad (1.1)$$

$$1x = x \qquad (x \text{ any vector}, 1 \text{ the identity of } F). \qquad (1.2)$$

For scalars α, β and vectors x, y, we demand

$$(\alpha + \beta)x = \alpha x + \beta x, \qquad (1.3)$$

$$\alpha(x + y) = \alpha x + \alpha y. \qquad (1.4)$$

We shall often describe the above situation by saying that X is a vector space *over F* and shall rather strictly adhere to using lower-case italic letters (especially $x, y, z, \ldots$) for vectors and lower-case Greek letters for scalars. The additive identity element of the field will be denoted by 0, and we shall also denote the additive identity element of the vector addition by 0. We feel that no confusion will arise from this practice, however. A third connotation that the symbol 0 will carry will be the transformation of one vector space into another and mapping every vector of the first space into the zero vector of the second.

The multiplicative identity of the field will be denoted by 1 and so will that transformation of a vector space into itself which maps every vector into itself.

A collection of vectors $x_1, x_2, \ldots, x_n$ is said to be *linearly independent* if the relation

$$\alpha_1 x_1 + \alpha_2 x_2 + \cdots + \alpha_n x_n = 0 \qquad (1.5)$$

implies that each $\alpha_i = 0$; in other words, no linear combination of linearly independent vectors whatsoever (nontrivial, of course) can yield the zero vector. An arbitrary collection of vectors is said to be *linearly independent* if every finite subset is linearly independent.

A linearly independent set of vectors with the property that any vector can be expressed as a linear combination of some subset is called a *basis* (for the vector space) or sometimes a *Hamel basis*. Note that a linear combination is always a *finite* sum, which means that, even though there may be an infinite number of vectors in the basis, we shall never express a vector as an *infinite* sum—the fact of the matter being that infinite sums are meaningless unless the notion of "limit of a sequence of vectors" has somehow been introduced to the space. Thus if B is a basis for X and if x and y are two distinct vectors, the typical situation would be the existence of basis vectors $z_1, z_2, ..., z_n$ and scalars $\alpha_1, \alpha_2, ..., \alpha_n$ such that

$$x = \alpha_1 z_1 + \alpha_2 z_2 + \cdots + \alpha_n z_n$$

and basis vectors $w_1, w_2, ..., w_m$ and scalars $\beta_1, \beta_2, ..., \beta_m$ such that

$$y = \beta_1 w_1 + \beta_2 w_2 + \cdots + \beta_m w_m.$$

A subset of a vector space that is also a vector space, with respect to the same operations, is called a *subspace*. Given any collection U of vectors, there is always a minimal† subspace containing them called the *subspace spanned by* U or the *subspace generated by* U and denoted by placing square brackets about the set: $[U]$.

In general, if a collection of vectors, linearly independent or not, is such that any vector in the space can be expressed as a linear combination of them, the collection is said to *span the space*.

A collection of subspaces $M_1, M_2, ..., M_n$ is said to be *linearly independent* if, for any $i = 1, 2, ..., n$,

$$M_i \cap (M_1 + M_2 + \cdots + M_{i-1} + M_{i+1} + \cdots + M_n) = \{0\},$$

where by $M_1 + M_2 + \cdots + M_{i-1} + M_{i+1} + \cdots + M_n$ we mean all sums $x_1 + \cdots + x_{i-1} + x_{i+1} + \cdots x_n$, where $x_j \in M_j$. Assuming that things take place in the vector space X if $M_1, M_2, ..., M_n$ are linearly independent and

$$X = M_1 + M_2 + \cdots + M_n,$$

the subspaces $\{M_i\}$ are said to form a *direct sum decomposition for* X, a situation we shall denote by writing

$$X = M_1 \oplus M_2 \oplus \cdots \oplus M_n. \tag{1.6}$$

Equation (1.6) is equivalent to requiring that every vector in X can be written *uniquely* in the form $x_1 + x_2 + \cdots + x_n$, where $x_j \in M_j$.

One will sometimes see another adjective involved in the description of the above situation—it is often called an *internal* direct sum decomposition. This is because a new vector space V can be formed from two others X and Y (over the same field); this new one is called the *external direct sum of X and Y*. What is done in this case

---

† By *minimal* we mean that any other subspace containing the vectors must also contain that subspace.

is to introduce vector operations to the cartesian product of X and Y, the set of all ordered pairs $\langle x, y \rangle$, where $x \in X$ and $y \in Y$. We define

$$\alpha_1 \langle x_1, y_1 \rangle + \alpha_2 \langle x_2, y_2 \rangle = \langle \alpha_1 x_1 + \alpha_2 x_2, \alpha_1 y_1 + \alpha_2 y_2 \rangle.$$

By identifying the vector $\langle x, 0 \rangle$ in V with the vector x in X, we can indulge in a common but logically inaccurate practice of viewing X as a subspace of V (similarly, Y can be viewed as a subspace of V) and throw this back into the case of the internal direct sum. Adopting this convention then, given a direct sum decomposition, we need not bother to enquire: Are the terms in the direct sum distinct vector spaces or are they subspaces of the same vector space?

Whenever the word "function," "mapping," or "transformation" is used, it will be tacitly assumed that it is well-defined (single-valued); symbolically, if f is a mapping defined on a set D (the *domain* of f), we require that $x = y$ implies $f(x) = f(y)$. If the converse is true, that is, if $x \neq y$ implies $f(x) \neq f(y)$ (distinct points have distinct images), we say that f is 1 : 1. In the case when f maps D into a set Y (denoted by $f : D \to Y$) and the *range* of f, $f(D) = Y$, we say that f is *onto* Y. Given a mapping A which maps a vector space X into a vector space Y (over the same field),

$$A : X \to Y,$$

with the property that for any scalars α and β and for any vectors x and y,

$$A(\alpha x + \beta y) = \alpha A(x) + \beta A(y),$$

A is called a *linear transformation*. Often when linear transformations are involved, we shall omit the parenthesis and write Ax instead of $A(x)$. In the special case when $Y = F$, the underlying field, A is said to be a *linear functional*. If the linear transformation is 1 : 1, it is called an *isomorphism* (of X into Y). If A is 1 : 1 and onto, the two spaces are said to be *isomorphic*. It can be shown that a linear transformation is 1 : 1 if and only if $Ax = 0$ implies $x = 0$ (see, for example, Halmos [1]). Two isomorphic spaces are completely indistinguishable as abstract vector spaces and in some cases, isomorphic spaces are actually identified. An instance where such an identification is convenient was mentioned in the preceding paragraph.

The collection of all linear transformations mapping X into Y can be viewed as a vector space by defining addition of the linear transformations A and B to be that transformation which takes x into $Ax + Bx$; symbolically, we have

$$(A + B)x = Ax + Bx.$$

As for scalar multiplication, let

$$(\alpha A)x = \alpha Ax.$$

Having given meaning to the formation of linear combinations of linear transformations, let us look at the special case of all linear transformations mapping a vector space into itself. Denoting the identity transformation by 1 (that transformation mapping x into x for all vectors x) and letting A be an arbitrary linear transformation, we now form $(A - \lambda 1)$, which we abbreviate simply $A - \lambda$, where λ is an arbitrary scalar. As noted above, if this transformation fails to be 1 : 1, there must be some *nonzero* vector x such that $(A - \lambda)x = 0$. When this situation occurs, one says that x is an *eigenvector* of A with *eigenvalue* λ. Equivalent terminology also used is *characteristic* vector and *proper* vector instead of eigenvector, and *characteristic* value and *proper* value for eigenvalue.

Still working in the linear space of all linear transformations mapping X into X, suppose the linear transformation E has the property that E is *idempotent*; that is, $E^2 = E$, where by E^2 we mean the transformation taking x into $E(Ex)$; $E^2x = E(Ex)$. Such a transformation is called a *projection*. As it happens, a projection always determines a direct sum decomposition of X into (cf. Ref. 1):

$$M = \{x \in X | Ex = x\} \quad \text{and} \quad N = \{x \in X | Ex = 0\},$$

and one usually says that E *is a projection on M along N*. On the other hand, given a direct sum decomposition of the space $X = M \oplus N$, every $z \in X$ can be written uniquely as $z = x + y$, where $x \in M$ and $y \in N$. In this case the linear transformation E, such that $Ez = x$, is a projection on M along N. One notes in this case that

$$E(M) = \{Ex | x \in M\} = M;$$

that is, the subspace M is left fixed by E. This is a special case of a more general property that certain pairs (A, L) possess, where A is a linear transformation and L is a collection of vectors, namely, the property of L *being invariant under A*, which means that $A(L) \subset L$. We do not mean by this that every vector in L is left fixed by A; it is just mapped into *some* vector in L.

Two other concepts that we shall make extensive use of are *greatest lower bounds* and *least upper bounds*. A set U of real numbers is said to be bounded *on the left* (*bounded from below*) if there exists some number a_l such that $a_l \leq x$ for all $x \in U$. Similarly, U is said to be bounded *on the right* (*bounded from above*) if there exists some number a_r such that $x \leq a_r$ for all $x \in U$. The numbers a_l and a_r have the properties that a_l can be replaced by anything smaller, whereas a_r can be replaced by anything larger, and we shall still obtain *lower* and *upper bounds*, respectively, *for U*. A fundamental property of the real numbers, however, is that among all the lower bounds, there is a *greatest* one in the sense that at least one point of U lies to the left of any number greater than the *greatest lower bound*. Symbolically, if we denote the greatest lower bound by λ and if we consider a number only slightly larger than λ, say $\lambda + \varepsilon$ where $\varepsilon > 0$, there must be some $x \in U$ such that $x < \lambda + \varepsilon$. The number λ that may or may not belong to U is called the *infimum of U* and we denote the relationship of λ to U by writing

$$\lambda = \inf U.$$

Similarly, of all the upper bounds there is a smallest, a number μ, such that a member of U lies to the right of any smaller number; for any $\varepsilon > 0$, there exists $x \in U$ such that $x > \mu - \varepsilon$. The number μ is called the *least upper bound of U* or *supremum of U* and is denoted by

$$\mu = \sup U.$$

Aside from this material pertaining to functions, there are many other notions from set theory that are absolutely indispensable to the study of functional analysis. We feel it is unnecessary to discuss any of the notions from set theory and consider it sufficient to list the pertinent notation we shall use in the List of Symbols. No doubt the reader has come in contact with them (for example, unions, intersections, membership) before. We make one exception to this rule with regard to the concept of an *equivalence relation*. Letting X be an arbitrary collection of elements, we suppose that $\sim$ is a relation between elements of X with the following three properties. For any $x, y, z \in X$,

(1) $x \sim x$ (reflexive);
(2) $x \sim y$ implies $y \sim x$ (symmetric);
(3) $x \sim y$ and $y \sim z$ imply $x \sim z$ (transitive).

In this case the relation is called an *equivalence relation*. As a mnemonic device for recalling the three defining properties of an equivalence relation, one sometimes sees it referred to as an *RST relation*. Examples of equivalence relations are equality of real numbers, equality of sets (subsets of a given set, say), congruence among the class of all triangles, and congruence of integers modulo a fixed integer. An equivalence relation has a very important effect on the set: It partitions it into disjoint equivalence classes of the form $\hat{y} = \{z \in X \mid z \sim y\}$. (For a proof and further discussion one might look at Ref. 3, pp. 11–13.)

1.2 Inner Product Spaces

Suppose X is a real or complex vector space; that is, suppose the underlying scalar field is either the real or complex numbers R or C. We now make the following definition.

DEFINITION 1.1. An *inner product* on X is a mapping from $X \times X$, the cartesian product space, into the scalar field, which we shall denote generically by F:

$$X \times X \to F,$$

$$\langle x, y \rangle \to (x, y)$$

[that is, (x, y) denotes the inner product of the two vectors, whereas $\langle x, y \rangle$ represents only the ordered pair in $X \times X$] with the following properties:

(I_1) let $x, y \in X$; then $(x, y) = \overline{(y, x)}$, where the bar denotes complex conjugation;

(I_2) if α and β are scalars and x, y, and z are vectors, then $(\alpha x + \beta y, z) = \alpha(x, z) + \beta(y, z)$;

(I_3) $(x, x) \geq 0$ for all $x \in X$ and equal to zero if and only if x is the zero vector. It is noted that by property (I_1) that (x, x) must always be real, so this requirement always makes sense.

A real or complex vector space with an inner product defined on it will be called an *inner product space*; one often sees this abbreviated as i.p.s. Inner product spaces are also called *pre-Hilbert spaces*.

Some immediate consequences of the definition will now be noted, the first of which is that if a vector y has the property that $(x, y) = 0$ for all $x \in X$, then y

must be the zero vector. To prove this, noting that this is true for every vector in the space, take x equal to y and apply part (I_3) of the definition.

If x, y, and z are vectors and α and β are scalars, then

$$(z, \alpha x + \beta y) = \bar{\alpha}(z, x) + \bar{\beta}(z, y).$$

The mapping of X into F via $(x, x)^{1/2}$ is a *norm* on X and will be denoted by $\|x\|$. The notion of a *norm* will be thoroughly elaborated upon later and, for the sake of the present discussion, it suffices to regard $\|x\|$ as merely a convenient shorthand representation for $(x, x)^{1/2}$.

Examples of Inner Product Spaces

In the following examples, the scalar field can be taken to be either the real numbers or the complex numbers. The verification that the examples have the properties claimed is left as an exercise for the reader.

EXAMPLE 1.1. Let $X = C^n$; that is all n-tuples of complex numbers. As the inner product of the two vectors

$$x = (\alpha_1, \alpha_2, \ldots, \alpha_n)$$

and

$$y = (\beta_1, \beta_2, \ldots, \beta_n),$$

where the α_i and β_i are complex numbers, we shall take

$$(x, y) = \sum_{i=1}^{n} \alpha_i \bar{\beta}_i.$$

C^n with this inner product is referred to as *complex euclidean n-space*. With R in place of C one speaks of *real euclidean n-space*.

EXAMPLE 1.2. Let $X = C[a, b]$ be complex-valued continuous functions on the closed interval $[a, b]$ with addition and scalar multiplication of such functions defined as follows: For continuous functions f and g, $(f + g)(x) = f(x) + g(x)$; for $\alpha \in C$ and f continuous, $(\alpha f)(x) = \alpha f(x)$. As the inner product of any two vectors $f(x)$ and $g(x)$ in this space, we shall take

$$(f, g) = \int_a^b f(x)\overline{g(x)} \, dx.$$

We shall sometimes use the same notation to denote the *real* space of *real-valued* continuous functions on $[a, b]$. We shall always specifically indicate which one is meant however.

EXAMPLE 1.3. Let $X = l_2$: all sequences of complex numbers $(a_1, \ldots, a_n, \ldots)$ with the property that

$$\sum_{i=1}^{\infty} |a_i|^2 < \infty. \tag{1.7}$$

As the inner product of the vectors

$$x = (\alpha_1, \ldots)$$

and

$$y = (\beta_1, \ldots),$$

we shall take

$$(x, y) = \sum_{i=1}^{\infty} \alpha_i \bar{\beta}_i,$$

which converges by property (1.7) and the Holder inequality for infinite series (cf. Theorem 8.2).

The following example is presented for those familiar with some measure theory. Since it is not essential to the later development, the reader should not be disturbed if he does not have this background.

EXAMPLE 1.4. Let Y be a set and let S be a collection of subsets of Y with the following properties:

(1) if $E, F \in S$, then $E - F \in S$;
(2) if $E_i \in S$ $(i = 1, 2, \ldots)$, then $\bigcup_{i=1}^{\infty} E_i \in S$;
(3) $Y \in S$.

In other words, we assume that S is a σ-algebra of subsets of Y. Let μ be a measure on S and let E be some particular set in S. Analogous to the case of $C[a, b]$ (Example 1.2), one can introduce vector operations on the class of all functions (complex-valued) which are measurable on E by defining

$$(f + g)(x) = f(x) + g(x) \qquad \text{and} \qquad (\alpha f)(x) = \alpha f(x).$$

Among the members of this collection, the relation "$f \sim g$ if and only if $f = g$ almost everywhere on E (on all but a subset of E with measure zero)" is easily demonstrated to be an equivalence relation (Sec. 1.1). As such, it partitions the set into disjoint equivalence classes, where a typical class is given by

$$\hat{f} = \{g, \text{ measurable on } E \mid g \sim f\},$$

and any $g \in \hat{f}$ would be called a *representative* of this class.

The vector operations can now be carried over to the classes by defining

$$\hat{f} + \hat{g} = \widehat{f + g} \qquad \text{and} \qquad \alpha \hat{f} = \widehat{(\alpha f)}.$$

It can easily be verified that these operations are well-defined—namely, that if $h \sim f$ and $k \sim g$; then $\widehat{h + k} = \widehat{f + g}$, and similarly for scalar multiplication.

Let us now restrict our attention to only those classes whose representatives are square-summable on E—those $\hat{f}$ such that

$$\int_E |f|^2 \, d\mu < \infty.$$

It is now easily demonstrated that

$$(\hat{f}, \hat{g}) = \int_E f\bar{g} \, d\mu$$

is an inner product on this space. The reason we went to the bother of dealing with equivalence classes and not the functions themselves is so that $(z, z) = 0$ only when z is the zero vector would not be violated. For if a measurable function f is equal to 0 almost everywhere on E, then $\int_E |f|^2 \, d\mu = 0$. (Note that the zero vector of the set of all square-summable *functions* on E is the function that is 0 everywhere on E.) Since the only nonnegative measurable functions on E satisfying the above equality are those that equal 0 almost everywhere on E, we see that this problem is circumvented when one deals with classes.

In the following we shall omit the formality of writing $\hat{f}$ to represent the class to which f belongs and shall write only f, it being understood that we are dealing with equivalence classes. This space of equivalence classes is denoted by $L_2(E, \mu)$. Sometimes we shall wish to deal with only real-valued functions and to view the collection as a *real* vector space. When this is done, the space will still be referred to as $L_2(E, \mu)$, but we shall specifically indicate how we are viewing things.

As a special case of the above result, the following example is presented.

EXAMPLE I.5. Let the set Y be the closed interval $[a, b]$, let S be the Lebesgue measurable sets in Y and let μ be Lebesgue measure. Then, for the equivalence classes of square-integrable functions (complex-valued) on $[a, b]$ we can take, as inner product between two classes f and g,

$$(f, g) = \int_a^b f(x)\overline{g(x)} \, dx,$$

where the integral is just the Lebesgue integral. This space is usually referred to as $L_2(a, b)$.

The proof of the following, very important result can be found in the references to this lecture and will only be stated here.

THEOREM I.1 (*Cauchy–Schwarz inequality*). Let X be an inner product space and let $x, y \in X$. Then

$$|(x, y)| \leq \|x\| \, \|y\|$$

with equality holding if and only if x and y are linearly dependent.

The following three results represent straightforward, although somewhat arduous, tasks to prove and will also only be stated.

THEOREM I.2 (*Polarization identity*). Let X be a real inner product space and let $x, y \in X$. Then

$$(x, y) = \tfrac{1}{4}\|x + y\|^2 - \tfrac{1}{4}\|x - y\|^2.$$

THEOREM I.3 (*Polarization identity*). Let X be a complex inner product space and let $x, y \in X$. Then

$$(x, y) = \frac{1}{4}\|x + y\|^2 - \frac{1}{4}\|x - y\|^2 + \frac{i}{4}\|x + iy\|^2 - \frac{i}{4}\|x - iy\|^2.$$

It should be noted from the above two results that if we know the norm in an inner product space, the inner product can be recovered.

THEOREM 1.4 (*Parallelogram law*). Let X be an inner product space, and let $x, y \in X$. Then

$$\|x + y\|^2 + \|x - y\|^2 = 2\|x\|^2 + 2\|y\|^2.$$

This theorem derives its name from the analogous statement one has in plane geometry about the opposite sides of a parallelogram.

Orthogonality and Orthogonal Sets of Vectors

If x and y represent two vectors from real euclidean 3-space, it is well known that the angle between these two lines in space has its cosine given by

$$\cos\left\langle \begin{matrix} x \\ y \end{matrix} \right. = \frac{\alpha_1\beta_1 + \alpha_2\beta_2 + \alpha_3\beta_3}{(\alpha_1^2 + \alpha_2^2 + \alpha_3^2)^{1/2}(\beta_1^2 + \beta_2^2 + \beta_3^2)^{1/2}} = \frac{(x, y)}{\|x\| \, \|y\|},$$

where $x = (\alpha_1, \alpha_2, \alpha_3)$, $y = (\beta_1, \beta_2, \beta_3)$, and the α_i and β_i are real numbers. Further, the two vectors will be perpendicular if and only if $(x, y) = 0$. (x and y are, of course, each assumed to be not the zero vector.) As a very natural generalization of this familiar result from the real vector space R^3, we now wish to state a definition.

DEFINITION 1.2. Let X be an inner product space and let $x, y \in X$. Then x is said to be *orthogonal* to y, written $x \perp y$, if $(x, y) = 0$. The next result follows immediately from just expanding the quantities involved and using the hypothesis.

THEOREM 1.5 (*Pythagorean theorem*). If $x \perp y$, then

$$\|x + y\|^2 = \|x\|^2 + \|y\|^2.$$

DEFINITION 1.3. Let X be an inner product space, let S be a subset of X, and suppose, for any two distinct vectors x and y in S, that $(x, y) = 0$. Then S is said to be an *orthogonal set of vectors*. Suppose now that S is an orthogonal set of vectors with the added property that, for all $x \in S$, $\|x\| = 1$. In this case S is said to be an *orthonormal set of vectors*.

Some examples of orthonormal sets of vectors will now be given.

EXAMPLE 1.6. Let the vector space be l_2 and consider the set of vectors from this space

$$(1, 0, 0, ..., 0, ...), (0, 1, 0, ..., 0, ...), (0, 0, 1, ..., 0, ...), ...$$

This collection or any subset of this collection, according to the inner product defined in Example 1.3, forms an orthonormal set in l_2.

EXAMPLE I.7. Consider the inner product space $L_2(-\pi, \pi)$ as defined in Example 1.5. The collection (or any subset thereof)

$$x_n = \frac{1}{\sqrt{2\pi}} e^{int} \qquad (n = 0, \pm 1, \dots)$$

constitutes an orthonormal set of vectors in this space.

EXAMPLE I.8. If we restrict our attention to only real-valued functions that are square-integrable on the closed interval $[-\pi, \pi]$, then the collection (or any subset thereof)

$$\frac{1}{\sqrt{2\pi}}, \frac{1}{\sqrt{\pi}} \cos t, \frac{1}{\sqrt{\pi}} \cos 2t, \dots$$

$$\frac{1}{\sqrt{\pi}} \sin t, \frac{1}{\sqrt{\pi}} \sin 2t, \dots$$

is an orthonormal set in this space.

We now wish to proceed to the intuitively plausible (from our geometric experience in R^3, real euclidean 3-space) result that a collection of mutually orthogonal vectors is linearly independent.

THEOREM I.6. Let X be an inner product space and let S be an orthogonal set of nonzero vectors. Then S is a linearly independent set.

Proof. Before passing to the proof, the reader is asked to recall that an arbitrary set of vectors is called linearly independent if and only if every finite subset is linearly independent. In light of this, choose any finite collection of vectors, $x_1, x_2, \dots, x_n$ from S and suppose that

$$\alpha_1 x_1 + \alpha_2 x_2 + \cdots + \alpha_n x_n = 0, \tag{1.8}$$

where the α_i are scalars. Equation (1.8) now assures us that

$$(\alpha_1 x_1 + \cdots + \alpha_n x_n, x_i) = 0 \qquad (\text{for any } i = 1, 2, \dots, n).$$

But, since S is an orthogonal set, the inner product on the left is just $\alpha_i(x_i, x_i)$. Hence, since S is a set of nonzero vectors, we can conclude that $\alpha_i = 0$, and this is true for any $i = 1, 2, \dots, n$. Thus the collection is linearly independent.

Knowing that there exists a basis for any finite-dimensional vector space, the following result assures that among other things, in such a space, there is an orthonormal basis.

THEOREM I.7 (*Gram–Schmidt process*). Let X be an inner product space and let $\{y_1, \dots, y_n, \dots\}$ be a linearly independent set of vectors. Then there exists an orthonormal set of vectors $\{x_1, \dots, x_n, \dots\}$ such that, for any n,

$$[\{y_1, y_2, \dots, y_n\}] = [\{x_1, x_2, \dots, x_n\}],$$

where the square brackets indicate the space spanned by the set of vectors enclosed; for example, if $x, y \in X$, then $[\{x, y\}] = \{\alpha x + \beta y \,|\, \alpha, \beta \in F\}$.

Proof. The proof will be by induction. First we verify that the theorem is true in the case when $n = 1$. Since $y_1 \neq 0$, it makes sense to consider $x_1 = y_1/\|y_1\|$, which, clearly, has norm 1. Further, the spaces that x_1 and y_1 span must be identical, because the two vectors are linearly dependent. Now that this is done, assume orthonormal vectors $\{x_1, x_2, \ldots, x_{n-1}\}$ have been constructed such that

$$[\{x_1, x_2, \ldots, x_t\}] = [\{y_1, y_2, \ldots, y_t\}]$$

for any t between 1 and $(n - 1)$ inclusive. Our job now is to construct the nth vector with the same properties. To this end, consider the vector

$$w = y_n - \sum_{i=1}^{n-1} (y_n, x_i)x_i,$$

and take, for $1 \leq j \leq n - 1$,

$$(w, x_j) = (y_n, x_j) - \sum_{i=1}^{n-1} (y_n, x_i)\,(x_i, x_j)$$

$$= (y_n, x_j) - (y_n, x_j) = 0.$$

Hence $w \perp x_i$ for $i = 1, 2, \ldots, n - 1$. Suppose it were possible for w to be zero. This would imply that

$$y_n = \sum_{i=1}^{n-1} (y_n, x_i)x_i$$

or, in words, that y_n could be written as a linear combination of $x_1, x_2, \ldots, x_{n-1}$, which, by the induction hypothesis, would imply that y_n could be written as a linear combination of $y_1, y_2, \ldots, y_{n-1}$, which is contrary to our assumption that the $\{y_1, y_2, \ldots, y_n, \ldots\}$ was a linearly independent set. Thus w cannot be zero. It now remains only to choose $x_n = w/\|w\|$, with the fact that the spaces spanned by $\{x_1, x_2, \ldots, x_n\}$ and $\{y_1, y_2, \ldots, y_n\}$ are the same being left as a simple exercise for the reader.

As applications of the above theorem the following examples are presented.

EXAMPLE 1.9. Consider the inner product space $L_2(a, b)$ mentioned in Example 1.5 and take the linearly independent set to be $\{1, t, t^2, \ldots, t^n, \ldots\}$. (It is easily verified that this is a linearly independent set by noting that any polynomial of degree n has at most n zeros.) Applying the Gram–Schmidt process would then lead us to the following orthonormal set in this space, where a_k is a scalar:

$$x_k = a_k \frac{d^k}{dt^k} (t - a)^k(t - b)^k \qquad (k = 0, 1, 2, \ldots). \tag{1.9}$$

In particular if $a = -1$ and $b = 1$, then a_k would be $\sqrt{k + \tfrac{1}{2}}/2^k k!$ and the functions $x_k/\sqrt{k + \tfrac{1}{2}}$ would be the Legendre polynomials.

In the next example, although the interval is changed, we shall consider essentially the same linearly independent set but multiplied by the weighting factor $e^{-t^2/2}$.

EXAMPLE 1.10. Consider $L_2(-\infty, \infty)$ and take the linearly independent set of vectors to be

$$e^{-t^2/2}, te^{-t^2/2}, ..., t^n e^{-t^2/2}, ...$$

Applying the Gram–Schmidt process in this case we would obtain, as the corresponding orthonormal set

$$a_k H_k(t)e^{-t^2/2} \qquad (k = 0, 1, 2, ...),$$

where a_k is a scalar and the $H_k(t)$ are the Hermite polynomials.

DEFINITION 1.4. Let X be an inner product space and let S be a subset of X. The collection of vectors

$$S^\perp = \{y \in X | y \perp x \quad \text{for all} \quad x \in S\}$$

is called the *orthogonal complement* of S. It is to be noted that, even though S is only a subset, $S^\perp$ is a subspace of X.

Our next theorem yields an important result about the orthogonal complement of a subspace that is finite-dimensional.

THEOREM 1.8. Let X be an inner product space and let M be a subspace of X such that dim $M < \infty$. Then we have the direct sum decomposition $X = M \oplus M^\perp$. [The notation $M \oplus N = X$ means that any vector $z \in X$ can be written uniquely as $z = x + y$, where $x \in M$ and $y \in N$, or, equivalently, that every $z \in X$ can be written as $z = x + y$, where $x \in M$ and $y \in N$ and $M \cap N = \{0\}$.]

Proof. First it will be shown that any vector in the space can be written as the sum of an element from M and an element from $M^\perp$.

Since dim $M < \infty$, we can choose an orthonormal basis for M : $x_1, x_2, ..., x_n$. Now it is clear that

$$\sum_{i=1}^{n} (z, x_i)x_i \in M \tag{1.10}$$

for any vector $z \in X$, because expression (1.10) is just a linear combination of the basis vectors in M. Now consider the vector

$$y = z - \sum_{i=1}^{n} (z, x_i)x_i.$$

Letting x_j be any one of the basis vectors in M, we have

$$(y, x_j) = (z, x_j) - (z, x_j) = 0 \qquad (j = 1, 2, ..., n)$$

or that y is orthogonal to each of the basis vectors in M. Hence it is orthogonal to every vector in M, which implies that $y \in M^\perp$. Therefore, we have the representation, for any vector $z \in X$,

$$z = \sum_{i=1}^{n} (z, x_i)x_i + y,$$

where the first term on the right is in M and the second is in $M^\perp$, and have completed the first part of the proof.

To prove the second part, suppose that $x \in M \cap M^{\perp}$. This immediately implies that $(x, x) = 0$, which implies that $x = 0$ and completes the proof.

1.3 Linear Functionals, the Riesz Representation Theorem, and Adjoints

Suppose X is an inner product space and that y is some fixed vector in X. From property I_2 of Definition 1.1, it is clear that the mapping

$$X \rightarrow F$$
$$x \rightarrow (x, y)$$

represents a linear functional on X. Thus every inner product gives rise to a collection of linear functionals on a space. The following representation theorem gives some answer as to the validity of the converse of the above statement.

THEOREM 1.9 (Riesz). If X is a finite-dimensional inner product space and f is a linear functional on X, $f: X \rightarrow F$, then there exists a unique vector $y \in X$ such that $f(x) = (x, y)$ for all $x \in X$.

Proof. Since dim $X < \infty$, we can choose an orthonormal basis for $X: x_1, x_2$, ..., x_n. Consider now the vector

$$y = \sum_{i=1}^{n} \overline{f(x_i)} x_i$$

and consider the mapping

$$f_y : X \rightarrow F$$
$$x \rightarrow f_y(x) = (x, y).$$

We now let f_y operate on any of the basis vectors x_j ($j = 1, 2, ..., n$) of X to get

$$f_y(x_j) = (x_j, y) = (x_j, \overline{f(x_j)} x_j) = f(x_j).$$

Thus the two linear functionals agree on each of the basis vectors, which implies that they must agree on any vector in the space, or that the two linear functionals are the same:

$$f_y(x) = (x, y) = f(x) \qquad \text{(for all } x \in X\text{)}.$$

Having established this, we now wish to show that the vector y is unique. Suppose there was a vector z such that

$$(x, y) = (x, z) \qquad \text{(for all } x \in X\text{)}.$$

This immediately implies $(x, y - z) = 0$ for every x which, in turn, implies that $y - z = 0$ or that $y = z$, and completes the proof of the theorem.

We shall now turn our attention from linear functionals to linear transformations. Suppose, again, that X is a finite-dimensional inner product space, and let A be a linear transformation on X into X:

$$A : X \rightarrow X.$$

Letting y be some particular vector from X, it is clear that the mapping

$$f^y : X \to F$$

$$x \to (Ax, y)$$

is a linear functional on X. Since dim $X < \infty$ and by the previous theorem, we can represent the linear functional $f^y(x)$ as

$$f^y(x) = (x, z)$$

for some unique z in X. Alternatively, we might say that there exists a unique z in X such that

$$(Ax, y) = (x, z) \qquad \text{(for all } x \in X).$$

In this manner then, for each y in X, a corresponding z in X is associated by invoking the previous representation theorem; that is, we have the mapping

$$A^* : X \to X$$

$$y \to z$$

or, writing z as A^*y, we have

$$(Ax, y) = (x, A^*y) \qquad \text{(for all } x, y \in X) \tag{1.11}$$

as the defining equation for the *adjoint*, A^*, *of A*. It is to be emphasized that, since we have relied upon the representation theorem so heavily, one can only speak of adjoints in *finite-dimensional* inner product spaces for the sake of the present discussion. Having defined the adjoint, we now show that it is also a linear transformation. It will be shown that it preserves sums here, the proof that it also preserves scalar multiples being left as an exercise for the reader.

Let $A^*y_1 = z_1$ and $A^*y_2 = z_2$, where A is a linear transformation on X and consider

$$\begin{aligned}
\left(x, A^*(y_1 + y_2)\right) &= (Ax, y_1 + y_2) \\
&= (Ax, y_1) + (Ax, y_2) \\
&= (x, A^*y_1) + (x, A^*y_2) \\
&= (x, A^*y_1 + A^*y_2).
\end{aligned}$$

Since this is true for every $x \in X$, we can conclude that

$$A^*(y_1 + y_2) = A^*y_1 + A^*y_2.$$

The proof that $A^*(\alpha x) = \alpha A^*x$, where α is a scalar, follows in a similar manner.

The following properties of adjoints may all be proved in a manner similar to the above proof (namely, using defining Eq. (1.11) strongly) and will not be proved here. With A and B as linear transformations and α a scalar, we have

(1) $(\alpha A)^* = \bar{\alpha} A^*$;
(2) $(A + B)^* = A^* + B^*$;
(3) $(AB)^* = B^*A^*$;
(4) $(A^*)^* = A^{**} = A$.

Before proceeding further, two new definitions are necessary and both should be viewed in the framework of a finite-dimensional (so that it makes sense to talk about adjoints) inner product space X.

DEFINITION 1.5. Let A be a linear transformation on X. If

(1) $A = A^*$, then A is said to be *self-adjoint*;
(2) $AA^* = A^*A$, then A is said to be *normal*.

One notes immediately that every self-adjoint linear transformation must also be normal. With these definitions in mind we can now proceed to the next theorem.

THEOREM 1.10. (1) If A is self-adjoint and λ is an eigenvalue of A, then λ is real. (2) Eigenvectors associated with distinct eigenvalues of a self-adjoint linear transformation are orthogonal.

Proof (1). Suppose x is an associated (nonzero) eigenvector with the eigenvalue λ; that is, $Ax = \lambda x$, $x \neq 0$. We then have

$$\lambda(x, x) = (\lambda x, x) = (Ax, x)$$
$$= (x, A^*x)$$
$$= (x, Ax)$$
$$= (x, \lambda x)$$
$$= \bar{\lambda}(x, x).$$

Since $x \neq 0$, we can conclude that $\lambda = \bar{\lambda}$.

Proof (2). Consider two distinct eigenvalues, λ and μ and associated eigenvectors, x and y and note immediately that, by part (1) of the theorem, both of the eigenvalues must be real. Consider now

$$\lambda(x, y) = (\lambda x, y) = (Ax, y) = (x, A^*y) = (x, Ay)$$
$$= (x, \mu y) = \bar{\mu}(x, y) = \mu(x, y).$$

Looking at the first and last terms, we now see that if $(x, y) \neq 0$, then λ must equal μ, which would contradict our original assumption that they were distinct. Hence $(x, y) = 0$ and completes the proof.

THEOREM 1.11. Let M be an invariant subspace of the finite-dimensional inner product space under the linear transformation A. Then $M^\perp$ is invariant under A^*.

Proof. Let $x \in M$ and let $y \in M^\perp$. Consider now

$$(x, A^*y) = (Ax, y),$$

which must be zero because $Ax \in M$ (by the invariance of M under A) and $y \in M^\perp$. Since the above would be true for any $x \in M$, we can conclude that $A^*y \in M^\perp$ for any $y \in M^\perp$.

THEOREM 1.12. Let X be an inner product space such that dim $X < \infty$ and let A be a linear transformation on X:

$$A : X \to X.$$

Then $R(A)^{\perp} = N(A^*)$, where $R(A)$ is the range of A, $\{Ax \mid x \in X\}$, and $N(A^*)$ is the null space of A^*, $\{x \mid A^*x = 0\}$.

Proof. Let $x \in X$ and let $y \in R(A)^{\perp}$. We now have

$$(Ax, y) = 0 = (x, A^*y) \qquad \text{(for any } x \in X),$$

which implies that $A^*y = 0$ or that $y \in N(A^*)$. Noticing that only definitions have been used in the proof, all the steps are seen to be reversible, or

$$R(A)^{\perp} = N(A^*).$$

EXERCISES I

1. Let X be a finite-dimensional inner product space and $A : X \to X$ a linear transformation. If A is self-adjoint and if $A^2x = 0$, show that $Ax = 0$.

2. Let $A : X \to X$ be a linear transformation, where X is a finite-dimensional complex inner product space. Show that A is self-adjoint if and only if (Ax, x) is real for all $x \in X$.

3. Let $A : X \to X$ be a linear transformation, where X is an inner product space. Show, if $\|Ax\| = \|x\|$ for all $x \in X$, that $(Ax, Ay) = (x, y)$ for all $x, y \in X$. Show also that if A is an onto map with the property that $(x, y) = (Ax, Ay)$ for all x, y then for any subset $U \subset X$, $A(U^{\perp}) = A(U)^{\perp}$.

4. Let S be a subset of an inner product space X. Show that $S^{\perp\perp} \supset [S]$. If X is finite-dimensional, show that $S^{\perp\perp} = [S]$.

5. Show that on any real or complex vector space an inner product can be introduced. (*Hint:* Use the fact that a basis exists for any vector space.)

6. Using the inner product of Example 1.1 in real euclidean 3-space, apply the Gram–Schmidt process to the vectors $x_1 = (1, 0, 1)$, $x_2 = (0, 1, 1)$, and $x_3 = (1, 1, 0)$ to obtain an orthonormal basis for this space.

7. Let $x_1, \ldots, x_n$ be a set of mutually orthogonal nonzero vectors in the space X. Suppose the only vector orthogonal to each of these vectors is the zero vector. Then prove that $x_1, \ldots, x_n$ form a basis for X.

8. Consider the (real) space $C(-1, 1)$ of all real-valued continuous functions on the closed interval $[-1, 1]$. Using the operations and inner product of Example 1.2, determine the orthogonal complement of the subspace of all odd functions. [By an *odd* function f, one means that $f(t) = -f(-t)$; the function $f(t) = t$ is an example of an odd function.]

9. Let X be a complex three-dimensional inner product space with the orthonormal basis x_1, x_2, x_3. The linear transformation $A : X \to X$ has the property that

$$Ax_1 = x_1 + 2x_2 + 3x_3,$$
$$Ax_2 = x_2 + x_3,$$

and

$$Ax_3 = x_1 + x_3.$$

Express A^*x_1, A^*x_2, A^*x_3 in terms of this basis.

10. Suppose the linear transformation $A : X \to X$, where X is finite-dimensional, has the property that $A^n = 0$ for some integer n. Determine the eigenvalues of A.

11. Consider the (real) space $C(-\infty, \infty)$ of real-valued functions continuous on all of R, with respect to the operations and inner product of Example 1.2. Show that the transformation A, where $(Af)(t) = \int_0^t f(x)\, dx$, has no eigenvalues.

REFERENCES†

References pertaining to linear algebra:

1. P. R. Halmos, "Finite-Dimensional Vector Spaces."
2. K. Hoffman and R. Kunze, "Linear Algebra."

The following references pertain more particularly to inner product spaces:

3. A. Kolmogorov and S. Fomin, "Elements of the Theory of Functions and Functional Analysis," Vol. 1.
4. A. Taylor, "Introduction to Functional Analysis."

General:

5. N. Jacobson, "Lectures in Abstract Algebra," Vol. II. Linear algebra is discussed here but from a somewhat more sophisticated viewpoint; for example, one should be familiar with some modern algebra in order to profit most from the book.
6. P. R. Halmos, "Measure Theory." Some further discussion of $L_2(E, \mu)$ for those familiar with measure theory.
7. I. P. Natanson, "Theory of Functions of a Real Variable." Discussion of $L_2(a, b)$.

† Here, as will be our custom in listing references throughout the book, only the author and title are listed. Other pertinent information about the reference appears in the Bibliography.

CHAPTER 2

Orthogonal Projections and the Spectral Theorem
for Normal Transformations

The first notion to be discussed in this chapter will be the *complexification* of a real vector space: a process by which we can extend a real vector space to a complex vector space. We shall be most interested in the case when the space to be extended is a real *inner product* space. After having defined what is meant by the complexification, the inner product on the original space will also be extended, as will linear transformations. It is to be noted that a strong motivational factor for making this extension is that the complex numbers constitute an algebraically closed field (that is, all polynomials completely factor into linear factors), whereas the field of real numbers does not have this property which is often so desirable when the eigenvalues of linear transformations are being sought. Next it will be shown that orthonormal bases for inner product spaces (finite-dimensional) can be constructed consisting purely of eigenvectors of certain types of linear transformations. After having done this, the notion of an orthogonal projection is introduced and certain properties of *orthogonal projections* are then demonstrated, along with the related notion of an *orthogonal direct sum decomposition* of the space.

In Sec. 2.2 it is shown that any normal transformation can be expressed in a unique way as a linear combination of orthogonal projections—a desirable representation, because orthogonal projections are comparatively simple to work with. However, from the way "adjoint" was defined in Chapter 1 and also from the nature of the proof, the result seems hopelessly tied to finite-dimensional spaces. This is not the case at all though, and by introducing suitable machinery, one can go far beyond this rather severe restriction. In particular, one might compare the spectral theorem for normal operators proved in Theorem 2.6 with the spectral theorem stated in Sec. 24.2 to see what sort of changes occur in the infinite-dimensional case. Loosely, one refers to the whole business of expressing a linear transformation *somehow* in terms of orthogonal projections as *spectral theory*. A nice application of the spectral theorem is given in Theorem 2.8: A transformation A is normal if, and only if, its adjoint A^* can be expressed as a polynomial in A.

2.1 The Complexification

Suppose X is a real vector space. The *complexification of X*, described below, will extend X to X^+, a complex vector space. It is now claimed that the set of elements

$$X^+ = \{\langle x, y \rangle | x, y \in X\}$$

with addition defined as

$$\langle x_1, y_1 \rangle + \langle x_2, y_2 \rangle = \langle x_1 + x_2, y_1 + y_2 \rangle,$$

and multiplication by a complex number (scalar) as

$$(\alpha + i\beta)\langle x, y \rangle = \langle \alpha x - \beta y, \beta x + \alpha y \rangle$$

constitutes a complex vector space. The verification of this fact is quite straightforward and will not be performed here.

It was claimed above, though, that X^+ was an "extension" of X; hence, one expects to find X embedded in X^+ somehow, and the embedding process consists of noting the following mapping:

$$X \to X^+$$

$$x \to \langle x, 0 \rangle$$

that is, we shall "identify" every element x of X with the element $\langle x, 0 \rangle$ of X^+ and in this way can locate every element of X in X^+. This identification process will be simplified even further. By our rules of addition and scalar multiplication, we can write

$$\langle x, y \rangle = \langle x, 0 \rangle + \langle 0, y \rangle$$
$$= \langle x, 0 \rangle + i\langle y, 0 \rangle$$
$$\text{"} = \text{"} \ x + iy,$$

where the last equal sign is in quotation marks because we have made the identification indicated above. It is to be noted, however, that the last representation is unique.

When one extends anything, it is usually desirable that the main structural properties of the original set be preserved in the larger one. Certainly, one will agree that a rather fundamental relationship in a vector space is that of linear independence, and it will be shown here that this relationship is invariant under the complexification.

(1) Suppose $x_1, \ldots, x_n$ are linearly independent in X. Then, making the identification mentioned above, they will be linearly independent in X^+, too. To verify this, suppose

$$\sum_{j=1}^{n} (\alpha_j + i\beta_j)x_j = 0 \qquad (= \langle 0, 0 \rangle \text{ here})$$

$$= \sum_{j=1}^{n} \alpha_j x_j + i \sum_{j=1}^{n} \beta_j x_j,$$

which immediately implies that

$$\sum_{j=1}^{n} \alpha_j x_j = 0 \quad \text{and} \quad \sum_{j=1}^{n} \beta_j x_j = 0.$$

To complete the proof, all we need do now is to note that the $x_1, x_2, \ldots, x_n$ were assumed to be linearly independent in X, which implies that all the α_j and β_j must be zero.

Our next result shows that a basis in X carries over as a basis in X^+.

(2) Suppose $B = \{x_1, x_2, \ldots, x_n\}$ is a basis for X. Then (again making the required identification) B is a basis for X also. Since B is a basis for X, then, for any two vectors x and y in X, we have real scalars such that

$$x = \sum_{j=1}^{n} \alpha_j x_j, \qquad y = \sum_{j=1}^{n} \beta_j x_j.$$

Thus we can write

$$x + iy = \sum_{j=1}^{n} (\alpha_j + i\beta_j)x_j$$

and, in this manner, can express any vector in X^+ as a linear combination of (using complex scalars, though) these vectors. We can also note from this result that the dimensions of X and X^+ must be the same.

Another property that one would certainly like to extend if the original space were an inner product space, is the inner product on that space, and the next result indicates this extension.

(3) If X is an inner product space, then the inner product (,) is extended to X^+ by defining the following mapping as an inner product on X^+:

$$(x_1 + iy_1, x_2 + iy_2)^+ = (x_1, x_2) + (y_1, y_2) - i((x_1, y_2) - (y_1, x_2)).$$

Two things should be verified now: (a) It should be verified that the mapping defined above is really an inner product and (b) the mapping defined above extends the original inner product. To verify the former, one need only perform the required expansions and check the axioms listed in Chapter 1. To verify the latter one need only set y_1 and y_2 to zero to see that this truly provides an extension.

(4) Suppose now that $A : X \to X$ is a linear transformation on X. We would like to extend A now, in a natural way, to a linear transformation on X^+, and this is accomplished by the following definition of the mapping A^+:

$$A^+ : X^+ \to X^+$$

$$A^+(x + iy) = Ax + iAy.$$

One notes immediately from this definition that A^+ is a linear transformation on X^+ with the following properties:

(a) $(\alpha A)^+ = \alpha A^+$;
(b) $(A + B)^+ = A^+ + B^+$;
(c) $(AB)^+ = A^+ B^+$;
(d) $(A^*)^+ = (A^+)^*$.

When one has linear transformations on a space, it becomes sensible to speak of the eigenvalues of those transformations. We should now like to investigate what relation, if any, exists between the eigenvalues of A^+ and the eigenvalues of A. To this end, suppose $\alpha + i\beta$ is an eigenvalue of A^+ with the associated eigenvector $x + iy$. We note immediately, since $x + iy$ is assumed to be eigenvector of A^+, that not both x and y can be zero. We now have

$$A^+(x + iy) = (\alpha + i\beta)(x + iy) = \alpha x - \beta y + i(\beta x + \alpha y),$$

which implies, by the uniqueness of the representation of $\langle x, y \rangle$ as $x + iy$, that

$$Ax = \alpha x - \beta y, \qquad Ay = \beta x + \alpha y,$$

in general. In the special case that $\beta = 0$, though, we have

$$Ax = \alpha x, \qquad Ay = \alpha y,$$

which in view of the fact that at least one of x, y must be nonzero, implies that α is a real eigenvalue of A. Hence if A^+ has a real eigenvalue α, then α is a real eigenvalue of A as well.

THEOREM 2.1. Let X be a finite dimensional inner product space and let A be a self-adjoint linear transformation on X: $A : X \rightarrow X$. Then there exists an orthonormal basis for X consisting of eigenvectors of A.

Proof. First it will be shown that the above statement is not vacuous; i.e., it will be shown that, since A is self-adjoint, it always possesses eigenvectors. There are two cases to consider: Either X is a complex space or it is a real space. In the former, the characteristic equation factors completely, and A must have eigenvalues. In the latter, however, it is not clear that A will always have eigenvalues and so we turn our attention toward the complexification of X and to the corresponding extension of A, A^+, it being clear that A^+ will have eigenvalues. Since A is self-adjoint, we have, by property (d) above,

$$A^+ = (A^*)^+ = (A^+)^*,$$

which shows that A being self-adjoint implies that A^+ is self-adjoint too. As shown at the end of the first chapter, however, the eigenvalues of a self-adjoint transformation must be real; hence all the eigenvalues of A^+ are real. Since they are all real and by the observation immediately preceding Theorem 2.1, they must also be eigenvalues of A. Thus we have shown that the consideration of the eigenvalues of a self-adjoint transformation is nonvacuous in either case, and we shall now proceed with the rest of the theorem.

Let λ be an eigenvalue of A and suppose x is a corresponding eigenvector. Thus we have $Ax = \lambda x$. Since x cannot be zero, we can choose $x_1 = x/\|x\|$ and if, in addition, dim $X = 1$, then we are done. If not, we shall proceed by induction; i.e., we shall assume that the theorem is true for all spaces of dimension less than X and then show that it follows for X from this assumption. Letting $M = [x_1] = \{\alpha x_1 \mid \alpha \in F\}$, the space spanned by x_1, we have the following direct sum decomposition of X:

$$X = M \oplus M^\perp,$$

and we can note immediately that we must have dim $M^\perp <$ dim X. As shown in Chapter 1, M being invariant under A implies that $M^\perp$ is invariant under A^*, which, in this case since A is self-adjoint, is just A. Consider now the restriction of A to $M^\perp$ denoted by $A|_{M^\perp}$ where $A|_{M^\perp}: M^\perp \rightarrow M^\perp$. It is clear that, if A is self-adjoint, then so must be the restriction of A to $M^\perp$. Now, since dim $M <$ dim X, we can apply the induction hypothesis to assert the existence of an orthonormal basis for $M^\perp$ consisting of eigenvectors for $A|_{M^\perp}$, $\{x_2, x_3, ..., x_n\}$. Eigenvectors of $A|_{M^\perp}$, however, must also be eigenvectors of A. Hence for the entire space we have $\{x_1, x_2, ..., x_n\}$ as orthonormal basis of eigenvectors of A.

Having made this statement for self-adjoint transformations, we now wish to make a similar statement about *normal* transformations; that is, linear transformations $A : A \rightarrow A$ that commute with their adjoints. Some preliminary results will be necessary first, however, and will be stated in the following two lemmas.

LEMMA 2.1. Let A be a normal transformation on X (finite-dimensional), an inner product space. Then $\|Ax\| = \|A^*x\|$ for all $x \in X$.

Proof. We note that Lemma 2.1 is certainly true in the event that $A = A^*$. In any case, though, we have

$$\|Ax\|^2 = (Ax, Ax) = (x, A^*Ax) = (x, A A^*x)$$
$$= (A^*x, A^*x)$$
$$= \|A^*x\|^2.$$

LEMMA 2.2. Let A be a normal transformation on a finite-dimensional inner product space X. Then if λ is an eigenvalue of A with eigenvector x, $\bar{\lambda}$ is an eigenvalue of A^* with eigenvector x.

Proof. Suppose x is an eigenvector of A and λ the corresponding eigenvalue which implies $(A - \lambda)x = 0$. By the results of Chapter 1 we have $(A - \lambda)^* = A^* - \bar{\lambda}$, which, clearly, must commute with $(A - \lambda)$. Thus if A is normal, then so must be $(A - \lambda)$, which, by Lemma 2.1 implies that

$$\|(A - \lambda)x\| = \|(A - \lambda)^*x\| = \|(A^* - \bar{\lambda})x\| = 0$$
$$\Rightarrow (A^* - \bar{\lambda})x = 0 \quad \text{or} \quad A^*x = \bar{\lambda}x,$$

or that x is an eigenvector of A^* with corresponding eigenvalue $\bar{\lambda}$.

THEOREM 2.2. Let X be a complex finite-dimensional inner product space and let A be a normal transformation on X. Then there exists an orthonormal basis for X consisting of eigenvectors of A.

Proof. Since we are working over the complex numbers, it is clear that A must have eigenvalues. Since this is the case, let x be an eigenvector of A such that $\|x\| = 1$. (This can always be accomplished because any scalar multiple of an eigenvector is still an eigenvector.) If the dimension of X happens to be 1, we are done. If not, we can proceed by induction as in the proof of Theorem 2.1. To this end, consider the subspace $M = [x]$, and the ensuing direct sum decomposition of X, $X = M \oplus M^\perp$. Since x is an eigenvector of A, it is clear that M is invariant under A which, by the result of Lemma 2.2, implies that M is invariant under A^*, too, because eigenvectors of A must also be eigenvectors of A^*. We can now apply the results of chapter 1 to assert that $M^\perp$ is invariant under $A^{**} = A$. We now have two statements:

(1) M is invariant under A;
(2) $M^\perp$ is invariant under A.

A look at the proof of Theorem 2.1 shows that the proof of this theorem, from this point on, can be carried out in exactly the same way.

2.2 Orthogonal Projections and Orthogonal Direct Sums

Before introducing the notion of an orthogonal projection, some basic ideas about projections in general will be reviewed. A projection on a vector space is a linear transformation of a special type. Suppose one has a direct sum decomposition of the space V, $V = M \oplus N$, say. Any $z \in V$ can now be uniquely written $z = x + y$, where $x \in M$ and $y \in N$. The mapping defined by $Ez = x$ is called the *projection on M along N*, and it is clear that M and N can now be written

$$M = \{z | Ez = z\}, \qquad N = \{z | Ez = 0\},$$

it now being clear that M and N represent the range and null space of E respectively.[+] An equivalent way of defining a projection is to require that the linear transformation be idempotent; that is, $E^2 = E$. An *orthogonal projection on M* is defined when one requires that the direct sum decomposition of X be $X = M \oplus M^\perp$, it being understood that whenever one says *orthogonal* projection on M, it is always *along $M^\perp$*. (For a further discussion of projections in general a good reference is: Halmos, *Finite-Dimensional Vector Spaces*, Sections 41–43.) With these basic notions about orthogonal projections in mind, we can proceed to our first theorem about them.

THEOREM 2.3. Let X be a finite-dimensional inner product space and let E be a projection. Then the following statements are equivalent.

(1) E is normal;
(2) E is self-adjoint;
(3) E is the orthogonal projection on its range.

Proof (1) $\Rightarrow$ (2):
By Lemma 2.1 we have, if E is normal,

$$\|E^* z\| = \|Ez\| \quad \text{(for all } z \in X),$$

which immediately tells us that $Ez = 0$ if and only if $E^* z = 0$.
Let z be any vector in X and consider the vector $w = z - Ez$. We now have

$$Ew = Ez - E^2 z = Ez - Ez = 0$$

$$\Rightarrow E^* w = 0.$$

Computing $E^* w$ directly, we have

$$E^* w = E^* z - E^* Ez = 0$$

or

$$E^* z = E^* Ez \quad \text{(for any } z \in X).$$

Hence, by the definition of what we mean by equal transformations, we conclude that

$$E^* = E^* E. \tag{2.1}$$

Computing the adjoint of each side, we get

$$E = (E^*)^* = (E^* E)^* = E^* E^{**} = E^* E$$

or

$$E = E^* E. \tag{2.2}$$

Combining (2.1) and (2.2), we have $E = E^*$, which proves our first contention.

[+] Projections are discussed briefly in Section 1.1.

Proof (2) $\Rightarrow$ (3). To prove this all we need show is that the null space of E is the orthogonal complement of the range of E, or, denoting the range of E by $R(E)$ and the null space of E by $N(E)$, that $R(E)^\perp = N(E)$. By Theorem 1.12 proved in Chapter 1, we have for any linear transformation A, that

$$R(A)^\perp = N(A^*).\tag{2.3}$$

Thus to complete the proof all we need do is replace A by E in Eq. (2.3) and note that $E = E^*$ by hypothesis.

Proof (3) $\Rightarrow$ (1). Let $x, y \in X$. Then, since E is idempotent,

$$x - Ex \in N(E) = R(E)^\perp$$

and

$$Ey \in R(E).$$

Thus

$$(x - Ex, Ey) = 0$$

or

$$(x, Ey) = (Ex, Ey) = (x, E^*Ey) \qquad \text{(for any } x, y \in X).$$

Therefore $E = E^*E$ and the proof becomes the same as in proving (1) $\Rightarrow$ (2); that is, it follows that E is self-adjoint, which certainly implies that E is normal, and completes the proof of the theorem.

The notation

$$X = M_1 \oplus M_2 \oplus \cdots \oplus M_k\tag{2.4}$$

read as "X is the *direct sum* of $M_1, M_2, ..., M_k$," where the M_i are subspaces in X means that the following two statements are true:

(1) $X = M_1 + M_2 + \cdots + M_k$. (Recall from Section 1.1 that this notation only denotes all possible linear combinations of the vectors in $M_1, M_2, ..., M_k$.)
(2) For any $i = 1, 2, ..., k$, $M_i \cap \{M_1 + M_2 + \cdots + M_{i-1} + M_{i+1} + \cdots + M_k\} = \{0\}$. Subspaces satisfying only (2) are called *linearly independent subspaces*. If in addition to (1) and (2) we also have $M_i \perp M_j$ for $i \neq j$ (i.e., $x \in M_i$, $y \in M_j \Rightarrow x \perp y$), we shall say that Eq. (2.4) represents an *orthogonal direct sum decomposition of the space*, and that the subspaces themselves are *orthogonal*.

THEOREM 2.4. Let X be a finite-dimensional inner product space, let $X = M_1 \oplus M_2 \oplus \cdots \oplus M_k$, and let E_j be the orthogonal projection on M_j for $j = 1, 2, ..., k$. Then the following statements are equivalent:

(1) $X = M_1 \oplus M_2 \oplus \cdots \oplus M_k$ (orthogonal);
(2) $1 = E_1 + E_2 + \cdots + E_k$ and $E_iE_j = 0$ for $i \neq j$;
(3) if B_j is an orthonormal basis for M_j, $j = 1, 2, ..., k$, then $\bigcup_{j=1}^{k} B_j$ is an orthonormal basis for X.

Proof. Since all three of these proofs run in a similar vein, only the fact that (1) implies (2) will be proved here. To prove this, suppose

$$X = M_1 \oplus M_2 \oplus \cdots \oplus M_k \qquad \text{(orthogonal)}.$$

This implies the existence, for any $x \in X$, of unique x_i ($i = 1, 2, ..., k$) such that

$$x = x_1 + x_2 + \cdots + x_k,$$

where $x_i \in M_i$. Applying E_i to both sides, we have

$$E_i x = E_i x_1 + E_i x_2 + \cdots + E_i x_i + \cdots + E_i x_k$$

$$= 0 + 0 + \cdots + E_i x_i + \cdots + 0 = x_i \quad \text{(for any } i\text{)}.$$

Hence we can write any x as

$$x = E_1 x + E_2 x + \cdots + E_k x,$$

and, by our rules for addition of linear transformations, we can paraphrase this last result as

$$x = (E_1 + E_2 + \cdots + E_k)x.$$

Since this is true for any $x \in X$, we conclude that

$$1 = E_1 + E_2 + \cdots + E_k,$$

where the symbol 1 represents the identity transformation—the transformation that takes everything into itself. It is clear that $M_j \subset M_i^\perp$, since the subspaces are orthogonal. Thus, since $E_j x \in M_j$, we have $E_i E_j x = 0$ for any $x \in X$ or $E_i E_j$ is the zero transformation for $i \neq j$.

THEOREM 2.5. Let X be a finite-dimensional inner product space, and let A be a normal transformation on X. In this case the eigenvectors belonging to distinct eigenvalues are orthogonal.

Proof. Suppose,

$$Ax = \lambda x \quad (x \neq 0),$$

$$Ay = \mu y \quad (y \neq 0),$$

and that $\lambda \neq \mu$. Now we can write

$$\lambda(x, y) = (\lambda x, y) = (Ax, y) = (x, A^* y).$$

Applying Lemma 2.2, we see that this is just

$$(x, \bar{\mu}y) = \mu(x, y).$$

Combining the first and last terms of this equality, we have

$$\lambda(x, y) = \mu(x, y),$$

which implies $(x, y) = 0$, and completes the proof.

We should now like to regroup and restate some results that have been obtained thus far in this chapter. We shall pick up a few new results that could easily have been deduced before, but we waited for this point as the more opportune moment to present them. Immediately thereafter we can proceed to one of our main results, namely, the *spectral decomposition theorem for normal transformations*.

As demonstrated by Theorem 2.2 in a complex finite-dimensional inner product space, given a normal transformation A, there exists an orthonormal basis for the space X consisting of the eigenvectors of A. Suppose $\lambda_1, \lambda_2, ..., \lambda_k$ are the distinct eigenvalues of A and let us group the members of the orthonormal basis, according to which of the eigenvectors they go with, as follows:

$$\underbrace{x_{11}, x_{12}, ..., x_{1j_1}}_{\lambda_1}; \quad \underbrace{x_{21}, x_{22}, ..., x_{2j_2}}_{\lambda_2}; \quad \cdots \quad \underbrace{x_{k1}, x_{k2}, ..., x_{kj_k}}_{\lambda_k}.$$

Introducing the notation

$$M_i = N(A - \lambda_i) \quad \text{[the null space of } (A - \lambda_i) \text{ for } i = 1, 2, ..., k\text{]},$$

we can write any vector $x \in X$ as

$$x = \underbrace{\alpha_{11}x_{11} + \cdots + \alpha_{1j_1}x_{1j_1}}_{\in M_1} + \underbrace{\alpha_{21}x_{21} + \cdots + \alpha_{2j_2}x_{2j_2}}_{\in M_2} + \cdots + \underbrace{\alpha_{k1}x_{k1} + \cdots + \alpha_{kj_k}x_{kj_k}}_{\in M_k},$$

and can conclude that $X = M_1 + M_2 + \cdots + M_k$. Suppose now that

$$x \in M_i \cap (M_1 + M_2 + \cdots + M_{i-1} + M_{i+1} + \cdots + M_k)$$

$$\Rightarrow x = x_1 + x_2 + \cdots + x_{i-1} + x_{i+1} + \cdots + x_k.$$

However, $x \in M_i$ implies that $(x, x_j) = 0$ for $j \neq i$; hence

$$(x, x) = (x_1 + x_2 + \cdots x_{i-1} + x_{i+1} + \cdots + x_k, x) = 0;$$

Therefore

$$x = 0.$$

Applying Theorem 2.5, we can conclude that the subspaces $M_1, M_2, ..., M_k$ constitute an *orthogonal* direct sum decomposition of the space.

Now suppose that $E_1, E_2, ..., E_k$ are the orthogonal projections on $M_1, M_2, ..., M_k$, respectively. By Theorem 2.4 of this lecture we see that

(1) $1 = E_1 + E_2 + \cdots + E_k$;
(2) $E_i \neq 0$ for any $i = 1, 2, ..., k$, because no $M_i = \{0\}$;
(3) $E_i E_j = 0$ for $i \neq j$.

Since we have a direct sum decomposition of the space, there must exist $x_i \in M_i$ $(i = 1, ..., k)$ such that, for any $x \in X$,

$$x = \sum_{i=1}^{k} x_i.$$

Applying A to x gives

$$Ax = \sum_{i=1}^{k} Ax_i$$

$$= \sum_{i=1}^{k} \lambda_i x_i$$

$$= \sum_{i=1}^{k} \lambda_i E_i x_i$$

$$= \sum_{i=1}^{k} \lambda_i E_i x$$

$$= \left(\sum_{i=1}^{k} \lambda_i E_i\right) x;$$

therefore

$$A = \sum_{i=1}^{k} \lambda_i E_i,$$ (2.5)

and we have "decomposed" the normal transformation A into a linear combination of orthogonal projections where the scalar multipliers are the distinct eigenvalues of A. The representation (2.5) for A is called the *spectral form* of A. We summarize these results by stating the next theorem.

THEOREM 2.6. To every normal transformation A on a complex finite-dimensional inner product space there correspond scalars $\lambda_1, \lambda_2, ..., \lambda_k$, the distinct eigenvalues of A, and orthogonal projections $E_1, E_2, ..., E_k$ (where k is a positive integer not greater than dim X) such that

(1) E_i is the orthogonal projection on $N(A - \lambda_i)$ $(i = 1, ..., k)$;
(2) $E_i \neq 0$ (for $i = 1, 2, ..., k$) and $E_i E_j = 0$ (for $i \neq j$);
(3) $\sum_{j=1}^{k} E_j = 1$;
(4) $\sum_{j=1}^{k} \lambda_j E_j = A$.

This theorem is called the *spectral decomposition theorem for normal transformations*. We note immediately that if the transformation was *self-adjoint*, we could even weaken the hypothesis to include the case of a *real* inner product space, the complex numbers being needed only to guarantee the existence of eigenvalues for normal transformations. If this were done, we would rename the theorem and call it the spectral decomposition theorem for *self-adjoint* transformations.

As a final word on the spectral form for any normal transformation, however, one can say that it is unique.

Before proving the uniqueness of the representation for a normal transformation, mentioned in the spectral decomposition, some facts about *Lagrange polynomials* will be stated (the interested reader might also check Ref. 2).

Let $\lambda_1, \lambda_2, ..., \lambda_k$ be distinct complex numbers. We define the *i*th *Lagrange polynomial* to be

$$p_i(\lambda) = \prod_{\substack{j=1 \\ j \neq i}}^{k} \frac{\lambda - \lambda_j}{\lambda_i - \lambda_j}$$

and note immediately that

$$p_i(\lambda_j) = \delta_{ij},$$

where δ_{ij} is the Kronecker delta. We now observe that if $P(\lambda)$ is any polynomial of degree less than or equal to $(k - 1)$, we have the following representation for $P(\lambda)$:

$$P(\lambda) = \sum_{i=1}^{k} p_i(\lambda)P(\lambda_i),$$

where the $p_i(\lambda)$ are just the Lagrange polynomials. With this in mind, we can now state the next theorem.

THEOREM 2.7. Suppose there exist distinct complex numbers $\lambda_1, \lambda_2, ..., \lambda_k$, and nonzero linear transformations $E_1, E_2, ..., E_k$ such that the normal transformation A can be written

(1)
$$A = \sum_{j=1}^{k} \lambda_j E_j,$$

(2)
$$E_i E_j = 0 \qquad (i \neq j),$$
and

(3)
$$1 = \sum_{j=1}^{k} E_j.$$

Then $\lambda_1, ..., \lambda_k$ are the distinct eigenvalues of A and E_i is the orthogonal projection on $N(A - \lambda_i)$ for $i = 1, ..., k$.

Proof. To show that each E_i is a projection, one need only apply E_i to both sides of (3) and apply (2). We now show that the $\lambda_1, \lambda_2, ..., \lambda_k$ are the distinct eigenvalues of A. Denoting the range of E_i by $R(E_i)$, let us consider some nonzero vector $x \in R(E_i)$. (Such a vector must exist because E_i is not identically zero.) Since $x \in R(E_i)$ and E_i is a projection, we have $E_i x = x$. Applying A to both sides of this last equality, we have

$$Ax = AE_i x$$
and we also contend that

$$AE_i x = E_i Ax,$$

that is, that A commutes with E_i for $i = 1, 2, ..., k$. To show that this last contention is so we note that

$$A = \sum_{j=1}^{k} \lambda_j E_j$$

and property (2) of the hypothesis imply that

$$A^2 = \sum_{j=1}^{k} \lambda_j^2 E_j. \tag{2.6}$$

We can now apply A to both sides in (2.6) repeatedly and reapply (2) to conclude that, for any positive integer n,

$$A^n = \sum_{j=1}^{k} \lambda_j^n E_j.$$

Hence, for any polynomial $f(\lambda)$, we can say that

$$f(A) = \sum_{j=1}^{k} f(\lambda_j) E_j,$$

and since it is true for any polynomial, it is certainly also true for the Lagrange polynomials mentioned before. In view of this we have, for the ith Lagrange polynomial $p_i(\lambda)$,

$$p_i(A) = \sum_{j=1}^{k} p_i(\lambda_j) E_j = \sum_{j=1}^{k} \delta_{ij} E_j$$

$$= E_i.$$

Thus L_i is a polynomial in A and must, therefore, commute with A, and we can now write, using (1) again,

$$Ax = E_i Ax = E_i \left(\sum_{j=1}^{k} \lambda_j E_j x \right)$$

$$= \lambda_i E_i x$$

$$= \lambda_i x$$

to conclude that each λ_i is an eigenvalue of A. We now wish to prove that these are the *only* eigenvalues that A has. To this end, suppose

$$Ax = \mu x \qquad (x \neq 0)$$

or

$$(A - \mu)x = 0.$$

Using (1) and (3), we can rewrite this as

$$\sum_{j=1}^{k} (\lambda_j - \mu) E_j x = 0.$$

Applying E_i to both sides, we obtain for any $i = 1, 2, ..., k$,

$$(\lambda_i - \mu) E_i x = 0.$$

Hence if $\mu \neq \lambda_i$ for any i, then $(\lambda_i - \mu) \neq 0$, and we can say that $E_i x = 0$ for $i = 1, 2, ..., k$, which, using (3) again, implies that

$$x = E_1 x + \cdots + E_k x = 0,$$

which is a contradiction because we assumed $x \neq 0$. Therefore μ must equal one of the λ_i.

We now wish to show that the E_i are orthogonal projections on the $N(A - \lambda_i)$, the null spaces of $A - \lambda_i$; hence we must first show that

$$R(E_i) = N(A - \lambda_i).$$

Suppose $x \in N(A - \lambda_i)$. This implies

$$(A - \lambda_i)x = 0.$$

Using (1) and (3), this can be rewritten as

$$\sum_{j=1}^{k} (\lambda_j - \lambda_i) E_j x = 0.$$

Applying E_m to both sides and using (2), we have

$$(\lambda_m - \lambda_i) E_m x = 0.$$

Thus $m \neq i$ implies that $E_m x = 0$. Hence

$$x = E_1 x + E_2 x + \cdots + E_k x = E_i x$$

or

$$x \in R(E_i).$$

Since we have already shown that $x \in R(E_i)$ implies that x is an eigenvector of A, we have demonstrated the equality of the two sets $R(E_i)$ and $N(A - \lambda_i)$.

To complete the proof we must show that E_i is the *orthogonal* projection on $N(A - \lambda_i)$. Since A is normal, $E_i = p_i(A)$ is normal too. Thus E_i is a projection and E_i is normal. By Theorem 2.3, we can conclude that E_i is the orthogonal projection on its range.

The following theorem yields an alternative way of characterizing the adjoint of a normal transformation and constitutes one nice application of the spectral theorem.

THEOREM 2.8.   Let X be a finite-dimensional inner product space and let A be a linear transformation on X. Then A is normal if and only if A^* can be expressed as some polynomial in A.

Proof. If A^* can be written as some polynomial in A, it must certainly commute with A, which establishes the sufficiency of the condition. To show that it is also necessary, suppose that A is normal. By the spectral decomposition theorem, then, we can write A as

$$A = \sum_{j=1}^{k} \lambda_j E_j,$$

where the λ_j are the eigenvalues of A, and the E_j are orthogonal projections. It was pointed out previously that an orthogonal projection must always be self-adjoint, which allows us to write

$$A^* = \sum_{j=1}^{k} (\lambda_j E_j)^*$$

$$= \sum_{j=1}^{k} \bar{\lambda}_j E_j.$$

But, as noted in Theorem 2.7, the E_j can be written as $p_j(A)$, where $p_j(\lambda)$ is just the jth Lagrange polynomial. Hence,

$$A^* = \sum_{j=1}^{k} \bar{\lambda}_j p_j(A),$$

which completes the proof.

2.3   Unitary and Orthogonal Transformations

We now wish to consider linear transformations $U : X \to X$ where X is a finite-dimensional inner product space X, with the property that

$$U^*U = 1. \tag{2.7}$$

If X is a complex space, then U is called *unitary*; if X is real, then U will be called *orthogonal*. Our first claim about transformations satisfying (2.7) is that they also satisfy

$$UU^* = 1.$$

To prove this, suppose $Ux = Uy$. Applying U^* to both sides, we have

$$U^*Ux = U^*Uy,$$

which, by (2.7), implies that $x = y$. Hence if U satisfies (2.7), it must be 1 : 1. On a finite-dimensional space, however, if a linear transformation is 1 : 1 it must also be onto, and we can conclude that U must have an inverse, which in this case must be U^*, to establish the contention. For most of our results about transformations which satisfy (2.7), we shall assume that X is complex (that is, that U is unitary) with appropriate modifications applying when one wishes to obtain analogous results about orthogonal transformations.

THEOREM 2.9. Let U be a linear transformation on the finite-dimensional inner product space X. In this event, the following statements are equivalent:

(1) $U^*U = 1$;
(2) $(Ux, Uy) = (x, y)$;
(3) $\|Ux\| = \|x\|$ for all $x \in X$.

Proof (1) $\Rightarrow$ (2). To prove this, all we need do is note that

$$(Ux, Uy) = (x, U^*Uy) = (x, y)$$

when (1) is satisfied.

To prove that (2) implies (3), we need only consider

$$\|Ux\|^2 = (Ux, Ux) = (x, x) = \|x\|^2.$$

Before establishing that (3) implies (1), however, we shall establish the following result: *if A is self-adjoint and $(Ax, x) = 0$ for all x, then $A = 0$*. Since $(Ax, x) = 0$ for any x, then for any x and y we also have

$$(A(x + y), x + y) = 0.$$

Expanding this, we get

$$0 = (Ax, x) + (Ax, y) + (Ay, x) + (Ay, y)$$
$$= (Ax, y) + (Ay, x)$$
$$= (Ax, y) + (y, A^*x)$$
$$= (Ax, y) + (y, Ax)$$
$$= (Ax, y) + \overline{(Ax, y)}.$$

Since y is arbitrary, let $y = Ax$. We then have $2\|Ax\|^2 = 0$ for all $x \in X$, which implies $Ax = 0$ for all x; hence $A = 0$.

We now show that (3) implies (1). By (3) we can write

$$(Ux, Ux) = (x, x)$$

or

$$(U^*Ux, x) = (x, x)$$

or

$$((U^*U - 1)x, x) = 0 \quad \text{(for any x)}.$$

All we need do now is note that $U^*U - 1$ is self-adjoint and apply the result just proved to complete the proof.

THEOREM 2.10. If U is a unitary transformation on the finite-dimensional inner product space X, then each of the eigenvalues of U must have an absolute value equal to 1.

Proof. Suppose $Ux = \lambda x$, $x \neq 0$. By the preceding theorem, then,

$$\|x\| = \|Ux\| = \|\lambda x\| = |\lambda| \, \|x\|,$$

which, since $\|x\| \neq 0$, implies that $|\lambda| = 1$.

Suppose now that A is a normal transformation on a complex finite-dimensional, inner product space. In view of the spectral decomposition theorem, we can write A as

$$A = \lambda_1 E_1 + \cdots + \lambda_k E_k,$$

where the λ_i and E_i are as in Theorem 2.6. We can also write

$$A^* = \bar{\lambda}_1 E_1 + \cdots + \bar{\lambda}_k E_k.$$

If all $\bar{\lambda}_i = \lambda_i$, then clearly $A^* = A$. Next we note that

$$A^*A = |\lambda_1|^2 E_1 + \cdots + |\lambda_k|^2 E_k.$$

Suppose now that each λ_i $(i = 1, 2, ..., k)$ has absolute value equal to 1. In this case then, clearly,

$$A^*A = 1.$$

On the other hand, suppose $A^*A = 1$. This implies that

$$1 = |\lambda_1|^2 E_1 + \cdots + |\lambda_k|^2 E_k,$$

which, in turn, implies that

$$E_i = |\lambda_i|^2 E_i$$

or

$$(1 - |\lambda_i|^2)E_i = 0.$$

Thus, since E_i cannot be zero for any $i = 1, 2, ..., k$, we can conclude that $|\lambda_i|^2 = 1$ for $i = 1, 2, ..., k$. We summarize these results in Theorem 2.11.

THEOREM 2.11. Let A be a normal transformation on a complex finite-dimensional inner product space. Then

(1) A is self-adjoint if and only if each eigenvalue of A is real;
(2) A is unitary if and only if each eigenvalue of A has absolute value equal to 1.

We noted before for any normal transformation A and any polynomial $p(\lambda)$, that $p(A) = \sum_{j=1}^{k} p(\lambda_j)E_j$, where, again, the λ_j and E_j are as in Theorem 2.6.

Suppose now that $f(\lambda)$ is any complex-valued function defined on the points $\lambda_1, ..., \lambda_k$, and consider the linear transformation one obtains by *defining* $f(A)$ to be

$$f(A) = \sum_{j=1}^{k} f(\lambda_j)E_j.$$

At first sight it seems reasonable to expect that the class of linear transformations obtained in this manner would be a somewhat "larger" class than the one gotten by considering just polynomials in A. This is not the case, however, and we shall show that the classes obtained by the two different methods of generation are identical. To this end, let $f(\lambda)$ be an arbitrary complex function defined on the distinct points $\lambda_1, \ldots, \lambda_k$ and let $\alpha_j = f(\lambda_j)$. Now consider the polynomial function

$$P(\lambda) = \sum_{j=1}^{k} \alpha_j p_j(\lambda),$$

where the $p_j(\lambda)$ are the jth Lagrange polynomials. In this case the linear transformation corresponding to this polynomial is just

$$P(A) = \sum_{j=1}^{k} \alpha_j p_j(A) = \sum_{j=1}^{k} f(\lambda_j) E_j = f(A).$$

Hence for any such complex-valued function, there exists a polynomial that will yield the same, linear transformation; for example, suppose $f(\lambda) = \bar{\lambda}$, which is not a polynomial. In this case we have $f(A) = A^* = \bar{\lambda}_1 E_1 + \cdots + \bar{\lambda}_k E_k$, which, after noting that $E_j = p_j(A)$, we see is a polynomial in A.

EXERCISES 2

1. Show that, if $\|Ax\| = \|A^*x\|$ for all x belonging to the finite-dimensional inner product space X, the linear transformation A is normal.

2. Let X be a finite-dimensional complex inner product space. Prove that the linear transformation A is normal if, and only if, $A(M) \subset M$ implies $A(M^\perp) \subset M^\perp$, where M is a subspace.

3. If A is a self-adjoint linear transformation on the finite-dimensional inner product space X such that $A^n = 1$ for some positive integer n, prove that $A^2 = 1$.

4. If A and B are normal linear transformations on the finite-dimensional complex inner product space X such that $AB = 0$, does it follow that $BA = 0$?

5. Let X be a complex three-dimensional vector space and let x_1, x_2, and x_3 be a basis for X. Let A be a linear transformation mapping X into X such that

$$Ax_1 = x_1 + x_2 - x_3;$$
$$Ax_2 = x_1 + 3x_2 + x_3;$$
$$Ax_3 = -x_1 + x_2 - x_3.$$

Determine the spectral decomposition of A.

6. Let A be a normal transformation on the finite-dimensional vector space X. If B is a transformation which commutes with A, show that B also commutes with A^*.

7. Let A and B be normal transformations on the finite-dimensional complex inner product space X such that A and B commute. Prove that there exists a normal transformation C such that both A and B are polynomials in C. (*Hint:* Let $A = \Sigma_i \lambda_i E_i$, $B = \Sigma_j \mu_j F_j$ be the spectral decompositions of A and B; let $C = \Sigma_{i,j} \gamma_{ij} G_{ij}$, where $G_{ij} = E_i F_j$ and $\{\gamma_{ij}\}$ are arbitrary complex numbers. Show that C is normal and then make use of the discussion after Theorem 2.11.)

8. Show that, if A commutes with AA^*, then A is normal.

9. Let A be a linear transformation defined on the complex inner product space X such that $(Ax, x) \geq 0$ for all $x \in X$. A is then called a *positive linear transformation*. Show that a positive linear transformation is self-adjoint and that a normal transformation is positive if and only if each eigenvalue is nonnegative.

10. If A is a positive linear transformation (see Exercise 9), show that A possesses a unique positive square root, that is, there exists a unique positive linear transformation B such that $A = B^2$.

REFERENCES

1. P. R. Halmos, "Finite-Dimensional Vector Spaces." A good discussion of projections and direct sums can be found here as well as the complexification (p. 151) and a proof of the spectral theorem (p. 156).

2. K. Hoffman and R. Kunze, "Linear Algebra." A discussion of Lagrange polynomials and Lagrange's interpolation formula appears on pp. 114–115; the spectral theorem appears on p. 275.

CHAPTER **3**

Normed Spaces and Metric Spaces

Probably the biggest difference between analysis and algebra is that in the former one utilizes the concept of limits. With limiting notions, one can discuss such things as differentiation and integration and can deal with infinite processes in general; or, rather, one can avoid infinite processes by reducing them to a consideration of some finite process. If one looks at anything pertaining to limits of real or complex numbers one sees that they all rest on the fact that some measure of "closeness" can be ascribed to any pair of points. In order to wrest linear spaces from purely the realm of algebra, we now wish to introduce some sort of distance-measuring device to those spaces and ultimately introduce limiting notions. Specifically, the concept of a *norm* on a vector space will be defined, and indeed almost all our subsequent results will involve *normed spaces*.

A more general notion than norm will also be dealt with in this chapter—that of a *metric*, a distance-measuring device defined on an arbitrary set. After defining these things, many familiar notions about real or complex numbers, such as limits of sequences, continuous functions, and Cauchy sequences, will be generalized to this new setting.

3.1   Norms and Normed Linear Spaces

In Chapter 1 we defined the notation $\|x\|$ to mean $(x, x)^{1/2}$ in an inner product space, and we would now like to note some further properties of this mapping. It is simple to verify that the following assertions about $\|x\|$ are valid:

$$\| \ \| : X \to R,$$

$$x \to \|x\|.$$

(1) $\|x\| \geq 0$ and $= 0$ if and only if $x = 0$;
(2) for a scalar α and a vector x, $\|\alpha x\| = |\alpha| \|x\|$;
(3) $\|x + y\| \leq \|x\| + \|y\|$ (triangle inequality).

We shall verify here that the third assertion is indeed correct. Consider

$$\|x + y\|^2 = (x + y, x + y)$$
$$= (x, x) + (x, y) + (y, x) + (y, y)$$
$$= \|x\|^2 + 2 \ \text{Re}(x, y) + \|y\|^2.$$

But, since $2 \ \text{Re}(x, y) \leq 2|(x, y)|$, we can say

$$\|x + y\|^2 \leq \|x\|^2 + 2|(x, y)| + \|y\|^2.$$

Appealing to the Cauchy–Schwarz inequality, noted in Chapter 1, we can now say

$$\|x + y\|^2 \leq \|x\|^2 + 2\|x\| \ \|y\| + \|y\|^2 = (\|x\| + \|y\|)^2,$$

and it remains only to take the square root of both sides to complete the proof.

We now wish to view mappings of the above type abstractly; that is, we now single out the three properties cited above and define mappings of this type, on a real or complex space, to be *norms*. Note that to have a norm it is not necessary to have an inner product at all. Conversely, though, whenever one has an inner product, one can always define a norm (that is, a mapping that will satisfy the above three conditions) right from the inner product as we have already indicated. Real or complex spaces over which a norm is defined are called *normed linear spaces* (n.l.s.), and we remark that there are normed linear spaces over which one cannot define an inner product that will "generate" [that is, no inner product such that $(x, x)^{1/2} = \|x\|$] the norm on the space. Thus the essential thing is to make the norm and inner product generated norm agree. As one might expect, if further conditions are imposed on the norm, one can guarantee the existence of an inner product that will generate the norm. One such further condition is the parallelogram law mentioned in Chapter 1. There are many other conditions that one might equally well impose upon the norm to guarantee an inner product representation but other interests prevent us from listing them all here. It is now necessary to leave normed spaces temporarily for the sake of a more general structure—the *metric space*. Normed spaces will be returned to and dwelt upon at great length from Sec. 8.2 on.

3.2   Metrics and Metric Spaces

If X is a normed linear space and $x, y \in X$, it is easily verified that the function $d(x, y) = \|x - y\|$ satisfies the following conditions:

(1) $d(x, y) \geq 0$ and is equal to zero if and only if $x = y$;
(2) $d(x, y) = d(y, x)$;
(3) $d(x, z) \leq d(x, y) + d(y, z)$ for any $x, y, z \in X$ (triangle inequality).

Again, in a manner analogous to the way the norm was defined abstractly, one can now make the following definition.

DEFINITION 3.1. X, an arbitrary set, will be called a *metric space* if there is a function $d : X \times X \to R$ satisfying properties (1)–(3) above. The mapping d itself will be called a *metric*.

One thing that should be pointed out immediately is that, when one refers to metric spaces, it does not suffice to mention just the set X involved; indeed, one must speak of the pair (X, d), for it is certainly possible that two different metrics could be defined on the same set and would, thus, give rise to two different metric spaces. To clarify these notions, several examples will now be given, noting that any normed space is, of course, a metric space.

EXAMPLE 3.1. Although the metric we shall define now is not too interesting for its own sake, it comes in quite handy in counterexamples. In addition, it illustrates that one can define a metric over any set whatsoever.

Let X be an arbitrary set and consider the function

$$d(x, y) = \begin{cases} 1 & (x \neq y), \\ 0 & (x = y), \end{cases}$$

where x and y are members of X. It is exceedingly simple to verify that properties (1)–(3) are satisfied by this mapping. This particular metric is called the *trivial metric*.

EXAMPLE 3.2. Take the set X to be the real numbers R, and $d(x, y)$ to be $|x - y|$, the usual absolute value.

EXAMPLE 3.3. Take X to be euclidean 2-space R^2, and for any two members of R^2, $x = (\alpha_1, \alpha_2)$ and $y = (\beta_1, \beta_2)$, take

$$d(x, y) = [(\alpha_1 - \beta_1)^2 + (\alpha_2 - \beta_2)^2]^{1/2}.$$

Alternatively, one might choose $d(x, y)$ to be

$$d(x, y) = |\alpha_1 - \beta_1| + |\alpha_2 - \beta_2|$$

or

$$d(x, y) = \max(|\alpha_1 - \beta_1|, |\alpha_2 - \beta_2|)$$

with similar results holding for euclidean n-space. Although these metrics are quite different, they do have a feature that unites them: They all lead to equivalent topologies. This notion will be discussed in more detail later.

EXAMPLE 3.4. Let $X = C[a, b]$ (continuous functions on the closed interval $[a, b]$) and let $f, g \in X$. We can now take, as a metric on this set,

$$d(f, g) = \max_{x \in [a,b]} |f(x) - g(x)|.$$

EXAMPLE 3.5. Let $X = Q$, the rational numbers, let c be some real number such that $0 < c < 1$, and let p be any prime. It is clear that any $x \in Q$ can be written

$$x = p^\alpha \frac{a}{b} \qquad (\text{where } p \nmid a \quad \text{and} \quad p \nmid b)$$

where $p|a$ means "p divides a" and $p \nmid a$ means "p does not divide a." We can now define the *p-adic valuation* of any $x \in Q$ as

$$|x|_p = c^\alpha \qquad (x \neq 0),$$

$$|0|_p = 0,$$

where $|\ \ |_p$ has the following properties:

(1) $|x|_p \geq 0$ and equals zero if and only if $x = 0$.
(2) $|xy|_p = |x|_p |y|_p$. To prove this consider

$$x = p^\alpha(a/b), \qquad y = p^\beta(a'/b'),$$

where $p \nmid a, b, a', b'$. Hence,

$$xy = p^{\alpha+\beta} \frac{aa'}{bb'} \qquad (p \nmid aa', p \nmid bb'),$$

and

$$|xy| = c^{\alpha+\beta} = c^\alpha c^\beta = |x|_p |y|_p.$$

(3) $|x + y|_p \leq \max(|x|_p, |y|_p)$. Proving this directly is a bit of chore, but it turns out that this condition is equivalent to the following condition [Eq. (3.1)], the verification of which is not as difficult:

$$|x|_p \leq 1 \Rightarrow |1 + x|_p \leq 1. \tag{3.1}$$

Let us prove that (3.1) is really equivalent to (3). Suppose (3) is true and suppose $|x|_p \le 1$. By (3), then,

$$|1 + x|_p \le \max(1, |x|_p) = 1,$$

it being clear that $|1|_p = 1$. Conversely, suppose now that (3.1) holds, and consider any two $x, y \in Q$. We may assume that x and y are not zero for the assertion is clearly valid in that case. Further, let us suppose that

$$|x|_p \le |y|_p.$$

This implies that

$$\frac{|x|_p}{|y|_p} = \left|\frac{x}{y}\right|_p \le 1.$$

Using (3.1), we have

$$\left|1 + \frac{x}{y}\right|_p \le 1,$$

which implies

$$|x + y|_p \le |y|_p = \max(|x|_p, |y|_p)$$

and establishes the equivalence of (3.1) and (3). We now show that (3.1) is satisfied. It follows immediately that, for $x \ne 0$, if $|x|_p = c^\alpha \le 1$, then $\alpha \ge 0$ and x can be written in the form $x = e/d$, where $(e, d) = 1$ (e and d are *relatively prime*, in other words the greatest common divisor of e and d is 1) and $p \nmid d$. In this case, then

$$1 + x = 1 + \frac{e}{d} = \frac{e + d}{d},$$

and $1 + x$ also has a denominator relatively prime to p, which implies $|1 + x|_p \le 1$ and completes the proof.

With these basic properties of the p-adic valuation established, consider the function $d(x, y) = |x - y|_p$. Clearly, $d(x, y) \ge 0$ and equals zero if and only if $x = y$. Also

$$d(y, x) = |y - x|_p = |-1|_p |x - y|_p = |x - y|_p = d(x, y)$$

so

$$d(x, y) = d(y, x).$$

Finally,

$$d(x, z) = |x - z|_p = |(x - y) + (y - z)|_p$$

$$\le \max(|x - y|_p, |y - z|_p) = \max(d(x, y), d(y, z)).$$

Thus

$$d(x, z) \le \max(d(x, y), d(y, z)), \tag{3.2}$$

from which it immediately follows that

$$d(x, z) \le d(x, y) + d(y, z),$$

and we have established that d is a metric or that (Q, d) is a metric space. Inequality (3.2) is often called the *ultrametric* inequality, it being clearly stronger than the triangle inequality. We note that the trivial metric satisfies the ultrametric inequality also.

EXAMPLE 3.6. Given a finite collection of metric spaces (X_1, d_1), (X_2, d_2), ..., (X_n, d_n), it is possible to assign a distance function d to the set of all ordered n-tuples $(x_1, x_2, ..., x_n)$, where $x_i \in X_i$ $(i = 1, 2, ..., n)$, that is, to the cartesian product $X_1 \times X_2 \times \cdots \times X_n$. Certainly one possibility is to assign the trivial metric. Other possibilities, each of which defines a metric of a less trivial nature are given below, the verification that they are indeed bona fide distance functions being left to the reader.

Let $x = (x_1, x_2, ..., x_n)$ and $y = (y_1, y_2, ..., y_n)$ be points of $X_1 \times X_2 \times \cdots \times X_n$. Each of the following defines a metric on $X_1 \times X_2 \times \cdots \times X_n$:

(1) $$d(x, y) = d_1(x_1, y_1) + d_2(x_2, y_2) + \cdots + d_n(x_n, y_n);$$

(2) $$d(x, y) = \max_i d_i(x_i, y_i);$$

(3) $$d(x, y) = \left[\sum_{i=1}^{n} d_i(x_i, y_i)^2 \right]^{1/2}.$$

Last, we remark that, if A is a subset of a metric space (X, d), A itself can be viewed as a metric space by merely restricting d to A. A with this restricted metric is called a *subspace* of the metric space.

3.3 Topological Notions in Metric Spaces

Suppose (X, d) is a metric space, $x \in X$, and r is a positive real number. We now define a *spherical neighborhood of radius r about x (with center x)* to be the set of points

$$S_r(x) = \{y \in X | d(y, x) < r\}.$$

For example, suppose X is a normed linear space. In this case we would have, as a neighborhood about a point,

$$S_r(x) = \{y \in X \mid \|y - x\| < r\}.$$

If the metric space was (R, d) as in Example 3.2, then

$$S_r(x) = \{y \in R \mid |y - x| < r\}.$$

If we considered (X, d), where X was any set and d was the trivial metric, then

$$S_r(x) = \begin{cases} \{x\} & (0 < r \leq 1), \\ X & (r > 1). \end{cases}$$

DEFINITION 3.2. Suppose now that $A \subset X$. Then $x \in A$ is called an *interior point of A* if there exists an $r > 0$ such that $S_r(x) \subset A$, the collection of all interior points of A being called the *interior of A* and denoted by A^0. It is clear that $A^0 \subset A$.

If every point of A is an interior point of A (that is, $A^0 = A$), then A is called an *open* set. We would now like to examine some properties that the class of all open sets in a given metric space (X, d) have. Denoting the class of all open sets by O, we have

(O_1) Arbitrary unions of open sets are also open. Symbolically, if $O_\alpha \in O$ $(\alpha \in \Lambda, \Lambda$ some index set), then $\bigcup_{\alpha \in \Lambda} O_\alpha \in O$.

(O_2) If $O_1, O_2, ..., O_n \in O$ for any finite n, then

$$\bigcap_{i=1}^{n} O_i \in O.$$

(O_3) X and $\emptyset$ are each open sets.

DEFINITION 3.3. A point x is called an *adherence point of the set* $A \subset X$ if, for any $S_r(x)$, $S_r(x) \cap A \neq \emptyset$. The set of all adherence points of A is called the *closure of A* and is denoted by $\bar{A}$.

Three observations that follow immediately from the definition are the following:

(1) $A \subset \bar{A}$;
(2) $A \subset B \Rightarrow \bar{A} \subset \bar{B}$;
(3) $\overline{A \cup B} = \bar{A} \cup \bar{B}$.

To illustrate how we prove observations such as these, we shall prove (3). It is clear that B and A are each contained in $A \cup B$. Since this is true then (2) implies that

$$\bar{A} \subset \overline{A \cup B},$$

$$\bar{B} \subset \overline{A \cup B},$$

which, together, imply that

$$\bar{A} \cup \bar{B} \subset \overline{A \cup B}.$$

Conversely, suppose that $x \in \overline{A \cup B}$ and, in addition, suppose that $x \notin \bar{A}$ and $x \notin \bar{B}$. This assumption means that there exist neighborhoods of x, $S_{r_1}(x)$ and $S_{r_2}(x)$, such that $S_{r_1}(x) \cap A = \emptyset$ and $S_{r_2}(x) \cap B = \emptyset$. Let r be the smaller of r_1 and r_2. In this case, we have

$$S_r(x) \cap (A \cup B) = \emptyset,$$

which is contradictory. Therefore $x \in \bar{A}$ or $x \in \bar{B}$, which implies that $x \in \bar{A} \cup \bar{B}$. One notes in passing here that the above argument cannot be extended to include the case of infinitely many sets. An essential feature in the proof is our ability to select the smaller of the two real numbers r_1 and r_2. If there were an infinite number of sets under consideration, no such selection would be possible. A concept closely related to, but somewhat stronger than, an adherence point is given in Definition 3.4. The reader should have the distinction between the two clearly fixed in his mind.

DEFINITION 3.4. The point x is said to be a *limit point* of $A \subset X$ if and only if, for every r, $S_r(x) \cap A$ contains infinitely many points of A, the collection of all limit points of A being called the *derived set* of A and denoted by A'. One notes immediately that no finite set can possess a limit point.

DEFINITION 3.5. Consider the sequence of points in X, $x_1, x_2, ..., x_n, ... = \{x_n\}$. The sequence of points $\{x_n\}$ is said to *converge* to the point x, denoted by

$x_n \rightarrow x$, if for any $S_r(x)$ there exists an N (which will depend on r, in general) such that

$$n > N \Rightarrow x_n \in S_r(x).$$

Clearly, $x_n \rightarrow x$ is equivalent to requiring that the sequence of real numbers, $d(x, x_n)$, converge to zero.

Now consider the pair (X, d), where d is the trivial metric. In this space $x_n \rightarrow x$ implies that, for some N, for $n > N$, all x_n must equal x.

As another example consider the pair (Q, d), where $d(x, y) = |x - y|_p$, and the sequence $\{p, p^2, ..., p^n, ...\}$. Now,

$$d(0, p^n) = |p|_p^{\,n} = c^n,$$

and, since $c < 1$, we have $p^n \rightarrow 0$, a result which rather confounds one's intuition.

3.4   Closed and Open Sets, Continuity, and Homeomorphisms

Let (X, d) be a metric space and let A be a subset of X. It is clear that, if x is an adherence point of A, every neighborhood of x, in particular each $S_{1/n}(x)$, where n runs through the positive integers, must contain a point of A. Calling a point of A selected from $S_{1/n}(x)$, x_n, it follows that the sequence of points $\{x_n\}$ must converge to $x : x_n \rightarrow x$. In light of this we can say:

$$x \in \bar{A} \text{ if and only if there exists a sequence } \{x_n\},$$

where $x_n \in A$ for every n, such that $x_n \rightarrow x$.

It is immaterial whether the points in the sequence are distinct or not; that is, the sequence $(x, x, x, ..., x, ...)$ is perfectly acceptable. We note that, if x is actually a point of A, this sequence will always satisfy the conditions in the above assertion. If, however, x is not a member of A but belongs to $\bar{A}$ (x must be a limit point of A in this case), then a sequence of distinct points (see Exercise 3.1) can be selected.

We define a *closed* set to be one that contains all its limit points. It can be shown that this is equivalent to defining a set A to be closed if and only if $A = \bar{A}$ (the reader is asked to verify that $\bar{A}$ itself is a closed set in Exercise 3.3), and it is this definition that we shall use in proving the following relationship between closed and open sets.

THEOREM 3.1. In a metric space (X, d), for a subset of X, A,

(1) A being open implies that CA, the complement of A, is closed;
(2) A closed implies that CA is open.

Proof. The reader, having seen the proof of assertion (1), can easily prove (2) by little more than just interchanging the words open and closed; for this reason, only (1) will be proved here.

Suppose A is open and $x \in A$. Since A is open, there exists some neighborhood of x, $S_r(x)$, such that $S_r(x) \subset A$. This implies that $S_r(x) \cap CA = \emptyset$ with analogous statements holding for every $x \in A$. Thus, for any $x \in A$, we can say that

$x \notin \overline{CA}$ (if it did, every neighborhood of x would have to contain a point of CA)

or

$$x \notin CA \Rightarrow x \notin \overline{CA}.$$

Taking the contrapositive of this statement, we get

$$x \in \overline{CA} \Rightarrow x \in CA$$

or

$$\overline{CA} \subset CA.$$

But, since any set is contained in its closure, we also have

$$CA \subset \overline{CA}.$$

Hence

$$CA = \overline{CA}$$

and CA is closed.

The proof of Theorem 3.2 is straightforward and will be omitted.

THEOREM 3.2. In a metric space (X, d), if $\{A_\alpha\}$ where $\alpha \in \Lambda$, Λ an index set, is a collection of closed sets, then

$$(\text{F}_1) \qquad\qquad \bigcap_{\alpha \in \Lambda} A_\alpha$$

is closed too. (The intersection of closed sets is always a closed set.) Further, if B_i $(i = 1, 2, ..., n)$ are closed sets, then

$$(\text{F}_2) \qquad\qquad \bigcup_{i=1}^{n} B_i$$

is also a closed set. (The union of a finite number of closed sets is a closed set.)

(F_3) X and $\emptyset$ are closed.

We would now like to extend the familiar concept of a function of a real (or complex) variable being continuous to a more general setting.

DEFINITION 3.6. Let (X, d) and (Y, d') be metric spaces and let $f: X \to Y$. The function f is said to be *continuous at the point* $x_0 \in X$ if, for any neighborhood $S_\varepsilon(f(x_0))$ there exists a $\delta > 0$ such that

$$f(S_\delta(x_0)) \subset S_\varepsilon(f(x_0)).$$

The function f is said to be a *continuous function* if it is continuous at every point of X.

It is easily verified that this definition matches up exactly with the usual notion of continuity of a real-valued function of a real variable, when viewed in that framework.

The following theorem gives us an equivalent way of defining continuity in a metric space.

THEOREM 3.3. If (X, d) and (Y, d') are metric spaces where $f: X \rightarrow Y$, then f is continuous at the point x if and only if for every sequence $\{x_n\}$ converging to x, $f(x_n) \rightarrow f(x)$.

Proof (necessity). Suppose f is continuous and let $x_n \rightarrow x$. For any ε there exists a δ such that

$$f(S_\delta(x)) \subset S_\varepsilon(f(x)).$$

For this δ there exists an integer N_δ, such that $n > N_\delta$ implies that $x_n \in S_\delta(x)$. By virtue of the above inclusion, then

$$f(x_n) \in S_\varepsilon(f(x)) \qquad \text{(for } n > N_\delta).$$

Hence $f(x_n) \rightarrow f(x)$.

Proof (sufficiency). Let $\varepsilon > 0$ be given and suppose for any δ there is an $x' \in S_\delta(x)$ such that

$$d'(f(x'), f(x)) \geq \varepsilon. \tag{3.3}$$

Consider now any sequence of real numbers $\{\delta_n\}$ such that $\delta_n \rightarrow 0$. For each n select the x' that satisfies Eq. (3.3) and call this x_n'. It is clear that $x_n' \rightarrow x$, but $f(x_n') \nrightarrow f(x)$, because the distance between $f(x_n')$ and $f(x)$ cannot be made smaller than ε for any n. Hence we have shown that, if f is not continuous, then not every sequence $\{x_n\}$ converging to x will yield a sequence $\{f(x_n)\}$ that converges to $f(x)$. Taking the contrapositive of this statement demonstrates that the condition is sufficient.

Continuing on in the same vein, we now state another condition that is equivalent to continuity.

THEOREM 3.4. If (X, d) and (Y, d') are metric space and $f: X \rightarrow Y$, then f is continuous if and only if $f^{-1}(F)$, where F is any closed set in Y, is a closed set in X.

Proof. First we shall show that the condition is necessary. To this end, suppose that f is continuous, F is a closed set in Y, and let $A = f^{-1}(F)$. What we want to show now is that every limit point of A must actually belong to A or that $x \in A'$ implies that $x \in A$. Thus, let us consider any limit point of A, x. Since $x \in A'$, there must exist a sequence of points of A, $\{x_n\}$ such that $x_n \rightarrow x$. Since f is continuous, though, this implies that

$$f(x_n) \rightarrow f(x).$$

Each of the terms $f(x_n)$, however, must be in F, which implies that $f(x) \in \bar{F}$. Since F is closed, though, $F = \bar{F}$, and we can say that $f(x) \in F$, which is equivalent to saying that $x \in f^{-1}(F) = A$ and completes this part of the proof.

Using the fact that the inverse mapping takes closed sets into closed sets, we must now show that f is continuous. For any $x \in X$ it is clear that $S_\varepsilon(f(x))$ is open, which implies that $CS_\varepsilon(f(x))$ is closed. Since this is a closed set in Y, its pre-image, $f^{-1}[CS_\varepsilon(f(x))]$, must be a closed set in X. We can rewrite the above†

$$f^{-1}[CS_\varepsilon(f(x))] = Cf^{-1}[S_\varepsilon(f(x))]$$

is closed. Taking the appropriate complement, we can say that

$$f^{-1}[S_\varepsilon(f(x))]$$

is open. Since the set is open and x is a member of the set, there must exist some neighborhood of x lying wholly within the set (x must be an interior point) or, there must exist a δ such that

$$S_\delta(x) \subset f^{-1}[S_\varepsilon(f(x))],$$

which implies

$$f(S_\delta(x)) \subset S_\varepsilon(f(x)),$$

which is the statement of continuity at the point x and completes the proof.

COROLLARY. If (X, d) and (Y, d') are metric spaces and $f: X \to Y$, then f is continuous if and only if $f^{-1}(B)$, where B is any open set in Y, is an open set in X.

EXERCISES 3

The context for the following exercises is assumed to be a metric space (X, d).

1. Show that, if x is an adherence point of a set A but is not a member of A, a sequence of distinct points x_n ($n = 1, 2, \ldots$) can be selected from A such that $x_n \to x$.

2. Prove that $\overline{S_\varepsilon(x)} \subset \{y \mid d(x, y) \leq \varepsilon\}$, and give an example for which the inclusion is proper.

3. For an arbitrary subset A of X, show that $\overline{\overline{A}} = \overline{A}$.

4. For an arbitrary subset A of X, prove that $C\overline{A} = (CA)^0$ and $C(A^0) = \overline{CA}$.

5. For any subsets A and B, prove that $(A \cap B)^0 = A^0 \cap B^0$.

6. Give examples for which (a) $\overline{A \cap B} \neq \overline{A} \cap \overline{B}$; (b) $(A \cup B)^0 \neq A^0 \cup B^0$.

7. Give an example of a metric space and subsets A_i ($i = 1, 2, \ldots$) such that $\overline{\left(\bigcup_{i=1}^\infty A_i\right)} \neq \bigcup_{i=1}^\infty \overline{A_i}$.

† For any function f and any two sets X and Y, where $f: X \to Y$, we can say that $f^{-1}(CA) = Cf^{-1}(A)$ for any $A \subset Y$; a rather nice property of inverse functions for which there is no corresponding statement about f itself.

8. Let A be a subset of X. For $x \in X$, the *distance from x to A*, denoted by $d(x, A)$ is defined as follows: $d(x, A) = \inf_{y \in A} d(x, y)$. Show that $d(x, A) = 0$ if and only if $x \in \bar{A}$.

9. Show that, if d satisfies the ultrametric inequality, then if $d(x, y) \neq d(y, z)$,

$$d(x, z) = \max(d(x, y), d(y, z)).$$

10. Show that, if d satisfies the ultrametric inequality, every point of the sphere $S_\varepsilon(x)$ is a center.

11. In the definition of $|\ \ |_p$ on Q, take $c = 1/p$. Then show that, if $x \in Q$ and $x \neq 0$, $\Pi_p |x|_p = 1/|x|$, where the product on the left is taken over all primes p.

12. (a) Show that $d(x, x_0)$ is a continuous function of x. (b) Show that $d(x, y)$ is a continuous function of x and y when $X \times Y$ carries the metric mentioned in (1) on p. 43.

13. Let f, g be two continuous mappings of (X, d) into (Y, d'). Show that the set $\{x \in X \mid f(x) = g(x)\}$ is a closed subset of X.

14. Let $X = R$ and define $d(x, y) = |x - y|^\alpha$, where $\alpha \in R$ $(0 < \alpha \leq 1)$. Show that (R, d) is a metric space.

15. Let $D_k(a, b)$ be the space of all complex-valued functions on $[a, b]$ having continuous derivatives up to order k. For $f \in D_k(a, b)$, define

$$\|f\| = \sum_{i=0}^{k} \frac{1}{i!} \max\{|f^{(i)}(t)| \mid t \in [a, b]\}.$$

One can easily show, with respect to the way linear operations were introduced to the space $C[a, b]$ of continuous functions on $[a, b]$ that $D_k(a, b)$ is a linear space. For the above operation, show that the norm axioms are satisfied.

REFERENCE

1. A. Komolgoroff and S. Fomin, "Elements of the Theory of Functions and Functional Analysis," Vol. 1. Discussion and examples of metric spaces, as well as properties of metric spaces, Sections 8–12.

CHAPTER **4**

Isometries and the
Completion of a Metric Space

A vital concept for the consideration of metric spaces is that of an *isometry*—a mapping between metric spaces which "preserves metric properties." After introducing an isometry and another similar mapping, the *homeomorphism*, in Sec. 4.1, we put it to almost immediate use in discussing the *completion of a metric space*. It will be shown that any metric space can be embedded, in a certain sense, in a complete metric space or, loosely speaking, that an arbitrary metric space can be incorporated into a larger space that is complete. The rest of the chapter will be devoted to the completion procedure which, aside from its usefulness in the sequel, is an elegant mathematical construction.

4.1   Isometries and Homeomorphisms

Whenever certain specific properties of a set are being examined, one usually wants to see what sort of mappings will preserve the properties under consideration. For example, in considering vector spaces where the basic operation is that of forming linear combinations, the isomorphism is of interest because it preserves this basic property. Aside from placing two linear spaces in 1 : 1 correspondence, it also sets up a correspondence of linear combinations in the domain with linear combinations of corresponding vectors (that is, images under the isomorphism) in the range. When an isomorphism exists between two linear spaces, there is no harm done in adopting the logically inaccurate convention of actually *identifying the two spaces and calling them the same*. Even though the two spaces may be very different concretely (for example, one space may be a collection of polynomials, whereas the other may be a set of n-tuples of numbers), there is no way of abstractly telling them apart as vector spaces.

So much for vector spaces. Since our present interest is in metric spaces, we now wish to define a mapping called an *isometry* that will preserve *metric* properties. After this is done, we shall feel free to identify *isometric* spaces.

Apropos to this discussion and to the preceding paragraph we now wish to make two definitions.

DEFINITION 4.1. Let (X, d) and (Y, d') be metric spaces and let f be a 1 : 1, onto mapping such that $f : X \to Y$. If f and f^{-1} are each continuous functions (mapping of this type are called *bicontinuous*), then f is said to be a *homeomorphism*. Further, X is said to be *homeomorphic* to Y if such a mapping exists.

Homeomorphisms preserve all the essential topological properties; in particular, open sets in X are mapped into open sets in Y, and, if x is a limit point of $A \subset X$, then $f(x)$ will be a limit point of $f(A)$. The next definition defines a mapping that preserves metric properties.

DEFINITION 4.2. If f maps (X, d) in 1 : 1 onto fashion onto (Y, d') such that, for any $x, y \in X$, $d(x, y) = d'(f(x), f(y))$, then f is said to be an *isometry* and the two spaces are said to be *isometric*.

It is clear that an isometry is also a homeomorphism. As an example of an isometry, one might take the two metric spaces to be the real numbers with the usual metric under the mapping $x \to x + a$, where a is a fixed real number. Another mapping that is also an isometry in this context is the one that takes any real number x into $-x$. As it happens, these are the only isometries of the real line onto itself.

4.2 Cauchy Sequences and Complete Metric Spaces

DEFINITION 4.3. If, in a metric space (X, d), the sequence $\{x_n\}$ has the property that, for any $\varepsilon > 0$, there exists an integer N such that for all n and m greater than N, $d(x_n, x_m) < \varepsilon$, the sequence is said to be a *Cauchy sequence*. In the case of the real line, every Cauchy sequence converges; that is, being a Cauchy sequence is sufficient to guarantee the existence of a limit. In the general case, however, this is not so. If a metric space does have the property that every Cauchy sequence converges, the space is called a *complete* metric space. Several examples will now be considered.

EXAMPLE 4.1. Consider R^n, all n-tuples of real numbers. This collection, along with any of the distance functions mentioned in Example 3.6, is complete.

EXAMPLE 4.2. Let X be any set with the trivial metric d assigned to it. The only way $\{x_n\}$ can be a Cauchy sequence in this space is if there is some integer N such that $x_N = x_{N+1} = \cdots$. Hence (X, d) is complete.

EXAMPLE 4.3. Consider the space $C[a, b]$ with the distance between any two functions $f, g \in C[a, b]$ defined to be

$$d(f, g) = \max_{x \in [a, b]} |f(x) - g(x)|,$$

and suppose $\{f_n\}$ is a Cauchy sequence. This implies that

$$\max_{x \in [a, b]} |f_n(x) - f_m(x)| < \varepsilon$$

for n and m bigger than some N, say. Hence we can say by a theorem of advanced calculus that the sequence must converge uniformly. We have a uniformly convergent sequence of continuous functions and, by another well-known theorem from advanced calculus, we can assert the continuity of the limit function. Thus this space is complete.

EXAMPLE 4.4. Consider $C[a, b]$ but with the metric

$$d(f, g) = \left(\int_a^b |f(x) - g(x)|^2 \, dx \right)^{1/2} \qquad (f, g \in C[a, b]).$$

We shall show, by a counterexample, that this space is not complete. Let a and b be -1, and 1, respectively and consider the Cauchy sequence (Fig. 4.1)

$$f_n(x) = \begin{cases} 0 & (-1 \le x \le 0), \\ nx & (0 < x \le 1/n), \\ 1 & (1/n < x \le 1). \end{cases}$$

It is clear that the limit function in this case is a discontinuous function and, hence, the space is not complete.

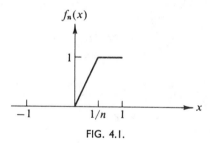

FIG. 4.1.

EXAMPLE 4.5. Let $X = (0, 1)$ and take the distance between any two numbers x and y belonging to X to be

$$d(x, y) = |x - y|.$$

The sequence $\{1/n\}$ is a Cauchy sequence in this space but converges to 0, which is not in the space. Thus this space is not complete.

EXAMPLE 4.6. Let $X = Q$, the rational numbers with distance between any two rational numbers x and y, be defined as

$$d(x, y) = |x - y|.$$

One can easily make up Cauchy sequences of rational numbers that converge to irrational numbers. Hence this space is not complete.

EXAMPLE 4.7. Again let $X = Q$, but take the distance between any two rational numbers to be $|x - y|_p$; that is, the p-adic valuation of their difference as discussed in Chapter 3. In the case when $p = 5$, it can be shown that the Cauchy sequence

$$x_n = \tfrac{1}{2} \sum_{k=0}^{n} (-1)^k \binom{\frac{1}{2}}{k} 5^k$$

converges to $\sqrt{-1}$ which is not a rational number. Similar examples can be constructed for each prime p. Thus Q with respect to any p-adic valuation is not complete.

The property of completeness is very important when one wishes to make an *existence* statement, and a fact of fundamental importance, which we shall prove later, is that every metric space can be regarded as a subset of a complete metric space. In a certain sense an incomplete metric space can be completed, the precise meaning of being completed being given by the following definition.

DEFINITION 4.4. Let (X, d) be an arbitrary metric space. The complete metric space (X^*, d^*) is said to be a *completion* of (X, d) if

(1) (X, d) is isometric to a subspace (X_0, d^*) of (X^*, d^*);
(2) the closure of X_0, $\overline{X}_0$, is all of $X^* : \overline{X}_0 = X^*$.

An equivalent way of stating condition (2) is saying that X_0 is *everywhere dense*, or simply *dense*, in X^*. It means that every point of X^* is either a point or a limit point of X_0, which means that, given any point x of X^*, there exists a sequence of points in X_0 that converge to x.

We are now in a position to demonstrate that every metric space has a completion.

THEOREM 4.1. Every metric space (X, d) has a completion (X^*, d^*) and, furthermore, if (X^{**}, d^{**}) is also a completion of (X, d), then (X^*, d^*) is isometric to (X^{**}, d^{**}); that is, the completion of a space is unique to within an isometry.

Proof. Consider the collection of all equivalence classes of Cauchy sequences in (X, d) under the following equivalence relation: If $\{x_n\}$ and $\{y_n\}$ are Cauchy sequences in (X, d), then $\{x_n\}$ is said to be equivalent to $\{y_n\}$, denoted by $\{x_n\} \sim \{y_n\}$, if and only if

$$\lim_{n \to \infty} d(x_n, y_n) = 0.$$

The collection of all these equivalence classes will be called X^*. We now wish to define a metric on X^* with the properties stated in the theorem. To this end, consider the following function: Let x^* and y^* be equivalence classes in X^*. We define

$$d^*(x^*, y^*) = \lim_{n \to \infty} d(x_n, y_n), \tag{4.1}$$

where $\{x_n\} \in x^*$ and $\{y_n\} \in y^*$. Before showing that this is even a metric, we must consider two other points. We must show that the limit [Eq. (4.1)] exists, and we must also show that it does not depend on which particular member of the equivalence class was chosen; that is, we must show that it is well-defined. First it will be shown that the limit (4.1) exists.

Let $\{s_n\}$ denote the sequence of real numbers $\{d(x_n, y_n)\}$, where the x_n and y_n are as in (4.1). Now we can say

$$|d(x_n, y_n) - d(x_m, y_m)| = |d(x_n, y_n) - d(x_n, y_m) + d(x_n, y_m) - d(x_m, y_m)|$$

$$\leq |d(x_n, y_n) - d(x_n, y_m)| + |d(x_n, y_m) - d(x_m, y_m)|$$

$$\leq d(y_n, y_m) + d(x_n, x_m),$$

where the last inequality follows from the fact that, if d is any metric, then $|d(x, y) - d(x, z)| \leq d(y, z)$. The last inequality, since $\{x_n\}$ and $\{y_n\}$ are Cauchy sequences, implies that $\{s_n\}$ is a Cauchy sequence. Since the real numbers are complete, the limit exists.

Now we show that d^* is well-defined. To this end, suppose

$$\{x_n\}, \{x_n'\} \in x^* \quad \text{and} \quad \{y_n\}, \{y_n'\} \in y^*.$$

We must show that the limit (4.1) is the same regardless of which sequences were chosen. Hence consider

$$|d(x_n, y_n) - d(x_n', y_n')|$$

$$\leq |d(x_n, y_n) - d(x_n', y_n)| + |d(x_n', y_n) - d(x_n', y_n')|$$

$$\leq d(x_n, x_n') + d(y_n, y_n').$$

Since $\{x_n\} \sim \{x_n'\}$ and $\{y_n\} \sim \{y_n'\}$, both terms on the right go to zero as n increases and d^* is, indeed, well-defined. We are now at least "in the ball park," for we have shown that it makes sense to consider d^* a function. We now show that it is a metric, by verifying that the axioms are satisfied.

(1) It is clear that $d^*(x^*, y^*) \geq 0$. We must also show that equality prevails if and only if $x^* = y^*$. To this end suppose $d^*(x^*, y^*) = 0$. This means that $\lim_{n \to \infty} d(x_n, y_n) = 0$, which implies that $\{x_n\} \sim \{y_n\}$ or $x^* = y^*$.

(2) Symmetry of d^* is clear because $d(x_n, y_n) = d(y_n, x_n)$ for every n.

(3) Last, we must show that the triangle inequality is satisfied. For this purpose let $x^*, y^*, z^* \in X^*$ and suppose that $\{x_n\}$, $\{y_n\}$, and $\{z_n\}$ are members of these classes, respectively. For every n, we can make the statement:

$$d(x_n, z_n) \leq d(x_n, y_n) + d(y_n, z_n).$$

Since the statement holds for every n, it is also true in the limit, and we have shown that (X^*, d^*) is a metric space.

We now wish to show that X^* contains a subspace X_0 that is isometric to X. In light of this, consider the subset of X^* consisting of those equivalence classes x', which contain the sequences $(x, x, ..., x, ...)$. Calling this collection X_0, we now contend that X is isometric to X_0, where the isometry is

$$X \to X_0,$$

$$x \to x'.$$

Since it is clear that there cannot be two different Cauchy sequences, of the type mentioned above, in the same equivalence class and that the mapping is onto by virtue of the way it was defined, we need only verify that it preserves distances to prove that it is isometry. For this purpose, let us compute the distance between any two points in X_0, x' and y':

$$d^*(x', y') = \lim_{n \to \infty} d(x, y) = d(x, y).$$

With this done we shall now show that $\overline{X}_0 = X^*$. To prove this, we show that, for any point in X^*, there is sequence of points in X_0 converging to it. Let $x^* \in X^*$ and let $\{x_n\} \in x^*$. For any $\varepsilon > 0$, there exists an N such that $n, m > N$ implies that $d(x_n, x_m) < \varepsilon$. Now let us examine the distance [in (X^*, d^*)] between the two points of X^*, x^* and x_n'; select the representatives

$$(x_1, x_2, ..., x_n, ...) \in x^*,$$

$$(x_n, x_n, ..., x_n, ...) \in x_n',$$

where $n > N$. Since $d(x_n, x_m) < \varepsilon$ for $n, m > N$, it follows that

$$d^*(x^*, x_n') = \lim_{m \to \infty} d(x_m, x_n) \leq \varepsilon.$$

Thus given any neighborhood of a point $x^* \in X^*$, there is some member of X_0 in there and, by virtue of this, it is possible to construct a sequence of points from X_0 converging to any desired point in X^*. We have now shown that $\overline{X}_0 = X^*$.

To show that X^* is complete, we must demonstrate that any Cauchy sequence of points in X^* converges to a point in X^*. Before doing this, however, we consider a more special type of sequence in X^* and show that it always converges to something in X^*: Consider the Cauchy sequence in X^*,

$$x_1', x_2', ..., x_n', ...,\qquad(4.2)$$

where

$$(x_i, x_i, ..., x_i, ...) \in x_i' \qquad (\text{for } i = 1, 2, ...).$$

Thus the distinguishing feature of this sequence is that the terms have been taken from X_0. We now claim that assuming (4.2) to be a Cauchy sequence in X^* implies that the sequence obtained by taking the isometric preimages of the entries in (4.2) is a Cauchy sequence in X (and, therefore, determines a point of X^*) and is also the limit that (4.2) approaches. Once again we are trying to show that the sequence " of sequences " [(4.2)] has a limit in X^*. Consider now the sequence obtained from (4.2) by taking the isometric images of each of its entries in X: Consider

$$\{x_n\} = (x_1, x_2, ...) \leftrightarrow (x_1', x_2', ...) = \{x_n'\}.$$

We wish to show first that $\{x_n\}$ is a Cauchy sequence in X. Since Eq. (4.2) is a Cauchy sequence and the mapping $X \to X_0$ is an isometry, we have that

$$d^*(x_m', x_n') = d(x_m, x_n)$$

is arbitrarily small for m and n bigger than some N. It is now seen that our original assumption that (4.2) was a Cauchy sequence implies that the sequence $\{x_n\}$ is a Cauchy sequence of points in X. Let us suppose that this Cauchy sequence belongs to the equivalence class x^*. We now show that the sequence (4.2) approaches x^* as a limit in X^*. Thus consider

$$\lim_{n \to \infty} d^*(x_n', x^*) = \lim_{n \to \infty} \lim_{m \to \infty} d(x_n, x_m) = 0,$$

because $\{x_n\}$ is a Cauchy sequence.

Now consider an arbitrary Cauchy sequence of points in X^*:

$$x_1^*, x_2^*, ..., x_n^*, ...$$

Since $\overline{X}_0 = X^*$, there exist points in X_0

$$x_1', x_2', ..., x_n', ...$$

such that $d^*(x_n^*, x_n') < (1/n)$. Consider now the sequences

$$x_1', x_2', ..., x_n', ...;$$

it is easily seen that, for any $\varepsilon > 0$,

$$d^*(x_n', x_m') \le d^*(x_n', x_n^*) + d^*(x_n^*, x_m^*) + d^*(x_m^*, x_m')$$

$$\le \frac{1}{n} + \varepsilon + \frac{1}{m}$$

for n and m large enough. Therefore,

$$\{x_n'\} \text{ is a Cauchy sequence.}$$

As noted above, then, $\{x_n'\}$ must converge to some $y^* \in X^*$; we now claim that $\{x_n^*\}$ must converge to y^*, too, and to prove it we only need note that

$$d^*(y^*, x_n^*) \le d^*(y^*, x_n') + d^*(x_n', x_n^*),$$

where, as previously noted, each of the terms on the right goes to zero as n increases.

Finally, we must show that we are justified in saying that (X^*, d^*) is *the* completion; that any other completion must be isometric to (X^*, d^*). To do this we must show that, for any other completion (X^{**}, d^{**}), there exists a $1:1$ onto mapping φ, taking X^* onto X^{**} with the property that if $\varphi(x^*) = x^{**}$ and $\varphi(y^*) = y^{**}$, then

$$d^{**}(x^{**}, y^{**}) = d^*(x^*, y^*).$$

In view of this, suppose there are two completions of X, X^* and X^{**}, with, of course, corresponding metrics. We first note that X itself must appear (actually an isometric image of X) in each of these spaces if each is to be a completion of X. This means that there is already a certain isometry established between the subsets of X^* and X^{**} that represents the image of X under that isometry. We shall call the images of X in X^* and X^{**}, X_0^* and X_0^{**}, respectively. For any point x^* of X^*, we can say that there must be a sequence of points in X_0^*, $\{x_n^*\}$, that must converge to x^*. Each of these points has an isometric image in X^{**}, though; thus the original (Cauchy) sequence in X^* gives rise to another Cauchy sequence in X^{**}, $\{x_n^{**}\}$. Since X^{**} is complete, this sequence must have a limit x^{**} in X^{**} and it is x^{**} that we shall take to be the image of x^*. It is clear that this mapping is $1:1$, onto, and does not depend on which member of the equivalence class was chosen. It only remains for us to show that distances are preserved under this mapping. Pictorially, we have, reading the arrows outside parentheses as "gives rise to"

$$(x^* \in X^*) \to (x_n^* \in X_0^*, \ x_n^* \to x^*) \to (\{x_n^*\} \text{ a Cauchy sequence})$$

$$\to (\{x_n^{**}\}, \text{ where } x_n^{**} \text{ is the isometric image of } x_n^* \text{ in } X_0^{**})$$

$$\to (x_n^{**} \to x^{**} \in X^{**})$$

as the mapping φ. We immediately note that, if one considers the isometric image x' of any point x, from X, in X^*, a sequence of points from X_0^* that converges to x' is just

$$(x', x', \ldots).$$

Thus $\varphi(x')$ is just the isometric image of x itself in X_0^{**}. If we had adopted the practice of just identifying an element x of a space X with its isometric image in the completion of X, we would say that φ reduces to the identity mapping on X_0^* and X_0^{**}, that is, takes an element into itself. Making this identification and also identifying d with d^* or d^{**} (when it is sensible to do so), we let

$$x_n \to x^* \in X^*,$$
$$y_n \to y^* \in X^* \qquad (x_n, y_n \in X_0^*),$$

and

$$x_n \to x^{**} \in X^{**},$$
$$y_n \to y^{**} \in X^{**}.$$

We now have

$$d^*(x^*, y^*) \le d^*(x^*, x_n') + d(x_n, y_n) + d^*(y_n', y^*),$$

for every n, where $x_n' = (x_n, x_n, \ldots)$ and similarly for y_n. Since it is true for every n, it is true in the limit as n becomes infinite, which yields

$$d^*(x^*, y^*) \le \lim_{n \to \infty} d(x_n, y_n).$$

But $d(x_n, y_n) \le d^*(x_n', x^*) + d^*(x^*, y^*) + d^*(y^*, y_n')$ which yields the reverse inequality. In a completely analogous manner, we can also show that

$$d^{**}(x^{**}, y^{**}) = \lim_{n \to \infty} d(x_n, y_n).$$

Consequently,

$$d^*(x^*, y^*) = d^{**}(x^{**}, y^{**}),$$

which completes the proof.

EXERCISES 4

Unless otherwise indicated, operations are assumed to take place in a metric space (X, d).

1. If $x_n \to x$, show that $\{x_n\}$ is a Cauchy sequence; show also that every subsequence of $\{x_n\}$ also converges to x.

2. Show that, if a Cauchy sequence $\{x_n\}$ has a convergent subsequence, it is convergent.

3. Give an example of a homeomorphism which is not an isometry.

4. Give an example of two metric spaces which are homeomorphic and where one is complete but the other is not.

5. Prove that the only isometries of R with the usual metric are $f(x) = x + a$ and $f(x) = -x + a$.

6. Assuming (X, d) to be a complete metric space, let A be a closed subset of X. Show that A considered a metric space by itself is complete.

7. If (X, d) and (Y, d') are metric spaces and if $f : X \to Y$, show that f is continuous if and only if for every $A \subset X, f(\bar{A}) \subset \overline{f(A)}$, and construct an example to show that the closure operation is not necessarily preserved by a continuous mapping. Moreover, if f is $1 : 1$ and onto, show that X and Y are homeomorphic if and only if $f(\bar{A}) = \overline{f(A)}$ for all $A \subset X$.

8. Let Q_p be the completion of Q with respect to $|\ \ |_p$ (see Example 3.5); actually Q_p can be made into a field called the *field of p-adic numbers*). Defining convergence in the natural fashion, show that if $\{a_n\}$ is a sequence of elements of Q_p, then $\sum_{n=1}^{\infty} a_n$ converges if and only if $a_n \to 0$.

9. Consider a mapping $f: X \to X$. f is said to be a *contraction mapping* if there exists a positive number a $(a < 1)$, such that $d(f(x), f(y)) \le a d(x, y)$ for any $x, y \in X$. Show that every contraction mapping is continuous and that every contraction mapping on a *complete* metric space has a unique fixed point, that is, a point y such that $f(y) = y$. [*Hint:* Consider for arbitrary $x_0 \in X$ the sequence $x_1 = f(x_0)$, $x_2 = f(x_1), \ldots, x_n = f(x_{n-1}), \ldots$.]

10. A mapping $f: X \to Y$, where (X, d) and (Y, d') are metric spaces, is said to be *uniformly continuous on* X if for every $\varepsilon > 0$ there exists a $\delta > 0$ such that $d(x_1, x_2) < \delta$ implies that $d'(f(x_1), f(x_2)) < \varepsilon$.

Prove (a) If (Y, d') is complete and $f: A \to Y$ is uniformly continuous on A and $\bar{A} = X$, then f has a unique continuous extension $g: X \to Y$, and this extension g is uniformly continuous on X. (b) Show that (a) is not true in general if Y is not complete.

11. Show that the cartesian product of a finite number of complete metric spaces, with respect to any of the product distances mentioned in Example 3.6, is complete.

REFERENCE

1. A. Komolgoroff and S. Fomin, "Elements of the Theory of Functions and Functional Analysis," Vol. 1, pp. 39–43.

CHAPTER 5

Compactness in Metric Spaces

We shall begin to exploit the concept of completeness, introduced in the last chapter, in this one. In many cases it will be possible to construct Cauchy sequences, and completeness will be used when we wish to assert that those sequences have limits. It will thus become quite evident why completeness is such a desirable property. At the outset, an equivalent way of describing a complete space is given in our first theorem. We shall then introduce the extremely important concept of *compactness* and other notions bearing a close relation to compactness in metric space such as ε-nets and *totally bounded sets*. Other related notions will then be introduced—namely, *countable compactness* and *sequential compactness*, and we then show that the notions of compactness, countable compactness, and sequential compactness are all equivalent in a metric space.

5.1 Nested Sequences and Complete Spaces

DEFINITION 5.1. By a *nested sequence* of sets $\{F_i\}$ ($i = 1, 2, \ldots$), we shall mean that

$$F_1 \supset F_2 \supset F_3 \supset \cdots.$$

DEFINITION 5.2. The *diameter of a set* $A \subset X$ where (X, d) is a metric space, $d(A)$, is defined to be $\sup_{a, b \in A} d(a, b)$.

Many important results can be derived, using the above notions, from the following theorem.

THEOREM 5.1. A necessary and sufficient condition that the metric space (X, d) be complete is that every nested sequence of nonempty closed sets $\{F_i\}$ ($i = 1, 2, \ldots$), with diameters tending to zero, have a nonempty intersection:

$$\bigcap_{i=1}^{\infty} F_i \neq \emptyset.$$

Proof. First suppose (X, d) is complete. We shall show that this implies that the above condition is satisfied. Suppose that $\{F_i\}$ ($i = 1, 2, \ldots$) is a nested sequence of nonempty closed sets such that

$$d(F_i) \to 0.$$

By selecting a point $x_n \in F_n$ for each $n = 1, 2, \ldots$, we can generate the sequence $\{x_n\}$. This sequence must be a Cauchy sequence though, because, assuming $m > n$, we have

$$d(x_m, x_n) \leq d(F_n),$$

which tends to zero. Since (X, d) is complete, $\{x_n\}$ must have a limit x. Now, for any given m, we have the sequence of points $\{x_m, x_{m+1}, \ldots\} \subset F_m$. In view of this,

$$x \in \bar{F}_n = F_n \quad \text{(for every } n\text{)}.$$

Hence

$$x \in \bigcap_{n=1}^{\infty} F_n,$$

and the intersection is seen to be nonempty.

Conversely, suppose that the condition is satisfied, and let $\{x_n\}$ be a Cauchy sequence. Letting $H_n = \{x_n, x_{n+1}, ...\}$, we can say that, since $\{x_n\}$ is a Cauchy sequence, $d(H_n) \to 0$. It is a simple exercise to verify that this implies that $d(\overline{H}_n) \to 0$. Further, since

$$H_n \supset H_{n+1} \Rightarrow \overline{H}_n \supset \overline{H}_{n+1},$$

we can say that $\{\overline{H}_n\}$ constitutes a closed, nested sequence of nonempty sets in X whose diameters tend to zero. By the hypothesis, then, we can assert the existence of an x such that

$$x \in \bigcap_{n=1}^{\infty} \overline{H}_n.$$

Therefore, since $d(x, x_n) \le d(\overline{H}_n) \to 0$, we can say

$$x_n \to x$$

to complete the proof.

Bounded Sets and Compactness

DEFINITION 5.3. A subset A of the metric space (X, d) is said to be *bounded* if there exists a real number M such that $d(A) \le M$. We define the *distance between the point x and the set B* of (X, d) to be $d(x, B) = \inf_{y \in B} d(x, y)$ and, in an analogous manner, define the *distance between the two sets B and C* of (X, d) to be

$$d(B, C) = \inf_{\substack{x \in B \\ y \in C}} d(x, y).$$

We now proceed to define the extremely important notion of compactness.

DEFINITION 5.4. Let Λ be an indexing set and further suppose that $\{G_\alpha\}$, where α runs through Λ, has the property that $A \subset \bigcup_{\alpha \in \Lambda} G_\alpha$. Then $\{G_\alpha\}$ is said to be a *covering* of the set A. If the set Λ is finite, it is said to be a *finite covering* of A. If each G_α is open, the covering is said to be an *open covering* of A. If the set A has the property that, from *every* open covering, one can select a finite subcovering, A is said to be *compact*.

Since we must be able to select a finite subcovering from *every* open covering of the set A, if A is to be compact, then it suffices to exhibit one open covering from which no finite subcovering can be selected to prove that A is not compact. For example, if we happen to be working on the real line with the usual metric and consider the open interval $(0, 1)$, it suffices to note that no finite subcovering can be selected from the open covering $\{(1/n, 1)\}$, where n runs through the positive integers, to prove that $(0, 1)$ is not compact. Similarly, working in the same space, if we consider the set $\{n\}$, where n runs through the integers, we can note that no finite subcovering can be selected from the open covering $\{(n - \frac{1}{2}, n + \frac{1}{2}) \mid n = 0, \pm 1, \pm 2, ...\}$ to prove that this set is not compact. In retrospect, we note that the first set was not closed and the second set was not bounded. Theorem 5.2 shows that both of these conditions must be satisfied for a set to have a chance of being compact, in a metric space.

THEOREM 5.2. Let (X, d) be a metric space. Then A is compact only if A is closed and bounded.

Proof. In the proof that follows, it will first be shown that compactness of a set implies that the set is closed. Then we shall show that compactness of a set implies that the set is bounded. A feature to be noted is that, in proving the first contention, we do not employ the metric directly; instead, we make use of the property of the metric that, given any two distinct points, we can find disjoint open sets containing each of the points. To prove the second contention, however, the metric itself plays a vital role.

Compactness of a Set Implying the Set Be Closed

Suppose A is a compact subset of (X, d). Let $x \in A$ and let $y \in CA$. For some real number ε_x, there exist neighborhoods of x and y, $S_{\varepsilon_x}(x)$ and $S_{\varepsilon_x}(y)$ such that

$$S_{\varepsilon_x}(x) \cap S_{\varepsilon_x}(y) = \emptyset. \tag{5.1}$$

In addition, it is clear that

$$A \subset \bigcup_{x \in A} S_{\varepsilon_x}(x).$$

Since A is compact, though, there must exist $x_1, x_2, \ldots, x_n$, such that

$$A \subset \bigcup_{i=1}^{n} S_{\varepsilon_{x_i}}(x).$$

For each of the x_i ($i = 1, 2, \ldots, n$), there is some neighborhood of the fixed point y, $S_{\varepsilon_{x_i}}(y)$, satisfying (5.1). Now we let

$$B = \bigcap_{i=1}^{n} S_{\varepsilon_{x_i}}(y).$$

Suppose now that $A \cap B$ was not null. This would mean that

$$B \cap S_{\varepsilon_{x_i}}(x) \neq \emptyset$$

for some i. Therefore, $S_{\varepsilon_{x_i}}(x) \cap S_{\varepsilon_{x_i}}(y) \neq \emptyset$, and contradicts the way in which the ε_{x_i} were chosen. Hence $A \cap B = \emptyset$, where B is an open set containing y, and we have shown that no point in the complement of A can be a limit point of A. Thus all the limit points of A must belong to A or $A' \subset A$, which implies that $A = \bar{A}$ or that A is closed.

Proof that Compactness Implies Boundedness

If A is not bounded, there must exist $x, y \in A$ such that, for any pre-assigned number, the distance between x and y is greater than that number. With this in mind, suppose that A is compact and consider an open covering of A consisting of 1-neighborhoods of each of its points:

$$A \subset \bigcup_{x \in A} S_1(x).$$

Since A is assumed compact, there must exist $x_1, x_2, ..., x_n$ such that

$$A \subset \bigcup_{i=1}^{n} S_1(x_i).$$

Now let $a = \max d(x_i, x_j)$, $1 \leq i \leq j \leq n$ and assume there exist $x, y \in A$ such that $d(x, y) > a + 2$. Since $x, y \in A$, there must exist elements x_i and x_j from the above-mentioned set such that $x \in S_1(x_i)$ and $y \in S_1(x_j)$. Applying the triangle inequality, we can now say that

$$d(x, y) \leq d(x, x_i) + d(x_i, x_j) + d(x_j, y) \leq 1 + a + 1$$

and contradict the way in which the x and y were chosen. Hence the assumption that they could be chosen in the first place must be wrong, and we have proved that A must be bounded.

A pitfall to be avoided is assuming that the converse of the above theorem holds in general, as is demonstrated by the following counterexample.

Suppose X is any infinite set and the trivial metric d has been assigned to it. For $\varepsilon < \frac{1}{2}$, we can say that

$$S_\varepsilon(x) = \{x\} \subset \{x\}$$

to conclude that even one-point sets are open sets in this space. Certainly, X itself is a bounded set, because the distance between any two of its points is at most 1 and, as noted in Chapter 4, this space is even complete. In addition, since X is the whole space, X is also closed. Now consider the following open covering of X:

$$X \subset \bigcup_{x \in X} \{x\}.$$

It is clear that no finite subcovering can be selected from this covering, and we conclude that the closed and bounded set X is not compact.

There are certain special spaces however in which the converse does hold, a notable example of which is R^n, all n-tuples of real numbers with any of the metrics mentioned in Example 3.6, and one need only slightly modify the ordinary Heine–Borel theorem to prove that this is so.

We now wish to proceed to some other, equivalent, ways of describing compact sets in metric spaces.

5.2 Relative Compactness, ε-Nets, and Totally Bounded Sets

DEFINITION 5.5. In the metric space (X, d) a subset A is said to be *relatively compact* if $\bar{A}$ is compact.

It is immediately evident that, in the euclidean spaces R^n (or C^n), relative compactness is equivalent to boundedness by virtue of the paragraph immediately preceding this section. In general, however, since compact sets must always be closed by Theorem 5.2, we can say that compact sets are always relatively compact.

DEFINITION 5.6. The set of points N of the metric space (x, d) is said to be an *ε-net with respect to the set A* if, for any $x \in A$, there exists a $y \in N$ such that $d(x, y) < \varepsilon$.

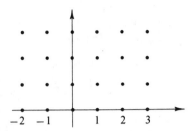

FIG. 5.1. An ε-net for R^2 for $ε > \sqrt{2}/2$.

For example, suppose A is R^2. Then the set $N = \{(m, n) \mid m, n = 0, \pm 1, ...\}$ constitutes an ε-net for R^2 (see Fig. 5.1), provided ε is greater than $\sqrt{2}/2$.

DEFINITION 5.7. A set A of the metric space (x, d) is said to be *totally bounded* if for any $ε > 0$ there exists a finite ε-net with respect to A.

We shall now prove that total boundedness in the general metric space immediately implies boundedness. To this end, let A be a totally bounded set, and suppose $ε > 0$ has been given. Since A is assumed to be totally bounded, there exists a finite ε-set for A, N. Since N is a finite set of points, the diameter of N must be finite also: $d(N) < \infty$. Now let a_1 and a_2 be any two points of A. There must be two corresponding points of N, y_1 and y_2, such that $d(a_1, y_1) < ε$ and $d(a_2, y_2) < ε$. Applying the triangle inequality, we can now say that

$$d(a_1, a_2) \le d(a_1, y_1) + d(y_1, y_2) + d(y_2, a_2)$$

$$\le 2ε + d(N).$$

But, since a_1 and a_2 were *any* two elements of A, this implies that

$$\sup_{a_1, a_2 \in A} d(a_1, a_2) = d(A) \le 2ε + d(N),$$

which proves that A must be bounded.

In the case of euclidean spaces, it can be shown that not only does total boundedness imply boundedness, but that the implication also goes the other way around. In the general case, however, the converse does not hold, as is evidenced by the following counterexample.

Consider the metric space l_2 (see Example 1.3), and let $x = (\alpha_1, \alpha_2, ..., \alpha_n, ...)$ and $y = (\beta_1, \beta_2, ..., \beta_n, ...)$ be two points of this space with distance between them $d(x, y)$ defined to be

$$d(x, y) = \left[\sum_{i=1}^{\infty} |\alpha_i - \beta_i|^2 \right]^{1/2}$$

Consider the subset of l_2 now,

$$A = \{x \in l_2 \mid d(x, 0) = 1\},$$

that is, the points on the surface of the unit sphere in l_2. For any two points x and y in A, we can say that

$$d(x, y) \le d(x, 0) + d(0, y) = 2$$

to conclude that A is indeed bounded. To prove that A is not totally bounded, consider the following collection of points from A:

$$e_1 = (1, 0, 0, \ldots)$$
$$e_2 = (0, 1, 0, \ldots)$$
$$e_3 = (0, 0, 1, \ldots)$$
$$\vdots \qquad\qquad \vdots$$

In this case, for $n \neq m$, we can say that $d(e_n, e_m) = \sqrt{2}$. We now contend that, for $\varepsilon < \sqrt{2}/2$, no finite ε-net can exist. Suppose $\varepsilon < \sqrt{2}/2$ and

$$d(e_i, y_i) < \varepsilon \qquad \text{and} \qquad d(e_j, y_j) < \varepsilon.$$

Is it possible now for y_i to equal y_j? If $y_i = y_j$, then

$$d(e_i, e_j) \leq d(e_i, y_i) + d(y_i, e_j) < 2\varepsilon \leq \sqrt{2},$$

which is impossible. Thus if we assume that a finite ε-net existed, at least one point in the net would have to be within $\sqrt{2}/2$ of infinitely many of the e_i, and the above contradiction would prevail. Therefore, there can be no finite ε-net for $\varepsilon < \sqrt{2}/2$, and A cannot be totally bounded.

The following theorem gives us an important criterion for determining when a set is totally bounded. It is identical in form to the converse of the Bolzano–Weierstrass theorem for real numbers. (See also Exercise 5.6.)

THEOREM 5.3. Let A be a subset of the metric space (X, d). If from every sequence of points from A one can select a convergent subsequence, then A is totally bounded.

Proof. We shall prove the contrapositive of the above statement. To this end, suppose that A is not totally bounded. This means that, for some ε, no finite ε-net exists for A. In particular, it also means that there can be no finite ε-net *in A itself*. Further if x_1 is any point in A, there must be another point of A, x_2, such that $d(x_1, x_2) \geq \varepsilon$, for otherwise the set $\{x_1\}$ would constitute an ε-net for A. Similarly, there must exist a point $x_3 \in A$ such that

$$d(x_1, x_3) \geq \varepsilon \qquad \text{and} \qquad d(x_2, x_3) \geq \varepsilon,$$

otherwise the set $\{x_1, x_2\}$ would constitute an ε-net for A. Proceeding in this fashion, we generate the sequence of points

$$x_1, x_2, x_3, \ldots,$$

with the property that, for $n \neq m$, $d(x_n, x_m) \geq \varepsilon$. In view of this, it is clearly impossible to extract any convergent subsequence from this sequence of points. (A necessary condition that a sequence be convergent is that the sequence be a Cauchy sequence.) Thus we have proved a little more than we originally set out to prove; indeed, we have shown that not only is it possible to assert that A is totally bounded if a convergent subsequence can be selected from every sequence, but also that the finite ε-net is *wholly contained in A*. We shall have occasion to make use of this fact very shortly.

5.3 Countable Compactness and Sequential Compactness

DEFINITION 5.8. A set A is said to be *countably compact* if every infinite subset of A has a limit point in A.

Our next theorem shows that a compact set must always be countably compact.

THEOREM 5.4. All compact sets are countably compact.

Proof. We shall, as in the previous theorem, prove the contrapositive; hence we shall assume that A is not countably compact and prove that this implies A is not compact. Since A is not countably compact, there must be some infinite subset of A, B with no limit point in A. Since B is infinite, it must be possible to select a denumerable set of distinct points from B. Let $M = \{x_1, x_2, \ldots\}$ be such a set. It is clear that the set M can have no limit point in A either. Let $E_1, E_2, \ldots, E_n, \ldots$ be a sequence of ε_n-neighborhoods of $x_1, x_2, \ldots, x_n, \ldots$, respectively, where the ε_n have been chosen to guarantee that

$$E_n \cap M = \{x_n\} \qquad (n = 1, 2, \ldots).$$

If it were not possible to construct such neighborhoods for every point, this would mean that for some x_n, every neighborhood of x_n would contain some other point of M distinct from x_n. Thus x_n would be a limit point of the set M and, hence, of B.

Now let x be a point of $A - M$. Since B can have no limit points in A, there must exist some neighborhood of x, $E(x)$, such that $E(x) \cap M = \emptyset$. Letting $U = \bigcup_{x \in A - M} E(x)$, we see that we now have an open covering for A, for

$$A \subset \left(\bigcup_{n=1}^{\infty} E_n \right) \cup U.$$

Since U contains no points of M and each of the E_n contain only one point of M, we see that it is impossible to select any finite subcovering of A from this covering. Thus A is not compact. Taking the contrapositive of this result yields the theorem.

DEFINITION 5.9. A set A is said to be *sequentially compact* if, from every sequence in A, one can select a convergent subsequence with limit in A.

Our next theorem, in conjunction with some other results of this chapter, will show that the notions of compactness, countable compactness and sequential compactness are all equivalent in a metric space.

THEOREM 5.5. A subset A of a metric space (X, d) is sequentially compact if and only if it is compact.

Proof (sufficiency). Suppose that A is compact. By Theorem 5.4, we note that this implies A is countably compact. We shall now show, just using the fact that A is countably compact, that A is sequentially compact. For this purpose let $\{x_n\}$

be a sequence of points in A. There are now two categories into which this sequence might fall:

(1) It might be that, for $n_1 < n_2 < n_3 \cdots$,

$$x_{n_1} = x_{n_2} = x_{n_3} = \cdots = x;$$

that is, infinitely many terms of the sequence may be equal. In this case the theorem immediately follows, because we can take the subsequence consisting of the repeated point. Clearly, then

$$x_{n_m} \to x \in A.$$

(2) The other possibility is that infinitely many points of the sequence are distinct. In this case the *set* consisting of the points in the sequence,

$$\{x_1, x_2, \ldots\}$$

is an infinite set of points in A. Since A is countably compact, this set must have a limit point in A, x, which implies that there is some subsequence of points $\{x_{n_k}\}$ converging to $x \in A$. Thus, in either case, from every sequence of points of A, it is possible to extract a convergent subsequence which converges to something in A, which is the statement of sequential compactness.

Before proceeding to the proof that the condition is also necessary, let us examine more carefully what has already been done. We have not only shown that compactness implies sequential compactness, but also that countable compactness implies sequential compactness. In view of the fact that we have already shown that compactness implies countable compactness, when, in the next part of this theorem, we show that sequential compactness implies compactness, we shall have the following statement:

(Compactness) $\Leftrightarrow$ (countable compactness) $\Leftrightarrow$ (sequential compactness)

in a metric space.

Proof (necessity). Let A be sequentially compact and suppose $\{G_\alpha\}$ is an open covering of A. For any point $x \in A$ we now define† $\delta(x)$ to be

$$\delta(x) = \sup\{a \mid S_a(x) \subset G_\alpha \text{ for some } G_\alpha\}.$$

Thus, loosely speaking, $\delta(x)$ is just about the radius of the biggest spherical neighborhood of x that can still be contained in some open set of the covering.

We first wish to show that, for any x, $\delta(x)$ is positive. Since $x \in A$, there is some G_α from the covering such that $x \in G_\alpha$. By the definition of an open set, we can say that there must be some spherical neighborhood of x, $S_p(x)$, such that $S_p(x) \subset G_\alpha$. Now, since $\delta(x)$ is the supremum of all such numbers p, we can say that $\delta(x) \geq p > 0$.

Thus, for every $x \in A$, we have assigned a real number $\delta(x)$. We can now define

$$\delta_0 = \inf\{\delta(x) \mid x \in A\}.$$

---

† We may assume that each $\delta(x)$ exists in view of the following easily verified result: A set A is compact if and only if from every cover of A by open spheres $S_{\varepsilon_\mu}(x_\mu)$ where, for all μ, $0 \leq \varepsilon_\mu \leq 1$, a finite subcover may be extracted.

From the way δ_0 was defined (i.e., since it is a greatest lower bound), we can assert the existence of a sequence $\delta(x_n)$, where all $x_n \in A$, such that $\delta(x_n) \to \delta_0$. Our next goal is to show that the number δ_0 is strictly positive.

Since A is assumed to be a sequentially compact set, we can say that there must exist a subsequence of the $\{x_n\}$, $\{x_{n_k}\}$, such that

$$x_{n_k} \to x_0 \in A.$$

Since this is true, then, for all n_k bigger than some N, say, we have

$$d(x_{n_k}, x_0) < \delta(x_0)/4,$$

and we now claim that this implies

$$\delta(x_{n_k}) \geq \delta(x_0)/4 \qquad (n_k \geq N).$$

To prove this, suppose,

$$z \in S_{\delta(x_0)/4}(x_{n_k}) \qquad \text{(for any } n_k \geq N).$$

This implies that

$$d(x_0, z) \leq d(x_0, x_{n_k}) + d(x_{n_k}, z) < \frac{\delta(x_0)}{4} + \frac{\delta(x_0)}{4} = \frac{\delta(x_0)}{2}.$$

Therefore,

$$z \in S_{\delta(x_0)/2}(x_0).$$

Now, since $\delta(x_0)/2 < \delta(x_0)$, there must exist an open set from the covering G_β such that

$$S_{\delta(x_0)/2}(x_0) \subset G_\beta.$$

But, since

$$d(x_{n_k}, x_0) < \delta(x_0)/4 \Rightarrow S_{\delta(x_0)/2}(x_0) \supset S_{\delta(x_0)/4}(x_{n_k}) \qquad \text{(for } n_k \geq N),$$

we can also say that

$$S_{\delta(x_0)/4}(x_{n_k}) \subset G_\beta,$$

which implies that

$$\delta(x_0)/4 \leq \delta(x_{n_k}) \qquad (n_k \geq N)$$

by the definition of $\delta(x_{n_k})$. Since any subsequence of a convergent sequence must have the same limit as the sequence, taking the limit of both sides of the last inequality as $k \to \infty$, we get

$$\lim_{k \to \infty} \delta(x_{n_k}) = \delta_0 \geq \delta(x_0)/4.$$

Since $\delta(x_0)$ is strictly greater than zero, there must exist some positive real number ε such that

$$\delta(x_0)/4 > \varepsilon > 0.$$

Therefore,

$$\delta_0 > \varepsilon > 0. \qquad (5.2)$$

Applying Theorem 5.3, we can say that A must be totally bounded and, by the observation right after the proof of that theorem, can further say that the ε-net must wholly be contained in A. Let the ε-net, where the ε is as in (5.2), be given by

$$y_1, y_2, \ldots, y_n \in A.$$

Since $\delta_0 > \varepsilon$, we are assured of the existence of open sets $G_1, G_2, ..., G_n$ from the covering such that

$$S_\varepsilon(y_i) \subset G_i \qquad (i = 1, 2, ..., n),$$

because certainly, for each i, $\delta(y_i) \geq \delta_0 > \varepsilon > 0$.

Now, for any $x \in A$, there is some y_i such that

$$d(x, y_i) < \varepsilon,$$

which implies that

$$x \in S_\varepsilon(y_i) \subset G_i.$$

Hence the family $\{G_i\}$ $(i = 1, 2, ..., n)$ constitutes a finite subcovering of A, and we conclude that A must be compact.

Our next theorem yields an equivalent way of describing a relatively compact set.

THEOREM 5.6. A set A is relatively compact if and only if a convergent subsequence can be selected from every sequence of points in A. (Note that we do not claim that the limit point is a member of A.)

Proof (necessity). Suppose A is relatively compact. By the definition of relative compactness, $\bar{A}$ must be compact. Using Theorem 5.5, we can say that, since $\bar{A}$ is compact, it must also be sequentially compact. Hence, from every sequence of points in $\bar{A}$, we can select a convergent subsequence with limit in $\bar{A}$. Since $A \subset \bar{A}$, every sequence of points of A is a sequence of points in $\bar{A}$, as well, and we have just noted that every such sequence does have a convergent subsequence.

Proof (sufficiency). Suppose every sequence in A has a convergent subsequence. To show that this implies that $\bar{A}$ is compact, we shall prove the equivalent result that $\bar{A}$ is sequentially compact. To this end, let $\{y_n\}$ be a sequence of points in $\bar{A}$. Since A is dense in $\bar{A}$, there must exist a sequence of points from A, $\{x_n\}$, such that, for each n,

$$d(x_n, y_n) < 1/n.$$

Using the hypothesis, we can now assert that a convergent subsequence $\{x_{n_k}\}$ can be selected from $\{x_n\}$ such that

$$x_{n_k} \to x \in \bar{A}.$$

This implies that

$$y_{n_k} \to x \in \bar{A}.$$

Therefore, $\bar{A}$ is sequentially compact which, by virtue of Theorem 5.5, implies that $\bar{A}$ is compact.

THEOREM 5.7.

(1) If a set A is relatively compact, it is also totally bounded, in the metric space (X, d).

(2) If (X, d) is complete and A is totally bounded, then A is relatively compact.

Proof.

(1) Suppose A is relatively compact. We can apply the preceding theorem to assert that, from any sequence in A, a convergent subsequence can be extracted. We need only apply Theorem 5.3 now to conclude that A is totally bounded.

(2) Let (X, d) be a complete metric space and let A be a totally bounded subset of X. By Theorem 5.6 we must show that it is possible to extract a convergent subsequence from every sequence of points in A. To this end, let $\{x_n\}$ be a sequence of points from A. Since A is assumed to be totally bounded, for each $\varepsilon_k = 1/k$, where k is a positive integer, there must be a finite ε_k-net:

$$\{y_1^{(k)}, y_2^{(k)}, \ldots, y_{n_k}^{(k)}\}.$$

Considering the collection of spheres of radius 1 about each point in the $k = 1$ net, we see that this collection must cover all of A. In particular, the x_n ($n = 1, 2, \ldots$) must be contained in a finite number of these spheres. Further, one of these spheres must contain an infinite number of the x_n. Hence we have some sphere, S_1 say, containing the subsequence of $\{x_n\}$:

$$x_1^{(1)}, x_2^{(1)}, x_3^{(1)}, \ldots,$$

where

$$x_i^{(1)} \in \{x_1, x_2, \ldots\} \quad \text{(for } i = 1, 2, \ldots\text{)}.$$

Reasoning along the same lines for the collection of neighborhoods of radius $\frac{1}{2}$ about the points in the $k = 2$ net, we can get some sphere of radius $\frac{1}{2}$ from this collection containing infinitely many of the $x_n^{(1)}$. Thus, calling this sphere S_2 and denoting the points it contains by $x_1^{(2)}, x_2^{(2)}, \ldots$, we have

$$S_2: x_1^{(2)}, x_2^{(2)}, x_3^{(2)}, \ldots,$$

where

$$x_i^{(2)} \in \{x_1, x_2, \ldots\} \quad (i = 1, 2, \ldots).$$

Proceeding in this fashion, we arrive at Fig. 5.2.

$$S_1 : x_1^{(1)}, x_2^{(1)}, \ldots$$
$$S_2 : x_2^{(2)}, x_2^{(2)}, \ldots$$
$$\vdots$$
$$S_n : x_1^{(n)}, x_2^{(n)}, \ldots$$

FIG. 5.2.

It is now clear that selecting the "diagonal" terms (see arrows in Fig. 5.2, the sequence $\{x_i^{(i)}\}$) yields a Cauchy sequence which, since the space is complete, must have a limit in X. Since each of the sequences $\{x_n^{(i)}\}$ (n is the index here, while i is fixed) is a subsequence of $\{x_n^{(i-1)}\}$ it is clear that $\{x_i^{(i)}\}$ is a subsequence of the original sequence $\{x_n\}$. Hence we need only apply the preceding theorem to complete the proof.

In conclusion it is noted that, in (X, d),

$$\text{theorems 5.2, 5.7(1)}$$
$$(A \text{ compact}) \Longrightarrow (A \text{ closed and totally bounded}).$$

Further, if (X, d) is complete, then A being closed and totally bounded implies that A is compact by Theorem 5.7, part (2).

The next theorem shows that continuous mappings preserve the property of compactness. It should be noted that the proof is not of a highly metric nature; that is, we really need only the concept of open set in each space.

THEOREM 5.8. Let f be a continuous function such that $f : X \to Y$ where X and Y are metric spaces. In this case, if A is a compact set in X, then $f(A)$ will be a compact set in Y. (The continuous image of a compact set is compact.)

Proof. We consider an arbitrary open covering of $f(A)$ and show that a finite subcovering can be selected. To this end, let $\{G_\alpha\}$ be a collection of open sets in Y such that

$$f(A) \subset \bigcup_\alpha G_\alpha.$$

Now we can say that

$$A \subset f^{-1}(f(A)) \subset f^{-1}\left(\bigcup_\alpha G_\alpha\right) = \bigcup_\alpha f^{-1}(G_\alpha).$$

Since f is continuous, each of the $f^{-1}(G_\alpha)$ are open sets (the inverse image of an open set must, itself, be open by the corollary to Theorem 3.4, and thus the family $\{f^{-1}(G_\alpha)\}$ constitutes an open covering of A. Since A is assumed to be compact, there must exist $G_1, G_2, \ldots, G_n$ such that

$$A \subset \bigcup_{i=1}^n f^{-1}(G_i),$$

which implies

$$f(A) \subset \bigcup_{i=1}^n f(f^{-1}(G_i)) \subset \bigcup_{i=1}^n G_i.$$

We have now succeeded in selecting a finite subcovering from the original open covering for $f(A)$, and therefore $f(A)$ is compact.

EXERCISES 5

1. Let (X, d) be a metric space and suppose A is a subset of X. Show that $d(A) = d(\bar{A})$.

2. In the metric space (X, d) show that, if $\{x_n\}$ is a Cauchy sequence, there must be some positive number M such that $d(x_n, x_m) \leq M$ for all n and m; that is, the *set* $\{x_1, x_2, \ldots, x_n, \ldots\}$ must be bounded.

3. Let A and B be compact and closed subsets, respectively, of the metric space (X, d) such that $A \cap B = \emptyset$. Prove that $d(A, B) > 0$. Construct an example to show that this need not be true if A and B are only closed sets.

4. Let A be a compact subset of a metric space. Show that A considered as a metric space by itself is complete.

5. Show that a subset of R, with the usual metric, is relatively compact if and only if it is bounded.

6. Show that in R^n (with respect to any of the metrics listed in Example 3.6) boundedness and total boundedness are equivalent.

7. Show that if (X, d) is complete, then $A \subset X$ is relatively compact, if for every $\varepsilon > 0$ there exists a relatively compact ε-net for A.

8. Show that the union of a finite number of relatively compact sets is relatively compact.

9. Let A be a subspace of the metric space (X, d). Prove that the completion A^* of A is compact if and only if A is totally bounded.

10. Let (X, d) be an arbitrary metric space and (Y, d') a compact metric space. Letting $C(X, Y)$ denote the set of all continuous functions from X into Y, we introduce the following metric ρ to this set: For $f, g \in C(X, Y)$, take

$$\rho(f, g) = \sup_{x \in X} d'(f(x), g(x)).$$

Show that $(C(X, Y), \rho)$ is a complete metric space.

11. Let f be a continuous function mapping X into Y, where (X, d) and (Y, d') are arbitrary metric spaces. Show that f maps relatively compact sets into relatively compact sets.

12. Let f be a continuous mapping of a compact subset A of the metric space X into R. Show that f assumes its maximum and minimum on A; that is, there exist points $x_0, y_0 \in A$ such that

$$f(x_0) = \inf_{x \in A} f(x) \quad \text{and} \quad f(y_0) = \sup_{x \in A} f(x).$$

13. Show that Theorem 5.1 ceases to be true if the assumption that $d(F_n) \to 0$ is removed from the hypothesis.

14. Show that the *Hilbert cube* [the set of all $x = (\alpha_1, \alpha_2, \ldots, \alpha_n, \ldots) \in l_2$ such that $|\alpha_n| \leq 1/n \ (n = 1, 2, \ldots)$] is a compact subspace of l_2.

REFERENCES

1. A. Komolgoroff and S. Fomin, "Elements of the Theory of Functions and Functional Analysis," Vol. 1, pp. 51–53. A rather short discussion covering relative compactness and total boundedness. Note that the term "compact" as used there denotes what we have designated as relative compactness.

2. J. Dieudonné, "Foundations of Modern Analysis," pp. 55–62. The word "precompact" used there denotes what we have referred to as totally bounded.

CHAPTER **6**

Category and Separable Spaces

Proofs in which the existence of some entity is to be shown are usually extremely difficult no matter how they are approached, and of the possible approaches, the constructive type of proof is usually the most difficult to achieve. But what about other possibilities? Certainly one thought would be to assume that the entity in question does not exist and arrive at a contradiction as is done, for example, in Euclid's proof of the existence of an infinite number of primes.

In this section, another tool that will be useful in existence proofs will be introduced. More specifically, it is the notion of *category* that we shall be concerned with. We define what is meant by saying "a set is of category I" or "a set is of category II." The notion is defined in such a way that a set must be of one of these two categories. It will then be observed that any set of category II cannot be empty. Thus, if we can show that a set is in category II, we can be assured that it is not empty, that is, we shall have proved the existence of members of the set. As an illustration of the way such arguments go, we shall prove the existence of functions continuous everywhere but differentiable nowhere. Another useful item for existence proofs is introduced in the Appendix to Chapter 9—Zorn's lemma.

The final object of our attention in this chapter is the *separable* metric space. The notion is defined, and several examples are given.

6.1 F_σ Sets and G_δ Sets

In this discussion, the following notation will be used: (X, d) will be an underlying metric space, $\mathcal{O}$ will designate the open sets of (X, d), and $\mathcal{F}$ will be used to denote the class of closed sets in (X, d).

DEFINITION 6.1. We shall mean by G_δ the class of all sets expressible as a countable (denumerable or finite) intersection of open sets. Any member of this class will be said to be *of type G_δ*.

DEFINITION 6.2. F_σ will denote the class of all sets that can be written as a countable union of closed sets. A member of this family of sets is said to be *of type F_σ*.

EXAMPLE 6.1. Consider the real numbers R, with the usual distance function, and consider the subset Q of all rational numbers. Since the rational numbers are a denumerable set, they can be enumerated, that is, set in 1 : 1 correspondence with the positive integers. (It is noted that the set Q is neither open nor closed.) Thus we can list all the rational numbers as follows:

$$Q: x_1, x_2, \ldots.$$

We can now write Q as

$$Q = \bigcup_{i=1}^{\infty} \{x_i\}.$$

Since each of the one-point sets $\{x_i\}$ is closed, it follows that Q is a set of type F_σ.

We note that, if

$$E \in F_\sigma,$$

then, by using the definition of an F_σ set and taking complements, we have

$$CE \in G_\delta.$$

Applying this result to the example, we can say that the irrational numbers, the complement of Q, must be a set of type G_δ. Thus, we have proved that the irrational numbers can be written as a countable intersection of open sets.

It is clear that any open set is of type G_δ, and any closed set is of type F_σ. We can write this fact symbolically by saying

$$\mathcal{O} \subset G_\delta \quad \text{and} \quad \mathcal{F} \subset F_\sigma.$$

We now contend that we can also say:

$$\mathcal{O} \subset F_\sigma, \tag{6.1}$$

$$\mathcal{F} \subset G_\delta. \tag{6.2}$$

It is clear, if we can show that (6.2) holds, that (6.1) will follow by taking complements. Now it will be shown that (6.2) is valid.

Let $F \in \mathcal{F}$. We shall show that $F \in G_\delta$. For each positive integer n, we define the open set O_n to be

$$O_n = \bigcup_{x \in F} S_{1/n}(x),$$

it being clear that $O_n \supset F$ for any n. Since this is so,

$$\bigcap_{n=1}^{\infty} O_n \supset F.$$

We now demonstrate that the inclusion can be reversed. Let

$$y \in \bigcap_{n=1}^{\infty} O_n.$$

This implies

$$y \in O_n = \bigcup_{x \in F} S_{1/n}(x)$$

for every n. By this result we can say that, for some $x \in F$,

$$y \in S_{1/n}(x),$$

which means that $d(y, x) < 1/n$. Let ε be an arbitrary positive number. It is clear that an n can be chosen so that $1/n < \varepsilon$, which means, for any ε, that there is some $x \in F$ such that

$$x \in S_\varepsilon(y).$$

Therefore

$$y \in \bar{F} = F,$$

because every neighborhood of y has been shown to contain a point of F. This completes the proof.

6.2 Nowhere-Dense Sets and Category

DEFINITION 6.3. A set A, a subset of the metric space (X, d), is said to be *nowhere dense* if $\bar{A}^o = \emptyset$. It can be shown (see Exercise 1) that this is equivalent to requiring that $C\bar{A}$ be dense in X or that $\overline{C\bar{A}} = X$.

REMARK 1. If d is the trivial metric, the only nowhere-dense set is the null set.

REMARK 2. The notion of being nowhere dense is not the opposite of being everywhere dense; that is, not being nowhere dense does not imply that the set is everywhere dense. As an example of a set which is neither, let R be the metric space with the usual metric, and consider the subset consisting of the open interval (1, 2). The interior of the closure of this set is nonempty, whereas the closure of (1, 2) is certainly not all of R.

DEFINITION 6.4. A set is said to be of *category I* if it can be written as a countable union of nowhere-dense sets. Otherwise it is said to be of *category II*.

REMARK 3. The above relation is mutually exclusive; a set is either of category I or category II.

It is clear that the null set is of category I. Another example of a set that is of category I is the rational numbers Q for, as in Example 6.1, we can write

$$Q = \bigcup_{i=1}^{\infty} \{x_i\},$$

it being clear that, for every i,

$$\overline{\{x_i\}} = \{x_i\} \quad \text{and} \quad \{x_i\}^o = \emptyset.$$

Since a denumerable union of denumerable sets is also a denumerable set, it follows that, if $A_1, A_2, ..., A_n, ...$ are each in category I, then so must be

$$\bigcup_{i=1}^{\infty} A_i.$$

In view of this, if

$$C = A \cup B,$$

where it is known that A is of category I and C is of category II, then B must be of category II. To prove this, one need only make use of Remark 3 and consider the two possibilities for B.

In light of this, denoting the irrational numbers by I, if it is known that R, the real numbers, is of category II, where

$$R = Q \cup I,$$

since Q is of category I, we can say that I is of category II. The following theorem will allow us to confirm this conjecture.

THEOREM 6.1 (*Baire*). Any complete metric space is of category II.

Proof. Letting X denote a complete metric space, we shall now show that no set of category I can possibly be all of X.

Suppose A is any set, in X, of category I. We can write A as

$$A = \bigcup_{n=1}^{\infty} E_n,$$

where each of the E_n is nowhere dense. Since this is so, $C\bar{E}_n$, for any n, is everywhere dense, and we can assert the existence of points in each of these sets: they cannot be empty. In the case of $C\bar{E}_1$ let x_1 be a member of the open set $C\bar{E}_1$. Since $C\bar{E}_1$ is open, there must be some neighborhood of x_1, $S_\rho(x_1)$, lying wholly within the set. We now have

$$S_\rho(x_1) \subset C\bar{E}_1.$$

For $\varepsilon_1 < \rho$ we can say

$$\overline{S_{\varepsilon_1}(x_1)} \subset S_\rho(x_1) \subset C\bar{E}_1 \subset CE_1,$$

which implies

$$\overline{S_{\varepsilon_1}(x_1)} \cap E_1 = \emptyset.$$

In view of this construction we now wish to make the following

Induction hypothesis. There exist spheres $S_{\varepsilon_k}(x_k)$ for $k = 1, 2, \ldots, n-1$ such that

$$\overline{S_{\varepsilon_k}(x_k)} \cap E_k = \emptyset \qquad \text{where } x_k \in C\bar{E}_k$$

and

$$\overline{S_{\varepsilon_k}(x_k)} \subset \overline{S_{\varepsilon_{k-1}}(x_{k-1})},$$

where

$$\varepsilon_k \leq \varepsilon_{k-1}/2.$$

Using this information, our wish now is to construct an nth sphere that also has these properties and, for this purpose, we now claim that there must be an element, x_n, in $S_{\varepsilon_{n-1}}(x_{n-1}) - \bar{E}_n$; that is,

$$x_n \in S_{\varepsilon_{n-1}}(x_{n-1}) - \bar{E}_n.$$

If no such element exists, then

$$S_{\varepsilon_{n-1}}(x_{n-1}) \subset \bar{E}_n \Rightarrow x_{n-1} \in \bar{E}_n^o$$

and contradicts the fact that E_n is nowhere dense. Thus we have

$$x_n \in S_{\varepsilon_{n-1}}(x_{n-1}) \cap C\bar{E}_n = B.$$

Since B is the intersection of two open sets, it must be open, as well, and there must be an ε such that

$$S_\varepsilon(x_n) \subset B.$$

We now choose

$$\varepsilon_n < \min(\varepsilon, \varepsilon_{n-1}/2),$$

which implies

$$\overline{S_{\varepsilon_n}(x_n)} \subset S_\varepsilon(x_n) \subset B$$

or

$$\overline{S_{\varepsilon_n}(x_n)} \cap E_n = \emptyset.$$

But, since we also have

$$\overline{S_{\varepsilon_n}(x_n)} \subset \overline{S_{\varepsilon_{n-1}}(x_{n-1})}$$

and

$$\varepsilon_n \leq \varepsilon_{n-1}/2,$$

we have constructed the desired nth sphere. We now have a nested sequence of nonempty closed spheres in a complete metric space with diameters tending to 0. Applying Theorem 5.1, we can say there exists some x_0 such that

$$x_0 \in \bigcap_{n=1}^{\infty} \overline{S_{\varepsilon_n}(x_n)}.$$

Further, since $\overline{S_{\varepsilon_n}(x_n)} \cap E_n = \emptyset$ for any n, we can say

$$x_0 \notin E_n \quad \text{(for any } n\text{)},$$

or

$$x_0 \in CE_n \quad \text{(for all } n\text{)}.$$

Thus, since

$$CA = \bigcap_{n=1}^{\infty} CE_n,$$

we have

$$x_0 \in CA,$$

which implies

$$A \neq X,$$

and we conclude that any set of category I cannot be the whole space. This completes the proof.

We note that, just because a set is of category I, the set itself need not be nowhere dense. For example, the rational numbers Q were shown to be in category I, whereas the closure of the rational numbers in R is all of R; that is, the set itself is everywhere dense even though it can be written as a countable union of nowhere-dense sets.

Our next theorem shows some connection between the notion of a set being of type G_δ and the category it is in.

THEOREM 6.2. Let A be an everywhere-dense subset of the complete metric space (X, d). If A is of type G_δ, then A is of category II.

Proof. Since A is in G_δ, there exist open sets O_n, such that

$$A = \bigcap_{n=1}^{\infty} O_n.$$

Taking the complement of both sides and calling the closed set $CO_n = F_n$, we have

$$CA = \bigcup_{n=1}^{\infty} CO_n = \bigcup_{n=1}^{\infty} F_n.$$

It is clear that, for every n, $CA \supset F_n$ or $A \subset CF_n$. Now, since

$$X = \bar{A} \subset \overline{CF_n} = \overline{C\overline{F}_n},$$

which yields

$$X = \overline{C\overline{F}_n},$$

we can say that each of the F_n is nowhere dense and can assert that CA is of category I. Now, since

$$X = A \cup CA$$

and X is of category II by virtue of Theorem 6.2, we can say that A is of category II, by applying the results directly following Remark 3 of this Section.† The proof is now complete.

6.3 The Existence of Functions Continuous Everywhere, Differentiable Nowhere

As a nice application of category arguments, it will now be shown that on the closed interval $[0, 1]$ there are functions continuous throughout this interval, while they possess a derivative at no point in this interval. The proof here will not be letter perfect but it will be brought to a point where it is felt that the reader can readily supply any further information that is needed.

It has been noted previously that the null set is in category I, which implies that it cannot be of category II. Thus if it can be shown that a set is of category II, we can be sure that it is nonempty. In the material that follows, it will be shown that a subset of the functions that are continuous everywhere but possess a derivative nowhere is of category II, and the existence of continuous functions, differentiable nowhere (throughout $[0, 1]$) will have been proved.

Consider the complete metric space $C[0, 1]$ with metric d given by

$$d(f, g) = \max_{x \in [0,1]} |f(x) - g(x)| \qquad (f, g \in C[0, 1]).$$

First consider the subset of this space

$A : f \in C[0, 1]$ such that, for some $x \in [0, 1]$, f has a finite right-hand derivative.

Next, let n be a positive integer and consider

$$E_n = \left\{ f \in C[0, 1] \, \middle| \, \frac{|f(x + h) - f(x)|}{h} \le n, 0 < h < 1/n, \text{ for some } x \in [0, 1 - 1/n] \right\}.$$

It is certainly true that $A \subset \bigcup_{n=1}^{\infty} E_n$. If it can be shown that $\bigcup_{n=1}^{\infty} E_n$ is of category I, it will immediately follow that A is of category I. If we can show this then, since we can write

$$C[0, 1] = A \cup CA$$

and it is known by Baire's theorem that $(C[0, 1], d)$ is of category II, we can apply the results after Remark 3 of the preceding section to assert that CA is of category II. CA, however, consists of those functions that do not possess a finite right-hand derivative and is certainly a subset of those $f \in C[0, 1]$ that do not possess any derivatives. But, if it can be shown that this set is in category II, it cannot be empty.

---

† In this case we have shown that A is of category II, whereas CA is of category I. Sets of this general type are called *residual* sets.

With this in mind let us proceed to showing that $\bigcup_{n=1}^{\infty} E_n$ is of category I. Certainly, if we can show that each of the E_n is nowhere dense or that, for every n,

$$\overline{CE}_n = X,$$

we shall have shown that $\bigcup_{n=1}^{\infty} E_n$ is of category I. Our first contention is that $\overline{E}_n = E_n$ or that $g \in \overline{E}_n$ implies $g \in E_n$.

Suppose that $g \in \overline{E}_n$. Since this is so, there must exist a sequence of functions from E_n, $\{f_j\}$, converging (with respect to d) to the continuous function g. Since each of the f_j is in E_n, there must be some point x_j associated with each of the f_j such that

$$\frac{|f_j(x_j + h) - f_j(x_j)|}{h} \leq n \qquad (0 < h < 1/n, \quad x_j \in [0, 1 - 1/n]). \qquad (6.3)$$

The $\{x_j\}$ now constitutes a bounded sequence (they all come from the interval $[0, 1]$) of real numbers and, by the Bolzano–Weierstrass theorem, we must be able to extract a convergent subsequence from them, $\{x_{j_k}\}$; $x_{j_k} \to x_0$. Since any subsequence of a convergent sequence must converge to the same limit to which the sequence converges, the limit in (6.3) can be passed to as j runs through the values j_k to yield†

$$\frac{|g(x_0 + h) - g(x_0)|}{h} \leq n,$$

which, since x_0 must belong to $[0, 1 - 1/n]$, implies $g \in E_n$.

Having proved that the E_n are closed sets, we wish now to return to proving that they are nowhere dense. As noted before, we must show

$$\overline{CE}_n = X.$$

First let us consider

$$CE_n = \left\{ f \in C[0, 1] \, \middle| \, \frac{|f(x + h) - f(x)|}{h} > n, \right.$$
$$\left. \text{for some } h \in (0, 1/n) \text{ and all } x \in [0, 1 - 1/n] \right\}.$$

To show that $\overline{CE}_n = X$, we must show that any neighborhood of any point of X contains a point of CE_n, or, for $h \in C[0, 1]$, we must show that there exists a $g \in CE_n$ such that

$$d(g, h) < \varepsilon,$$

where ε is any preassigned, positive number.

Suppose now that $h \in C[0, 1]$. By the theorem due to Weierstrass,‡ there exists some polynomial $p(x)$ such that

$$\max_{x \in [0,1]} |h(x) - p(x)| < \varepsilon/4.$$

† The full justification lies in observing $f_{j_k}(x) \to g(x)$ uniformly, that $g(x)$ is continuous so that $\lim g(x_{j_k}) = g(\lim x_{j_k}) = g(x_0)$ and applying the triangle inequality.

‡ *Weierstrass' theorem:* Let T be a closed bounded set of real numbers. Then every function $f(x)$, continuous there, is the limit of a sequence of polynomials in x with real coefficients which is uniformly convergent on that set. For a proof, see Naimark, *Normed Rings*, pp. 30–33, or Buck, *Advanced Calculus.*

We now draw a saw-tooth function (Fig. 6.1) whose slope will be greater than $n + \max\{|p'(x)| \mid x \in [0, 1]\}$ and whose magnitude is less than $\varepsilon/4$. Calling the saw-tooth function $s(x)$, it is clear that adding $p(x)$ to $s(x)$ will yield a continuous function that is in CE_n and that is within ε of the given, continuous function.

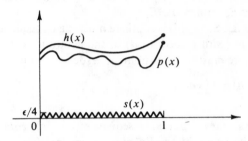

FIG. 6.1. $h(x)$ is the given continuous function, $p(x)$ is the polynomial, and $s(x)$ is the saw-tooth function.

Now, since we can write $C[0, 1] = A \cup CA$, as noted earlier, we can say that CA must be of category II, because the complete space $C[0, 1]$ is of category II. Denoting the subset of $C[0, 1]$ that is differentiable nowhere by N, it is clear that, if a function has no finite right-hand derivative, it must also be in N or

$$CA \subset N.$$

CA being of category II implies that N must be of category II, which means that it is a nonempty set; indeed one rather expects " more " elements in sets of category II than in sets of category I. In view of this, it is almost surprising that we can find continuous functions that are differentiable everywhere, the other variety being the more common.

6.4 Separable Spaces

DEFINITION 6.5. Let (X, d) be a metric space. X is said to be *separable* or to be a *separable space* if it contains a denumerable dense subset. It is easy to see that subsets of a separable space are separable.

To make the idea of a separable space a little clearer, we give several examples.

EXAMPLE 6.2. The real numbers, with the usual metric, constitute a separable space because we can exhibit the rational numbers Q as a denumerable, dense subset.

EXAMPLE 6.3. Consider $C[a, b]$ with distance between any two functions (points) f and g taken to be

$$d(f, g) = \max_{x \in [a, b]} |f(x) - g(x)|.$$

We know, by virtue of the Weierstrass theorem cited on page 81 that a dense subset of this space is given by the collection of all polynomials. This collection, however, is not denumerable. The collection of all polynomials with rational coefficients is a denumerable subset, however, that is also dense in $C[a, b]$ and therefore $(C[a, b], d)$ is a separable space.

EXAMPLE 6.4. Let (X, d) be any trivial space (d is the trivial metric), and consider any subset of this space

$$A = \{x_\alpha\}_{\alpha \in \Lambda}$$

It is easily seen that the closure of this set is just the set itself:

$$\bar{A} = A.$$

Thus we can conclude that a trivial space will be separable if and only if X is a denumerable set.

EXAMPLE 6.5. Consider l_2, the collection of all infinite sequences with real entries

$$(\alpha_1, \alpha_2, ..., \alpha_n, ...)$$

such that $\sum_{i=1}^{\infty} |\alpha_i|^2 < \infty$ with distance between $x = (\alpha_1, \alpha_2, ..., \alpha_n, ...)$ and $y = (\beta_1, \beta_2, ..., \beta_n, ...)$ taken to be $(\sum_{i=1}^{\infty} |\alpha_i - \beta_i|^2)^{\frac{1}{2}}$.

A denumerable dense subset for this space is the set of all sequences with rational entries such that all but a finite number of the entries are zero. We can easily extend this to the case of complex entries by an analogous procedure. In this case, we would have, as a denumerable dense subset, entries whose real and imaginary parts are rational and all but a finite number of the entries are zero.

Our next example is an example of a space that is not separable.

EXAMPLE 6.6. Consider the space l_∞; all bounded sequences $(\alpha_1, \alpha_2, \alpha_n, ...)$. As distance between any two points of l_∞, $x = (\alpha_1, \alpha_2, ..., \alpha_n, ...)$ and $y = (\beta_1, \beta_2, ..., \beta_n, ...)$, we take

$$d(x, y) = \sup_k |\alpha_k - \beta_k|.$$

Now consider a point x of this space such that all the entries are either 0 or 1. The collection of all such points is not a countable set, because this set can be put in 1 : 1 correspondence with the real numbers between 0 and 1 [each member of this set can be regarded as the image of the dyadic (base 2) expansion of a real number between 0 and 1] and, as such, must have the power of the continuum. It is clear that if any two members of this set are distinct, they must be of distance 1 apart.

Suppose now that we surrounded each member of this set by a sphere of radius $\frac{1}{2}$. By virtue of the preceding paragraph, these spheres do not intersect. Now for any dense subset in this space we can say that at least one member of the dense subset must be in each of these nonintersecting spheres. Hence there can be no denumerable dense subset in this space.

EXERCISES 6

1. Show that A is nowhere dense if and only if $C\overline{A}$ is dense in X; that is, $\overline{C\overline{A}} = X$.

2. Show that if O is any open set, $\overline{O} - O$ is nowhere dense.

3. Construct a function $f(x)$ which is discontinuous at every rational point of $[0, 1]$ and continuous at every irrational point of $[0, 1]$.

4. Prove the impossibility of constructing a function on $[0, 1]$ which is continuous at every rational point and is discontinuous at every irrational point.

5. Show the Cantor ternary set to be nowhere dense and perfect (where a set A is *perfect* if and only if $A' = A$).

6. Let s be the set of all infinite sequences of real numbers. For any two points of s, $x = (\alpha_1, \alpha_2, \ldots)$ and $y = (\beta_1, \beta_2, \ldots)$, define

$$d(x, y) = \sum_{i=1}^{\infty} \frac{1}{2^i} \frac{|\alpha_i - \beta_i|}{1 + |\alpha_i - \beta_i|}.$$

Show that (s, d) is a separable, complete metric space.

7. Show that a set A is nowhere dense if and only if any nonempty open set U has a nonempty open subset V such that $V \cap A = \phi$.

REFERENCES

1. R. Boas, "A Primer of Real Functions." Further category arguments similar in spirit to the one presented in Section 6.3.

2. A. Komolgoroff and S. Fomin, "Elements of the Theory of Functions and Functional Analysis," Vol. 1, p. 25. A few more examples of separable spaces.

CHAPTER **7**

Topological Spaces

In the last few chapters we have been considering properties of metric spaces. Many of the concepts introduced there, such as adherence point, continuity, and compactness, can be characterized purely in terms of open sets (see e.g., Theorem 3.1 and Corollary to Theorem 3.4). We shall now introduce a more general structure than a metric space, namely, a topological space. Many of our previous results for metric spaces will carry over to topological spaces, directly or with at most minor modifications, whereas others (the equivalence of the various notions of compactness, for example, discussed in Sec. 5.3) will completely fail to be true in this more general setting.

We do not discuss topological spaces just for the sake of generalization, however. Many of the results developed here will be vital for our subsequent development of Banach algebras and operator topologies in Chapter 19. However, the reader primarily interested in Banach spaces or Hilbert spaces, and the spectral theory, may omit this chapter and go directly to the next chapter on Banach spaces.

7.1 Definitions and Examples

We observed after Definition 3.2 that the open sets as defined in a metric space (X, d) satisfied the three basic properties O_1, O_2, and O_3. Those properties are precisely what we shall single out and make axioms of in the foundation of a topological space. There are other equivalent axiomatic characterizations of topological spaces, some of which are referred to in the exercises, but we choose the one via open sets for its simplicity and applicability to the structures that will be considered later. We now proceed to the formal definition of a topological space.

DEFINITION 7.1. A topological space is a set X (assumed to be nonempty throughout) together with a nonempty collection of subsets $\mathcal{O}$ of X called *open sets* such that:

(O$_1$) The union $\bigcup_\alpha O_\alpha$, of any number of sets of $\mathcal{O}$ is again a member of $\mathcal{O}$.
(O$_2$) The intersection of a finite number of sets of $\mathcal{O}$ is in $\mathcal{O}$.
(O$_3$) $\emptyset$ and X are members of $\mathcal{O}$.

We observe that (O$_3$) could be omitted if we include in (O$_1$) and (O$_2$) the possibility of an empty union and empty intersection and agree that an empty union designates the null set and an empty intersection, the whole set X.

The pair $(X, \mathcal{O})$ satisfying the conditions of the definition is called a *topological space*, and the collection $\mathcal{O}$ is called a *topology for X*. Usually, we shall just refer to X itself as a topological space—it being understood that we are considering a fixed topology $\mathcal{O}$. But we warn the reader that, on a given set X, it is possible to have more than one collection of subsets $\mathcal{O}$ satisfying the above conditions (see Examples 7.1 and 7.2) and constituting, therefore, a topology for X.

As mentioned in Definition 7.1, the members of $\mathcal{O}$ are referred to as *open sets*, whereas the elements of X are frequently called *points* of the topological space.

We now consider a number of examples that will be useful for illustrative purposes and for counterexamples later. They will also be useful in many of the exercises.

EXAMPLE 7.1. Let X be an arbitrary set and take $\mathcal{O} = \{X, \emptyset\}$. It is evident that $\mathcal{O}$ satisfies the conditions of Definition 7.1, and it is called the *trivial topology*.

EXAMPLE 7.2. Let X be an arbitrary set. We now go to the opposite extreme of Example 7.1 and take $\mathcal{O} = P(X)$, the power set of X; that is, the collection of all subsets of X. $\mathcal{O}$ is, of course, a topology for X and is called the *discrete topology*, because every subset, even discrete points, are open sets in this topology.

EXAMPLE 7.3. Let (X, d) be a given metric space and define a set $O \subset X$ to be open according to Definition 3.2. The observations following Definition 3.2 show that this collection of sets can still be called open sets in the new sense, that is, they do indeed form a topology for X called the *metric topology*.

EXAMPLE 7.4. Let X be an arbitrary infinite set. We define a subset O to be open if either O is equal to $\emptyset$, X or if CO is finite. The reader may verify easily that this collection of sets gives a topology for X called the *cofinite topology*.

DEFINITION 7.2. A subset F of the topological space X is called *closed* if CF, the complement of F, is open.

Thus the notion of a closed set is thrown back to the basic collection of open sets $\mathcal{O}$. Similarly, we reduce the concept of an adherence point of an arbitrary subset E of a topological space X to one involving just the open sets, namely Definition 7.3.

DEFINITION 7.3. A point $x \in X$ is called an *adherence point* of the subset E if every open set containing x contains a point of E.

It should be observed that every point of E is clearly an adherence point of E. As in Definition 3.3, the set of all adherence points of E is called the *closure* of E and is designated, as in the metric space situation, by $\bar{E}$. We have, by the preceding observation, that $E \subset \bar{E}$.

Finally we characterize the notions of continuity and homeomorphisms in terms of open sets. Recalling the Corollary to Theorem 3.4 we make the following definition.

DEFINITION 7.4. Let $(X_1, \mathcal{O}_1)$ and $(X_2, \mathcal{O}_2)$ be two topological spaces and let $f: X_1 \to X_2$ be a mapping of X_1 into X_2. The mapping f is called *continuous* if, for every $O \in \mathcal{O}_2$, $f^{-1}(O) \in \mathcal{O}_1$ (we require that the inverse image of every open set be an open set).

If f is $1:1$, continuous, and such that f^{-1} is also continuous, then f is called a *homeomorphism*. Finally, if f is a homeomorphism mapping X_1 onto X_2, the topological spaces X_1 and X_2 are said to *homeomorphic*.

If X_1 and X_2 are homeomorphic spaces, there exists a $1:1$ correspondence f between the elements of the two spaces which preserves open sets. Since topological properties of a space can eventually be characterized solely in terms of open sets, two homeomorphic spaces are indistinguishable from the topological point of view, although concretely they might consist of vastly different types of elements. For this reason, considering them exclusively with regard to their topological properties, we can frequently identify them (see the corresponding discusssion related to metric spaces and isometries in Sec. 4.1).

We shall now establish a number of theorems, in the setting of the general topological space, which were proved for the special case of the metric topology. In all these cases, the modifications are minor, and because of this many proofs are left as exercises for the reader. Before proceeding, we must once again warn the reader that many results proved for the case of the metric topology simply do not remain true in the more general setting. This will be especially apparent in Sec. 7.2 and, in particular, in the exercises for that section. We can already point out one marked contrast: It is clear that, if X is a metric space and $x \in X$ then $\overline{\{x\}} = \{x\}$; whereas $\overline{\{x\}} = X$ in the topological space of Example 7.1! With this word of warning, we proceed to the theorems.

THEOREM 7.1. Let $(X, \mathcal{O})$ be a topological space and let A and B be subsets of X. Then

(1) $A \subset B \Rightarrow \bar{A} \subset \bar{B}$;
(2) $\overline{A \cup B} = \bar{A}.\cup \bar{B}$;
(3) $\overline{\bar{A}} = \bar{A}$.

(Compare these results with those following Definition 3.3.)

Proof. The proof of (1) is clear. Using this result, it follows immediately that $\bar{A} \cup \bar{B} \subset \overline{A \cup B}$.

If, on the other hand, an element $x \in \overline{A \cup B}$, any open set O containing x has a nonempty intersection with $A \cup B$. If $x \notin \bar{A} \cup \bar{B}$, there exists an open set O_1 containing x such that $O_1 \cap A = \emptyset$ and an open set O_2 containing x such that $O_2 \cap B = \emptyset$. Thus $x \in O_1 \cap O_2$, which is an open set by the second axiom, and $O_1 \cap O_2 \cap (A \cup B) = \emptyset$, a contradiction. Therefore, $\overline{A \cup B} \subset \bar{A} \cup \bar{B}$, and this completes the proof of (2).

Finally, we prove (3). $\bar{A} \subset \overline{\bar{A}}$, as noted earlier. Thus, suppose that $x \in \overline{\bar{A}}$ and let O be any open set containing x; then $O \cap \bar{A} \neq \emptyset$. Let $y \in O \cap \bar{A}$; hence $y \in \bar{A}$, and therefore any open set containing y has a nonempty intersection with A, but $y \in O$, which is open. Thus, $O \cap A \neq \emptyset$. Since O was any open set containing x, we can conclude that $x \in \bar{A}$, so $\overline{\bar{A}} \subset \bar{A}$, which completes the proof of part (3).

The next theorem gives an alternate characterization of the closedness of a set (cf. Theorem 3.1).

THEOREM 7.2. Let $(X, \mathcal{O})$ be a topological space. A subset F of X is closed if and only if $F = \bar{F}$.

Proof (1). If F is closed, CF is open. Let x be any element of CF. Thus $x \in CF$, which is open, and of course $F \cap CF = \emptyset$. Therefore, no element of CF is an adherence point of F, so that we must have $\bar{F} \subset F$. But we always have $F \subset \bar{F}$; hence, $F = \bar{F}$.

Proof (2). Conversely, suppose that $F = \bar{F}$. Then if $x \in CF$, x is certainly not an adherence point of F. Thus there exists an open set O_x which contains x and such that $O_x \cap F = \emptyset$. Let $O = \bigcup_{x \in CF} O_x$. O is open by the first axiom for a topological space, and, moreover, it is clear that $CF = O$, since $O_x \cap F = \emptyset$ for each $x \in CF$. Thus CF is open, and consequently F is closed. This completes the proof of the theorem.

We observe that, as in the case of the metric topology, we can define the notion of a limit point of a set. Namely, if $(X, \mathcal{O})$ is a topological space and A is a subset of X, a point $x \in X$ is called a *limit point of A* if every open set containing x contains a point of A distinct from x. This is not quite the definition that was taken in the metric space situation; as the reader can readily show, however, they are equivalent in the case of metric space (cf. Exercise 7.1). This, however, is no longer the case for the general topological space as illustrated by Exercise 7.22.

Again, the set of all limit points of A is denoted by A' and is called the *derived set of A*. Clearly, $\bar{A} = A \cup A'$ and, in view of Theorem 7.2, A is closed if and only if $A' \subset A$.

If $\{x_n\}$ is a sequence of points in a topological space $(X, \mathcal{O})$ we can introduce the notion of *convergence* as follows: The sequence $\{x_n\}$ is said to converge to the point $x \in X$, denoted by $x_n \to x$, if, for every open set O containing x, there exists an index N (depending in general on O) such that $x_n \in O$ for all $n > N$. Or, to paraphrase this requirement, every open set containing x must contain almost all (all but a finite number) of the x_n.

The notion of convergence of a sequence of points of a general topological space does not play as important a role in the general setting as it does in the special case of a metric space, as will be apparent from many of the exercises. Actually to get certain analogous theorems in the general case, one must introduce some sort of notion of *generalized* sequence.

We shall now establish a few theorems related to continuous mappings.

THEOREM 7.3. Let X_1, X_2, and X_3 be three topological spaces. Suppose that f is a continuous mapping of X_1 into X_2 and g a continuous mapping of X_2 into X_3. Then the composite mapping gf of X_1 into X_3 is continuous.

Proof. Pictorially, we have the following situation:

$$X_1 \xrightarrow{f} X_2 \xrightarrow{g} X_3.$$

Let O_3 be any open set in X_3 and let $h = gf$ be the composite map. Then

$$h^{-1}(O_3) = f^{-1}(g^{-1}(O_3))$$

and, since g is continuous, $g^{-1}(O_3)$ is open in X_2; since f is continuous, $f^{-1}(g^{-1}(O_3))$ is open in X_1. This completes the proof.

THEOREM 7.4. Let X_1 and X_2 be two topological spaces and $f: X_1 \to X_2$ be a mapping of X_1 into X_2. f is continuous if and only if for every subset A of X, $f(\bar{A}) \subset \overline{f(A)}$.

Proof (1). First assume that f is continuous and let A be an arbitrary subset of X_1. Let $x \in \bar{A}$ and consider $f(x)$. If O_2 is any open set of X_2 such that $f(x) \in O_2$, then $x \in f^{-1}(O_2)$, which is an open set of X, since f is assumed continuous. Therefore, $f^{-1}(O_2) \cap A \neq \emptyset$, since $x \in \bar{A}$. Consequently, $O_2 \cap f(A) \neq \emptyset$, but O_2 was any open set containing $f(x)$. Thus $f(x) \in \overline{f(A)}$, or $f(\bar{A}) \subset \overline{f(A)}$.

Proof (2). Conversely, we now assume that $f(\bar{A}) \subset \overline{f(A)}$, where A is any subset of X. Let O_2 be an open set in X_2; we must show that $U = f^{-1}(O_2)$ is open in X_1. It is easy to see that

$$f(CU) = f(X_1) \cap CO_2.$$

Now, by hypothesis, $f(\overline{CU}) \subset \overline{f(CU)}$; that is,

$$f(\overline{CU}) \subset \overline{f(X_1) \cap CO_2}. \tag{7.1}$$

Let $x \in \overline{CU}$. It is clear that $x \notin U$, for otherwise $f(x) \in O_2$ and by Eq. (7.1), $f(x) \in \overline{f(X_1) \cap CO_2}$, which would imply, since O_2 is open and $f(x) \in O_2$, that $O_2 \cap CO_2 \neq \emptyset$, which is clearly absurd. Thus, if $x \in \overline{CU}$, $x \notin U$ or $\overline{CU} \subset CU$, so CU is closed and $U = f^{-1}(O_2)$ is open, completing the proof.

Strengthening the inclusion of Theorem 7.4 to an equality, we have the following result.

THEOREM 7.5. Let X_1 and X_2 be two topological spaces and $f: X_1 \to X_2$ be a $1:1$, onto mapping. Then f is a homeomorphism of X_1 onto X_2 if and only if, for all subsets A of X_1, $f(\bar{A}) = \overline{f(A)}$.

Proof. The proof follows readily from the definition of a homeomorphism and the preceding theorem and is left as an exercise for the reader.

7.2 Bases

Again, let $(X, \mathcal{O})$ be a topological space and let A be a subset of X. We want to make A into a topological space in a natural fashion; for this purpose we must define the open sets of A. The collection of all sets of the form $A \cap O$, where $O \in \mathcal{O}$, is easily seen to form a topology for A and is called the topology *induced by X* or the *relative topology of A*, and A together with this topology is called a *subspace* of the given topological space $(X, \mathcal{O})$.

A word of caution is necessary in considering a subspace A of the topological space X. It is entirely possible for a set to be open in A but not in X, for if E is open in A, it simply means that E is of the form $A \cap O$, where O is open in X. Clearly, $A \cap O$ need not be open in X; it surely will be, however, if A itself is open in X.

We note that, if (X, d) is a metric space and if we consider the metric topology, for any subset A of X, then (A, d) is a metric space (Sec. 3.2) and therefore A can be considered a topological space with metric topology determined by the restriction of d to A. In addition, however, A inherits the induced topology of X. It is not difficult to see, and we have left it as an exercise (Exercise 7.3), that these two topologies on A coincide.

In specifying a topological space $(X, \mathcal{O})$, it is frequently unnecessary to give the entire collection $\mathcal{O}$ of open sets, but instead a certain subcollection. We now consider precisely how to specify such a subcollection as we turn to the notion of a *basis for a topology*—a notion which, incidentally, will turn out to be quite useful in constructing topologies.

DEFINITION 7.5. Let $(X, \mathcal{O})$ be a topological space. A collection $\mathcal{B}$ of open sets such that every open set can be written as a union of sets from $\mathcal{B}$ is called a *basis*. We observe that a basis always exists; if worse comes to worst, just take $\mathcal{B} = \mathcal{O}$. Next it is clear that, if $\mathcal{B} = \{B_\alpha\}_{\alpha \in \Lambda}$ (Λ is some index set of arbitrary cardinality) is a basis for X, then $\{A \cap B_\alpha\}_{\alpha \in \Lambda}$ is a basis for the subspace A. Also, many of our previous results and concepts can be characterized in terms of the basis sets. For example, the reader can easily verify (Exercise 7.4) that, if $f: X_1 \to X_2$, where X_1 and X_2 are two topological spaces, then f is continuous if and only if $f^{-1}(B)$ is open, where B is any set of a basis for X_2.

We note the next theorem.

THEOREM 7.6. A collection $\mathcal{B} = \{B_\alpha\}_{\alpha \in \Lambda}$ of open sets of the topological space X is a basis if and only if, for every open set O and every $x \in O$, there exists a $B_x \in \mathcal{B}$ such that $x \in B_x \subset O$.

Proof. If the condition is true, then, for any open set O,

$$O = \bigcup_{x \in O} B_x,$$

and $\mathcal{B}$ is therefore a basis.

Conversely, if $\mathcal{B}$ is a basis, let O be any open set. Thus $O = \bigcup_\alpha B_\alpha$, where $B_\alpha \in \mathcal{B}$, and if $x \in O$, then $x \in B_\alpha$ for some α and $B_\alpha \subset O$. This completes the proof.

In the metric space (R^2, d) with d given by the first metric of Example 3.3, it is readily seen that the open spheres of radius ε constitute a basis. (It is further evident that the collection of all open triangles, open crescents, etc., will also form a basis for this space.) More generally, if (X, d) is a metric space with metric topology, the collection of all $S_\varepsilon(x_0)$ is a basis. Indeed, recalling Definition 3.2, if A is open then $A = A^o$, the interior of A; therefore $x \in A$ implies that x is an interior point, and so there exists an $S_{\varepsilon(x)}(x) \subset A$. Hence,

$$A = \bigcup_{x \in A} S_{\varepsilon(x)}(x),$$

and thus any open set is a union of sets of the form $S_\varepsilon(x)$, the collection of which is consequently a basis.

We return again to the case of R^2, only now with the metric given either by $d(x, y) = |x_1 - y_1| + |x_2 - y_2|$ or by $d(x, y) = \max\{|x_1 - y_1|, |x_2 - y_2|\}$, where $x = (x_1, x_2)$ and $y = (y_1, y_2)$. Although all three metrics we have considered on R^2 are distinct, the resulting metric topologies are all the same. This is a special case of the following, more general, situation, which we shall now discuss.

Suppose X is an arbitrary set and $\mathscr{B}$ is a nonempty collection of subsets of X which satisfies the conditions:

(B_1) For each $x \in X$ there exists a $B_x \in \mathscr{B}$ such that $x \in B_x$.

(B_2) Given any two sets $B_1, B_2 \in \mathscr{B}$, if $x \in B_1 \cap B_2$, then there exists another set $B_3 \in \mathscr{B}$ such that $x \in B_3 \subset B_1 \cap B_2$.

If $(X, \mathcal{O})$ is a topological space and $\mathscr{B}$ is a basis of open sets, then $\mathscr{B}$ clearly satisfies condition (B_1). As for (B_2), if $B_1, B_2 \in \mathscr{B}$, since $B_1 \cap B_2$ is open, it can be written as a union of sets of $\mathscr{B}$ and therefore (B_2) must be satisfied.

Conversely, we now show that, if X is an arbitrary set and $\mathscr{B}$ a collection of subsets satisfying the above conditions, then a topology $\mathcal{O}$ can be introduced on X for which $\mathscr{B}$ is a basis. We achieve this by defining the collection $\mathcal{O}$ to consist of arbitrary unions of sets from $\mathscr{B}$ together with the null set (or use the convention mentioned after Definition 7.1). $\mathcal{O}$ is clearly closed under arbitrary unions and $X \in \mathcal{O}$, for $X = \bigcup_{x \in X} B_x$, where B_x is as in (B_1). To show that $\mathcal{O}$ is closed under finite intersections and is consequently a topology, it suffices by induction to show that, if $O_1, O_2 \in \mathcal{O}$, then $O_1 \cap O_2 \in \mathcal{O}$. Now, since $O_1, O_2 \in \mathcal{O}$, $O_1 = \cup B_\alpha^1$ and $O_2 = \cup B_\gamma^2$, where $B_\alpha^1, B_\gamma^2 \in \mathscr{B}$, and

$$O_1 \cap O_2 = \bigcup_{\alpha, \gamma} B_\alpha^1 \cap B_\gamma^2.$$

If we can show that an arbitrary $B_\alpha^1 \cap B_\gamma^2$ can be written as a union of sets from $\mathscr{B}$, then we are, of course, done. Thus, let $x \in B_\alpha^1 \cap B_\gamma^2$. By ($B_2$) there exists a $B_{\alpha,\gamma}^{1,2} \in \mathscr{B}$ that contains x, and such that $B_{\alpha,\gamma}^{1,2} \subset B_\alpha^1 \cap B_\gamma^2$ which clearly implies that $B_\alpha^1 \cap B_\gamma^2$ can be written as a union of sets from $\mathscr{B}$. Thus $\mathcal{O}$ is a topology and, by its very definition, $\mathscr{B}$ is a basis of open sets for this topology. Consequently, one also speaks of a collection of sets $\mathscr{B}$ satisfying conditions (B_1) and (B_2) as a *basis* and the topology defined as above as the topology *determined by* or *generated by* $\mathscr{B}$.

If we are given two bases $\mathscr{B}_1$ and $\mathscr{B}_2$ of X, that is, two collections of subsets of X satisfying (B_1) and (B_2), they are called *equivalent* if they determine the same topologies. With further regard to equivalence we have the next theorem.

THEOREM 7.7. Two bases $\mathscr{B}_1$ and $\mathscr{B}_2$ are equivalent if and only if, for each $B_1 \in \mathscr{B}_1$ and each $x \in B_1$, there exists a $B_2 \in \mathscr{B}_2$ such that $x \in B_2 \subset B_1$. Conversely, for each $\hat{B}_2 \in \mathscr{B}_2$ and each $y \in \hat{B}_2$, there exists a $\hat{B}_1 \in \mathscr{B}_1$ such that $y \in \hat{B}_1 \subset \hat{B}_2$.

Proof. Suppose the condition is satisfied. Let $\mathcal{O}_1$ be the topology determined by $\mathscr{B}_1$ and $\mathcal{O}_2$ that are determined by $\mathscr{B}_2$. If $O_1 \in \mathcal{O}_1$, then

$$O_1 = \bigcup_\alpha B_\alpha^1,$$

where the $B_\alpha^1 \in \mathscr{B}_1$. But $x \in O_1$ implies that $x \in B_\alpha^1$ for some α, whence there exists a $B_\alpha^2 \in \mathscr{B}_2$ such that $x \in B_\alpha^2 \subset B_\alpha^1$. It follows that O_1 is certainly a union of sets from $\mathscr{B}_2$ and so belongs to $\mathcal{O}_2$. Similarly, any set in $\mathcal{O}_2$ is in $\mathcal{O}_1$, so $\mathcal{O}_1 = \mathcal{O}_2$.

If the bases are equivalent, $\mathcal{O}_1 = \mathcal{O}_2$, so, if $B_1 \in \mathcal{O}_1$, then $B_1 \in \mathcal{O}_2$. Hence,

$$B_1 = \bigcup B_\alpha^2$$

where the $B_\alpha^2 \in \mathscr{B}_2$. Therefore, if $x \in B_1$, then $x \in B_\alpha^2$ for some α and $B_\alpha^2 \subset B_1$. Similarly, the other half of the condition is satisfied and this completes the proof.

In the particular case of R^n or, more particularly, R^2 (with any of the three metrics considered in Example 3.3) and taking in each instance the bases to be the ones given by the respective ε-spheres, we see that these bases are equivalent and the resulting metric topologies the same.

Let $(X, \mathcal{O})$ be a topological space. Besides a basis of open sets, we also consider a basis of open sets at a point x, defined as follows.

DEFINITION 7.6. A collection $\mathscr{B}_x$ of open sets of the topological space X is called a *basis at the point* $x \in X$ if, for any open set O containing x, there exists a set $B \in \mathscr{B}_x$ such that $x \in B \subset O$.

In the metric topology of a metric space (X, d), the collection of all $S_\varepsilon(x_0)$, where ε runs through the positive real numbers, constitutes a basis at the given point $x_0 \in X$. However, the collection of all $S_r(x_0)$, where r runs through the set of positive rational numbers is already a basis at the point x and this collection is countable, which brings us to our next point.

DEFINITION 7.7. A topological space $(X, \mathcal{O})$ is said to satisfy the *first axiom of countability* if there exists a countable basis at every point. It is said to satisfy the *second* axiom of countability if there is a countable basis for $\mathcal{O}$. Clearly, the second axiom of countability implies the first.

We have just seen that every metric space satisfies the first axiom of countability. In general, an arbitrary metric space does not satisfy the second axiom of countability. The particular metric space (R^n, d), where d is any of the three metrics of Example 3.6 does, however, satisfy the second axiom of countability. As a basis we may choose the collection of all ε-spheres with rational radii and rational centers, that is, with centers $x = (\alpha_1, \ldots, \alpha_n)$, where the α_i are all rational numbers. This collection is, of course, countable and so this space satisfies the second axiom of countability. As an example of a space that does not satisfy the second axiom of countability, consider an infinite nondenumerable set X with trivial metric, which leads clearly to the discrete topology as the metric topology.

7.3 Weak Topologies

The next situation that we shall consider is a given set X with two topologies $\mathcal{O}_1$ and $\mathcal{O}_2$ defined on X. If $\mathcal{O}_1 \subset \mathcal{O}_2$, we say that the topology $\mathcal{O}_1$ is *weaker* (*coarser*) than $\mathcal{O}_2$ or that the topology $\mathcal{O}_2$ is *stronger* (*finer*) than $\mathcal{O}_1$. If X is an arbitrary set, then the trivial topology is clearly weaker than any other topology on X and the discrete topology is stronger than every other.

Since the intersection of an arbitrary collection $\{\mathcal{O}_\alpha\}$ of topologies defined on X is a topology, it is clear that given an arbitrary collection $\mathscr{S}$ of subsets of X, there is a unique topology containing $\mathscr{S}$ that is weaker than all other topologies containing $\mathscr{S}$. This topology is obtained by taking the intersection of all topologies containing $\mathscr{S}$. We note that there is at least one topology containing $\mathscr{S}$, namely, the discrete topology. Another way to characterize this unique weakest topology containing $\mathscr{S}$ is by taking the collection of all arbitrary unions of finite intersections of sets in $\mathscr{S}$. It is not difficult to see that this collection of sets forms a topology adopting the customary conventions concerning empty intersections and unions. The proof follows directly from our previous discussion on bases, if we observe that the set of all finite intersections of sets from $\mathscr{S}$ forms a basis, that is, satisfies conditions (B_1) and (B_2) of Sec. 7.2. (B_2) is clearly true here, whereas (B_1) follows from the fact that X is in our collection being an empty intersection of sets from $\mathscr{S}$. We take now the topology determined by this basis and call it the topology generated by $\mathscr{S}$.

DEFINITION 7.8. A nonempty collection $\mathscr{S}$ of open subsets of a topological space X is called a *subbasis* if the collection of all finite intersections of sets from $\mathscr{S}$ forms a basis.

The previous discussion shows that the following theorem is true.

THEOREM 7.8. Let X be an arbitrary set and let $\mathscr{S}$ be any collection of subsets of X. Then there exists a topology for X in which $\mathscr{S}$ is a subbasis.

A subbasis for the metric space R with the usual metric is the collection of all (a, ∞) and $(-\infty, b)$, where $a, b \in R$.

Now let X be an arbitrary (nonempty, as usual) set and let $\{(X_\alpha, \mathcal{O}_\alpha)\}_{\alpha \in \Lambda}$ be a collection of topological spaces. We assume that for each $\alpha \in \Lambda$ there is a given mapping

$$f_\alpha : X \to X_\alpha.$$

If we consider X with the discrete topology, then each f_α is, of course, continuous. Thus there is certainly one topology on X with respect to which all the mappings f_α are continuous. Consider the weakest topology with respect to which all the f_α are continuous. This topology is called the *weak topology generated by the* f_α. We could view this topology as follows: If the f_α are to be continuous, we must have $f_\alpha^{-1}(O_\alpha)$ open for any O_α open in X_α and this must be true for all α. Thus we can view the weak topology as the topology generated by the collection $\mathscr{S}$ of all sets of the form $f_\alpha^{-1}(O_\alpha)$, that is, take this collection of sets as a subbasis. We observe that, instead of all open sets O_α in X_α, it suffices to take the topology generated by all $f_\alpha^{-1}(B_\alpha)$, where B_α runs over a basis of X_α. We must do this for each α in order to obtain the weak topology generated by the f_α.

In the particular case in which all $X_\alpha = C$ or all $X_\alpha = R$, a subbasis $\mathscr{S}$ for the weak topology on X generated by all f_α consists of all sets of the form

$$\{x \in X \mid |f_\alpha(x) - f_\alpha(x_0)| < \varepsilon\}, \qquad (7.2)$$

where ε is a positive real number. A finite intersection of sets of this sort with a fixed $x_0 \in X$ and ε will be denoted by $V(x_0; f_1, f_2, ..., f_n, \varepsilon)$. Hence,

$$V(x_0; f_1, ..., f_n, \varepsilon) = \{x \in X \mid |f_i(x) - f_i(x_0)| < \varepsilon, i = 1, ..., n\}. \qquad (7.3)$$

We now establish several properties pertaining to sets of the form of Eq. (7.3).

(1) $x_0 \in V(x_0; f_1, ..., f_n, \varepsilon)$.

Proof. Clear.

(2) Given $V(x_0; f_1, ..., f_n, \varepsilon_1)$ and $V(x_0; g_1, ..., g_m, \varepsilon_2)$, then

$$V(x_0; f_1, ..., f_n, g_1, ..., g_m, \min(\varepsilon_1, \varepsilon_2)) \subset V(x_0; f_1, ..., f_n, \varepsilon_1) \cap V(x_0; g_1, ..., g_m, \varepsilon_2)$$
(7.4)

Proof. Suppose that x belongs to the left member of Eq. (7.4). Then

$$|f_i(x) - f_i(x_0)| < \min(\varepsilon_1, \varepsilon_2) \le \varepsilon_1 \qquad (i = 1, ..., n)$$

and

$$|g_i(x) - g_i(x_0)| < \min(\varepsilon_1, \varepsilon_2) \le \varepsilon_2 \qquad (i = 1, ..., m).$$

Therefore $x \in V(x_0; f_1, ..., f_n, \varepsilon_1)$ and $x \in V(x_0; g_1, ..., g_m, \varepsilon_2)$ and this establishes the desired inclusion.

(3) If $x_1 \in V(x_0; f_1, ..., f_n, \varepsilon)$, there exists a $\delta > 0$ such that $V(x_1; f_1, ..., f_n, \delta) \subset V(x_0; f_1, ..., f_n, \varepsilon)$.

Proof. Let $|f_i(x_1) - f_i(x_0)| = v_i < \varepsilon$ $(i = 1, ..., n)$, and let $v = \max_i v_i$. Finally, choose δ such that $\varepsilon - v > \delta > 0$, and consider $V(x_1; f_1, ..., f_n, \delta)$. If $x \in V(x_1; f_1, ..., f_n, \delta)$, then

$$|f_i(x) - f_i(x_1)| < \delta \qquad (i = 1, ..., n).$$

Hence,

$$|f_i(x) - f_i(x_0)| \le |f_i(x) - f_i(x_1)| + |f_i(x_1) - f_i(x_0)|$$

$$< \delta + v$$

$$< \varepsilon.$$

Thus we see that

$$V(x_1; f_1, ..., f_n, \delta) \subset V(x_0; f_1, ..., f_n, \varepsilon).$$

Now we recall that the weak topology generated by the family of functions consists of arbitrary unions of finite intersections of sets of the form (7.2). Considering finite intersections of sets of the form (7.2) with a fixed x_0 and ε gave us sets of the form (7.3). Now, for different x_0, say x_0 and y_0, and different ε, say ε_1 and ε_2, let us consider

$$V(x_0; f_1, ..., f_n, \varepsilon_1) \cap V(y_0; g_1, ..., g_m, \varepsilon_2),$$

and suppose that x_1 belongs to this intersection. Then by property (3), there exist δ_1 and δ_2 such that

$$V(x_1; f_1, ..., f_n, \delta_1) \subset V(x_0; f_1, ..., f_n, \varepsilon_1)$$

and

$$V(x_1; g_1, ..., g_m, \delta_2) \subset V(y_0; g_1, ..., g_m, \varepsilon_2),$$

so by properties (1) and (2),

$$x_1 \in V(x_1; f_1, ..., f_n, g_1, ..., g_m, \min(\delta_1, \delta_2))$$

$$\subset V(x_1; f_1, ..., f_n, \delta_1) \cap V(x_1; g_1, ..., g_m, \delta_2)$$

$$\subset V(x_0; f_1, ..., f_n, \varepsilon_1) \cap V(y_0; g_1, ..., g_m, \varepsilon_2).$$

It follows immediately from this that the sets of the form (7.2) form a basis and that the weak topology generated by the f_α consists in this case of arbitrary unions of sets of the form $V(x_0; f_1, ..., f_n, \varepsilon)$.

We make one final observation concerning the weak topology in this situation. Suppose $x_0, y_0 \in X$ and $x_0 \neq y_0$. In addition, assume that there exists an $f \in \{f_\alpha\}_{\alpha \in \Lambda}$ such that $f(x_0) \neq f(y_0)$. Then choose ε so that $0 < \varepsilon < |f(x_0) - f(y_0)|$. It is now evident that

$$V(x_0; f, \varepsilon/2) \cap V(y_0; f, \varepsilon/2) = \emptyset \tag{7.5}$$

for, if x belongs to this intersection, then $|f(x) - f(x_0)| < \varepsilon/2$ and $|f(x) - f(y_0)| < \varepsilon/2$, so

$$|f(x_0) - f(y_0)| \leq |f(x_0) - f(x)| + |f(x) - f(y_0)|$$

$$< \frac{\varepsilon}{2} + \frac{\varepsilon}{2} = \varepsilon,$$

a contradiction, proving (7.5).

We shall return to the situation at hand later in a more special setting in which X is not just an arbitrary set but has certain algebraic and analytic structure defined on it. The results established here in the more general situation will then be carried over to the special case and will prove quite useful.

We continue with the general notion of weak topology generated by a family of functions. Given again a family $\{X_\alpha\}_{\alpha \in \Lambda}$ of topological spaces, we wish to introduce a topology on the cartesian product, $X = \prod_{\alpha \in \Lambda} X_\alpha$, of the spaces X_α. We remind the reader that X consists of all functions, $x : \Lambda \to \bigcup_{\alpha \in \Lambda} X_\alpha$ such that $x(\alpha) \in X_\alpha$. For each $\alpha \in \Lambda$, there is associated a mapping

$$pr_\alpha : X \to X_\alpha$$

given by $pr_\alpha(x) = x(\alpha)$, called the *projection of X onto X_α*. We now take for a topology of X the weak topology generated by the family $\{pr_\alpha\}_{\alpha \in \Lambda}$.

Since $pr_\gamma^{-1}(O_\gamma)$, where O_γ is open in X_γ is just $\prod_{\alpha \in \Lambda} U_\alpha$, where all $U_\alpha = X_\alpha$, $\alpha \neq \gamma$, and $U_\gamma = O_\gamma$, and since a finite intersection of sets of this type gives a set of the form

$$\prod_{\alpha \in \Lambda} V_\alpha, \tag{7.6}$$

where all but a finite number of the $V_\alpha = X_\alpha$ and the others are open sets in the respective spaces, we can say that the weak topology generated by the pr_α is just the topology determined by taking all sets of the form (7.6) as a basis. This topology on the cartesian product is frequently referred to as the *Tychonoff topology* or sometimes as just the *product* topology. We shall return to this topology shortly. For the moment, we shall leave the discussion of weak topologies and go on to a consideration of separation properties in general topological spaces.

7.4 Separation

We have seen that, if X is a topological space with trivial topology and if $x \in X$, then $\overline{\{x\}} = X$. Also, it is easy to see, using the notion of convergence of a sequence in a topological space introduced in Section 7.1, that any sequence $\{x_n\}$ of points in this space converges to any and every point in the space. Needless to say, this type of chaos is a bit undesirable, at least from certain points of view, and for this reason we start by restricting our attention to spaces where there is more " separation " between the points of the space.

A topological space X is said to satisfy axiom T_0 or to be a T_0 *topological space* if it satisfies the following condition:

(T_0). If x and y are any two distinct points of X, there exists an open set which contains one of them but not the other.

It is clear that, in general, a space X with trivial topology does not satisfy axiom T_0 ; all other examples given in Sec. 7.1 do.

A stronger separation demand on a topological space X is the following:

(T_1). If x and y are any two distinct points of the topological space X, there exists an open set containing x and not y and an open set containing y but not x.

Again, all our spaces given in Sec. 7.1 except the first satisfy axiom (T_1). Such spaces are called T_1 *topological spaces*. However, there are examples of spaces that satisfy (T_0) but not (T_1). For example, take $X = [0, 1)$, the topology $\mathcal{O}$ as $\emptyset$, and all sets of the form $[0, a)$, where $0 < a \leq 1$. It is readily seen that this is a T_0 topological space but certainly not T_1.

The next separation axiom (T_2) is still stronger than (T_1) and is frequently called the *Hausdorff separation axiom*.

(T_2). If x and y are two distinct points of the topological space X, there exist disjoint open sets containing x and y, respectively.

A topological space satisfying this axiom is called either a T_2 *topological space* or a *Hausdorff space*. Clearly, any Hausdorff space is a T_1 space; however, the cofinite topology (Example 7.4) is an example of a T_1 space that is not Hausdorff. We saw in the previous section that, under special conditions, the weak topology generated by a family of functions $\{f_\alpha\}_{\alpha \in \Lambda}$ where each f_α maps X into C say, is a Hausdorff space. It is also a simple matter to verify that every metric space is a Hausdorff space.

We show next that the situation which occurred in the space with trivial topology, namely $\overline{\{x\}} = X$, cannot occur in a T_1 space, assuming that X to begin with does not consist of just a single point.

THEOREM 7.9. A topological space X satisfies axiom (T_1) if and only if every subset consisting of a single point is closed.

Proof (1). Suppose X is a T_1 space. Let $x \in X$; then for any $y \in X - \{x\}$, there exists an open set containing y but not x. Thus $y \notin \overline{\{x\}}$ for any $y \in X - \{x\}$; whence $\overline{\{x\}} = \{x\}$, and $\{x\}$ is a closed set.

Proof (2) Conversely, suppose that, for every $x \in X$, $\{x\}$ is a closed subset. If x and y are distinct points of X, then $X - \{x\}$ and $X - \{y\}$ are open sets, and $y \in X - \{x\}$ but $x \notin X - \{x\}$; also $x \in X - \{y\}$ but $y \notin X - \{y\}$. Hence X is a T_1 space, and this finishes the proof.

We observe that if X satisfies any of the axioms (T_0), (T_1), or (T_2), then so does any subspace of X. Also, if $\{X_\alpha\}_{\alpha \in \Lambda}$ is a family of topological spaces each of which is, say, Hausdorff, then so is the product space. We leave this last fact as a simple exercise for the reader. We also observe that in a Hausdorff space a sequence $\{x_n\}$ cannot converge to more than one point.

Actually, this completes the material that we need for our purposes concerning separation properties. However, for completeness, we list the most important remaining separation axioms for the reader's convenience and leave some matters related to them for the exercises. Suitable references for this material will also be listed at the end of this chapter.

DEFINITION 7.9. A topological space X is called *regular* if for each point x and each closed set F such that $x \notin F$, there exist disjoint open sets containing x and F, respectively. A regular T_1 space is called a T_3 *topological space.*

It is clear, in view of Theorem 7.9, that a T_3 topological space is a Hausdorff space.

DEFINITION 7.10. A topological space X is said to be *completely regular* if, for each $x \in X$ and each open set O containing x, there exists a real-valued continuous function f defined on X such that $0 \le f(y) \le 1$ for all $y \in X$, $f(x) = 0$, and $f(y) = 1$ for $y \in X - O$. A completely regular T_1 space is called a $T_{3\frac{1}{2}}$ space or a *Tychonoff space.*

It is not difficult to show (see Exercise 7.31) that a completely regular space is regular. It therefore follows that every $T_{3\frac{1}{2}}$ space is a T_3 space.

DEFINITION 7.11. A topological space X is called *normal* if, for any two disjoint closed sets F_1 and F_2 of X, there exist disjoint open sets O_1 and O_2 such that $F_1 \subset O_1$ and $F_2 \subset O_2$. A normal T_1 space is called a T_4 *space.*

Since every T_4 space is completely regular (see Exercise 7.31), it follows that every T_4 space is a $T_{3\frac{1}{2}}$ space.

DEFINITION 7.12. Two subsets A and B of a topological space X are said to be *separated* if and only if $\bar{A} \cap B = \emptyset$ and $A \cap \bar{B} = \emptyset$. A topological space X is called *completely normal* if, for any two separated sets A and B, there exist disjoint open sets O_1 and O_2 such that $O_1 \supset A$ and $O_2 \supset B$. A completely normal T_1 space is called a T_5 *space.*

Since completely normal clearly implies normal, it follows that any T_5 space is a T_4 space. Thus we have a hierarchy of T spaces. None of the implications that we have mentioned go in the reverse way; a few of these were illustrated by examples for the lower T spaces. Other examples may be found in the exercises, some with references given.

7.5 Compactness

In the case of metric spaces, we introduced several notions of compactness and showed that they were all equivalent. This, as we already warned the reader, is not the case in general topological spaces. We proceed now to discuss these concepts in a general topological space. We leave, for the exercises, examples illustrating the nonequivalence of these notions. The definition of covering and compactness, given in Definition 5.4, can be carried over verbatim to general topological spaces and we list it for convenience.

DEFINITION 7.13. Let Λ be an indexing set and further suppose that $\{G_\alpha\}$, where α runs through Λ, has the property that $X \subset \bigcup_{\alpha \in \Lambda} G_\alpha$. Then $\{G_\alpha\}$ is said to be a *covering of X*. If the set Λ is finite, it is said to be a *finite covering*. If X is a topological space and if each G_α is open, the covering is said to be an *open covering* of X. Finally, if the space X has the property that from every open covering one can select a finite subcovering, then X is said to be *compact*. A subspace A of X is compact if it is compact in the induced topology.

In a metric space we saw (Theorem 5.2) that every compact set is closed. In general, we can prove the next theorem.

THEOREM 7.10. A compact subspace of a Hausdorff space is closed.

Proof. Let X be a Hausdorff space and A a compact subspace. Let $x \in A$ and let y be a fixed but arbitrary point of CA. Since X is Hausdorff, there exist disjoint open sets O_x and $O_y{}^x$, depending on x, such that $x \in O_x$ and $y \in O_y{}^x$. It is clear that

$$A = \bigcup_{x \in A} (O_x \cap A)$$

and, since A is assumed compact,

$$A = \bigcup_{i=1}^{n} (O_{x_i} \cap A).$$

Consider now the corresponding $O_y{}^{x_i}$ ($i = 1, 2, ..., n$). Each of these open sets contains y; therefore,

$$V = \bigcap_{i=1}^{n} O_y{}^{x_i} \neq \emptyset.$$

In addition $V \cap A = \emptyset$, for, if $V \cap A \neq \emptyset$, then $V \cap O_{x_i} \neq \emptyset$ for some i ($1 \leq i \leq n$); hence,

$$O_{x_i} \cap O_y{}^{x_i} \neq \emptyset \qquad \text{(for some } i, 1 \leq i \leq n),$$

a contradiction. Thus any point $y \in CA$ lies in an open set V having an empty intersection with A so $y \notin \bar{A}$. Hence $\bar{A} \subset A$ and A is closed, which completes the proof.

THEOREM 7.11. A closed subspace of a compact space is compact.

Proof. Let X be a compact topological space and F a closed subspace. Let $\{U_\alpha\}$ be an open covering of F in the relative topology on F. Then each U_α is of the form $O_\alpha \cap F$, where O_α is open in X. Clearly, the collection $\{O_\alpha\} \cup \{CF\}$ of open sets of X covers X and, since X is compact, a finite subcovering $O_1, O_2, ..., O_n, CF$

can be found. Then the collection $\{O_i \cap F\}$, $(i = 1, \ldots, n)$ a subcollection of the original collection $\{U_\alpha\}$, yields a covering of the space F; hence F is compact.

We recall that the proof of Theorem 5.8 made no use of the metric properties of the spaces X and Y but only of the characterization of compactness in terms of open sets, as well as the characterization of the continuity of f via open sets. Thus the proof of that theorem can be carried over completely to the case of topological spaces and we have:

THEOREM 7.12. Let f be a continuous mapping of the topological space X into the topological space Y. If A is a compact subspace of X, then $f(A)$ is a compact subspace of Y.

A related result is the following theorem.

THEOREM 7.13. If $f : X \to Y$ is a continuous 1 : 1 mapping of a compact space X into a Hausdorff space Y, then f is a homeomorphism.

Proof. Let F be an arbitrary closed subspace of X. Since X is compact, it follows from Theorem 7.11 that F is compact and therefore, by Theorem 7.12, $f(F)$ is compact. However, Y is Hausdorff; hence by Theorem 7.10 $f(F)$ is closed. Thus f maps closed sets into closed sets, so f^{-1} is continuous, and since f is assumed 1 : 1 and continuous, the desired conclusion that f is a homeomorphism follows.

Earlier we introduced into a general topological space X the notion of limit point of a set and convergence of a sequence of points. Thus, as in Sec. 5.3 for metric spaces, we can introduce in an identical manner into topological spaces the notions of countably compact and sequentially compact. We list them for reference for the reader, with the word of warning that a number of authors take a different definition for countably compact than we do; it is not equivalent to ours. The relationship between these two definitions are given in the exercises (see especially Exercise 7.19).

DEFINITION 7.14. A set A is said to be *countably compact* if every infinite subset of A has a limit point in A.

DEFINITION 7.15. A set A is said to be *sequentially compact* if, from every sequence in A, one can select a convergent subsequence with limit in A.

As noted earlier, these concepts of compactness are not equivalent in a general topological space, as they are in metric spaces. The full interconnection can be found in the exercises (see Exercises 7.14–7.21) and is not needed for our later applications, where we shall be concerned exclusively with ordinary compactness as given in Definition 7.13. Here we simply note and prove with minor modifications of the proof of Theorem 5.4 that compactness implies countable compactness.

THEOREM 7.14. Every compact subspace A of a topological space X is countably compact.

Proof. Let B be an infinite subset of A and suppose that B has no limit point in A. Choose an infinite sequence of distinct points of B, $x_1, x_2, ..., x_n, ...$, say, and let $F = \{x_1, x_2, ..., x_n, ...\}$. Clearly, F has no limit point in A. Hence no x_n ($n = 1, 2, ...$) is a limit point of F. Thus, for each n, there exists an open set O_n such that $x_n \in O_n$ and $O_n \cap F = \{x_n\}$. Also, for each $x \in A - F$, there exists an open set O_x such that $x \in O_x$ and $O_x \cap F = \emptyset$. Consider $U = \bigcup_{x \in A - F} O_x$; U is, of course, an open set, and the collection $\{O_n | n = 1, 2, ...\} \cup \{U\}$ is an open covering of A, which clearly has no finite subcovering. This contradicts the hypothesis that A is compact and completes the proof.

Whenever we work with sets, there is always a certain duality present stemming from the fact that if two sets are equal, their complements must also be equal. One manifestation of this duality in topological spaces is that, for every statement about open sets, there is a corresponding statement about closed sets. Compactness is no exception for, if X is compact,

$$\bigcup_\alpha O_\alpha = X \Rightarrow \text{there exists } \bigcup_{i=1}^n O_{\alpha_i} = X.$$

The corresponding dual statement is

$$\bigcap_\alpha F_\alpha = \emptyset \Rightarrow \text{there exists } \bigcap_{i=1}^n F_{\alpha_i} = \emptyset,$$

where the F_α are closed, which is equivalent to:

$$\text{if no } \bigcap_{i=1}^n F_{\alpha_i} = \emptyset, \quad \text{then } \bigcap_\alpha F_\alpha \neq \emptyset.$$

We have thus obtained an equivalent characterization of compactness which we summarize in the next theorem.

THEOREM 7.15. A topological space X is compact if and only if, for any collection of closed sets $\{F_\alpha\}_{\alpha \in \Lambda}$ having the *finite intersection property* (that is, the intersection of any finite number of sets from the collection is nonempty), then $\bigcap_{\alpha \in \Lambda} F_\alpha \neq \emptyset$.

COROLLARY. A topological space X is compact if and only if every collection $\{E_\alpha\}_{\alpha \in \Lambda}$ that has the finite intersection property has at least one common adherence point: $\bigcap_\alpha \bar{E}_\alpha \neq \emptyset$.

Proof. The proof follows by considering the family of closed sets $\{\bar{E}_\alpha\}_{\alpha \in \Lambda}$.

The final theorem we shall consider pertaining to the notion of compactness is the important Tychonoff theorem. We shall only apply this result later in our discussion of Banach algebras, but it has many other important applications in analysis and topology.

THEOREM 7.16 (*Tychonoff*). The cartesian product X of an arbitrary collection $\{X_\alpha\}_{\alpha \in \Lambda}$ of compact spaces is compact (with respect to the topology discussed in Section 7.3 for cartesian products).

Proof. Let $\{E^\gamma\}_{\gamma \in \Gamma}$ be a collection of sets of X having the finite intersection property. By the preceding Corollary, we shall be done if we can show that $\bigcap_{\gamma \in \Gamma} \bar{E}^\gamma \neq \emptyset$. The collection $\{E^\gamma\}_{\gamma \in \Gamma}$ may be assumed to be a maximal collection of sets having the finite intersection property, for, if this is not the case, we observe that the totality of all collections of sets of X containing the collection $\{E^\gamma\}_{\gamma \in \Gamma}$ and having the finite intersection property is inductively ordered by set inclusion (see Appendix to Chapter 9). Consequently, by Zorn's lemma, it has a maximal element. If the intersection of the closures of all sets from this maximal collection, which is contained in the intersection of the closures of our original collection, is nonempty, then of course the same is true for the original collection. Thus we assume henceforth that $\{E^\gamma\}_{\gamma \in \Gamma}$ is a maximal collection of sets having the finite intersection property.

Let $E_\alpha^\gamma = pr_\alpha(E^\gamma)$. For fixed α and any finite collection $\gamma_1, \ldots, \gamma_k \in \Gamma$, consider the sets $E_\alpha^{\gamma_1}, E_\alpha^{\gamma_2}, \ldots, E_\alpha^{\gamma_k}$ in X_α. We contend that their intersection is nonempty, for, suppose

$$\emptyset = \bigcap_{i=1}^{k} E_\alpha^{\gamma_i} = \bigcap_{i=1}^{k} pr_\alpha(E^{\gamma_i}).$$

Then clearly, since $\bigcap_{i=1}^{k} pr_\alpha(E^{\gamma_i}) \supset pr_\alpha(\bigcap_{i=1}^{k} E^{\gamma_i})$, we must have $\bigcap_{i=1}^{k} E^{\gamma_i} = \emptyset$—a contradiction; therefore, $\bigcap_{i=1}^{k} E_\alpha^{\gamma_i} \neq \emptyset$, and we see that the family $\{E_\alpha^\gamma\}_{\gamma \in \Gamma}$ has the finite intersection property but X_α is, by hypothesis, compact. Hence, by the previous corollary, $\bigcap_{\gamma \in \Gamma} \bar{E}_\alpha^\gamma \neq \emptyset$. Let $\hat{x}_\alpha \in \bigcap_{\gamma \in \Gamma} \bar{E}_\alpha^\gamma$. We now contend that the element $\hat{x}$ of X, where $\hat{x}(\alpha) = \hat{x}_\alpha$, all $\alpha \in \Lambda$, is a common adherence point of the collection $\{E^\gamma\}_{\gamma \in \Gamma}$. To prove this, we must show that every basis open set at the point $\hat{x}$ has a nonempty intersection with each E^γ.

Recall (see Section 7.3) that a typical basis set at the point $\hat{x}$, $V(\hat{x})$ of the product topology consists of those $x \in X$ such that

$$x(\alpha_i) = x_{\alpha_i} \in V(\hat{x}_{\alpha_i}) \qquad (i = 1, 2, \ldots, k),$$

where $V(\hat{x}_{\alpha_i})$ is an open set in X_{α_i} containing $\hat{x}_{\alpha_i}$ ($i = 1, 2, \ldots, k$). If we just consider, for a fixed i,

$$V_i(\hat{x}) = \{x \in X \mid x(\alpha_i) = x_{\alpha_i} \in V(\hat{x}_{\alpha_i})\},$$

then $V_i(\hat{x})$ is a subbasis set (Section 7.3) of the product topology, and $V(\hat{x}) = \bigcap_{i=1}^{k} V_i(\hat{x})$. However, we know that $\hat{x}_{\alpha_i} \in \bigcap_{\gamma \in \Gamma} \bar{E}_{\alpha_i}^\gamma$, and therefore, $V(\hat{x}_{\alpha_i}) \cap E_{\alpha_i}^\gamma \neq \emptyset$ for all $\gamma \in \Gamma$ and each $i = 1, 2, \ldots, k$. Thus

$$V_i(\hat{x}) \cap E^\gamma \neq \emptyset \tag{7.7}$$

for all $\gamma \in \Gamma$ and each $i = 1, 2, \ldots, k$.

Next, by the maximality of the collection $\{E^\gamma\}_{\gamma \in \Gamma}$ with respect to the finite intersection property, it follows that the intersection of any finite number of sets in this collection is again in the collection (the adjunction of a finite intersection of sets from $\{E^\gamma\}_{\gamma \in \Gamma}$ to the collection $\{E^\gamma\}_{\gamma \in \Gamma}$ will again yield a collection satisfying the finite intersection property). Equation (7.7) therefore implies that each $V_i(\hat{x})$ has a nonempty intersection with any finite intersection of sets from the collection

$\{E^\gamma\}_{\gamma \in \Gamma}$. However, again by the maximality of the collection $\{E^\gamma\}_{\gamma \in \Gamma}$, it follows that each $V_i(\hat{x}) \in \{E^\gamma\}_{\gamma \in \Gamma}$, $i = 1, 2, \ldots, k$. Hence

$$V(\hat{x}) = \bigcap_{i=1}^{k} V_i(\hat{x}) \in \{E^\gamma\}_{\gamma \in \Gamma},$$

and since by assumption the collection possesses the finite intersection property, we have, of course, that

$$V(\hat{x}) \cap E^\gamma \neq \emptyset \qquad \text{(for all } \gamma \in \Gamma).$$

This is what we desired to show and completes the proof of Tychonoff's theorem.

We round out this section with two important notions, one of which we considered in the special case of metric spaces. These concepts are added mainly for completeness and will not play a role in our subsequent development. We shall, however, refer to them in the exercises.

DEFINITION 7.16. A topological space X is called *locally compact* if for each $x \in X$ there exists an open set O_x containing x and such that $\bar{O}_x$ is compact.

The real axis with the customary metric topology is an example of a locally compact space that is not compact. It is clear that any compact space is certainly locally compact.

DEFINITION 7.17. A topological space X is called *separable* if there exists a countable dense subset A of X; that is, $\bar{A} = X$.

THEOREM 7.17. Let X be a topological space that satisfies the second axiom of countability; then X is separable.

Proof. By hypothesis, X has a countable basis $\mathscr{B}$ of open sets. We form a set A consisting of one element from each of the sets of the basis $\mathscr{B}$. A is of course countable and $\bar{A} = X$, for if not, then $X - \bar{A} \neq \emptyset$, and since $X - \bar{A}$ is open, it must contain some set $O \in \mathscr{B}$. But by the construction of A, $A \cap O \neq \emptyset$, which is clearly a contradiction, and completes the proof.

We note that in general topological spaces the converse of this theorem is false. It is even possible (see Exercise 7.29) for a space to satisfy the first axiom of countability and be separable but still fail to satisfy the second axiom of countability. However, in a metric space the converse is true (see Exercise 7.30), namely, every separable metric space satisfies the second axiom of countability.

EXERCISES 7

1. Show that the general topological definition of limit point of a set given in this chapter agrees with that given earlier (Definition 3.4) in the special case of metric spaces.

2. Prove Theorem 7.5.

3. Let (X, d) be a metric space. Show that the topology defined by the induced metric on a subset A agrees with the induced topology on A from the metric topology of X.

4. Let X_1 and X_2 be topological spaces and $f: X_1 \to X_2$. Show that f is continuous if and only if $f^{-1}(B)$ is open, where B is any basis open set of X.

5. Let X and Y be topological spaces and $f: X \to Y$. f is said to be *continuous at* $x \in X$ if, for each open set U of Y containing $f(x)$, there exists an open set O of X containing x such that $f(O) \subset U$. Show that f is continuous if and only if it is continuous at every point $x \in X$.

6. Let X be an arbitrary set and suppose that, to each subset E of X, there is associated another subset $\hat{E}$ such that
   (a) $E \subset \hat{E}$,
   (b) $\hat{\emptyset} = \emptyset$,
   (c) $\widehat{E \cup F} = \hat{E} \cup \hat{F}$,
   (d) $\hat{\hat{E}} = \hat{E}$.
   Define E to be closed if and only if $\hat{E} = E$ and E open if and only if CE is closed. Show that the system of open sets defined in this way is a topology for X and that the resulting closure $\bar{E}$ defined in terms of this topology agrees with $\hat{E}$, that is, $\bar{E} = \hat{E}$. [Axioms (a)–(d) are called the *Kuratowski closure axioms*.]

7. *The neighborhood topology.* Let X be an arbitrary set and suppose that, to each $x \in X$ there corresponds a collection $N(x)$ of subsets of X called *the set of neighborhoods of* x, such that:
   (a) if $A \in N(x)$ and B is any subset of X such that $A \subset B$, then $B \in N(x)$; that is, any superset of a set of $N(x)$ belongs to $N(x)$;
   (b) the point x belongs to all sets of $N(x)$;
   (c) all finite intersections of sets of $N(x)$ belong to $N(x)$;
   (d) if $V \in N(x)$, there exists a $U \in N(x)$ such that, for all $y \in U$, $V \in N(y)$— in other words, a neighborhood of x is a neighborhood of all points y "sufficiently close" to x.
   Define a set O to be open if and only if for each $x \in O$ there exists a neighborhood $V(x)$ of x such that $V(x) \subset O$ (that is, O is open if and only if O is a neighborhood of each of its points). Show that the collection of open sets so defined is a topology.
   Conversely, if $(X, \mathcal{O})$ is a topological space, define U to be a neighborhood of x $[U \in N(x)]$ if and only if there exists an $O \in \mathcal{O}$ such that $x \in O \subset U$. Show that $N(x)$ satisfies the neighborhood axioms (a)–(d) and that the resulting collection of open sets defined in terms of neighborhoods, as given in the previous paragraph, coincides with the given topology.

8. Construct an example to show that the closure operation is not necessarily preserved by a continuous mapping.

9. Let $X = Z$, the set of integers, and let $p \in Z$ be a fixed prime. Define U to be a neighborhood of n if U contains all integers $n + mp^k$ for some k and all $m = 0$, $\pm 1, \pm 2, \ldots$. Show that the neighborhood axioms given in Exercise 7 are satisfied.

10. Let X_1 and X_2 be two topological spaces and $f: X_1 \rightarrow X_2$. Show that f is continuous at $x \in X_1$ (see Exercise 5) if and only if, to every neighborhood (as in Exercise 7) V_2 of $f(x)$, there exists a neighborhood V_1 of x such that $f(V_1) \subset V_2$.

11. Show that the cartesian product of T_1 spaces is a T_1 space; same problem for T_2 spaces. (It is assumed, of course, that the Tychonoff topology has been assigned to the cartesian product, an assumption that will be made every time cartesian products of topological spaces are mentioned.)

12. Prove that a compact Hausdorff space is normal.

13. Show that every metric space is a T_5 space.

14. Take $X = Z^+$, the positive integers, and let $U_n = \{2n, 2n - 1\}$. Take, now, the family $\{U_n \mid n \in Z^+\}$ as a subbasis (actually a basis in this instance) for a topology on Z^+. Show that X is *countably compact, but not compact*.

15. Show that every sequentially compact space is countably compact.

16. Give an example of a topological space that is *countably compact but not sequentially compact*. (*Hint:* Consider the space of Exercise 14.)

17. Give an example of a topological space that is *sequentially compact but not compact*. (*Hint:* Let Ω_0 be the set of all ordinal numbers less than the first uncountable ordinal with the order topology—see Ref. 4 with regard to this exercise.)

18. Give an example of a topological space which is *compact but not sequentially compact*. (*Hint:* Take X to be the set of all functions $f: [0, 1] \rightarrow [0, 1]$, where $[0, 1]$ has the induced topology of the real axis. Note that this collection of functions is nothing more than the cartesian product ΠX_α, where each $X_\alpha = [0, 1]$ and $\alpha \in [0, 1]$. In this (compact) space consider the following sequence of functions: Divide $[0, 1]$ into intervals of length $1/10^n$ (there is nothing special about 10 here) and consider the sawtooth function of Fig. 7.1. Consider any

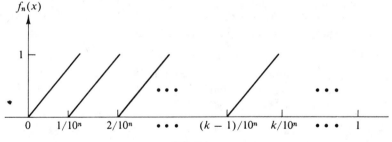

FIG. 7.1.

subsequence $\{f_{n_k}\}$ of $\{f_n\}$ now. Recalling that a typical basis set is of the form $V(g; x_1, ..., x_n, \varepsilon)$, show that no function g mapping $[0, 1]$ into $[0, 1]$ can exist such that $f_{n_k} \to g$. Loosely speaking, the sequence $\{f_n\}$, or any subsequence thereof, is approaching something that is not single-valued, that is, not a function.)

19. Some authors define a topological space X to be *countably compact* if, from every countable covering of X by open sets, a finite subcovering can be obtained. Show that this definition agrees with the one given in the text if X is a T_1 topological space, and in general (for arbitrary topological spaces) implies the one in the text. Construct an example which is countably compact according to the definition of the text but not countably compact according to the definition given above. (*Hint:* See Exercise 14.)

20. Assuming that the topological spaces satisfies the first axiom of countability, show that compactness implies sequential compactness.

21. Show that all the notions of compactness are equivalent in a space satisfying the second axiom of countability and axiom (T_1).

22. Give an example of a topological space and a limit point of a subset such that any open set about the limit point does not contain infinitely many points of the set. (*Hint:* Consider any finite set with the trivial topology.) Show that this cannot happen in a Hausdorff space.

23. Give an example of a topological space X and a limit point of a subset of X for which no sequence of points of the set converges to the limit point. (*Hint:* Consider $X = [0, 1]$ with the cocountable topology and the subset $[\frac{1}{2}, 1]$ and note that the point $\frac{1}{4}$, for example, is a limit point of this subset. The cocountable topology is defined such that a set O is open if $O = \phi$ or if CO is countable.) Show that this cannot occur if X satisfies the first axiom of countability.

24. Show that the continuous image of a countably compact set, as defined in the text, need not be countably compact. [*Hint:* Consider the space of positive integers with the topology defined in Exercise 14 as X_1 and take as X_2 the set of positive integers only with the discrete topology. Let $f: X_1 \to X_2$ be defined by $f(2n) = f(2n - 1) = n$.] Show, however, that the continuous image of a T_1 countably compact space is countably compact.

25. Let X_1 and X_2 be two topological spaces and $f: X_1 \to X_2$. Show that, if f is continuous, then $x_n \to x$ implies $f(x_n) \to f(x)$. Construct an example to show that the converse is false. [*Hint:* Let X_1 be the space in Hint to Exercise 23 and let X_2 be the same set with the usual topology; let $f: X_1 \to X_2$ be defined by taking $f(x) = x$.]

26. Under the same assumptions as the preceding exercise plus the assumption that X_1 satisfies the first axiom of countability, show that f is continuous if and only if $x_n \to x$ implies $f(x_n) \to f(x)$.

27. Let $(X, \mathcal{O}_1)$ be a Hausdorff topological space and let $(X, \mathcal{O}_2)$ be a compact topological space. If $\mathcal{O}_1 \subset \mathcal{O}_2$, show that $\mathcal{O}_1 = \mathcal{O}_2$.

28. Let X be a topological space and let $\{X_\alpha\}_{\alpha \in \Lambda}$ be a collection of topological spaces. If $f: X \to \Pi_\alpha X_\alpha$, show that f is continuous if and only if each $pr_\alpha(f)$ is continuous.

29. Construct an example of a topological space X which does not satisfy the second axiom of countability, but is separable and satisfies the first axiom of countability. [*Hint:* Let $X = \{\langle x, y \rangle \in R \times R \mid y \geq 0\}$ and let $E = \{\langle x, y \rangle \in X \mid y = 0\}$. For $p = \langle x, y \rangle \in X$ and $\varepsilon > 0$, define $U_\varepsilon(p) = S_\varepsilon(p) \cap X$ if $p \in X - E$ and $U_\varepsilon(p) = S_\varepsilon(p) \cap (X - E) \cup \{p\}$ if $p \in E$ and take the topology generated by the $U_\varepsilon(p)$.]

30. Show that every separable metric space satisfies the second axiom of countability.

31. Prove that a completely regular space is regular. Prove that every T_4 space is completely regular. (*Hint:* Use Urysohn's lemma; see, for example, Ref. 3 or 4.)

32. Show that a T_2 space is not necessarily T_3. (*Hint:* Consider the example of Exercise 29; for other examples showing the nonequivalence of the T spaces, see Ref. 2 or 5.)

REFERENCES

1. D. Bushaw, "Elements of General Topology." An introductory treatment.

2. D. Hall and G. Spencer, "Elementary Topology." An elementary treatment with numerous examples and counterexamples.

3. N. Bourbaki, "Topologie Genérale," Chap. I. A more sophisticated approach.

4. J. Kelley, "General Topology."

5. R. Vaidyanathaswamy, "Set Topology."

CHAPTER **8**

Banach Spaces, Equivalent Norms, and Factor Spaces

In this chapter the emphasis will be on completeness. In Sec. 3.1 the concept of a normed linear space was defined. In Sec. 3.2 we observed that any normed linear space was certainly a metric space, and, as such, all the subsequent notions about metric spaces can be applied to it. In particular, the concept of completeness is applicable as is the notion of a completion. Since the general metric space need not have any algebraic structure, merely completing a normed linear space is hardly satisfactory. What is much more desirable is to be able to complete it (as a metric space) and also extend the vector operations to the completion—which will be demonstrated in Sec. 8.3. Calling a complete normed linear space a *Banach space*, we shall show that any normed linear space can be embedded in a Banach space. In addition, several examples pertaining to Banach spaces will be given.

Our main interest in norms is for their metric qualities, that is, when a norm is present we can deal with such things as Cauchy sequences and open sets. Hence it is desirable to investigate when two norms have similar or identical metric qualities. More specifically, we wish to obtain a criterion for determining when a Cauchy sequence with respect to a first norm will also be a Cauchy sequence with respect to a second norm and when an open set with respect to a first norm will be an open set with respect to a second norm. Naturally enough when this situation occurs we shall call the two norms *equivalent*.

Indeed an old friend from linear algebra is that of the smallest subspace containing a given set of vectors—the *subspace spanned* or *generated* by that set. Viewing this process in the context of a normed space, we now wish to consider the smallest *closed* subspace containing a given set of vectors. In a finite-dimensional space, this subspace is always a closed set; in fact any subspace of a finite-dimensional space is closed.

Last, we wish to consider the notion of a *factor space*—another vector space that is readily available once a vector subspace has been chosen. In the event that the original space is a normed space, we shall be particularly concerned with the case when a *closed* subspace has been chosen, because then a new *normed* space can be formed.

8.1 The Hölder and Minkowski Inequalities

Having a good supply of inequalities to draw from is a handy supplement to any mathematical toolbox, and the ones that will be demonstrated here are particularly useful in functional analysis. Knowing them is often vital when we encounter an unfamiliar metric or normed linear space and are trying to verify the triangle inequality. However, this is not their only use by far.

Although the first result that we wish to prove might be put in a more rigorous setting, the proof offered here is retained because of its intuitive appeal.

THEOREM 8.1. Let a and b be nonnegative real numbers and let p and q be real numbers, each of which is greater than 1 and with the added property that

$$\frac{1}{p} + \frac{1}{q} = 1.$$

Then

$$ab \le \frac{a^p}{p} + \frac{b^q}{q}.$$

Proof. Clearly, three possibilities about the relative size of a and b exist: Either $a < b, b < a$, or $a = b$. The function $y = x^{p-1}$ is sketched in Fig. 8.1 for the cases of the first two possibilities. Appealing to the diagram, we see that the area of the rectangle, ab, will be less than $A_1 + A_2$. These two areas will now be computed by integration:

$$A_1 = \int_0^a x^{p-1}\, dx = \frac{a^p}{p}$$

and, since $(1/p) + (1/q) = 1$ implies that $1/(p-1) = q - 1$, we can say that

$$x = y^{1/(p-1)} = y^{q-1}.$$

Consequently,

$$A_2 = \int_0^b y^{q-1}\, dy = \frac{b^q}{q}.$$

Thus,

$$ab \le A_1 + A_2 = \frac{a^p}{p} + \frac{b^q}{q}.$$

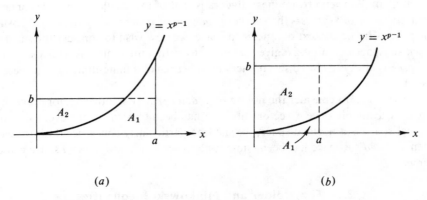

(a)   (b)

FIG. 8.1. The function x^{p-1}. (a) The case $a > b$. (b) The case $b > a$. In each case A_1 represents the area between the curve and the x-axis and A_2 represents the area between the curve and the y-axis.

We can now proceed to the more important inequality, namely, Hölder's inequality.

THEOREM 8.2 (*Hölder*). Let p and q be real numbers greater than 1, with the added property that

$$\frac{1}{p} + \frac{1}{q} = 1.$$

Then for any complex numbers x_j and y_j $(j = 1, 2, ..., n)$,

$$\sum_{j=1}^n |x_j y_j| \le \left(\sum_{j=1}^n |x_j|^p \right)^{1/p} \left(\sum_{j=1}^n |y_j|^q \right)^{1/q}.$$

REMARK 1. There are some extensions of this inequality that are also of interest. One extension is to infinite sums, in which case one must, of course, be assured of the convergence of each of the infinite series in the inequality. Another important extension is to integrals, in which case the Hölder inequality takes the form

$$\int_a^b |f(x)g(x)|\,dx \le \left(\int_a^b |f(x)|^p\,dx \right)^{1/p} \left(\int_a^b |g(x)|^q\,dx \right)^{1/q},$$

where f and g are assumed to be pth and qth power summable, respectively, on $[a, b]$.

Proof. We claim that, if we can show the following condition to hold, we shall be done:

$$\left\{ \sum_{j=1}^n |x_j|^p = 1 \quad \text{and} \quad \sum_{j=1}^n |y_j|^q = 1 \right\} \Rightarrow \sum_{j=1}^n |x_j y_j| \le 1. \tag{8.1}$$

First it will be shown that Eq. (8.1) suffices, and then it will be shown that it holds.
Suppose (8.1) is satisfied and suppose

$$\left(\sum_{j=1}^n |x_j|^p \right)^{1/p} = \alpha \quad \text{and} \quad \left(\sum_{j=1}^n |y_j|^q \right)^{1/q} = \beta.$$

We now define

$$x_j' = \frac{x_j}{\alpha}, \quad y_j' = \frac{y_j}{\beta}$$

and can, in this case, say

$$\left(\sum_{j=1}^n |x_j'|^p \right)^{1/p} = 1 \quad \text{and} \quad \left(\sum_{j=1}^n |y_j'|^q \right)^{1/q} = 1.$$

It immediately follows that

$$\sum_{j=1}^n |x_j'|^p = 1 \quad \text{and} \quad \sum_{j=1}^n |y_j'|^q = 1$$

and we can apply (8.1) to obtain

$$\sum_{j=1}^n |x_j' y_j'| \le 1$$

or

$$\sum_{j=1}^n |x_j y_j| \le \alpha\beta.$$

Thus (8.1) is clearly sufficient. To show that it holds, we use the inequality proved immediately before this theorem.
Let

$$\sum_{j=1}^n |x_j|^p = 1 = \sum_{j=1}^n |y_j|^q$$

and take

$$a = |x_j|, \quad b = |y_j|.$$

By the preceding theorem,

$$ab = |x_j y_j| \le \frac{|x_j|^p}{p} + \frac{|y_j|^q}{q}.$$

Summing both sides with respect to j, we get

$$\sum_{j=1}^{n} |x_j y_j| \le \frac{1}{p} + \frac{1}{q} = 1,$$

which completes the proof.

Another quite useful inequality in the same vein is the Minkowski inequality. To prove this result, we use the Hölder inequality; the proof is straightforward and only the result will be stated here.

THEOREM 8.3 (*Minkowski*). Let p be a real number such that $p \ge 1$. Then, for any complex numbers x_j and y_j ($j = 1, 2, ..., n$),

$$\left(\sum_{j=1}^{n} |x_j + y_j|^p \right)^{1/p} \le \left(\sum_{j=1}^{n} |x_j|^p \right)^{1/p} + \left(\sum_{j=1}^{n} |y_j|^p \right)^{1/p}.$$

REMARK 2. This result, as was the case for the preceding result, has some important generalizations to infinite sums and integrals. In each case one need only impose those restrictions, which guarantee that the quantities (infinite sums, integrals) in question make sense, to obtain the extension. In particular, for integrals one has

$$\left(\int_a^b |f(t) + g(t)|^p \, dt \right)^{1/p} \le \left(\int_a^b |f(t)|^p \, dt \right)^{1/p} + \left(\int_a^b |g(t)|^p \, dt \right)^{1/p},$$

where f and g are assumed to be the pth power summable on $[a, b]$.

8.2 Banach Spaces and Examples

As noted earlier, in any normed linear space one always has a convenient metric immediately available. For any two vectors x and y in the space, one can take as the distance between them

$$d(x, y) = \|x - y\|.$$

With this particular metric in mind, we can now speak of normed linear spaces as metric spaces, and a question of interest about any such space is whether the space is complete or not. Spaces that are complete are very important, and a great deal of study is devoted to them.

DEFINITION 8.1. Let X be a normed linear space. As the distance between any two points $x, y \in X$, we shall take $d(x, y) = \|x - y\|$. If (X, d) is a complete metric space, X is said to be a *Banach space* or *B-space*. Spaces of this type are also called *l.n.c. spaces* (linear, normed, complete spaces).

We first wish to show that, given a particular vector space and a metric on this space, one cannot always come up with a norm that will agree (in the sense of the definition of Banach space) with the metric already defined on that space.

Consider the vector space of all infinite-tuples of complex numbers over the scalar field of complex numbers, with linear operations taken pointwise and with distance between any two points of the space

$$x = (\alpha_1, \alpha_2, ..., \alpha_n, ...) \quad \text{and} \quad y = (\beta_1, \beta_2, ..., \beta_n, ...)$$

defined to be

$$d(x, y) = \sum_{i=1}^{\infty} \frac{1}{2^i} \frac{|\alpha_i - \beta_i|}{1 + |\alpha_i - \beta_i|}.$$

One can verify that this actually is a metric space. In this case, however, it is easily seen that, if $\lambda \in C$, then

$$d(\lambda x, \lambda y) \neq |\lambda| \, d(x, y),$$

which makes it impossible to come up with any norm that will "yield" this metric, for, for any norm whatsoever,

$$\|\lambda x - \lambda y\| = |\lambda| \cdot \|x - y\|.$$

We now wish to demonstrate another property of normed linear spaces.

THEOREM 8.4. Let X be a normed linear space and let F denote R or C, the real or complex numbers. Then the mappings

$$X \times X \to X,$$

$$\langle x, y \rangle \to x + y,$$

(8.2)

$$F \times X \to X,$$

$$\langle \alpha, x \rangle \to \alpha x,$$

(8.3)

are continuous.

Proof. We shall assign a product metric (any of the three mentioned in Example 3.6) to the spaces $X \times X$ and $F \times X$ so that continuity of the mappings mentioned above may be considered. It is in this sense that we shall prove they are continuous.

To prove that the mapping (8.2) is continuous, we need only note that

$$\|x + y - (x_0 + y_0)\| \leq \|x - x_0\| + \|y - y_0\|$$

for any x_0 and y_0. To prove (8.3) is continuous, we note that

$$\|\alpha x - \alpha_0 x_0\| = \|\alpha(x - x_0) + (\alpha - \alpha_0)x_0\|$$

$$\leq |\alpha| \cdot \|x - x_0\| + |\alpha - \alpha_0| \cdot \|x_0\|$$

for any fixed α_0 and x_0.

It is clear in each case that appropriate restrictions in the respective domains can give rise to any desired "closeness" in the range.

Another fact, in the same vein, is that the mapping

$$\| \quad \| : X \to R,$$

$$x \to \|x\|$$

is continuous. To prove this we need only note that

$$\big|\,\|x\| - \|x_0\|\,\big| \le \|x - x_0\|$$

and apply the reasoning used above. We now wish to exhibit some Banach spaces.

Examples of Banach Spaces

In most of the following examples we list only the space and the norm, and shall not verify that each of the spaces listed is actually a Banach space. As usual, the scalar field in each case will be either R or C, whichever is appropriate.

EXAMPLE 8.1. Consider, as a linear space, R or C with norm taken to be the usual absolute value in either case.

EXAMPLE 8.2. Consider all n-tuples of real or complex numbers R^n or C^n. As the norm of

$$x = (\alpha_1, \alpha_2, \ldots, \alpha_n),$$

we take

$$\|x\| = \left(\sum_{i=1}^{n} |\alpha_i|^p \right)^{1/p} \qquad (p \ge 1).$$

One can easily verify that this is actually a norm, the only difficulty being in verifying the triangle inequality at which point the Minkowski inequality is useful.

EXAMPLE 8.3. Consider the linear space $C[a, b]$ and as norm of $f \in C[a, b]$, we take

$$\|f\| = \max_{x \in [a, b]} |f(x)|.$$

To show that $C[a, b]$ is complete with respect to this norm, we now consider any Cauchy sequence of elements $\{f_n\}$ of $C[a, b]$. Since it is a Cauchy sequence with respect to the *sup* norm, then, for any $\varepsilon > 0$, there must be an N such that $n, m > N$ imply

$$|f_n(x) - f_m(x)| < \varepsilon$$

for all $x \in [a, b]$. But this means that the sequence $\{f_n\}$ is a uniform Cauchy sequence of continuous functions and therefore converges uniformly to some function f. Whenever a sequence of continuous functions converges uniformly, however, the limit function must also be continuous. Moreover, invoking the uniform convergence again, we see clearly that

$$\|f_n - f\| \to 0,$$

which completes the proof.

EXAMPLE 8.4. Consider $l_p(p \ge 1)$, all infinite sequences of complex numbers

$$x = (\alpha_1, \alpha_2, \ldots, \alpha_n, \ldots)$$

such that

$$\sum_{i=1}^{\infty} |\alpha_i|^p < \infty.$$

As norm of x we take

$$\|x\| = \left(\sum_{i=1}^{\infty} |\alpha_i|^p \right)^{1/p}.$$

Comparing this to Example 8.2, one might suspect that the Minkowski inequality will come in handy when one is trying to verify that the triangle inequality holds for the proposed norm. In this case, however, the Minkowski inequality for infinite sums is needed. Completeness will follow as a special case of the space in Example 8.6.

EXAMPLE 8.5. Consider the linear space l_∞ mentioned in Example 6.6. As norm of $x \in l_\infty$,

$$x = (\alpha_1, \alpha_2, ..., \alpha_n, ...),$$

we take

$$\|x\| = \sup_i |\alpha_i|.$$

In this case, to verify the triangle inequality, one needs only the fact that the sup of a sum is less than or equal to the sum of sups.

EXAMPLE 8.6. Suppose X is any set, S is a σ-algebra (Example 1.4) of subsets of X, and μ is a measure on S. Let E be any fixed set in S. In this case we define the linear space

$$L_p(E, \mu) \qquad (1 \le p < \infty)$$

to be all equivalence classes† of functions that are measurable and pth-power summable on E. The Minkowski inequality for integrals is used here in verifying that

$$\|f\|_p = \left(\int_E |f|^p \, d\mu \right)^{1/p} \qquad [f \in L_p(E, \mu)] \tag{8.4}$$

satisfies the triangle inequality. Here it is to be noted that we are identifying functions that are equal almost everywhere on E. This is necessary to guarantee that the first norm axiom will be satisfied. With these notions in mind it is easily verified that $\| \quad \|_p$ is, indeed, a norm on $L_p(E, \mu)$.

We now prove that $L_p(E, \mu)$ is complete with respect to this norm. Let $\{f_n\}$ be a Cauchy sequence in $L_p(E, \mu)$ so that $\|f_n - f_m\|_p \to 0$ as $n, m \to \infty$; that is,

$$\lim_{n, m} \left(\int_E |f_n - f_m|^p \, d\mu \right)^{1/p} = 0.$$

This, however, implies that $\{f_n\}$ is a Cauchy sequence in measure (cf. Ref. 1) in regard to convergence in measure and other convergence relationships) which, in turn, implies that there exists a subsequence $\{f_{n_k}\}$ which converges almost everywhere. Thus there exists a measurable function f such that

$$\lim_k f_{n_k} = f \qquad \text{(almost everywhere)}.$$

---

† See also Example 1.4 for a discussion of this space and mention of the need for considering equivalence classes of functions instead of the functions themselves.

For a fixed j, we then have

$$\lim_k |f_{n_j} - f_{n_k}|^p = |f_{n_j} - f|^p \qquad \text{(almost everywhere)},$$

and, consequently, by Fatou's lemma,

$$\int_E \underline{\lim_k} |f_{n_j} - f_{n_k}|^p \, d\mu = \int_E |f_{n_j} - f|^p d\mu \le \underline{\lim_k} \int_E |f_{n_j} - f_{n_k}|^p d\mu. \qquad (8.5)$$

Let $\varepsilon > 0$ now be given. Since $\{f_n\}$ is a Cauchy sequence in $L_p(E, \mu)$ to begin with, there must exist an N such that, for $j, k > N$,

$$\int_E |f_{n_j} - f_{n_k}|^p \, d\mu \le (\varepsilon/2)^p$$

so

$$\underline{\lim_k} \int_E |f_{n_j} - f_{n_k}|^p d\mu \le (\varepsilon/2)^p, \qquad \text{(for } j > N)$$

which, together with (8.5) implies that

$$\int_E |f_{n_j} - f|^p d\mu \le (\varepsilon/2)^p \qquad \text{(for } j > N). \qquad (8.6)$$

It follows from (8.6) that $(f_{n_j} - f) \in L_p(E, \mu)$ and, since $f_{n_j} \in L_p(E, \mu)$, we get that $f \in L_p(E, \mu)$. From the Minkowski inequality, we now observe that

$$\left(\int_E |f_n - f|^p \, d\mu \right)^{1/p} \le \left(\int_E |f_n - f_{n_j}|^p \, d\mu \right)^{1/p} + \left(\int_E |f_{n_j} - f|^p \, d\mu \right)^{1/p}.$$

This together with (8.6) and the fact that the given sequence is a Cauchy sequence in $L_p(E, \mu)$ implies that, for n sufficiently large,

$$\left(\int_E |f_n - f|^p \, d\mu \right)^{1/p} \le \varepsilon$$

or that $\|f_n - f\|_p \to 0$ as $n \to \infty$; since we have already seen that $f \in L_p(E, \mu)$, this completes the proof that $L_p(E, \mu)$ is complete.

Using this result, we now show that the space considered in Example 8.4, l_p, is complete. Consider the class of all positive integers Z^+. Clearly $P(Z^+)$, the class of all subsets of Z^+ (the power set of Z^+) is a σ-algebra of subsets of Z^+. As a measure μ, defined on this collection, we consider the following function: For $E \subset Z^+$ (or, equivalently, $E \in P(Z^+)$), take

$$\mu(E) = \begin{cases} \text{number of elements in } E \text{ if } E \text{ is a finite set,} \\ \infty \qquad \text{(otherwise).} \end{cases}$$

with the usual conventions about ∞ prevailing.

It is easily verified that μ is a measure on $P(Z^+)$. For the set E, mentioned above in the more general framework, we now take all of Z^+; thus we consider the space $L_p(Z^+, \mu)$. Since all subsets of Z^+ are in the σ-algebra of the power set, all functions defined on E are measurable; note too that a complex-valued function defined on the integers is nothing more than a sequence. By the definition of abstract integral, the integral of such a function, f on Z^+, is just (using the countable additivity of the integral)

$$\int_{Z^+} f \, d\mu = \int_{\underset{n \in Z^+}{\cup \{n\}}} f \, d\mu = \sum_{n \in Z^+} \int_{\{n\}} f \, d\mu = \sum_{n=1}^{\infty} f(n)\mu(\{n\}) = \sum_{n=1}^{\infty} f(n).$$

Hence, to say that the function f belongs to $L_p(Z^+, \mu)$ is equivalent to saying that the sequence $\{f(n)\}$ is a member of l_p; l_p, therefore, since we can view it in this framework, is complete.

EXAMPLE 8.7. Again we consider a set X, a σ-algebra of subsets of X, S, and μ, a measure on S. Now let E be some particular set from S. We define the space $L_\infty(E, \mu)$ to be all those functions (real- or complex-valued), f measurable on E such that, for any given f, there is some real number M such that

$$|f(x)| \leq M \qquad \text{(for almost all } x \in E).$$

(Note that we do not require the same M to work for all f.) It is quite simple to verify that $L_\infty(E, \mu)$ is indeed a linear space with respect to pointwise addition of functions and pointwise multiplication by a scalar.

We now consider all equivalence classes of functions in this space defined by the equivalence relation:

$$f \sim g \qquad \text{if and only if } f = g \qquad \text{almost everywhere on } E.$$

We let any member of a given class denote the whole class; that is, f will now denote not only f itself, but all other functions that are equal to f almost everywhere on E. Having adopted this convention, we now wish to consider the mapping from the equivalence classes of $L_\infty(E, \mu)$ into the real numbers defined for any class f, as follows:

$$\text{ess sup } f = \inf \{M \,||\, |f(x)| \leq M \text{ almost everywhere on } E\}.$$

This is read as the *essential sup of* f and, denoting ess sup f by α, it can be shown that†

$$|f(x)| \leq \alpha \qquad \text{(almost everywhere on } E).$$

Furthermore, it is now easy to verify that ess sup f is actually a norm on the equivalence classes of $L_\infty(E, \mu)$. Once again, it is vital that we work with the equivalence classes to guarantee that

$$\|f\| = \text{ess sup } f = 0 \qquad (\text{if and only if } \; f = 0).$$

It can also be shown that the above space is complete.

† To prove this, we note that a denumerable union of sets of measure zero is itself of measure zero and observe that we can write

$$E(|f| > \alpha) = \bigcup_{n=1}^{\infty} E\left(|f| > \alpha + \frac{1}{n}\right),$$

where $E(|f| > \alpha)$ denotes those points of E where $|f| > \alpha$.

Thus, in the sequel, if we refer to $L_\infty(E, \mu)$ as a Banach space, it is understood that we are actually considering the equivalence classes of functions from $L_\infty(E, \mu)$ and not the functions themselves.

EXAMPLE 8.8. Consider the closed interval $[a, b]$ and all functions (real- or complex-valued) of bounded variation on $[a, b]$. This class will be denoted by $BV[a, b]$. With respect to the operations of ordinary addition of functions and multiplication by a real or complex number, it can be shown that this space is a vector space, and we wish to introduce the following norm on $BV[a, b]$:

$$\|f\| = |f(a)| + V(f) \qquad (f \in BV[a, b]),$$

where $V(f)$ denotes the total variation of f on $[a, b]$. We state without proof that the above is truly a norm and that, with respect to this norm, $BV[a, b]$ is a Banach space.

EXAMPLE 8.9. Let X, S, and E be as in Example 8.6. If we introduce the following notation:

$$S_E: \quad \text{all measurable subsets of } E,$$
$$B: \quad \text{all finite signed measures } (s\text{-measures}) \text{ on } S_E,$$

we can show that:

(1) B is a real linear space with respect to addition of the signed measures (set functions) defined in the usual way and multiplication by a real number;

(2) B is a Banach space with respect to the norm

$$\|\mu\| = |\mu|(E) \qquad (\mu \in B),$$

where $|\mu|$ denotes the total variation of μ on E.

8.3   The Completion of a Normed Linear Space

Let X be a normed linear space and consider the distance function d defined by taking

$$d(x, y) = \|x - y\| \qquad (x, y \in X).$$

For the sake of this proof, we shall refer to this distance function as the *norm-derived* metric. Relying upon the results of Sec. 4.2 and treating (X, d) as just a metric space, we know that there is a complete metric space (X^*, d^*) such that X is isometric to a dense subset of this space. We shall adopt the convention of identifying the element, x of X with its isometric image in X^* and shall also often not distinguish between d and d^*.

Our goal now is to show that, after defining vector addition and scalar multiplication appropriately, X^* will be a complete normed linear space with the property that not only is X isometric to a dense subset of X^* but is also isomorphic to this dense subset. Further, we shall show that the norm on X^* will actually extend the norm on X, the extension being made with the above identification in mind. Thus we wish to exhibit a subspace of X^*, X_0, such that

(1) $\overline{X}_0 = X^*$;

(2) X is isomorphic and isometric to X_0.

Another way of stating (2) is to say that X and X_0 are *congruent*: In general, a mapping $A : X \to Y$, where X and Y are normed linear spaces, is said to be a *congruence* if it is simultaneously an isometry and an isomorphism.

Using the notation and conventions mentioned above we now proceed to the proof.

Let x^* and y^* be elements (equivalence classes of Cauchy sequences of X) of X^* and let

$$\{x_n\} \in x^*, \qquad \{y_n\} \in y^*. \tag{8.7}$$

We now define $x^* + y^*$ to be the equivalence class containing $\{x_n + y_n\}$ and shall refer to this class as z^*. We must first show that $\{x_n + y_n\}$ is a Cauchy sequence, and to prove this, we need only note that

$$\|x_n + y_n - (x_m + y_m)\| \leq \|x_n - x_m\| + \|y_n - y_m\|.$$

Having noted this, we must now demonstrate that the mapping does not depend on the particular representatives of the classes x^* and y^* that were chosen; that is, we must show that the operation is well-defined. Suppose

$$\{\hat{x}_n\} \sim \{x_n\} \qquad \text{and} \qquad \{\hat{y}_n\} \sim \{y_n\}.$$

Recalling from Sec. 4.2 what is meant for two sequences to be equivalent, we can now show that $\{x_n + y_n\} \sim \{\hat{x}_n + \hat{y}_n\}$ by noting that

$$\|x_n + y_n - (\hat{x}_n + \hat{y}_n)\| \leq \|x_n - \hat{x}_n\| + \|y_n - \hat{y}_n\|.$$

Now let $\alpha \in F$, where F, as usual, denotes the real or complex numbers. With $\{x_n\} \in x^*$ as in (8.7) we now define αx^* to be the class containing $\{\alpha x_n\}$. It is easily verified that $\{\alpha x_n\}$ is a Cauchy sequence and that the operation of scalar multiplication, as we have defined it, is well-defined. It is equally simple to verify that X^* with respect to these two operations is, indeed, a linear space. We now wish to introduce a norm on X^* that will have the properties claimed above. With $\{x_n\} \in x^*$ as in (8.7) we now define

$$\|x^*\| = \lim \|x_n\|. \tag{8.8}$$

Strictly speaking, the norm on the left side of (8.8) might be better denoted by some different symbol, $\|\ \|_*$, say. In view of what will be shown below, however, we shall dispense with this formality and shall use the same symbol for the norm on X^* as for the norm on X.

In order for (8.8) to be meaningful, it must first be shown that the limit in that equation exists. But, since

$$\big|\ \|x_m\| - \|x_n\|\ \big| \leq \|x_m - x_n\|,$$

it is seen that the sequence of real numbers $\{\|x_n\|\}$ is a Cauchy sequence and, hence, the limit exists.

Suppose now that $\{x_n\} \sim \{\hat{x}_n\}$. Since

$$\big|\ \|x_n\| - \|\hat{x}_n\|\ \big| \leq \|x_n - \hat{x}_n\|$$

and the term on the right must go to zero, the norm defined in (8.8) is well-defined.

Knowing that the mapping defined in (8.8) is meaningful, we now show that it is truly a norm. In the following statements $\{x_n\}$ and $\{y_n\}$ are as in (8.7).

(1) The mapping is clearly nonnegative and equals zero if $x^* = 0^*$. Now suppose

$$\|x^*\| = 0.$$

This implies

$$\lim \|x_n\| = 0,$$

which, in turn, implies that

$$x_n \to 0.$$

Thus,

$$\{x_n\} \sim (0, 0, 0, \ldots) \quad \text{or} \quad \{x_n\} \in 0^* \quad \text{and } x^* = 0^*.$$

(2) Consider

$$\|\alpha x^*\| = \lim \|\alpha x_n\| = |\alpha| \lim \|x_n\| = |\alpha| \|x^*\|.$$

(3) Last, the triangle inequality must be verified. For this purpose we note that

$$\|x^* + y^*\| = \lim \|x_n + y_n\| \leq \lim \|x_n\| + \lim \|y_n\|$$

or

$$\|x^* + y^*\| \leq \|x^*\| + \|y^*\|.$$

Thus (8.8) determines a norm on X^*. However, the fact that we now have a normed linear space is still not enough. We want X^* to be complete with respect to the distance function determined by this norm, which we shall denote by d_N. If we can show that d_N and d^*, where d^* is as in Sec. 4.2, agree, however, we shall have achieved this result. With $\{x_n\}$ and $\{y_n\}$ as in (8.7), consider

$$d_N(x^*, y^*) = \|x^* - y^*\| = \lim \|x_n - y_n\| = \lim d(x_n, y_n)$$
$$= d^*(x^*, y^*);$$

hence we can conclude that X^* is a complete normed linear space.

In Sec. 4.2 we had an isometric mapping A from the metric space X onto the dense subset X_0, of X^*: The set of all equivalence classes of X^* containing all elements of the form

$$(x, x, x, \ldots) \in x',$$

where $x \in X$.

We now show that the mapping A also establishes an isomorphism between X and X_0. It is already known that the mapping is onto X_0 and, since it is an isometry, it must also be $1:1$. Thus all we need show now is that it preserves linear combinations. To this end, suppose $x, y \in X$ and let

$$Ax = x' \quad \text{and} \quad Ay = y'.$$

Now consider

$$A(x + y) = (x + y)'.$$

Since

$$(x + y, x + y, \ldots) \in (x + y)'$$

and

$$(x + y, x + y, \ldots) = (x, x, \ldots) + (y, y, \ldots),$$

we can say, by our rule for addition, that

$$(x + y)' = x' + y'$$

and A thus preserves sums. Scalar multiplication follows in a similar fashion, and we have proved Theorem 8.5.

THEOREM 8.5. For every normed linear space X, there is a complete normed linear space X^* such that X is congruent to a dense subset of X^*, X_0, and the norm on X^* extends the norm on X.

8.4 Generated Subspaces and Closed Subspaces

Suppose X is a linear space and S is an arbitrary subset of X. Since the intersection of any collection of subspaces is still a subspace and X itself is a subspace containing S, we can define the *subspace generated by* S, or the *subspace spanned by* S, to be the intersection of all subspaces containing S. One can easily show that the subspace spanned by S is the same as the set of all linear combinations of elements in S, $[S]$.

Suppose now that X is a *normed* linear space. In this context it is meaningful to ascertain whether a given subspace is closed or not. In any finite-dimensional normed linear space, all subspaces are closed. In the infinite-dimensional case, subspaces that are not closed can easily be exhibited and, for this reason, we now wish to consider the smallest *closed* subspace containing an arbitrary subset of the space. To aid this investigation we first prove the following, simple, theorem.

THEOREM 8.6. In a normed linear space X, if M is a subspace, the closure of M, $\overline{M}$ is a subspace too.

Proof. To prove this, we must show that any linear combination of elements in $\overline{M}$ is also in $\overline{M}$. For any element to be in $\overline{M}$, however, it suffices to show that, for any ε, there is some element of M that is within ε distance from it. Thus let $\varepsilon > 0$ be given, let x and y be elements of $\overline{M}$, and consider $\alpha x + \beta y$, where α and β are scalars. Since x and y are in $\overline{M}$, there must exist elements x_1 and y_1 in M such that

$$\|x - x_1\| < \varepsilon \quad \text{and} \quad \|y - y_1\| < \varepsilon,$$

which implies

$$\|\alpha x + \beta y - (\alpha x_1 + \beta y_1)\| \le |\alpha| \, \|x - x_1\| + |\beta| \, \|y - y_1\|$$

$$< \varepsilon(|\alpha| + |\beta|).$$

Noting that $\alpha x_1 + \beta y_1 \in M$, the proof is seen to be complete.

We can describe the smallest closed subspace containing a given subset of a normed linear space as the intersection of all closed subspaces containing the given set. If S is the subset and M denotes the smallest closed subspace containing S, we can say that, since $\overline{[S]}$ is a closed subspace and

$$[S] \supset S,$$

then since M is the *smallest* closed subspace containing S

$$\overline{[S]} \supset M.$$

But

$$[S] \subset M,$$

which implies

$$\overline{[S]} \subset \overline{M} = M.$$

Therefore,

$$M = \overline{[S]}.$$

Thus the smallest closed subspace containing a given set of vectors is just the closure of the subspace generated by that set.

8.5 Equivalent Norms and a Theorem of Riesz

DEFINITION 8.2. Suppose X is a vector space over F and suppose that $\| \ \|_1$ and $\| \ \|_2$ are each norms on X. $\| \ \|_1$ is said to be *equivalent* to $\| \ \|_2$, $\| \ \|_1 \sim \| \ \|_2$, if there exist positive numbers a and b such that

$$a\|x\|_1 \leq \|x\|_2 \leq b\|x\|_1 \qquad \text{(for all } x \in X). \tag{8.9}$$

It is not difficult to show that this relation is an equivalence relation on the set of all norms over a given space.

If two norms are equivalent then certainly if $\{x_n\}$ is a Cauchy sequence with respect to $\| \ \|_1$, it must also be a Cauchy sequence with respect to $\| \ \|_2$ and vice versa. Another fact that is evident about equivalent norms is that the class of open sets defined by one is the same as the class defined by the other. To prove this one need only note that in any ε-neighborhood induced by $\| \ \|_1$, a neighborhood induced by $\| \ \|_2$ is wholly contained, and conversely.

Whenever one introduces an equivalence relation on a certain set, the set becomes partitioned into equivalence classes; something to look for anytime such a relation has been introduced is what the classes look like. We shall provide a partial answer to this question with the following theorem.

THEOREM 8.7. On a finite-dimensional space all norms are equivalent.

Proof. We shall show that all norms are equivalent by showing that any norm is equivalent to the particular norm defined below and called the *zeroth norm*:

Let a basis for the finite-dimensional space X be given by

$$x_1, x_2, ..., x_n.$$

For any $x \in X$, there exist unique scalars $\alpha_1, \alpha_2, ..., \alpha_n$ such that

$$x = \sum_{i=1}^{n} \alpha_i x_i. \tag{8.10}$$

One can easily show that

$$\|x\|_0 = \max_i |\alpha_i|$$

is indeed a norm.

Now let $\| \ \|$ be any norm on X. Our job now is to find real numbers $a, b > 0$ such that (8.8) is satisfied, where $\| \ \|_2$ is replaced by $\| \ \|$ and $\| \ \|_1$ is replaced by $\| \ \|_0$.

The right side of (8.9) is easily seen to be satisfied by noting, for x as in (8.10)

$$\|x\| = \|\sum_{i=1}^{n} \alpha_i x_i\| \leq \sum_{i=1}^{n} |\alpha_i| \, \|x_i\| \leq \|x\|_0 \sum_{i=1}^{n} \|x_i\|$$

because, since the basis is fixed, we can take as the number b,

$$b = \sum_{i=1}^{n} \|x_i\|$$

to get, for any $x \in X$,

$$\|x\| \leq b\|x\|_0.$$

The left side of (8.9) does not follow quite as simply. Consider the simple case of a one-dimensional space with basis x_1. Any vector in the space x can be written uniquely as

$$x = \alpha_1 x_1$$

for some $\alpha_1 \in F$. Hence,

$$\|x\| = |\alpha_1| \, \|x_1\| = \|x\|_0 \, \|x_1\|.$$

Thus, in this case, the number a on the left side of (8.9) can be taken to be just $\|x_1\|$. Having verified this, we shall now proceed by induction; suppose the theorem is true for all spaces of dimension less than or equal to $n - 1$. We can now say that, if $\dim X = n$, with basis $\{x_1, x_2, ..., x_n\}$ and

$$M = [\{x_1, x_2, ..., x_{n-1}\}],$$

the subspace spanned by the first $n - 1$ basis vectors, then

$$\| \cdot \| \sim \| \ \|_0$$

on M. Since this is so, if $\{y_n\}$ is a Cauchy sequence of elements from M with respect to $\| \ \|$, then $\{y_n\}$ is also a Cauchy sequence with respect to $\| \ \|_0$. Consider the ith term of this sequence now:

$$y_i = \alpha_1^{(i)} x_1 + \alpha_2^{(i)} x_2 + \cdots + \alpha_{n-1}^{(i)} x_{n-1}.$$

By the above,

$$\|y_n - y_m\|_0 \to 0 \qquad (\text{as} \quad n, m \to \infty). \tag{8.11}$$

But

$$\|y_n - y_m\|_0 = \max_{j} |\alpha_j^{(n)} - \alpha_j^{(m)}|,$$

which, by (8.11) implies

$$|\alpha_j^{(n)} - \alpha_j^{(m)}| \to 0 \qquad (\text{as} \quad n, m \to \infty) \tag{8.12}$$

for $j = 1, 2, ..., n - 1$. Since $F = R$ or C and each is complete, and (8.12) states that each of the $\{\alpha_j^{(m)}\}$ is a Cauchy sequence, there must exist $\alpha_1, \alpha_2, ..., \alpha_{n-1} \in F$ such that

$$\alpha_j^{(m)} \to \alpha_j \qquad (j = 1, 2, ..., n - 1).$$

In view of this, it is clear that

$$y_m \to y = \sum_{j=1}^{n-1} \alpha_i x_i,$$

with respect to the zeroth norm. Further,

$$y_m \xrightarrow{\ \|\ \|_0\ } y \Rightarrow y_m \xrightarrow{\ \|\ \|\ } y.$$

Thus, under the induction hypothesis, we have shown that the subspace M is complete with respect to an arbitrary norm which immediately implies that it is closed. Furthermore, from the above we see that this statement will be true for any finite-dimensional subspace of a normed space.

Consider the nth basis vector x_n now, and form the set

$$x_n + M = \{x_n + z \mid z \in M\}.$$

Since, for any $y, z \in M$,

$$\|x_n + z - (x_n + y)\| = \|z - y\|$$

$x_n + M$ is seen to be isometric to M under the mapping $z \to x_n + z$. Hence, since M is closed, $x_n + M$ must be closed as well, which implies, denoting the complement of $x_n + M$ by $C(x_n + M)$, that $C(x_n + M)$ is open.

We now contend that

$$0 \notin x_n + M,$$

for, if it did, we would be able to write, for some $\beta_1, \beta_2, \ldots, \beta_{n-1} \in F$,

$$0 = x_n + \beta_1 x_1 + \beta_2 x_2 + \cdots + \beta_{n-1} x_{n-1},$$

which is ridiculous. We now have that 0 is a point of the open set $C(x_n + M)$; hence there must be a whole neighborhood of 0 lying entirely within $C(x_n + M)$. In other words, there must exist $c_n > 0$ such that, for any $x \in x_n + M$,

$$\|x\| \geq c_n.$$

(Note that all we are saying here is that the distance from any point in $x_n + M$ to 0 is positive.) Thus, for all $\alpha_i \in F$ $(i = 1, \ldots, n-1)$,

$$\|\alpha_1 x_1 + \alpha_2 x_2 + \cdots + \alpha_{n-1} x_{n-1} + x_n\| \geq c_n,$$

which implies, for any $\alpha_n \in F$, that

$$\|\alpha_1 x_1 + \alpha_2 x_2 + \cdots + \alpha_n x_n\| \geq |\alpha_n| c_n$$

because we can write, for $\alpha_n \neq 0$,

$$\left\| \frac{\alpha_1}{\alpha_n} x_1 + \cdots + \frac{\alpha_{n-1}}{\alpha_n} x_{n-1} + x_n \right\| \geq c_n.$$

Suppose now that we had not taken

$$M = [\{x_1, x_2, \ldots, x_{n-1}\}]$$

but had taken instead

$$[\{x_1, x_2, \ldots, x_{i-1}, x_{i+1}, \ldots, x_n\}].$$

Since the only fact about M that was essential to our discussion was that its dimension was $n - 1$, it is clear that, in an analogous fashion, we could have arrived at some $c_i > 0$ such that

$$\|\alpha_1 x_1 + \cdots + \alpha_n x_n\| \geq c_i |\alpha_i|,$$

for any $i = 1, 2, \ldots, n$. In view of this, we can say for any $x = \sum_{i=1}^{n} \alpha_i x_i$,

$$\|\alpha_1 x_1 + \alpha_2 x_2 + \cdots + \alpha_n x_n\| \geq \min_i c_i \max_i |\alpha_i| = \min_i c_i \|x\|_0,$$

which completes the proof, since $a = \min_i c_i$ is positive.

There are some immediate corollaries to this theorem; indeed, they are more than corollaries for they have actually been proved in the course of proving the theorem.

COROLLARY 1. If X is any finite-dimensional normed linear space, X is complete.

COROLLARY 2. If X is a normed linear space and M is any finite-dimensional subspace, M is closed.

We turn now to another interesting fact about closed subspaces.

THEOREM 8.8 (*Riesz*). Let M be a closed proper subspace of the normed linear space X, and let a be a real number such that $0 < a < 1$. Then there exists a vector $x_a \in X$ such that

$$\|x_a\| = 1$$

and

$$\|x - x_a\| \geq a \qquad \text{(for all } x \in M).$$

Proof. Since M is a proper subspace, there must exist some vector x_1 such that

$$x_1 \in X - M.$$

Now let

$$d = \inf_{x \in M} \|x - x_1\| = d(x_1, M).$$

It is clear that d must be strictly greater than zero, for otherwise we would have

$$d = 0 \Rightarrow d(x_1, M) = 0 \Rightarrow x_1 \in \overline{M} = M,$$

which contradicts the way in which x_1 was chosen. Letting a be the a mentioned in the hypothesis, it is obvious that

$$\frac{d}{a} > d.$$

This now implies that there is some $x_0 \in M$ such that

$$\|x_0 - x_1\| < \frac{d}{a},$$

because, if $\|x_0 - x_1\|$ were greater than or equal to $a^{-1}d$ for all $x_0 \in M$, this would contradict the fact that d was the greatest lower bound of $\{d(x_0, x_1) \mid x_0 \in M\}$. Consider now

$$x_a = \frac{x_1 - x_0}{\|x_1 - x_0\|}.$$

It is clear that $\|x_a\| = 1$. To complete the proof we now let x be any vector in M and consider

$$\|x - x_a\| = \left\| x - \frac{x_1}{\|x_1 - x_0\|} + \frac{x_0}{\|x_1 - x_0\|} \right\| = \frac{1}{\|x_1 - x_0\|} \left\| (\|x_1 - x_0\|)x + x_0 - x_1 \right\|.$$

Since $(\|x_1 - x_0\|)x + x_0 \in M$, however, the above quantity must be greater than or equal to

$$\frac{d}{\|x_1 - x_0\|}.$$

Now, noting that

$$\frac{d}{\|x_1 - x_0\|} \geq \frac{d}{a^{-1}d} = a,$$

we see that the proof is complete.

We wish to turn to a new concept now—the notion of a *factor space*.

8.6   Factor Spaces

Let X be a vector space and let M be a subspace of X. We shall now define a new vector space which will, in some sense, allow us to focus attention on the complementary subspace of M; loosely speaking, the vectors in M will all correspond to the zero vector in the space that we are about to define.

Consider the relation $\sim$ now for any two vectors in X, x and y, defined by $x \sim y$ if and only if $x - y \in M$. We now show that this is an equivalence relation on X:

(1) Since $0 \in M$, it is clear that $x \sim x$ for

$$x - x = 0 \in M.$$

(2) If $x \sim y$, then $x - y \in M$. Since M is a subspace, though,

$$-(x - y) = y - x \in M,$$

which means that $y \sim x$. Thus the relation $\sim$ is symmetric.

(3) Suppose that $x \sim y$ and $y \sim z$. This implies that

$$x - y \in M \qquad \text{and} \qquad y - z \in M.$$

Since M is a subspace, the sum of any two of its members must also be in M; hence

$$x - y + y - z = x - z \in M \Rightarrow x \sim z,$$

and the relation is seen to be transitive. Therefore, $\sim$ is an equivalence relation.

If two vectors x and y are equivalent under $\sim$, this means that

$$x \in y + M = \{y + z \,|\, z \in M\}, \tag{8.13}$$

and any other vector equivalent to x under $\sim$ must also belong to $y + M$. In view of this, a typical equivalence class under $\sim$ must have the same form as (8.13). Hence, given any two equivalence classes

$$x + M \qquad \text{and} \qquad y + M,$$

either

$$x + M = y + M$$

or

$$(x + M) \cap (y + M) = \emptyset.$$

The set of all such equivalence classes $\{x + M \,|\, x \in X\}$ will be referred to as X/M, and is read X modulo M or, more simply, X mod M. X/M is called the *factor space* or *quotient space* of X with respect to M.

We now introduce operations of addition and scalar multiplication on the elements of X/M and show that, under these operations, X/M is a vector space.

We define

$$(x + M) + (y + M) = (x + y) + M;$$

that is, we add the particular representative of the equivalence classes and take the equivalence class to which their sum belongs. Similarly, if $\alpha \in F$, we define

$$\alpha(x + M) = (\alpha x) + M.$$

It must now be shown that the operations of addition and scalar multiplication do not depend on the particular representatives chosen. To this end, let $z_1, z_2 \in M$. (It is easily demonstrated that $y \in M$ implies that $y + M = M$ by just using the definition of $y + M$; moreover, $y + M = M$ if and only if $y \in M$.) Noting now that

$$x + M = x + z_1 + M \qquad \text{and} \qquad y + M = y + z_2 + M,$$

we consider the sum

$$(x + z_1 + M) + (y + z_2 + M) = x + y + (z_1 + z_2) + M.$$

Since $z_1 + z_2 \in M$, by virtue of the fact that M is a subspace, we can now write

$$x + y + (z_1 + z_2) + M = x + y + M$$

to conclude that the operation of addition is well-defined. Scalar multiplication can be shown to be well-defined in an exactly analogous manner. It can now easily be verified that X/M is a vector space. In the event that X is finite-dimensional, it can readily be shown (see Exercise 8.16) that, if dim $X = n$ and dim $M = m$, then dim $(X/M) = n - m$, by merely choosing an appropriate basis for X.

Example of a Factor Space

We now apply the notions introduced above to the particular setting of R^2, all ordered pairs of real numbers over R, so that the entire process can be viewed in the euclidean plane. Any one-dimensional subspace of this space must be a straight line passing through the origin. Denoting some particular subspace of R^2 by M (see Fig. 8.2), we now wish to see what the equivalence classes in this space look like. Proceeding as in Fig. 8.2, we draw a vector z_1 from the origin. We now wish to

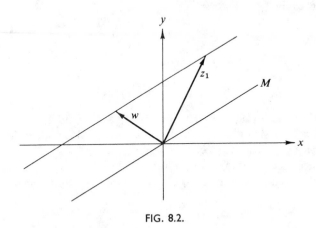

FIG. 8.2.

determine the set of all vectors in the plane that are equivalent to z_1 under the equivalence relation $z_1 \sim z_2$ if and only if $z_1 - z_2 \in M$. We leave it as an exercise for the reader to verify that this set consists of all vectors drawn from the origin such that their terminal point intersects the line drawn parallel to M and passing through the terminal point of z_1 as shown in the figure; that is, a typical vector in $z_1 + M$ is w, as indicated in Fig. 8.2.

Having noted that X/M is a linear space with respect to the operations defined above, we now wish to suppose that X is a *normed* linear space and exhibit a norm for X/M. To do this, however, we must restrict ourselves to considering only *closed* subspaces of X.

Using the same notation as above, we now define for any element of X/M, $x + M$,

$$\|x + M\| = \inf_{y \in x+M} \|y\| = \inf_{v \in M} \|x + v\|.$$

Since this operation is clearly well-defined, it remains for us only to verify that the norm axioms are satisfied; first scalar multiplication will be investigated. To this end, let $\alpha \in F$ and consider

$$\|\alpha x + M\| = \inf_{v \in M} \|\alpha x + v\|$$

$$= \inf_{v \in M} \|\alpha x + \alpha v\|,$$

provided $\alpha \neq 0$, the desired result being trivially obtained when $\alpha = 0$. The last term is equal to

$$|\alpha| \inf_{v \in M} \|x + v\| = |\alpha| \, \|x + M\|,$$

which is the desired result.

Next, we wish to show that the triangle inequality is satisfied for any two elements of X/M, $x + M$, and $y + M$. By the definition of addition in X/M, we have

$$\|x + M + y + M\| = \|x + y + M\|,$$

but since $\|x + y + M\|$ has been defined as a greatest lower bound, we can assert the existence of sequences of points in M, $\{z_n\}$ and $\{w_n\}$, such that

$$\lim_n \|x + z_n\| = \|x + M\| \quad \text{and} \quad \lim_n \|y + w_n\| = \|y + M\|,$$

which implies, for any n, that

$$\|x + y + M\| \le \|x + y + z_n + w_n\|$$

$$\le \|x + z_n\| + \|y + w_n\|.$$

Since the above statement is true for any n, it must also be true in the limit, which yields

$$\|x + y + M\| \le \|x + M\| + \|y + M\|$$

and, thus, the triangle inequality is seen to be satisfied.

Since $0 \in M$, it is clear that if $x + M = M$ (the zero vector in X/M) that $\|x + M\| = 0$. We now wish to show that the converse of this statement also holds; hence suppose

$$\|x + M\| = 0.$$

By the definition of the above norm as a greatest lower bound, we can assert the existence of a sequence of points in M, $\{z_n\}$ such that

$$\lim_n \|x + z_n\| = \|x + M\| = 0,$$

which implies

$$\|x + z_n\| \to 0$$

or

$$x + z_n \to 0$$

or

$$z_n \to -x.$$

Now, using for the first time the fact that M is closed, we can say that $(-x) \in M$, because it is the limit of a Cauchy sequence of points of M, which implies $x + M = M$ and completes the proof.

8.7 Completeness in the Factor Space

Suppose that X is a normed linear space and that M is a closed subspace of X. We now wish to show that, if X is a Banach space, then X/M is a Banach space too; or that completeness of X implies the completeness of X/M. To achieve our result, we must show that any Cauchy sequence of points in X/M must converge to some other point in X/M using, somewhere along the line, the fact that X was complete. To this end suppose that $\{x_n + M\}$ is a Cauchy sequence of points of X/M. Since this is assumed to be Cauchy sequence, if a convergent subsequence of the above sequence can be extracted, it follows that the entire sequence must converge and, furthermore, must converge to the same limit as the subsequence. With this in mind, we shall now try to extract a convergent subsequence.

By the definition of a Cauchy sequence, there must exist some integer n_1 such that the distance between any pair of terms whose indices exceed n_1 will be less than $\frac{1}{2}$. Thus we have

$$x_{n_1} + M, \qquad x_{n_1+1} + M, \ldots . \tag{8.14}$$

We can now choose an n_2 such that a subsequence of (8.14) can be extracted with the property that the distance between any two terms (with respect to the norm on X/M) is less than $\frac{1}{4}$; thus we have

$$x_{n_2} + M, \qquad x_{n_2+1} + M, \ldots . \tag{8.15}$$

Proceeding in the same way, we next extract a subsequence of (8.15) such that, for all pairs of indices greater than n_3, the distance between corresponding terms is less than $\frac{1}{8}$:

$$x_{n_3} + M, \qquad x_{n_3+1} + M, \ldots, \tag{8.16}$$

and so on. We now consider the "diagonal" terms of the subsequences so obtained and denote them by $\{y_i + M\}$; that is,

$$y_i = x_{n_i+i}.$$

We further note that

$$\|y_{i+1} + M - (y_i + M)\| = \|(y_{i+1} - y_i) + M\| < 1/2^i$$

and, by virtue of the definition of the above norm, assert the existence of

$$u_i \in (y_{i+1} - y_i) + M,$$

such that

$$\|u_i\| < 1/2^i.$$

Choose $z_1 \in y_1 + M$ now, and suppose $z_1, z_2, \ldots, z_n$ have been chosen such that

$$z_{i+1} \in y_{i+1} + M \tag{8.17}$$

and

$$z_{i+1} - z_i = u_i \tag{8.18}$$

for $i = 1, 2, 3, \ldots, n-1$. We wish to demonstrate that an $(n+1)$st vector can be added to the list having the same properties. To this end, we write

$$u_n = y_{n+1} - y_n + v,$$

where $v \in M$, and also

$$z_n = y_n + w,$$

where w is some element of M. The first equation implies that

$$u_n + y_n = y_{n+1} + v.$$

Adding w to both sides and letting $v + w = w'$, we have

$$z_n + u_n = y_{n+1} + w',$$

and we now define

$$z_{n+1} = y_{n+1} + w'$$

which completes the induction process concerning the existence of vectors $z_1, z_2, \ldots$, having properties (8.17) and (8.18). It will now be shown that $\{z_n\}$ is a Cauchy sequence. Suppose $n > m$, and consider

$$\|z_n - z_m\| \leq \|z_n - z_{n-1}\| + \cdots + \|z_{m+1} - z_m\|,$$

the terms on the right being just

$$\|u_{n-1}\| + \|u_{n-2}\| + \cdots + \|u_m\| \leq \frac{1}{2^{n-1}} + \cdots + \frac{1}{2^m} < \frac{1}{2^{m-1}}.$$

It is now clear that $\{z_n\}$ is a Cauchy sequence of elements from X which, since X is a Banach space, must converge to some point z of X. Next, we wish to show that the sequence of elements from X/M, $\{y_i + M\}$, approaches as a limit $(z + M)$. Noting, by the way the z_n were defined, that

$$y_i + M = z_i + M,$$

we consider

$$\|z + M - (y_i + M)\| = \|z + M - (z_i + M)\|.$$

But

$$\|z - z_i + M\| \leq \|z - z_i\|$$

and

$$\|z - z_i\| \to 0,$$

which implies

$$y_i + M \to z + M.$$

We need only note now that, if a convergent subsequence can be extracted from a Cauchy sequence, the sequence itself must converge to this limit, to complete the proof.

8.8 Convexity

In this section we assume throughout that X is a real or complex vector space. We want to introduce and discuss here a few results related to the notion of convexity; some of these results will be used later in our discussion of linear transformations on Hilbert spaces. However, we are essentially only introducing the reader to a rather extensive body of results related to convexity. For further results, the reader may consult some of the books listed in the references—in particular Ref. 4.

DEFINITION 8.3. The *segment* connecting the vectors $x, y \in X$ is the set of all vectors $z = (1 - \alpha)x + \alpha y$, where $0 \leq \alpha \leq 1$. Alternatively, we could write the segment connecting $x, y \in X$ as the set of all vectors $z = \alpha x + \beta y$, where $\alpha + \beta = 1$ ($\alpha \geq 0, \beta \geq 0$). A subset K of X is called *convex* if, whenever $x, y \in K$, it follows that the segment connecting them also belongs to K. (The concept of a segment in the euclidean plane is actually a straight-line segment as is easily verified.)

We proceed at once to establish certain elementary properties pertaining to convex sets.

THEOREM 8.9. Let K be a convex set and let $x_1, x_2, ..., x_n \in K$. Then all elements of the form $\alpha_1 x_1 + \alpha_2 x_2 + \cdots + \alpha_n x_n$, where all $\alpha_i \geq 0$ and $\alpha_1 + \alpha_2 + \cdots + \alpha_n = 1$, belong to K too.

Proof. The proof will be by induction. The contention for $n = 2$ is satisfied by the definition of a convex set. We therefore assume the result to be true for $n - 1$ and proceed to n.

If $\alpha_1 + \alpha_2 + \cdots + \alpha_{n-1} = 0$, then each α_i $(i = 1, 2, ..., n - 1)$ is zero and $\alpha_n = 1$, which implies

$$\alpha_1 x_1 + \alpha_2 x_2 + \cdots + \alpha_n x_n \in K.$$

Thus we assume that $\beta = \alpha_1 + \alpha_2 + \cdots + \alpha_{n-1} > 0$, and $\alpha_1 + \alpha_2 + \cdots + \alpha_n = 1$. Then,

$$\frac{\alpha_1}{\beta} x_1 + \frac{\alpha_2}{\beta} x_2 + \cdots + \frac{\alpha_{n-1}}{\beta} x_{n-1} \in K$$

by the induction assumption. Consequently,

$$\alpha_1 x_1 + \alpha_2 x_2 + \cdots + \alpha_n x_n = \beta \left(\frac{\alpha_1}{\beta} x_1 + \frac{\alpha_2}{\beta} x_2 + \cdots + \frac{\alpha_{n-1}}{\beta} x_{n-1} \right) + \alpha_n x_n \in K,$$

since K is convex, which completes the proof.

THEOREM 8.10. Let X be a normed linear space and let K be a convex subset of X; then $\overline{K}$ is convex.

Proof. Let $x, y \in \overline{K}$ and let ε be an arbitrary positive real number. There exist elements $x_1, y_1 \in K$ such that

$$\|x - x_1\| < \varepsilon \quad \text{and} \quad \|y - y_1\| < \varepsilon.$$

Let $\alpha, \beta \geq 0$ and $\alpha + \beta = 1$. Then

$$\|\alpha x + \beta y - (\alpha x_1 + \beta y_1)\| \leq \alpha \|x - x_1\| + \beta \|y - y_1\|$$

$$\leq (\alpha + \beta)\varepsilon = \varepsilon.$$

But $\alpha x_1 + \beta y_1 \in K$, since K is convex. Thus, since $\varepsilon > 0$ is arbitrary, it follows that $\alpha x + \beta y \in \overline{K}$, so $\overline{K}$ is convex also.

THEOREM 8.11. The intersection of any number of convex subsets of the vector space X is a convex subset.

Proof. Clear.

DEFINITION 8.4. Let S be a subset of the vector space X. The *convex hull* of S is the intersection of all convex sets containing S.

We note that the definition makes sense; namely, there is a convex set containing S: X itself. For that matter, it is clear that every subspace of X is a convex set. An alternate characterization of convex hull is given in the next theorem.

THEOREM 8.12. The convex hull of the subset S of the vector space X consists of all vectors of the form $\alpha_1 x_1 + \alpha_2 x_2 + \cdots + \alpha_n x_n$, where the $x_i \in S$, $\alpha_i \geq 0$, for $i = 1, 2, \ldots, n$ and $\sum_{i=1}^{n} \alpha_i = 1$.

Proof. Let K be the set of all vectors of the form given in the statement of the theorem. K is clearly a convex set from its very definition. Moreover, $K \supset S$. It follows that $K \supset S_c$, the convex hull of S. However, it is clear by Theorem 8.9 that any convex set containing S must contain K; in particular, $S_c \supset K$. Therefore $S_c = K$, and this completes the proof.

DEFINITION 8.5. Let X be a normed linear space and let S be any subset of X. The closure of the convex hull of S is called the *closed convex hull* of S.

THEOREM 8.13. Let $E = \bigcap \{K \mid K \supset S, K \text{ convex and closed}\}$, that is, E is the intersection of all closed and convex sets containing S. Let S_c equal the convex hull of S. Then $E = \bar{S}_c$, the closed convex hull of S.

Proof. Since S_c is convex, $\bar{S}_c$ is convex by Theorem 8.10 and is, of course, closed and $\bar{S}_c \supset S$. Therefore $E \subset \bar{S}_c$. However, $E \supset S_c$, since S_c is the intersection of all convex sets containing S, whereas E is just the intersection of the closed and convex sets containing S. But $E \supset S_c$ implies that $E = \bar{E} \supset \bar{S}_c$. This, combined with the preceding inclusion, gives $E = \bar{S}_c$.

EXERCISES 8

1. Let X be a normed linear space. Show that

$$\overline{S_\varepsilon(x)} = \{y \mid \|y - x\| \leq \varepsilon\}.$$

(See also Exercise 3.2.)

2. Let X be a normed linear space, let x_0 be a fixed vector in X, and let $\alpha \neq 0$ be a fixed scalar. Show that the mappings $x \to x_0 + x$ and $x \to \alpha x$ are homeomorphisms of X onto itself.

3. Show that, if X is a normed linear space, then X is homeomorphic to $S_r(0)$ for any fixed $r > 0$. [*Hint:* Consider the mapping $x \to rx/(1 + \|x\|)$.]

4. Let A and B be two subsets of a normed linear space X. Let

$$A + B = \{x + y \mid x \in A, y \in B\}.$$

Prove: (a) If A or B is open, then $A + B$ is open.
  (b) If A and B are compact, then $A + B$ is compact.

(c) If A is compact and B is closed, then $A + B$ is closed.

(d) Give an example where A and B are closed but $A + B$ is not closed.

5. Prove that l_∞ is a complete space (cf. Example 6.6).

6. Show that l_p (Example 8.4) is separable for $1 \le p < \infty$.

7. Show that a set A in a normed linear space is bounded if and only if $\sup_{x \in A} \|x\| < \infty$.

8. If $f \in L_p(E, \mu)$ (Example 8.6), show that

$$\|f\|_p = \max \left| \int_E fg \, d\mu \right| = \max \int_E |fg| \, d\mu,$$

where the max is taken over all $g \in L_q$ such that $\|g\|_q \le 1$ and $(1/p) + (1/q) = 1$.

9. If $f \in L_p(E, \mu)$ and $f \in L_q(E, \mu)$, where $p < q$, show that $f \in L_r(E, \mu)$ for all r satisfying $p \le r \le q$.

10. Denote by c the set of all convergent sequences of complex numbers. By defining vector operations as in l_2, it is easily verified that c is a linear space. Letting $x \in c$, where $x = (\alpha_1, \alpha_2, \ldots)$, one can take $\sup_i |\alpha_i| = \|x\|$ as a norm on c. Show that c is a Banach space.

11. Prove that an infinite-dimensional Banach space cannot have a countable basis. (*Hint:* Use the Baire category theorem of Sec. 6.2.)

12. If X and Y are normed linear spaces over the same field $F(= C$ or $R)$, both having the same finite dimension n, then prove that X and Y are *topologically isomorphic*, where a topological isomorphism is defined to be a mapping that is simultaneously an isomorphism and a homeomorphism.

13. Let X be a normed linear space. Letting S be a subset of X, give an example for which $[S] \ne \overline{[S]}$.

14. Let X be an n-dimensional normed linear space. Let $x_1, x_2, \ldots, x_n \in X$ such that $\overline{[\{x_1, x_2, \ldots, x_n\}]} = X$. Show that every $x \in X$ can be written uniquely in the form $x = \sum_{i=1}^n \alpha_i x_i$.

15. Let X be a normed linear space. Show that X is finite-dimensional if and only if every closed and bounded set in X is compact.

16. If $\dim X = n$ and M is a subspace of X of dimension m, show that $\dim (X/M) = n - m$.

17. Show that if X is a normed linear space and if X is separable, then so is X/M.

18. Let M be a finite-dimensional subspace of the normed space X. If $x + M \in X/M$, show that there exists an element $y \in x + M$ such that $\|y\| = \|x + M\|$.

19. If X is a vector space and if $X = M \oplus N$, where M and N are subspaces, prove that N is isomorphic to X/M.

20. Let f be a continuous linear functional defined on the normed linear space X. Show that $M = \{x \in X \mid f(x) = 0\}$ is a closed subspace of X.

REFERENCES

1. P. Halmos, "Measure Theory." A more analytic proof of the Hölder and Minkowski inequalities.

Material on Banach spaces

2. N. Dunford and J. Schwartz, "Linear Operators. Part I: General Theory."

3. A. Zaanen, "Linear Analysis."

4. F. Valentine, "Convex Sets."

CHAPTER **9**

Commutative Convergence, Hilbert Spaces, and Bessel's Inequality

When one considers an infinite series or sum, the index of summation is only allowed to range through the integers. It is clear that letting the index range through any denumerable set will serve the purpose just as well. As a generalization of this, we now wish to consider the case for which the index of summation ranges through a nondenumerable set. We shall refer to this consideration as *summability* or *commutative convergence*. Unfortunately, as will be seen, the generalization provided by such considerations is not so grand as it seems.

When dealing with spaces of certain types, one is always interested in combining them in various ways to generate new spaces. Thus, in this vein, given two normed linear spaces X and Y, norms will be proposed for the cartesian product (external direct sum) $X \times Y$ in such a way that the norm on the product space will, in some sense, extend the norms on the original spaces. Similarly, if X and Y are inner product spaces, an inner product will be suggested for $X \times Y$.

Since an inner product space is a normed space and hence a metric space, it makes sense to consider when an inner product space is complete. According to the terminology of the last chapter, such a space, if complete, would be called a *Banach space*. Since the norm is derived from an inner product, however, we wish to give some special consideration to these spaces and, to facilitate referring to them, we shall call them *Hilbert* spaces. Just as we were able to embed any metric space in a complete metric space and any normed linear space in a complete normed linear space, we shall be able to embed any inner product space in a complete inner product space or Hilbert space.

Historically, when Hilbert spaces were first considered, an additional demand was made upon them: It was required that, in addition to being complete, they also be separable. In most modern treatments this requirement has been dropped, and in Sec. 9.4 an example of a nonseparable Hilbert space is given.

A rather important Hilbert space is $L_2(0, 2\pi)$—equivalence classes of square-summable functions on $[0, 2\pi]$—and we shall devote some time to it in this chapter. For example, it is observed that the collection

$$\frac{1}{\sqrt{2\pi}}, \quad \frac{\cos t}{\sqrt{\pi}}, \quad \frac{\sin t}{\sqrt{\pi}}, \quad \dots$$

is an orthonormal set in this space. Using this collection in the familiar context of ordinary Fourier series, the well-known Riemann–Lebesgue lemma is deduced. Another problem that will be answered by working in this space is the following: Given a collection of real numbers, under what conditions will there exist a function such that these numbers will be the Fourier coefficients of that function? This rather formidable sounding problem is completely answered in Sec. 9.6 by the Riesz–Fischer theorem and in such a way that its simplicity is the most singular feature.

When a basis for a vector space was defined, it was required that every vector in the space be expressible as a *finite* linear combination of basis elements; indeed, it was not required that every vector in the space be expressible in terms of the same finite collection of basis elements but just some finite collection of basis elements (for example, x might be expressed in terms of the basis vectors $x_1, x_2, \dots, x_n$, whereas y is expressible in terms of the basis vectors $y_1, y_2, \dots, y_m$). A new concept will be introduced in this chapter somewhat akin to the notion of a basis, namely, the notion of a *complete* orthonormal set in an inner product space. Instead of just forming linear combinations (that is, finite) of these vectors, we shall form commutatively convergent sums as defined in Sec. 9.1. It will then be shown that complete orthonormal sets exist in any inner product space, and also that such a set, when it is infinite, is never a basis in a Hilbert space.

As a concrete illustration of these notions, it will be shown that

$$\left\{ \frac{1}{\sqrt{2\pi}} e^{int} \middle| n = 0, \pm 1, \pm 2, \dots \right\}$$

is a complete orthonormal set in $L_2(0, 2\pi)$.

An appendix to this chapter is included in which the notions of partially ordered sets and Zorn's lemma are discussed.

9.1 Commutative Convergence

DEFINITION 9.1. Suppose $\{x_\alpha\}$, where α runs through an index set Λ, is a collection of elements from the normed linear space X. $\{x_\alpha\}$ is said to be *summable* (*commutatively convergent*) *to* $x \in X$, written

$$\sum_{\alpha \in \Lambda} x_\alpha = x \qquad \text{or sometimes just} \qquad \sum_\Lambda x_\alpha = x$$

if, for all $\varepsilon > 0$, there exists some finite set of indices $H \subset \Lambda$, such that, for any finite set of indices $J \supset H$,

$$\left\| \sum_{\alpha \in J} x_\alpha - x \right\| < \varepsilon.$$

Before proving some immediate consequences of the definition, we note the similarity of this notion to that of *unconditional convergence* of real numbers: a convergence strong enough to guarantee that no regrouping of the terms will have any effect. (The last observation rather suggests the term *commutatively* convergent.)

The first observation we wish to make is that, if $\sum_{\alpha \in \Lambda} x_\alpha = x$ and $\beta \in F$, then

$$\sum_{\alpha \in \Lambda} \beta x_\alpha = \beta x.$$

To prove this, let ε be given. It is clear that a finite set of indices H can be chosen so that, for all finite sets $J \supset H$,

$$\left\| \sum_{\alpha \in J} \beta x_\alpha - \beta x \right\| = |\beta| \left\| \sum_{\alpha \in J} x_\alpha - x \right\| < \varepsilon.$$

Next we claim that if

$$\sum_{\alpha \in \Lambda} x_\alpha = x \qquad \text{and} \qquad \sum_{\alpha \in \Lambda} y_\alpha = y,$$

then

$$\sum_{\alpha \in \Lambda} (x_\alpha + y_\alpha) = x + y.$$

To prove this we choose H_1 and H_2 such that, for all finite sets of indices J_1 and J_2, where $J_1 \supset H_1$ and $J_2 \supset H_2$,

$$\left\| \sum_{\alpha \in J_1} x_\alpha - x \right\| < \varepsilon/2 \qquad \text{and} \qquad \left\| \sum_{\alpha \in J_2} y_\alpha - y \right\| < \varepsilon/2$$

and note that for any finite set

$$J \supset H_1 \cup H_2 \Rightarrow J \supset H_1 \quad \text{and} \quad J \supset H_2.$$

Thus,

$$\left\| \sum_{\alpha \in J} x_\alpha + y_\alpha - (x + y) \right\| < \varepsilon.$$

Our next result shows that, if a series is to be summable in the sense defined above, all but a countable number of terms in the series must be zero.

THEOREM 9.1. In the normed linear space X, suppose that $\{x_\alpha\}$ is summable to $x \in X$, where α runs through an index set Λ. Then all but a countable number of the x_α must be zero.

Proof. Let $\varepsilon > 0$ be given and a corresponding finite set of indices H be chosen so that, for all $J_1 \supset H$,

$$\left\| \sum_{\alpha \in J_1} x_\alpha - x \right\| < \varepsilon/2.$$

Now consider any other finite set of indices J, such that $J \cap H = \emptyset$. In this case,

$$\left\| \sum_{\alpha \in J} x_\alpha \right\| = \left\| \sum_{\alpha \in J \cup H} x_\alpha - \sum_{\alpha \in H} x_\alpha \right\|,$$

which is less than or equal to

$$\left\| x - \sum_{\alpha \in J \cup H} x_\alpha \right\| + \left\| x - \sum_{\alpha \in H} x_\alpha \right\| < \varepsilon.$$

Now let $\varepsilon = 1/n$ $(n = 1, 2, \ldots)$. There exist finite sets of indices H_n such that, if J is a finite set of indices and

$$J \cap H_n = \emptyset,$$

then

$$\left\| \sum_{\alpha \in J} x_\alpha \right\| < 1/n.$$

Consider the countable set $\bigcup_{n=1}^{\infty} H_n$ and consider some index element α such that

$$\{\alpha\} \cap \left(\bigcup_{n=1}^{\infty} H_n \right) = \emptyset.$$

This implies

$$\{\alpha\} \cap H_n = \emptyset$$

for any n, which means

$$\|x_\alpha\| < 1/n$$

for any n. Consequently,

$$x_\alpha = 0,$$

and this must be the case for any x_α, where

$$\alpha \notin \bigcup_{n=1}^{\infty} H_n.$$

The proof is now complete.

As a final comment, note that nowhere in the above discussion did we attempt to sum up an infinite number of terms. In spite of what our notation might indicate, a closer examination reveals that only finite sums have been dealt with to characterize this special type of infinite sum.

9.2 Norms and Inner Products on Cartesian Products of Normed and Inner Product Spaces

In Example 3.6 it was shown that, if (X, d) and (Y, d') were metric spaces, then many metrics, each of which involved the original metrics in some way, could be constructed for $X \times Y$. If X and Y are each normed linear spaces (we shall not distinguish between the symbol for norm on X and on Y), we wish now to introduce a norm on $X \times Y$. First, however, we must introduce operations on $X \times Y$ so that it can be viewed as a linear space. To avoid confusion with inner products, members of $X \times Y$ will be denoted $\langle x, y \rangle$. Using this notation, we define addition of elements of $X \times Y$ as

$$\langle x_1, y_1 \rangle + \langle x_2, y_2 \rangle = \langle x_1 + x_2, y_1 + y_2 \rangle$$

and multiplication by a scalar α, as

$$\alpha \langle x, y \rangle = \langle \alpha x, \alpha y \rangle$$

and leave as an exercise for the reader to verify that $X \times Y$ with respect to these operations is truly a linear space.

We now claim that the following mappings are all norms on $X \times Y$ and, furthermore, are all equivalent on $X \times Y$:

$$\| \langle x, y \rangle \| = \|x\| + \|y\|, \tag{9.1}$$

$$\| \langle x, y \rangle \| = (\|x\|^p + \|y\|^p)^{1/p} \qquad (p \geq 1), \tag{9.2}$$

$$\| \langle x, y \rangle \| = \max(\|x\|, \|y\|). \tag{9.3}$$

The above norms all give rise to the same topology for $X \times Y$, and it can also be demonstrated that if X and Y are each Banach spaces, $X \times Y$ is a Banach space, too, with respect to any of the norms defined above.

Still working in the same notation, suppose X and Y are now inner product spaces over the same field. We shall define an inner product for $X \times Y$ as

$$(\langle x_1, y_1 \rangle, \langle x_2, y_2 \rangle) = (x_1, x_2) + (y_1, y_2).$$

That the above is really an inner product on $X \times Y$ can easily be verified. We further note that, if one considers the rather natural correspondence between elements x, of X (or Y), and the element of $\langle x, 0 \rangle$ of $X \times Y$ (or $\langle 0, y \rangle$), one can readily show that this is an isomorphism between X and $\{ \langle x, 0 \rangle \mid x \in X \}$ (with a corresponding statement for Y if Y had been chosen). Thus the inner product defined here and the norms defined above actually *extend* the original inner product and norm on each of the spaces; that is the newly defined quantities reduce to the original mappings when restricted to the (isomorphic image of) original space.

Now that we have introduced a norm or metric on the cartesian product of linear spaces, it becomes meaningful for us to speak of continuous mappings from such cartesian products into other spaces. The definition of continuity most appropriate for such discussions is Definition 3.6, namely, a function will be continuous at a point x if for every ε-neighborhood of $f(x)$ in the range, one can find a δ-neighborhood of x in the domain $S_\delta(x)$, such that $f(S_\delta(x)) \subset S_\varepsilon(f(x))$. In the particular situation when the domain is the cartesian product of two spaces, the point x mentioned above will be an ordered pair of elements from the component spaces, and an ε-neighborhood of the point will be a neighborhood in the cartesian product space.

It will now be shown that the inner product is a continuous mapping.

THEOREM 9.2. If X is an inner product space, the inner product (x, y) is a continuous function mapping $X \times X$ into F.

Proof. Consider the fixed point in the range (x_2, y_2). Now let

$$x_3 = x_1 - x_2 \quad \text{and} \quad y_3 = y_1 - y_2,$$

which implies

$$|(x_1, y_1) - (x_2, y_2)| = |(x_2 + x_3, y_2 + y_3) - (x_2, y_2)|.$$

Expanding the first inner product and appealing to the Cauchy–Schwarz inequality, noted in Theorem 1.1, we have

$$|(x_2, y_3) + (x_3, y_2) + (x_3, y_3)| \leq \|x_2\| \, \|y_1 - y_2\| + \|y_2\| \, \|x_1 - x_2\|$$
$$+ \|x_1 - x_2\| \, \|y_1 - y_2\|,$$

and the continuity of the above mapping is evident.

9.3 Hilbert Spaces

If X is an inner product space, we can immediately define a norm in terms of the inner product. Once we have a norm, this immediately gives rise to a metric on the space and, as usual, an important fact about the space is whether it is complete with respect to this metric. We now make the following definition.

DEFINITION 9.2. If an inner product space X is complete with respect to the metric derived from the inner product, X is said to be a *Hilbert space*.

In previous chapters it was shown that any metric space could be completed and that any normed linear space could be embedded in a complete normed linear space or Banach space. We would now like to show that any inner product space can be embedded in a complete inner product space or Hilbert space. To do this, suppose X is the given inner product space. Viewing it as a normed space where $\|x\|$ is taken to be $(x, x)^{1/2}$, we can apply the results of Sec. 8.3 to assert that the collection of all equivalence classes of Cauchy sequences from X is a complete normed linear space with norm of $x^* \in X^*$ (the completion of X) defined to be

$$\|x^*\| = \lim_n \|x_n\|,$$

where $\{x_n\}$ is any member of the equivalence class x^*. We now wish to define an inner product on X^* in such a way that the norm determined by this inner product is the same as the above norm. If this can be accomplished, it is clear that X^* will be a complete inner product space. Let $x^*, y^* \in X^*$. We propose that

$$(x^*, y^*) = \lim_n (x_n, y_n),$$

where $\{x_n\} \in x^*$ and $\{y_n\} \in y^*$, does indeed constitute an inner product on X^*.

As always, when such a mapping is proposed, one must verify that

(1) the limit exists;
(2) the mapping is well-defined;
(3) the axioms that an inner product must satisfy are obeyed.

Leaving these verifications as exercises, we now proceed to show that the norm derived from this inner product does indeed agree with the norm with respect to which the space is complete. We denote the norm defined by the inner product as $\| \ \|_N$; that is

$$\|x^*\|_N = (x^*, x^*)^{1/2} \qquad (\text{for } x^* \in X^*).$$

Letting $\{x_n\} \in x^*$, where $x^* \in X^*$, we can now write

$$\|x^*\|_N = (x^*, x^*)^{1/2}$$
$$= \lim_n (x_n, x_n)^{1/2}$$
$$= \lim_n \|x_n\|$$
$$= \|x^*\|$$

to conclude that $\| \ \|$ and $\| \ \|_N$ are the same; hence, X^* is a Hilbert space containing a congruent image of X that is everywhere dense in X^*. Further, as before, the space X^* is unique up to a congruence. In view of this, one can readily see why incomplete inner product spaces are often referred to as *pre-Hilbert spaces*. In the event that a Hilbert space is finite-dimensional, one sometimes sees the term *euclidean space* used to describe it. (One often sees real euclidean spaces referred as just euclidean spaces and complex euclidean spaces called *unitary* spaces.)

Some examples of Hilbert spaces will now be given. The space C^n, all n-tuples of complex numbers (or real numbers) with inner product between $x = (\alpha_1, \alpha_2, ..., \alpha_n)$ and $y = (\beta_1, \beta_2, ..., \beta_n)$ given by

$$(x, y) = \sum_{i=1}^{n} \alpha_i \bar{\beta}_i$$

is a Hilbert space.

The space l_2, mentioned in Example 1.3, is a Hilbert space.

The space of equivalence classes $L_2(E, \mu)$, discussed in Example 1.4, is another example of a Hilbert space of which the Hilbert space $L_2[a, b]$ discussed in Example 1.5 is a special case.

As an example of an inner product space that is not a Hilbert space, consider the linear space $C[a, b]$ (Example 1.2) with inner product of $f, g \in C[a, b]$ defined to be

$$(f, g) = \int_a^b f(x)\overline{g(x)} \, dx.$$

If a sequence of functions from $C[a, b]$, $\{f_n\}$ converges with respect to the norm defined by this inner product, they will only converge " in the mean." Since mean convergence does not necessarily imply uniform convergence, it is seen that this spaces does not have to be complete and the counterexample (Example 4.4) shows that this is actually so.

In Theorem 1.4 we noted that, if x and y are any two vectors in an inner product space X, then

$$\|x + y\|^2 + \|x - y\|^2 = 2\|x\|^2 + 2\|y\|^2.$$

This result is called the parallelogram law and is of interest to us now because it provides a necessary condition that any inner-product-determined norm must satisfy. Thus if we are given a normed linear space and are wondering whether an inner product can be introduced which will determine a norm that will agree with the norm already defined, we know that we need check no further if the norm already defined does not satisfy the parallelogram law; any norm derivable from an inner product must satisfy this equality. As an immediate application of this fact, consider the linear space $L_1[0, 1]$ consisting of equivalence classes of functions summable on $[0, 1]$ with respect to Lebesgue measure (see Example 8.6 for a discussion of the B-space, $L_p(E, \mu)$), with the norm of $f \in L_1[0, 1]$ taken as

$$\|f\| = \int_0^1 |f(x)| \, dx.$$

It will now be shown that this norm does not satisfy the parallelogram law and thus precludes the possibility of viewing this space as a Hilbert space (with respect to this norm).

Consider the sets $A = [0, \frac{1}{2}]$, $B = [\frac{1}{2}, 1]$ and the characteristic functions of these sets, k_A and k_B. (The *characteristic function of a set* is that function which is one everywhere on the set and zero elsewhere.) We now note that

$$\|k_A + k_B\|^2 = 1 \quad \text{and} \quad \|k_A - k_B\|^2 = 1,$$

whereas

$$2\|k_A\|^2 + 2\|k_B\|^2 = 2 \cdot \tfrac{1}{4} + 2 \cdot \tfrac{1}{4} = 1,$$

and, therefore, the parallelogram law is not satisfied.

9.4 A Nonseparable Hilbert Space

Recalling the notion of a separable metric space (see Sec. 6.4), we now wish to exhibit a Hilbert space that is not separable. The reasoning will go in a fashion similar to the way in which l_∞ was shown (Example 6.6) to be nonseparable.

Let X be any nondenumerable set, and consider those functions f on X (real or complex-valued) that are equal to zero on all but a countable subset of X and such that

$$\sum_{x \in X} |f(x)|^2 < \infty.$$

It can be shown that the collection of all such functions with respect to addition of functions in the usual manner and multiplication by scalars is a linear space and will be denoted by $L_2(X)$. Furthermore, the operation

$$(f, g) = \sum_{x \in X} f(x)\overline{g(x)},$$

where $f, g \in L_2(X)$ can be shown to be an inner product on $L_2(X)$. (The Hölder inequality for infinite sums mentioned in Remark 1 of Sec. 8.1 is useful in verifying this claim.) To show that $L_2(X)$ is nonseparable, consider the subset of $L_2(X)$ consisting of all characteristic functions defined for one-point sets in X; that is, $\{k_{\{x\}} \mid x \in X\}$. Suppose now that $x_1 \neq x_2$ and consider the distance between $k_{\{x_1\}}$ and $k_{\{x_2\}}$:

$$\|k_{\{x_1\}} - k_{\{x_2\}}\|^2 = \sum_{x \in X} |k_{\{x_1\}}(x) - k_{\{x_2\}}(x)|^2 = 2.$$

Thus

$$\|k_{\{x_1\}} - k_{\{x_2\}}\| = \sqrt{2}.$$

We now surround each of these points $k_{\{x\}}$ (characteristic functions) by spheres of radius $\sqrt{2}/2$, it being clear that none of the spheres so defined overlap (have a nonempty intersection). For any dense subset, each sphere must contain at least one distinct point of the dense subset. Since there are a nondenumerable number of such spheres, because X was assumed to be nondenumerable, the subset itself must also be nondenumerable and, hence, the space cannot be separable.

Actually the preceding example is a special case of $L_2(E, \mu)$ in which one takes E to be X, S to be the collection of all subsets of X, and μ a discrete measure on S; that is, $\mu(A)$ is the number of elements in A if A is finite; otherwise, $\mu(A) = \infty$, for any subset A of X.

9.5   Bessel's Inequality

THEOREM 9.3. Let X be an inner product space, A an orthonormal set of vectors in X, and y an arbitrary vector in X. Then:

(1) (Bessel's inequality) for all $x_1, x_2, \ldots, x_n \in A$,

$$\sum_{i=1}^{n} |(y, x_i)|^2 \leq \|y\|^2;$$

(2) the set $E = \{x \in A \mid (y, x) \neq 0\}$ is countable;

(3) if $z \in X$, then $\sum_{x \in A} |(y, x)\overline{(z, x)}| \leq \|y\| \|z\|$.

Proof. Let $\alpha_i = (y, x_i)$. It is clear that

$$0 \leq \left(y - \sum_{i=1}^{n} \alpha_i x_i, \; y - \sum_{j=1}^{n} \alpha_j x_j \right)$$

$$= \|y\|^2 - \sum_{j=1}^{n} \bar{\alpha}_j (y, x_j) - \sum_{i=1}^{n} \alpha_i (x_i, y) + \sum_{i=1}^{n} \sum_{j=1}^{n} \alpha_i \bar{\alpha}_j (x_i, x_j)$$

$$= \|y\|^2 - \sum_{j=1}^{n} |\alpha_j|^2 - \sum_{i=1}^{n} |\alpha_i|^2 + \sum_{i=1}^{n} |\alpha_i|^2$$

$$= \|y\|^2 - \sum_{j=1}^{n} |\alpha_j|^2$$

or

$$\|y\|^2 \geq \sum_{j=1}^{n} |(y, x_j)|^2,$$

which proves (1).

To prove (2), consider the set

$$E_n = \left\{ x \in A \;\middle|\; |(y, x)| \geq \frac{1}{n} \right\},$$

where n is some positive integer and suppose that $x_1, x_2, \ldots, x_k \in E_n$. In this case, by Bessel's inequality and the definition of E_n,

$$k \cdot \frac{1}{n^2} \leq \sum_{i=1}^{k} |(y, x_i)|^2 \leq \|y\|^2,$$

which implies

$$k \leq n^2 \|y\|^2 < \infty.$$

Thus the set

$$E = \{ x \in A \mid (y, x) \neq 0 \} = \bigcup_{n=1}^{\infty} E_n$$

is a denumerable union of finite sets and is therefore countable.

Last, applying part (2) of the theorem to assert that $(y, x_i) = 0$ and $(z, x_i) = 0$ for all but a countable number of $x_i \in A$, and using the Hölder inequality and Bessel's inequality, we can say, for any n whatsoever, that

$$\sum_{i=1}^{n} |(y, x_i)\overline{(z, x_i)}| \leq \left(\sum_{i=1}^{n} |(y, x_i)|^2 \right)^{1/2} \left(\sum_{i=1}^{n} |(z, x_i)|^2 \right)^{1/2}$$

$$\leq \|y\| \, \|z\|.$$

The series $\sum_{x \in A} |(y, x_i)\overline{(z, x_i)}|$ is now seen to converge absolutely and therefore, unconditionally, which, for series of complex numbers, means commutative convergence. Actually for series of complex numbers, commutative convergence is equivalent to absolute convergence as one can readily see.

9.6   Some Results from $L_2(0, 2\pi)$ and the Riesz–Fischer Theorem

Consider the real Hilbert space of real-valued square-summable functions $L_2(0, 2\pi)$† (Example 1.5) and the orthonormal set in this space

$$A: \frac{1}{\sqrt{2\pi}}, \frac{\sin t}{\sqrt{\pi}}, \frac{\cos t}{\sqrt{\pi}}, \frac{\sin 2t}{\sqrt{\pi}}, \frac{\cos 2t}{\sqrt{\pi}}, \ldots = x_0, x_1, x_2, \ldots;$$

that is, for $n > 0$,

$$x_n = \begin{cases} \dfrac{1}{\sqrt{\pi}} \cos\left(\dfrac{n}{2} t\right) & (n \text{ even}), \\[3mm] \dfrac{1}{\sqrt{\pi}} \sin\left(\dfrac{n+1}{2} t\right) & (n \text{ odd}). \end{cases}$$

We now define, for $f \in L_2(0, 2\pi)$,

$$\alpha_n = (f, x_n);$$

that is,

$$\alpha_0 = \frac{1}{\sqrt{2\pi}} \int_0^{2\pi} f(t)\, dt,$$

$$\alpha_1 = \frac{1}{\sqrt{\pi}} \int_0^{2\pi} f(t) \sin t\, dt,$$

$$\vdots$$

Consider now the ordinary Fourier series coefficients:

$$a_0 = \frac{1}{\pi} \int_0^{2\pi} f(t)\, dt,$$

and, for $n > 0$,

$$a_n = \frac{1}{\pi} \int_0^{2\pi} f(t)\cos\left(\frac{n}{2} t\right) dt \qquad (n \text{ even}),$$

and

$$b_n = \frac{1}{\pi} \int_0^{2\pi} f(t)\sin\left(\frac{n+1}{2} t\right) dt \qquad (n \text{ odd}),$$

We now have the relationships

$$\alpha_0 = a_0 \sqrt{\frac{\pi}{2}},$$

$$\alpha_1 = b_1 \sqrt{\pi},$$

$$\alpha_2 = a_2 \sqrt{\pi},$$

$$\vdots \qquad \vdots$$

† For a proof that this space is complete see Example 8.6.

Appealing to Bessel's inequality [Theorem 9.3(1)], we have

$$\sum_{n=0}^{\infty} |\alpha_n|^2 \leq \|f\|^2$$

and can substitute the preceding relationships to get

$$a_0^2 \frac{\pi}{2} + \pi \left(\sum_{n=1}^{\infty} a_n^2 + b_n^2 \right) \leq \int_0^{2\pi} f^2(t)\, dt$$

which implies

$$\lim_n a_n = 0 \quad \text{and} \quad \lim_n b_n = 0,$$

an important result usually called the *Riemann–Lebesgue lemma*.

Returning to the general situation once again, we now wish to prove the following theorem.

THEOREM 9.4. If X is a separable inner product space and A is any orthonormal set in X, then A is countable.

Proof. Let x and y be any two elements of A. We can say immediately

$$\|x - y\|^2 = (x - y, x - y) = \|x\|^2 - (x, y) - (y, x) + \|y\|^2$$
$$= \|x\|^2 + \|y\|^2$$
$$= 2,$$

which implies

$$\|x - y\| = \sqrt{2}.$$

Now consider the collection of $(\sqrt{2}/2)$-neighborhoods about each point in A. It is clear that no two of these neighborhoods can have a point in common which implies that each such neighborhood must contain a distinct point of the countable dense subset. Thus if A were not countable we would have to have a noncountable dense subset contrary to our hypothesis.

Our next result yields an important fact about convergence (in the sense of summability introduced in Sec. 9.1) in Hilbert space.

THEOREM 9.5. Let X be a Hilbert space and let

$$A: x_1, x_2, \ldots$$

be an orthonormal set in X. Then, if $\alpha_n \in F$ $(n = 1, 2, \ldots)$, we have (1) $\sum_{n=1}^{\infty} \alpha_n x_n$ converges if and only if $\sum_{n=1}^{\infty} |\alpha_n|^2 < \infty$, and (2) if $\sum_{n=1}^{\infty} \alpha_n x_n$ converges to x, then $\alpha_n = (x, x_n)$.

Proof (1). Consider the partial sum

$$S_n = \sum_{i=1}^{n} \alpha_i x_i$$

and suppose now that $n > m$; then

$$\|S_n - S_m\|^2 = \left\| \sum_{i=m+1}^{n} \alpha_i x_i \right\|^2$$

$$= \sum_{i=m+1}^{n} |\alpha_i|^2. \qquad (9.4)$$

Clearly, if the series mentioned in (1) converges $\sum_{i=m+1}^{n} |\alpha_i|^2$ must go to zero and the series $\sum_{n=1}^{\infty} |\alpha_n|^2$ must converge. Conversely suppose as $n, m \to \infty$,

$$\sum_{i=m+1}^{n} |\alpha_i|^2 \to 0.$$

This implies that the sequence of partial sums $\sum_{i=1}^{n} \alpha_i x_i$ is a Cauchy sequence and, since X is a Hilbert space,

$$\sum_{n=1}^{\infty} \alpha_n x_n$$

must converge to some $x \in X$.

Proof (2). Consider

$$x = \sum_{n=1}^{\infty} \alpha_n x_n$$

and the partial sum

$$S_n = \sum_{i=1}^{n} \alpha_i x_i .$$

For $n > j$, we define

$$\alpha_j = (S_n , x_j).$$

Since this relationship is true (that is, the value remains the same) for any $n > j$, it must also be true in the limit, and we can write

$$\lim_n (S_n , x_j) = \alpha_j .$$

Appealing to Theorem 9.2, we can assert the continuity of the inner product mapping which allows us to interchange the operations of limit and inner product in the above equation, which yields

$$(x, x_j) = \alpha_j ,$$

which is the desired result. Thus, knowing that the given series converges to some x, we have a strong relationship between the coefficients in the series and x.

Let us now return to the special setting of $L_2(0, 2\pi)$ mentioned at the beginning of this section with the same notation prevailing.

THEOREM 9.6 (*Riesz–Fischer*). Let $c_0 , d_1, c_2 , \ldots$ be a collection of real numbers such that

$$\left(c_0^2 + \sum_{n \text{ even}} c_n^2 + \sum_{n \text{ odd}} d_n^2 \right) < \infty.$$

Then there exists a function $f \in L_2(0, 2\pi)$ such that the numbers c_k and d_k are the Fourier coefficients of f.

Proof. To prove this we need only appeal to part (1) of Theorem 9.5, where the orthonormal set is

$$A: \frac{1}{\sqrt{2\pi}}, \frac{\sin t}{\sqrt{\pi}}, \frac{\cos t}{\sqrt{\pi}}, \ldots = x_0, x_1, x_2, \ldots$$

to assert that [with respect to the norm on $L_2(0, 2\pi)$]

$$\frac{c_0}{\sqrt{2\pi}} + \frac{1}{\sqrt{\pi}} \left(\sum_{n \text{ even}} c_n \cos nt + \sum_{n \text{ odd}} d_n \sin nt \right)$$

converges to some $f \in L_2(0, 2\pi)$ and $f/\sqrt{\pi}$ has c_n and d_n $(n = 0, 1, 2, \ldots)$ as the customary Fourier coefficients.

9.7 Complete Orthonormal Sets

DEFINITION 9.3. Let A be an orthonormal set in the inner product space X. A is said to be *complete* if there exists no other orthonormal set containing A; that is, A must be a maximal orthonormal set.

CRITERION. An orthonormal set A is complete if and only if, for any x such that $x \perp A$, x must be zero.

Proof. Suppose A is complete and x is a nonzero vector such that $x \perp A$. Clearly, this is contradictory because the orthonormal set

$$A \cup \left\{ \frac{x}{\|x\|} \right\}$$

contains A properly and contradicts the maximality of A.

Conversely, suppose the above condition is satisfied, that is, $x \perp A$ implies $x = 0$. If A is not complete, there must exist some orthonormal set B such that $B \supset A$ properly. In this case let $x \in B - A$. Since $\|x\| = 1$ and $x \perp A$, the assumption that such an orthonormal set B exists must be false and A must be complete.

Our next theorem shows that there are complete orthonormal sets in any inner product space.

THEOREM 9.7. Let X be an inner product space (not the trivial space consisting of just the zero vector). Then:

(1) there exist complete orthonormal sets in X;
(2) any orthonormal set can be extended to a complete orthonormal set.

Proof. It is clear that if (2) can be proved, this will imply (1) by virtue of the fact that, in any space, orthonormal sets must exist; for, for any nonzero vector x, the set $\{x/\|x\|\}$ is an orthonormal set. Hence we shall prove (2).

Let E be a given orthonormal set and let S be the collection of all orthonormal sets containing E. It is clear that the collection S is partially ordered by set inclusion (see Appendix to Chapter 9). We wish to show now that every totally ordered subset of S has an upper bound in S; that is, S is inductively ordered. We can then apply Zorn's lemma (see Appendix to Chapter 9), which states that every inductively ordered set must have a maximal element to obtain a maximal orthonormal set.

Let T be any totally ordered subset of S; let

$$T = \{A_\alpha\} \qquad (\alpha \in \Lambda).$$

It is clear that, for any α,

$$A_\alpha \subset \bigcup_\alpha A_\alpha;$$

furthermore,

$$E \subset \bigcup_\alpha A_\alpha.$$

Hence to show that $\bigcup_\alpha A_\alpha \in S$ it remains to show that it is an orthonormal set.

Let x and y be any two members of $\bigcup_\alpha A_\alpha$. This implies there exist A_α and A_β such that

$$x \in A_\alpha \qquad \text{and} \qquad y \in A_\beta.$$

Since T was totally ordered, however, either $A_\alpha \subset A_\beta$ or $A_\beta \subset A_\alpha$; supposing the former inclusion to hold, we can say

$$x, y \in A_\beta,$$

which implies

$$x \perp y \qquad (\|x\| = \|y\| = 1)$$

and, therefore,

$$\bigcup_\alpha A_\alpha \in S.$$

Thus the set $\bigcup_\alpha A_\alpha$ is an upper bound for T in S and S is seen to be inductively ordered. By Zorn's lemma there must exist a maximal element in S. It is clear that no other orthonormal set in the space could contain this maximal element, because this would contradict its maximality.

Recall now the definition of a basis for a vector space. This was a collection of linearly independent elements such that any vector in the space could be expressed as a *finite* linear combination of them; indeed, the general vector space is an entity without limiting notions attached to it, and, for this reason, consideration of anything but finite linear combinations is out of the question. With these notions in mind, we now wish to show that an infinite complete orthonormal set is never a basis in a Hilbert space.

To this end suppose X is a Hilbert space and A is an infinite complete orthonormal set in this space. Since A is assumed to be infinite, we can certainly extract a denumerable sequence of distinct points of A,

$$x_1, x_2, x_3, \dots.$$

Consider now the series

$$\sum_{k=1}^{\infty} \frac{1}{k^2} x_k.$$

Since $\sum_{k=1}^{\infty}(1/k^4)$ converges we can apply Theorem 9.5(1) to assert that

$$\sum_{k=1}^{\infty}\frac{1}{k^2}x_k$$

must converge to some $x \in X$. Suppose now that A were a basis for X. If so, we could write x as some finite linear combination of basis elements; that is, we could write

$$x = \gamma_\alpha x_\alpha + \cdots + \gamma_\nu x_\nu,$$

where

$$x_\alpha, \ldots, x_\nu \in A \qquad \text{and} \qquad \gamma_\alpha, \ldots, \gamma_\nu \in F.$$

Now let j be any other index value different from $\alpha, \ldots, \nu$ and compute

$$\frac{1}{j^2} = \left(x_j, \sum_{k=1}^{\infty}\frac{1}{k^2}x_k\right) = (x_j, \gamma_\alpha x_\alpha + \cdots + \gamma_\nu x_\nu) = 0,$$

which is ridiculous. Thus A cannot be a basis.

THEOREM 9.8. Let X be a Hilbert space, suppose that $A = \{x_\alpha\}_{\alpha \in \Lambda}$, is an orthonormal set in X, and let x be an arbitrary vector in X. Then the following statements are true.

(1) $y = \sum_{\alpha \in \Lambda}(x, x_\alpha)x_\alpha$ exists; that is, the series is summable;
(2) the vector y mentioned in (1) belongs to $\overline{[A]}$;
(3) $x \in \overline{[A]}$ if and only if $x = y$ (it can be written as the above series);
(4) $x - y \perp \overline{[A]}$.

Proof (1). We note that $\sum_\alpha \|(x, x_\alpha)x_\alpha\|^2$ converges, since

$$\sum_\alpha \|(x, x_\alpha)x_\alpha\|^2 = \sum_\alpha |(x, x_\alpha)|^2 \le \|x\|^2 \cdot$$

by the Bessel inequality [Theorem 9.3(1)].

Since this is so, using the same argument used in proving Theorem 9.1, for any $\varepsilon > 0$ there must exist a finite set $K \subset \Lambda$ such that, for any finite subset $J \subset \Lambda$, where $J \cap K = \emptyset$,

$$\sum_{\alpha \in J} \|(x, x_\alpha)x_\alpha\|^2 < \varepsilon.$$

But, since

$$\left\|\sum_{\alpha \in J}(x, x_\alpha)x_\alpha\right\|^2 = \sum_{\alpha \in J}\left\|(x, x_\alpha)x_\alpha\right\|^2$$

we can say

$$\left\|\sum_{\alpha \in J}(x, x_\alpha)x_\alpha\right\|^2 < \varepsilon.$$

Now by Exercise 9.1 we can say

$$\sum_\alpha (x, x_\alpha)x_\alpha$$

exists.

Proof (2). It is clear that any partial sum of $\sum_\alpha (x, x_\alpha)x_\alpha$ must belong to $[A]$. This implies that the limit y must belong to $\overline{[A]}$.

Proof (4). Let $x_\beta \in A$ and consider

$$(x - y, x_\beta) = (x, x_\beta) - (y, x_\beta)$$

$$= (x, x_\beta) - \left(\sum_\alpha (x, x_\alpha)x_\alpha, x_\beta \right).$$

$$= (x, x_\beta) - (x, x_\beta) \qquad \text{(See Exercise 9.3)}$$

$$= 0.$$

Thus

$$x - y \perp A,$$

which implies

$$x - y \perp [A].$$

For any $z \in \overline{[A]}$, there must exist a sequence of elements of $[A]$, $\{v_n\}$, such that

$$z = \lim_n v_n,$$

which means

$$(x - y, v_n) = 0 \qquad \text{(for every n)}.$$

Since this statement is true for any n, it is also true in the limit; thus

$$\lim_n (x - y, v_n) = 0.$$

By the continuity of the inner product mapping (Theorem 9.2), the operations of limit and inner product can be interchanged to yield

$$(x - y, z) = 0,$$

which is the statement of the fact that

$$x - y \perp \overline{[A]},$$

which proves (4).

Proof (3). It is clear that if $x = y$, then $x \in \overline{[A]}$ by virtue of (2). Conversely, suppose that $x \in \overline{[A]}$. Since, by (2), $y \in \overline{[A]}$, we must have

$$x - y \in \overline{[A]}.$$

But, by (4),

$$x - y \perp \overline{[A]},$$

which implies

$$x - y = 0$$

and completes the proof.

Note that in our next theorem we do not need the more special framework of a Hilbert space; an ordinary inner product space suffices. In words, the following theorem says that, if the space spanned by an orthonormal set is dense in the whole space, the orthonormal set must be complete.

THEOREM 9.9. Let X be an inner product space, and suppose the orthonormal set A has the property that $\overline{[A]} = X$. Then A is complete.

Proof. The proof of this result will be by contradiction. It will be shown that, assuming the hypothesis to be true while assuming the conclusion to be false, leads to incompatible results.

If A is not complete, there must be some nonzero vector x, such that $x \perp A$. This implies that

$$x \perp [A],$$

which, in turn, implies

$$x \perp \overline{[A]}.$$

But, since $\overline{[A]} = X$, this implies

$$x \perp x.$$

Consequently, x must be zero, and we have arrived at the desired contradiction.

Our next theorem shows that, if we are working in a Hilbert space, the converse of the above theorem holds.

THEOREM 9.10. Suppose $A = \{x_\alpha\}_{\alpha \in \Lambda}$, is a complete orthonormal set in the Hilbert space X. Then $\overline{[A]} = X$.

Proof. The contrapositive of this result will be proved here. We shall assume that A is an orthonormal set in X and that

$$\overline{[A]} \neq X,$$

which means there is some element $x \in X - \overline{[A]}$. Now, since X is a Hilbert space, we can apply part (1) of Theorem 9.8 to guarantee the existence of

$$\sum_\alpha (x, x_\alpha)x_\alpha = y.$$

Appealing now to part (4) of that theorem, we can say that

$$\frac{x - y}{\|x - y\|} \perp \overline{[A]} \Rightarrow \frac{x - y}{\|x - y\|} \perp A,$$

being assured, by part (3) of Theorem 9.8, that $\|x - y\| \neq 0$. We have now found a nonzero vector orthogonal to all of A. In view of this we can say that A is not complete for

$$A \subset \left\{ \frac{x - y}{\|x - y\|} \right\} \cup A.$$

9.8 Complete Orthonormal Sets and Parseval's Identity

Our next result gives us an equivalent description of a complete orthonormal set in a Hilbert space. Note, however, that to prove part (1) of the following theorem we need only an inner product space.

THEOREM 9.11. Let X be an inner product space and let $A = \{x_\alpha\}_{\alpha \in \Lambda}$, be an orthonormal set in X. Then:

(1) If for any $x \in X$,

$$\|x\|^2 = \sum_\alpha |(x, x_\alpha)|^2, \tag{9.5}$$

then A is complete.

(2) If X is a Hilbert space and A is a complete orthonormal set then Eq. (9.5) holds for any $x \in X$.

REMARK. The equality of Eq. (9.5) is called *Parseval's identity*.

Proof (1). If we assume A not to be complete, there must be some nonzero vector x such that

$$x \perp A.$$

Substituting this x into Parseval's identity, we get

$$\|x\|^2 = \sum_\alpha |(x, x_\alpha)|^2 = 0,$$

which is a contradiction; consequently A must be complete.

Proof (2). By Theorem 9.10, the closure of the subspace generated by a complete orthonormal set in a Hilbert space must be the whole space; hence

$$\overline{[A]} = X.$$

For any vector $x \in X$ we have that $x \in \overline{[A]}$. Applying Theorem 9.8(3), we can now say that

$$x = \sum_\alpha (x, x_\alpha) x_\alpha.$$

To show that Parseval's identity holds, we compute

$$\|x\|^2 = (x, x)$$

$$= \left(\sum_\alpha (x, x_\alpha) x_\alpha, \sum_\beta (x, x_\beta) x_\beta \right)$$

$$= \sum_\alpha \sum_\beta (x, x_\alpha) \overline{(x, x_\beta)} (x_\alpha, x_\beta) \qquad \text{(see Exercise 9.3)}$$

$$= \sum_\alpha |(x, x_\alpha)|^2.$$

Combining the first and last terms, we see that Parseval's identity is indeed satisfied.

We now wish to summarize all our results with the following theorem. Note that even though the special setting of a Hilbert space is assumed, it is not essential in every case. It is felt that, although some of the implications stated below have not been proved so far, the reader can easily supply any missing details.

THEOREM 9.12. Let $A = \{x_\alpha\}_{\alpha \in \Lambda}$ be any orthonormal set in the Hilbert space X. Then

(1) A is complete;

$\Leftrightarrow$ (2) $x \perp A \Rightarrow x = 0$;

$\Leftrightarrow$ (3) $x \in X \Rightarrow x = \sum_\alpha (x, x_\alpha) x_\alpha$;

$\Leftrightarrow$ (4) $\overline{[A]} = X$;

$\Leftrightarrow$ (5) $\|x\|^2 = \sum_\alpha |(x, x_\alpha)|^2$

$\Leftrightarrow$ (6)   for any $x, y \in X$, $(x, y) = \sum_\alpha (x, x_\alpha)(x_\alpha, y)$.

9.9   A Complete Orthonormal Set for $L_2(0, 2\pi)$

Returning once more to the special setting of the Hilbert space $L_2(0, 2\pi)$ (complex-valued functions), we wish to demonstrate that

$$A = \{e^{int}/\sqrt{2\pi} \mid n = 0, \pm 1, \pm 2, \ldots\}$$

constitutes a complete orthonormal set in this space. To do this, we shall select a function $f \in L_2(0, 2\pi)$, suppose that it is orthogonal to A, and show that this implies that f must be zero almost everywhere on the closed interval $[0, 2\pi]$; remembering that we are identifying functions equal almost everywhere, we shall have shown that f is the zero function.

To this end, let $f \in L_2(0, 2\pi)$ and suppose $f \perp A$. This means that

$$\int_0^{2\pi} f(t) e^{-int} \, dt = 0 \qquad \text{(for any } n\text{)}. \tag{9.6}$$

Now define

$$G(t) = \int_0^t f(x) \, dx - C,$$

where the constant C will be determined later. Since

$$\int_0^t f(x) \, dx$$

is an absolutely continuous function of t, we can say that

$$G'(t) = f(t) \qquad \text{(almost everywhere)},$$

which implies

$$\int_0^{2\pi} G'(t) e^{-int} \, dt = 0.$$

Integrating this last expression by parts, we have for $n = \pm 1, \pm 2, \ldots$

$$0 = e^{-int} G(t) \Big|_0^{2\pi} - \int_0^{2\pi} G(t)(-in) e^{-int} \, dt$$

$$= G(2\pi) - G(0) - \int_0^{2\pi} G(t)(-in) e^{-int} \, dt. \tag{9.7}$$

Now, since (9.6) holds for any n (in particular $n = 0$),

$$\int_0^{2\pi} f(x)\, dx = 0,$$

and we have

$$G(2\pi) = -C = G(0).$$

Thus, by (9.7), for all integers $n \neq 0$,

$$\int_0^{2\pi} G(t) e^{-int}\, dt = 0. \tag{9.8}$$

Letting

$$F(t) = \int_0^t f(x)\, dx$$

we see that (9.8) is the same as

$$\int_0^{2\pi} (F(t) - C) e^{-int}\, dt = 0 \tag{9.9}$$

for $n = \pm 1, \pm 2, \ldots$. Choose C now such that

$$\int_0^{2\pi} F(t)\, dt = 2\pi C,$$

that is, such that (9.9) holds for $n = 0$ too. Now by the theorem due to Weierstrass,† we can uniformly approximate $G(t)$ by the trigonometric polynomial

$$T(t) = \sum_{k=-m}^{m} a_k e^{ikt};$$

that is, we can now write, for any given $\varepsilon > 0$,

$$|G(t) - T(t)| < \varepsilon$$

for all $t \in [0, 2\pi]$ for an appropriate m. [Note also that since $G(t)$ is absolutely continuous, it is certainly continuous.] Now, by taking conjugates in (9.8) we can write

$$\int_0^{2\pi} \overline{G(t)} T(t)\, dt = \sum_{k=-m}^{m} a_k \int_0^{2\pi} \overline{G(t)} e^{ikt}\, dt = 0,$$

which implies

$$\int_0^{2\pi} |G(t)|^2\, dt = \int_0^{2\pi} G(t) \overline{G(t)}\, dt$$

$$= \int_0^{2\pi} \overline{G(t)} (G(t) - T(t))\, dt$$

$$\leq \varepsilon \int_0^{2\pi} |G(t)| \cdot 1\, dt$$

$$\leq \varepsilon \sqrt{2\pi} \left(\int_0^{2\pi} |G(t)|^2\, dt \right)^{1/2},$$

† *Weierstrass' Theorem.* Let $f(x)$ be a continuous periodic function of period 2π. For every $\varepsilon > 0$, there exists a trigonometric polynomial $T(x)$ such that $|f(x) - T(x)| < \varepsilon$ for all x.

where the last inequality is obtained by applying the Hölder inequality to the previous term.

Suppose that

$$\left(\int_0^{2\pi} |G(t)|^2 \, dt\right)^{1/2} \neq 0,$$

If so, we can divide through by this term in the previous inequality to get

$$\left(\int_0^{2\pi} |G(t)|^2 \, dt\right)^{1/2} \leq \varepsilon \sqrt{2\pi}$$

or

$$\int_0^{2\pi} |G(t)|^2 \, dt \leq \varepsilon^2 \, 2\pi.$$

Since ε is arbitrary, however, this implies

$$\int_0^{2\pi} |G(t)|^2 \, dt = 0,$$

which, since $G(t)$ is continuous, implies

$$G(t) = 0$$

or

$$\int_0^t f(x) \, dx = C.$$

Differentiating both sides, we get

$$f(t) = 0 \qquad \text{(almost everywhere)},$$

and, noting that we are identifying functions that are equal almost everywhere, the proof is complete.

APPENDIX

In this appendix the notions of partially ordered set, upper bounds, and maximal elements will be discussed. All these notions will then be lumped together to obtain the notion of an inductively ordered set and, once this has been defined, Zorn's lemma will be stated. For our purposes it suffices to consider Zorn's lemma an axiom.

DEFINITION A.1. Let X be a set and let the relation $\mathscr{R}$ be defined between some elements of this set. X is said to be *partially ordered under $\mathscr{R}$* if the following conditions are satisfied among the elements of X that are "comparable" with respect to $\mathscr{R}$.

(1) Let $a \in X$. Then $a\mathcal{R}a$. (reflexive);
(2) Let $a, b, c \in X$. Then $a\mathcal{R}b$ and $b\mathcal{R}c \Rightarrow a\mathcal{R}c$ (transitive);
(3) For $a, b \in X$, if $a\mathcal{R}b$ and $b\mathcal{R}a$, then $a = b$.

Furthermore if for any two elements of X one of the relations

$$a\mathcal{R}b \quad \text{or} \quad b\mathcal{R}a$$

holds, then X is said to be *totally ordered* under $\mathcal{R}$.

Some examples of partially ordered sets will now be considered.

EXAMPLE A.I. Let the set X be the real numbers, and let $\mathcal{R}$ be $\leq$. It is clear that, for any real numbers a, b and c,

(1) $a \leq a$,
(2) $a \leq b$ and $b \leq c \Rightarrow a \leq c$,
(3) $a \leq b$ and $b \leq a \Rightarrow a = b$.

Consequently, $\leq$ is a partially ordering for the real numbers. In this case we note that the real numbers are actually totally ordered.

EXAMPLE A.2. Let the set X consist of the points in Fig. 9.1. For any two elements x and y, of X, we shall say $x\mathcal{R}y$ if y can be reached from x by a strictly ascending path or standing still. Thus, referring to Fig. 9.1, $c\mathcal{R}a$ and $c\mathcal{R}d$. It is easily verified that the requirements (1)–(3) are met and it is further noted that the points b and e are not "comparable" with respect to $\mathcal{R}$; that is, neither $b\mathcal{R}e$ nor $e\mathcal{R}b$ holds. Consequently, X is not totally ordered with respect to $\mathcal{R}$.

FIG. 9.1.

EXAMPLE A.3. Let X be any set and S be any collection of subsets of X. It is clear that, taking set inclusion as $\mathcal{R}$,

(1) for any $A \in S$, we have $A \subset A$;
(2) if $A, B, C \in S$, $A \subset B$ and $B \subset C$, then $A \subset C$;
(3) for $A, B \in S$, if $A \subset B$ and $B \subset A$, then $A = B$. Indeed this is the definition of set equality.

Having seen that set inclusion is a partial ordering on S, we note that, if two sets are disjoint, they are not "comparable" with respect to $\mathcal{R}$; consequently S is not totally ordered.

If one takes as the prototype that partial ordering mentioned in Example A.1, it seems natural to wonder about such notions as the "largest" member of a given set and this motivates our next two definitions.

DEFINITION A.5. Let X be partially ordered under $\mathscr{R}$ and let A be a subset of X. The element $a \in X$ (but not necessarily to A) is said to be an *upper bound for A* if for *all $y \in A$*,

$$y \mathscr{R} a.$$

Note that we require an upper bound for a set to be "comparable" to every member of the set.

DEFINITION A.6. Let X be as in the previous definition. The element $a \in X$ is said to be a *maximal element of X* if $a \mathscr{R} y$ implies a must equal y.

Loosely speaking, we do not require a maximal element to be the biggest element in a given set; we require only that nothing be bigger than it.

It is further noted that nothing unique about the maximal element or the upper bound is required. In a given situation there may be many of each. For example, in Example A.2 the elements a and d are each maximal elements of the set.

In Example A.3, if we extend the partial ordering to the collection of all subsets of X, it is clear that the union of all the sets in S is an upper bound for S and that any other set in X containing S is also an upper bound for S or any subset thereof. This union may not be a maximal element for S though, because it may not be a member of S.

DEFINITION A.7. The partially ordered set X, under $\mathscr{R}$, is said to be *inductively ordered* if any totally ordered subset of X has an upper bound in X.

With these notions in mind we can now state the following theorem, which is equivalent to the axiom of choice.

ZORN'S LEMMA. Every nonempty inductively ordered set has a maximal element.

Needless to say, this statement is extremely useful when one wishes to make some statement about the *existence* of "maximal" elements in a given set. The key feature in successfully utilizing Zorn's lemma is to define a partial ordering in such a way as to fit the particular problem at hand. For example, suppose one wishes to prove the existence of maximal ideals in a ring with identity. By considering the set of all ideals of the ring and taking set inclusion to be the partial ordering, one can show that this set is inductively ordered and can, thus, assert the existence of maximal ideals without resorting to trying to construct one.

As another application of Zorn's lemma, suppose we wish to prove that every vector space has a basis. In this case we would consider the set of all linearly independent subsets of the vector space. By Example A.3, it is clear that set inclusion induces a partial ordering on this collection, and it can be further verified that any totally ordered subset has an upper bound in the collection. Zorn's lemma can now be applied to assert the existence of a maximal linearly independent subset of the space. If this maximal element were not a basis for the space, there would have to be some element x of the space that could not be expressed as a linear combination of the members of the maximal linearly independent subset; the union of x and this set would then constitute a "larger" (under set inclusion) linearly independent subset and would contradict the maximality of that subset. Hence the maximal element must be a basis for the space.

EXERCISES 9

1. Show that, if X is a Banach space and if for every $\varepsilon > 0$, there exists a finite set K of indices such that $\left\| \sum_{\alpha \in J} x_\alpha \right\| < \varepsilon$ for any finite set of indices J such that $J \cap K = \emptyset$, then $\{x_\alpha\}_{\alpha \in \Lambda}$ is summable. (Note how very much this is like the Cauchy criterion for convergence of a sequence in a complete space.)

2. Let X be a Banach space. Show that, if $\sum_{n=1}^{\infty} x_n$ is absolutely convergent (that is, $\sum_{n=1}^{\infty} \|x_n\|$ is convergent), then $\sum_{n=1}^{\infty} x_n$ converges and $\left\| \sum_{n=1}^{\infty} x_n \right\| \le \sum_{n=1}^{\infty} \|x_n\|$.

3. Show that, if X is an inner product space and if $\sum_\alpha x_\alpha = x$, then $\sum_\alpha (x_\alpha, y) = (x, y)$ and $\sum_\alpha (y, x_\alpha) = (y, x)$ for every $y \in X$.

4. If X is a real inner product space and if $\|x + y\|^2 = \|x\|^2 + \|y\|^2$, show that $x \perp y$. Show that this is not necessarily true in a complex inner product space.

5. If X is a complex inner product space and if $x, y \in X$, show that $x \perp y$ if and only if $\|\alpha x + \beta y\|^2 = \|\alpha x\|^2 + \|\beta y\|^2$ for all scalars α and β.

6. Let X be an inner product space and $\{x_n\}$ a sequence of elements of X. If $\|x_n\| \to \|x\|$ and $(x_n, x) \to (x, x)$, show that $x_n \to x$.

7. Let X be an inner product space and A be an orthonormal set of vectors of X such that $(x, y) = (x, z)$ for all $x \in A$ implies $y = z$. Prove that A is complete and conversely.

8. In the Hilbert space l_2, prove that $(1, 0, 0, \ldots), (0, 1, 0, 0, \ldots), (0, 0, 1, 0, 0, \ldots), \ldots$, is a complete orthonormal set.

9. Let X be a separable Hilbert space and M a closed subspace of X. Prove that M is a separable Hilbert space.

10. Show that if $y_1, y_2, \ldots, y_n, \ldots$ is a sequence of vectors in the Hilbert space X such that every $x \in X$ is a linear combination of finitely many of the y_i then X is finite-dimensional.

11. Give an example of a sequence $\{x_n\}$ such that $\sum_{n=1}^{\infty} \|x_n\|^2 < \infty$ but $S_n = \sum_{i=1}^{n} x_i$ is not a Cauchy sequence.

12. Show that, if X is a Hilbert space and A is a countable complete orthonormal set, then X is separable.

13. If X is an inner product space which contains a finite complete orthonormal set $\{x_1, \ldots, x_n\}$, show that X is finite-dimensional.

14. Let X be an inner product space and y a fixed vector of X. Prove that the functional $f(x) = (x, y)$ for all $x \in X$ is a continuous linear functional.

15. Prove that l_p, $p \neq 2$, with usual norm, is not an inner product space.

16. Let X be a Hilbert space and let $A = \{x_\alpha \mid \alpha \in \Lambda\}$ be an orthogonal family of vectors. Show that $\{x_\alpha\}$ is summable if and only if $\{\|x_\alpha\|^2\}$ is summable and, if $x = \sum_\alpha x_\alpha$, then $\|x\|^2 = \sum_\alpha \|x_\alpha\|^2$.

17. Let X be a Hilbert space and let S be a convex set in X. Let $d = \inf_{x \in S} \|x\|$. Prove that, if $\{x_n\}$ is a sequence of elements of S such that $\lim_n \|x_n\| = d$, then $\{x_n\}$ is a Cauchy sequence.

18. Using Exercise 17, show that any closed convex subset S of a Hilbert space X contains a unique element of minimal norm.

19. Let f be a linear functional defined on a subspace M of the linear space X. Prove that f can be extended to the whole space; that is, there exists a linear functional $\hat{f}$ defined on all X which extends f:

$$\hat{f} : X \to F$$
$$|$$
$$f : M \to F$$

(*Hint:* See Lemma and Theorem 11.1.)

20. Consider the class of continuous functions on the closed interval $[a, b]$, $C[a, b]$. Show that the relation, for $f, g \in C[a, b]$, $f \leq g$ if and only if $f(t) \leq g(t)$ for all $t \in [a, b]$ is a partial ordering on $C[a, b]$.

21. Prove that every vector space has a basis.

REFERENCES

1. P. Halmos, "Introduction to Hilbert Space."
2. A. Taylor, "Introduction to Functional Analysis."
3. A. Zaanen, "Linear Analysis."

Material on Zorn's lemma
4. H. Rubin and J. Rubin, "Equivalents of the Axiom of Choice."
5. J. Kelley, "General Topology."

CHAPTER 10

Complete Orthonormal Sets

The principal topics of discussion in this chapter will be complete orthonormal sets and Hilbert spaces. A feature to note is that the special setting of a Hilbert space is not necessary to all our theorems; in many the more general notion of inner product space suffices. For example in Theorem 9.11(1) it was shown that, in any inner product space, if Parseval's identity held for every vector in the space, then the orthonormal set in question must be complete. This result will be extended a little further in this chapter, where it will be shown that, with a particular orthonormal set in mind, if Parseval's identity is satisfied by every vector in a dense subset of the space, then the orthonormal set is complete. Using this fact, we shall see that an alternate proof that the functions

$$\frac{1}{\sqrt{2\pi}}, \frac{\cos t}{\sqrt{\pi}}, \frac{\sin t}{\sqrt{\pi}}, \cdots$$

constitute a complete orthonormal set for $L_2(0, 2\pi)$. If one compares this proof to the one given in the preceding chapter, it seems that the newer proof is the simpler. This is somewhat illusory, however, for this proof rests very heavily on the fact that the trigonometric polynomials are dense in $L_2(0, 2\pi)$—a somewhat deeper fact than anything used in the previous proof.

A familiar result from linear algebra is that any finite-dimensional vector space, of dimension n say, over a field F is isomorphic to F^n—all n-tuples of entries from that field. Here it will be shown that any complex n-dimensional inner product space is congruent to C^n; that is, it is isometric as well as isomorphic.

Toward the end of the chapter, the special setting of a Hilbert space becomes indispensable, and the last few theorems just could not be pushed through in the general inner product space. One result about Hilbert spaces of particular interest is the projection theorem for Hilbert spaces—the fact that if M is any closed subspace of a Hilbert space X, then X can be decomposed into the direct sum decomposition of M and its orthogonal complement $M^\perp$.

10.1 Complete Orthonormal Sets and Parseval's Identity

THEOREM 10.1. Let S be a dense subset of the inner product space X, and let $A = \{x_\alpha\}$, where α runs through the index set Λ, be an orthonormal set in X. If, for any $y \in S$

$$\|y\|^2 = \sum_\Lambda |(y, x_\alpha)|^2$$

(that is, Parseval's identity holds for every $y \in S$), then A is complete.

Proof. By part (1) of Theorem 9.11, we know that, if we can show Parseval's identity to hold for every vector in the space, we can conclude that A is complete, and this is what will be shown in this proof. To this end, let x be any vector in X. Since S is a dense subset, for any preassigned $\varepsilon > 0$, there is some $y \in S$ such that

$$\|y - x\| < \varepsilon.$$

Since $y \in S$, we can apply the hypothesis to assert that

$$\|y\|^2 = \sum_\Lambda |(y, x_\alpha)|^2. \tag{10.1}$$

Now let us note that, for any finite subset of Λ, J,

$$\left\| y - \sum_J (y, x_\alpha)x_\alpha \right\|^2 = \|y\|^2 - \sum_J |(y, x_\alpha)|^2 - \sum_J |(y, x_\alpha)|^2 + \sum_J |(y, x_\alpha)|^2$$

$$= \|y\|^2 - \sum_J |(y, x_\alpha)|^2$$

$$\leq | \|y\|^2 - \sum_J |(y, x_\alpha)|^2 |.$$

Since the series (10.1) is summable, the last term above can be made arbitrarily small, and by taking a suitable finite subset K of Λ, we can say that, for any finite subset $J \supset K$,

$$\left\| y - \sum_J (y, x_\alpha)x_\alpha \right\| < \varepsilon. \tag{10.2}$$

Now consider

$$\left\| \sum_J (y, x_\alpha)x_\alpha - \sum_J (x, x_\alpha)x_\alpha \right\|^2 = \left\| \sum_J (y - x, x_\alpha)x_\alpha \right\|^2$$

$$= \sum_J |(y - x, x_\alpha)|^2$$

$$\leq \|y - x\|^2 < \varepsilon^2,$$

the next to the last inequality being obtained by applying the Bessel inequality (Theorem 9.3) to the vector $x - y$. Combining all the above results after adding and subtracting some terms and applying the triangle inequality, we have

$$\left\| x - \sum_J (x, x_\alpha)x_\alpha \right\| \leq \|x - y\| + \left\| y - \sum_J (y, x_\alpha)x_\alpha \right\|$$

$$+ \left\| \sum_J (y - x, x_\alpha)x_\alpha \right\| < 3\varepsilon,$$

where J is as in (10.2). This implies now that $\{(x, x_\alpha)x_\alpha\}$ is summable to x or that

$$x = \sum_\Lambda (x, x_\alpha)x_\alpha.$$

We can now apply Exercise 9.3 to assert that

$$(x, x) = \|x\|^2 = \sum_\Lambda |(x, x_\alpha)|^2.$$

Thus we have shown that Parseval's identity holds for every vector in the space; A, therefore, is complete.

The beauty of the above result is that if we wish to check whether a given orthonormal set is complete, we need not check Parseval's identity for any vector in the space, we need only check a dense subset. Not only has this cut down immensely the number of things to check, but in many cases there will be a dense subset available that lends itself very readily to the particular situation, as shown by the following example.

A Complete Orthonormal Set in $L_2(0, 2\pi)$

It will be shown here that the set

$$\frac{1}{\sqrt{2\pi}}, \frac{\cos t}{\sqrt{\pi}}, \frac{\sin t}{\sqrt{\pi}}, \ldots = x_0, x_1, x_2, \ldots \tag{10.3}$$

is a complete orthonormal set in the real vector space $L_2(0, 2\pi)$. To facilitate the discussion we make the following definition.

DEFINITION 10.1. Any function of the form

$$T(t) = a + \sum_{k=1}^{n} (c_k \cos kt + d_k \sin kt)$$

is said to be a *trigonometric polynomial of degree n.*

It can be shown that the class of all trigonometric polynomials S is dense in $L_2(0, 2\pi)$†; thus if we can show that Parseval's identity holds for every vector in S, we can apply the preceding result to conclude that the proposed orthonormal set is complete. For a typical vector in S,

$$y = T(t) = a + \sum_{k=1}^{n} (c_k \cos kt + d_k \sin kt),$$

let us compute the square of the norm of y:

$$\|y\|^2 = (y, y) = 2\pi a^2 + \pi \sum_{k=1}^{n} (c_k{}^2 + d_k{}^2). \tag{10.4}$$

On the other hand, let us now compute

$$\sum_{n=1}^{\infty} |(y, x_n)|^2 = 2\pi a^2 + \sum_{k=1}^{n} \left(\left(\frac{c_k}{\sqrt{\pi}} \pi \right)^2 + \left(\frac{d_k}{\sqrt{\pi}} \pi \right)^2 \right). \tag{10.5}$$

Since expressions (10.4) and (10.5) are the same, we conclude that the orthonormal set is complete. Further, we can say that Parseval's identity holds for every vector in the space or that

$$\frac{a_0{}^2}{2} + \sum_{n=1}^{\infty} (a_n{}^2 + b_n{}^2) = \frac{1}{\pi} \int_0^{2\pi} |f(t)|^2 \, dt,$$

where the $a_0, a_1, a_2, \ldots, a_n, \ldots$ are the ordinary Fourier series coefficients mentioned in Sec. 9.6.

Now, since the orthonormal set is complete and $L_2(0, 2\pi)$ is a Hilbert space, we can appeal to Theorem 9.10 to say that the closure of the space spanned by the orthonormal set must be the whole space. Then, by part (3) of Theorem 9.8 for any $f \in L_2(0, 2\pi)$,

$$f = \sum_{n=1}^{\infty} (f, x_n) x_n,$$

† For a proof see I. P. Natanson, *Theory of Functions of a Real Variable*, Vol. 1, Theorem 6, p. 172. Actually it follows directly from this that the set A given in (10.3) is complete for $[A]$ equals the set of all trigonometric polynomials; therefore by the denseness, $\overline{[A]} = X$, whence A is complete by Theorem 9.9. We proceed as we do in order to show how Theorem 10.1 is frequently applied.

where the x_n are as in (10.3). In terms of the Fourier series coefficients, we can write

$$f(t) = \frac{a_0}{2} + a_1 \cos t + a_2 \cos 2t + \cdots$$

$$+ b_1 \sin t + b_2 \sin 2t + \cdots$$

and note that every function in the space possesses such an expansion. A fact of great importance is that the above series converges in the norm on $L_2(0, 2\pi)$; that is, in the mean. This does *not* imply that the Fourier series actually converges to the function point by point or even that it converges pointwise to the function almost everywhere, the latter still being an open question. The only thing about pointwise convergence that can be definitely inferred from the above is that there is a subsequence (of the sequence of partial sums, that is) that converges to the function f, pointwise almost everywhere. (See Exercise 10.7.)

10.2　The Cardinality of Complete Orthonormal Sets

In the following theorem the symbol a will denote the power of a denumerable set and $\bar{A}$ will denote the power of the set A. In addition, if A and B are two sets, the following two facts about the arithmetic of cardinal numbers will be used:

(1) $\bar{A}a = \bar{A}$ for any set A of infinite cardinality;
(2) If $\bar{A} \le \bar{B}$ and $\bar{B} \le \bar{A}$, then $\bar{A} = \bar{B}$.

We now show that all complete orthonormal sets in an inner product space have the same cardinality.

THEOREM 10.2. Let $A = \{x_\alpha\}_{\alpha \in \Lambda}$, and $B = \{y_\beta\}_{\beta \in \Gamma}$, be complete orthonormal sets in the inner product space X. Then $\bar{A} = \bar{B}$.

Proof. For any $x_\alpha \in A$ the set

$$B_{x_\alpha} = \{y_\beta \in B | (x_\alpha, y_\beta) \ne 0\}$$

must be countable [Theorem 9.3(2)]. We now claim that any $y_\beta \in B$ must belong to some B_{x_α} for if it did not, this would mean that

$$(x_\alpha, y_\beta) = 0 \qquad \text{(for all } \alpha).$$

In other words, the vector y_β is orthogonal to all of A. Since A is complete, however, this means that $y_\beta = 0$, which is impossible because $\|y_\beta\| = 1$. Hence $y_\beta \in B$ must belong to some B_{x_α}. In view of this, we can write B as

$$B = \bigcup_{\alpha \in \Lambda} B_{x_\alpha}$$

which implies that

$$\bar{B} \le \bar{A}a = \bar{A}.$$

(Note that we are assuming both A and B to be infinite, the theorem being trivial otherwise.) In an exactly analogous procedure, we arrive at

$$\bar{\bar{A}} \le \bar{\bar{B}}$$

and conclude that

$$\bar{\bar{A}} = \bar{\bar{B}},$$

the desired result.

10.3   A Note on the Structure of Hilbert Spaces

THEOREM 10.3. Let X be a complex inner product space of dimension n. Then X is congruent to C^n.

Proof. Let an orthonormal basis for X be given by $x_1, x_2, x_3, ..., x_n$. Any vector $x \in X$ can now be written as the linear combination

$$x = \sum_{i=1}^{n} \alpha_i x_i \qquad (\alpha_i \in C).$$

We now define the linear transformation A to be

$$A: X \to C^n,$$

$$x \to (\alpha_1, \alpha_2, ..., \alpha_n).$$

It is easy to verify that A is an isomorphism. We wish to show that A is a norm-preserving mapping as well (the norm on C^n being that mentioned in Example 1.1), and for this purpose we compute

$$\|x\|^2 = \left(\sum_{i=1}^{n} \alpha_i x_i, \sum_{i=1}^{n} \alpha_i x_i \right) = \sum_{i=1}^{n} |\alpha_i|^2 = \|Ax\|^2,$$

which proves that the mapping is an isometry and establishes the theorem.

REMARK. If X is a separable inner product space, then X contains a countable complete orthogonal set. To prove this we need only note that by virtue of Theorem 9.7 complete orthonormal sets must exist in X. But, by Theorem 9.4, an orthonormal set in a separable space must be countable. With this in mind we now prove Theorem 10.4.

THEOREM 10.4. Let X be a separable, complex Hilbert space of infinite dimension. Then X is congruent to l_2.

Proof. Because of the Remark above, we can safely assume the existence of a complete orthonormal set for $X: x_1, x_2, ..., x_n, ...$. For any $x \in X$ we now define

$$\alpha_n = (x, x_n).$$

Since the orthogonal set is complete and X is a Hilbert space, Parseval's identity must hold [Theorem 9.11(2)] for every x or

$$\|x\|^2 = \sum_{n=1}^{\infty} |(x, x_n)|^2 = \sum_{n=1}^{\infty} |\alpha_n|^2 < \infty.$$

Consider now the mapping

$$A: X \to l_2$$

$$x \to (\alpha_1, \alpha_2, ..., \alpha_n, ...).$$

Since the inner product is linear in the first argument, it is clear that A is a linear transformation. Since it is linear, to prove that it is $1 : 1$ we need only show that the zero vector of X is the only thing that maps into the zero vector of l_2. Hence, suppose

$$Ax = (\alpha_1, \alpha_2, ..., \alpha_n, ...) = 0,$$

which implies that

$$\alpha_n = (x, x_n) = 0$$

for every n. Since the orthonormal set is complete, this means that x must be zero and proves that the mapping is $1 : 1$.

To prove that the mapping is onto, consider an arbitrary element

$$(\beta_1, \beta_2, ..., \beta_n, ...) \in l_2.$$

Since

$$\sum_{n=1}^{\infty} |\beta_n|^2 < \infty,$$

then $\sum_{n=1}^{\infty} \beta_n x_n$ converges to some $x \in X$ and furthermore

$$\beta_n = (x, x_n)$$

by Theorem 9.5(2). Thus, since

$$Ax = (\beta_1, \beta_2, ..., \beta_n, ...),$$

we have established the ontoness of the mapping and shown that A is an isomorphism. Lastly, by Parseval's identity, we note that

$$\|Ax\|^2 = \sum_{n=1}^{\infty} |\alpha_n|^2 = \|x\|^2,$$

which proves that A is an isometry as well as an isomorphism and proves that the two spaces are congruent.

Now, in view of Theorems 10.3 and 10.4, we have the option of viewing any finite-dimensional inner product space (which is a separable Hilbert space) and any separable Hilbert space in a different setting. Although we shall rarely have occasion to do this, we note that when pondering whether some property holds for a separable Hilbert space, it might turn out that the answer to the problem may turn up in some already proved theorem about C^n or l_2 (see Exercise 10.1 for the extension of these results to inseparable Hilbert spaces).

10.4 Closed Subspaces and the Projection Theorem for Hilbert Spaces

THEOREM 10.5. Let M be a closed subspace of the Hilbert space X, and let $x \in X$. Then, denoting the distance between x and M by $d = \inf_{y \in M} \|x - y\|$ there exists a vector $y \in M$ such that $\|y - x\| = d$.

Proof. Since, by definition,

$$d(x, M) = d = \inf_{z \in M} \|z - x\|,$$

there must exist a sequence of points of M, $\{y_n\}$, such that

$$\|y_n - x\| \to d.$$

Consider the two vectors $y_n - x$ and $y_m - x$ now. By the parallelogram law, we can say

$$\|y_n + y_m - 2x\|^2 + \|y_n - y_m\|^2 = 2\|y_n - x\|^2 + 2\|y_m - x\|^2$$

or

$$\|y_n - y_m\|^2 = 2\|y_n - x\|^2 + 2\|y_m - x\|^2 - 4\left\|\frac{y_n + y_m}{2} - x\right\|^2.$$

We now note that since

$$\frac{y_n + y_m}{2} \in M,$$

then

$$\left\|\frac{y_n + y_m}{2} - x\right\| \geq d,$$

which implies

$$\|y_n - y_m\|^2 \leq 2\|y_n - x\|^2 + 2\|y_m - x\|^2 - 4d^2.$$

Taking the limit as $n, m \to \infty$ and noting that $\|y_n - x\|$ and $\|y_m - x\|$ each approach d, we conclude that

$$\|y_n - y_m\| \to 0$$

or that $\{y_n\}$ is a Cauchy sequence. Since X is a Hilbert space, there must be some $y \in X$ such that $y_n \to y$ and, further, since M is a closed subspace, y must belong to M. Since the norm is a continuous mapping, we can now say that

$$d = \lim_n \|y_n - x\| = \|\lim_n y_n - x\| = \|y - x\|.$$

The proof is now complete.

In the following theorems we shall have occasion to make use of the vector whose existence was proved in the above theorem many times. In particular, note the role it plays in our next theorem.

THEOREM 10.6. Let M be a closed subspace of the Hilbert space X, and let N be a subspace that properly contains M. Then there exists a nonzero vector $w \in N$ such that $w \perp M$.

Proof.† Since the inclusion $M \subset N$ is proper, there must exist an $x \in N - M$. For

---

† Actually we could give a shorter, less constructive proof of this theorem using the results of the preceding chapter; namely by Theorem 9.7 there exists a complete orthonormal set A in M and since M is a closed subspace of the Hilbert space X, and therefore a Hilbert space, $\overline{[A]} = M$ by Theorem 9.10. Now since $N \supset M$ properly, by Theorem 9.7 we can extend A to a complete orthonormal set B of N and $B \supset A$ properly. Let $y \in B - A$; then $y \perp A$ and therefore $y \perp \overline{[A]} = M$, y being of course not 0. We choose the proof above because of its more elementary nature.

this x let $d = d(x, M)$. Applying the preceding theorem, there must be some $y \in M$ such that $d = \|y - x\|$. Now we define

$$w = y - x.$$

Clearly, $w \neq 0$ for $y \in M$ and $x \in N - M$, which precludes the possibility of y being equal to x. We wish now to show that w is orthogonal to every vector in M. For this purpose let z be any element of M, let α be a scalar, and consider

$$\|w + \alpha z\| = \|y - x + \alpha z\| = \|y + \alpha z - x\| \geq d = \|w\|,$$

the last inequality following from the definition of $d = d(x, M)$ and the fact that $y + \alpha z \in M$. Thus,

$$0 \leq \|w + \alpha z\|^2 - \|w\|^2 = (w + \alpha z, w + \alpha z) - (w, w)$$
$$= \bar{\alpha}(w, z) + \alpha(z, w) + |\alpha|^2 \|z\|^2 \qquad (10.6)$$

for any scalar α. Now let α assume the particular value $\beta(w, z)$, where β is real. If we substitute this for α, the last expression becomes

$$2\beta|(w, z)|^2 + \beta^2|(w, z)|^2\|z\|^2 = \beta|(w, z)|^2(2 + \beta\|z\|^2).$$

If $(w, z) \neq 0$, we can clearly choose β so that

$$\beta|(w, z)|^2(2 + \beta\|z\|^2) < 0$$

and contradict (10.6). Hence (w, z) must be zero for any $z \in M$ or $w \perp M$.

We now wish to list some simple observations about Hilbert spaces.

(1) If X was any inner product space and S was an arbitrary subset, then $S^\perp$ was a subspace. We now note that it is more: It is a closed subspace. To prove this, let y be any member of the closure of $S^\perp$. This implies that there is a sequence of points of $S^\perp$, $\{y_n\}$, such that

$$y_n \to y.$$

Thus, for any $x \in S$ and any n

$$0 = (x, y_n).$$

Since this is true for any n, it must also be true in the limit and, since the inner product is a continuous mapping (Theorem 9.2), we have

$$0 = \lim_n(x, y_n) = (x, \lim_n y_n) = (x, y)$$

which proves assertion (1).

(2) For any inner product space X and any subset S,

$$S \cap S^\perp \subset \{0\},$$

the inclusion going the other way as well if S is a subspace.

(3) For any inner product space X and any subset S, $S \subset S^{\perp\perp}$.

(4) For any inner product space X and any subsets S_1 and S_2 such that $S_1 \subset S_2$,

$$S_1{}^\perp \supset S_2{}^\perp.$$

For our next result the framework of Hilbert space is needed.

(5) If M is a closed subspace of the Hilbert space X, then $M = M^{\perp\perp}$.

Proof. By (3) we know that $M \subset M^{\perp\perp}$. Suppose now that this inclusion were proper; that is, $M \neq M^{\perp\perp}$. Since M is a closed subspace of a Hilbert space, we can apply Theorem 10.6 to assert the existence of a nonzero vector $w \in M^{\perp\perp}$ such that $w \perp M$ or $w \in M^{\perp}$. But we now have

$$w \in M^{\perp} \cap M^{\perp\perp},$$

and since

$$M^{\perp} \cap M^{\perp\perp} = \{0\},$$

w must be zero—a contradiction. Thus the inclusion cannot be proper and we must have $M = M^{\perp\perp}$.

(6) If X is an inner product space and S is any subset, then $S^{\perp} = S^{\perp\perp\perp}$.

(7) If X is a Hilbert space and S is any subset, then $S^{\perp\perp} = \overline{[S]}$.

Proof. Clearly, by (3) and (1), $S^{\perp\perp}$ is a closed subspace containing S; hence it must certainly contain the smallest closed subspace containing S, $\overline{[S]}$.

It is clear that $S \subset \overline{[S]}$. By (4) this implies that

$$S^{\perp} \supset \overline{[S]}^{\perp}$$

and that

$$S^{\perp\perp} \subset \overline{[S]}^{\perp\perp}.$$

But, noting that $\overline{[S]}$ is a closed subspace of a Hilbert space, we can apply (5) to assert that $\overline{[S]} = [\overline{S}]^{\perp\perp}$ or that $S^{\perp\perp} \subset \overline{[S]}$ which completes the proof.

THEOREM 10.7. Let M and N be closed subspaces of the Hilbert space X and suppose that $M \perp N$ (this immediately implies that $M \cap N = \{0\}$). Then $M + N = \{x + y \mid x \in M, y \in N\}$ is a closed subspace.

Proof. It is clear that $M + N$ is a subspace; thus all we have to show is that it closed. Let

$$w \in \overline{M + N}.$$

There must be a sequence of points of $M + N$, $\{w_n\}$, such that $w_n \to w$. Since each $w_n \in M + N$, we have the representation

$$w_n = x_n + y_n,$$

where $x_n \in M$ and $y_n \in N$. Using this representation, let us examine the norm of

$$w_n - w_m = (x_n - x_m) + (y_n - y_m).$$

Since $(x_n - x_m) \perp (y_n - y_m)$, we can apply the Pythagorean theorem (see Theorem 1.5) to assert that

$$\|w_n - w_m\|^2 = \|x_n - x_m\|^2 + \|y_n - y_m\|^2.$$

Since the term on the left must go to zero as $n, m \to \infty$, so must each of the terms on the right; hence $\{x_n\}$ and $\{y_n\}$ are Cauchy sequences. Using simultaneously the facts that X is a Hilbert space and that M and N are closed subspaces, we can assert the existence of vectors $x \in M$ and $y \in N$ such that

$$x_n \to x \quad \text{and} \quad y_n \to y.$$

Now, since addition is a continuous operation, we can say that

$$x_n + y_n \to x + y.$$

Thus,

$$w = x + y \in M + N.$$

THEOREM 10.8 (*projection theorem*). If M is a closed subspace of the Hilbert space X, we have the direct sum decomposition of X, $X = M \oplus M^\perp$.

REMARK. Note that in Theorem 1.8 it was shown that if M was a finite-dimensional subspace of any inner product space, such a decomposition of the space was possible.

Proof. Since it is clear that $M \cap M^\perp = \{0\}$, we need only check that $X = M + M^\perp$. To this end, consider

$$N = M + M^\perp.$$

By Theorem 10.7, it is clear that N is a closed subspace; further, we have

$$M \subset N \quad \text{and} \quad M^\perp \subset N,$$

which, by (4), implies that

$$N^\perp \subset M^\perp \quad \text{and} \quad N^\perp \subset M^{\perp\perp}$$

or

$$N^\perp \subset M^\perp \cap M^{\perp\perp} = \{0\}.$$

Therefore,

$$N^\perp = \{0\}$$

and, by (5),

$$N = N^{\perp\perp} = \{0\}^\perp = X.$$

The proof is now complete.

Vital to the truth of this result is the completeness of X; a counterexample is offered in Exercise 10.8.

EXERCISES 10

1. Let X be a Hilbert space and $A = \{x_\alpha\}_{\alpha \in \Lambda}$ a complete orthonormal set. If the cardinality of A is γ, and, if S is any set with cardinality γ, show that X is congruent to $L_2(S)$. (The space $L_2(S)$ is first discussed in Sec. 9.4.)

2. Let X be an inner product space and S a subset of X. Show that $\overline{[S]}^{\perp} = S^{\perp}$.

3. If M is a closed subspace of the Hilbert space X, prove that X/M is isomorphic to $M^{\perp}$.

4. If M is a complete subspace of the inner product space X, prove that $X = M^{\perp} \oplus M$ and that $M^{\perp\perp} = M$.

5. Consider the real inner product space $L_2(-1, 1)$. Show that in this space the set of Legendre polynomials constitutes a complete orthonormal set.

6. Let M be a subspace of the Hilbert space X. Prove that M is dense in X if and only if $y \perp M \Rightarrow y = 0$ (that is, $M^{\perp} = \{0\}$).

7. As an example of a sequence of functions that converge in the mean to a function but do not converge pointwise almost everywhere to that function, consider the following: On the half-open interval $[0, 1)$ define the k functions

$$f_1^{(k)}(x), f_2^{(k)}(x), \ldots, f_k^{(k)}(x),$$

for every natural number k according to the following definition:

$$f_i^{(k)}(x) = \begin{cases} 1 & \text{for} \quad x \in \left[\dfrac{i-1}{k}, \dfrac{i}{k}\right), \\ 0 & \text{for} \quad x \notin \left[\dfrac{i-1}{k}, \dfrac{i}{k}\right) \end{cases}$$

[let $f_1^{(1)}(x) = 1$ on $[0, 1)$]. We now obtain

$$\varphi_1(x) = f_1^{(1)}(x), \quad \varphi_2(x) = f_1^{(2)}(x), \quad \varphi_3(x) = f_2^{(2)}(x), \quad \varphi_4(x) = f_1^{(3)}(x), \ldots .$$

The reader should verify that $\{\varphi_n(x)\}$ converges in the 2-norm on $L_2(0, 1)$ to the zero function and also that $\{\varphi_n(x)\}$ does not converge to 0 pointwise almost everywhere.

8. Let X be the following inner product space: The elements of X are all sequences of complex numbers $(\alpha_1, \alpha_2, \ldots)$ such that $\alpha_n = 0$ for all n greater than some N. (The number N is not assumed to be the same for each sequence.) The vector operations are as in l_2. For $x = (\alpha_1, \alpha_2, \ldots)$ and $y = (\beta_1, \beta_2, \ldots)$, one can show that

$$(x, y) = \sum_{i=1}^{\infty} \alpha_i \bar{\beta}_i$$

constitutes an inner product on X. Let M be the set of all $x \in X$ such that $\sum_{i=1}^{\infty}(1/i)\alpha_i = 0$. Show that M is a closed subspace of X but $X \neq M \oplus M^{\perp}$ and $M \neq M^{\perp\perp}$.

9. Show that the projection theorem ceases to be true if M is not closed.

10. Let S_1 and S_2 be arbitrary subspaces of the inner product space X. Prove that $S_1^{\perp\perp} + S_2^{\perp\perp} \subset (S_1 + S_2)^{\perp\perp}$.

REFERENCES

1. S. Berberian, "Introduction to Hilbert Space."
2. P. Halmos, "Introduction to Hilbert Space."
3. A. Taylor, "Introduction to Functional Analysis."

CHAPTER 11

The Hahn–Banach Theorem

In this chapter a mapping similar to a norm is introduced and called a *convex functional*. It is easily verified that every norm is a convex functional and, thus, any results proved for convex functionals will also hold for norms. The first result, the Hahn–Banach theorem, is exceedingly important in functional analysis and has many striking consequences. The theorem is first proved for real linear spaces and then, with a slight modification of the hypothesis, for complex spaces. It is interesting to note that the complex version of the theorem was not proved until some eight years after the version for real spaces.

When one studies finite-dimensional vector spaces, a considerable amount of time is usually devoted to the space of all linear functionals over a vector space. This space, the conjugate or dual space, is a linear space having the same dimension as the original space. In the infinite-dimensional case, however, there are just "too many" linear functionals on which to focus attention. In our investigation, we shall limit ourselves to the examination of only those linear functionals (over a normed linear space) that are *bounded*—the next notion introduced in this chapter. It is shown that the collection of all bounded linear functionals over a normed linear space after introducing suitable operations, is, itself, a normed linear space. Furthermore, it is shown that this space is a Banach space regardless of whether the original space is a Banach space.

II.I   The Hahn–Banach Theorem

DEFINITION II.I. Let X be a linear space and consider the mapping

$$p : X \rightarrow R$$

with the properties

(1) $p(x) \geq 0$ for all $x \in X$;
(2) for all $x, y \in X, p(x + y) \leq p(x) + p(y)$ (subadditive);
(3) for $\alpha \in F$ and $x \in X, p(\alpha x) = \alpha p(x)$ provided $\alpha \geq 0$ (positive homogeneous).

A mapping satisfying all three conditions is said to be a *convex functional*. If only the last two requirements are met (that is, the mapping is subadditive and positive homogeneous), the mapping is called a *sublinear functional*.

One should note the similarity of a convex functional to a norm. It is evident that every norm is a convex functional from the above definition.

Before proceeding to our main result, we first wish to prove the following lemma.

LEMMA. Let M be a proper subspace of the real vector space X and let $x_0 \in X - M$. Consider the subspace spanned by M and $\{x_0\}$; that is, consider

$$N = [M \cup \{x_0\}],$$

and suppose that f is a linear functional defined only on M, that p is a sublinear functional defined on all of X, and that

$$f(x) \leq p(x)   (\text{for } x \in M).$$

Then f can be extended to a linear functional F defined on N with the property that

$$F(x) \leq p(x)   (\text{for all } x \in N).$$

REMARK. Pictorially we are trying to establish the following:

$$N \xrightarrow{F} R \quad [F(x) \leq p(x)],$$

$$M \xrightarrow{f} R \quad [f(x) \leq p(x)].$$

Proof. Since $f(x) \leq p(x)$ on M then, for arbitrary $y_1, y_2 \in M$, we can say

$$f(y_1 - y_2) = f(y_1) - f(y_2) \leq p(y_1 - y_2) = p(y_1 + x_0 - y_2 - x_0)$$

$$\leq p(y_1 + x_0) + p(-y_2 - x_0).$$

Grouping the terms involving y_2 on one side and those involving y_1 on the other, we have

$$-p(-y_2 - x_0) - f(y_2) \leq p(y_1 + x_0) - f(y_1). \tag{11.1}$$

Suppose now that y_1 is held fixed while y_2 is allowed to vary through all of M. From (11.1), it is clear that the set of real numbers

$$\{-p(-y_2 - x_0) - f(y_2) \,|\, y_2 \in M\}$$

has an upper bound—hence a least upper bound. In view of this we define

$$a = \sup\{-p(-y_2 - x_0) - f(y_2) | y_2 \in M\}.$$

In a similar fashion, we are also assured of the existence of

$$b = \inf\{p(y_1 + x_0) - f(y_1) \,|\, y_1 \in M\}.$$

By (11.1), we have

$$a \leq b$$

and, hence, there must be a real number c_0, such that

$$a \leq c_0 \leq b.$$

In the event that $a = b$, c_0 is just the common value.

We now have, for any $y \in M$,

$$-p(-y - x_0) - f(y) \leq c_0 \leq p(y + x_0) - f(y). \tag{11.2}$$

Since $x_0 \notin M$, we can write any $x \in N$ as

$$x = y + \alpha x_0,$$

where α is a uniquely determined scalar and y is a uniquely determined vector in M. Because of this unique representation, the mapping

$$F : N \to R$$

via

$$F(y + \alpha x_0) = f(y) + \alpha c_0$$

is certainly well-defined and is, furthermore, clearly a linear functional on the subspace N. In addition, it is also evident that, if $y \in M$, then

$$F(y) = f(y)$$

or that F extends f. To complete the proof it only remains to show that

$$F(x) \le p(x)$$

for all $x \in N$. To prove this, we now consider three cases: For any $x \in N$, $x = y + \alpha x_0$, one of $\alpha = 0$, $\alpha > 0$, or $\alpha < 0$ must hold; hence consider

(1) $\alpha = 0$. In this case we need only note that $F(y + \alpha x_0) = F(y) = f(y)$ and apply the hypothesis.

(2) $\alpha > 0$. In this case we consider the right half of inequality (11.2) with y replaced by y/α. This yields

$$c_0 \le p\left(\frac{y}{\alpha} + x_0\right) - f(y/\alpha).$$

Multiplying through by α and using the fact that p is a sublinear functional, we have

$$f(y) + \alpha c_0 \le p(y + \alpha x_0)$$

or

$$F(x) \le p(x).$$

(3) Last, suppose $\alpha < 0$. We now appeal to the left half of (11.2) with y replaced by y/α. This yields

$$-p(-y/\alpha - x_0) - f(y/\alpha) \le c_0$$

or

$$-p(-y/\alpha - x_0) \le c_0 + f(y/\alpha).$$

Multiplying through by α changes the sense of the inequality and gives

$$(-\alpha)p(-y/\alpha - x_0) \ge \alpha c_0 + f(y)$$

or, since $-\alpha > 0$,

$$p(y + \alpha x_0) \ge \alpha c_0 + f(y),$$

which completes the proof.

We can now proceed to the theorem.

THEOREM 11.1 (*Hahn–Banach*). Let M be a subspace of the real linear space X, p a sublinear functional on X, and f a linear functional on M such that, for all $x \in M$,

$$f(x) \le p(x).$$

Then there exists a linear functional F defined on all of X and extending f such that

$$F(x) \le p(x) \qquad \text{(for all } x \in X\text{)}.$$

REMARK. Pictorially this amounts to the following:

$$X \xrightarrow{\ F\ } R \qquad [F(x) \le p(x)],$$

$$M \xrightarrow{\ f\ } R \qquad [f(x) \le p(x)].$$

Proof. To prove this result, a partial ordering will be introduced on a certain set. The set will be shown to be inductively ordered, and then Zorn's lemma will be applied. (Zorn's lemma is discussed in the Appendix to Chapter 9.)

Let S denote the class of all linear functionals $\{\hat{f}\}$, extending f, and such that

$$\hat{f}(x) \le p(x),$$

when $x \in D_{\hat{f}}$, where $D_{\hat{f}}$ represents the domain of $\hat{f}$. The set S must be nonempty, for certainly f itself is a member of S and, furthermore, by virtue of the lemma just proved, it is seen that there are other less trivial members of S. Being assured that our results will not be vacuous, we define, for $\hat{f}_1, \hat{f}_2 \in S$,

$$\hat{f}_1 < \hat{f}_2$$

if $\hat{f}_2$ extends $\hat{f}_1$ or, equivalently, if $D_{\hat{f}_2} \supset D_{\hat{f}_1}$ and

$$\hat{f}_{2|D_{\hat{f}_1}} = \hat{f}_1$$

where the notation $\hat{f}_2|\,D_{\hat{f}_1}$ is read "$\hat{f}_2$ restricted to $D_{\hat{f}_1}$." Using an argument very similar to that employed in Example A.3 in the Appendix to Chapter 9, it is easily verified that $<$ induces a partial ordering on S. We now wish to show that S is inductively ordered, namely, that every totally ordered subset of S has an upper bound in S. To this end let $T = \{\hat{f}_\alpha\}$ be a totally ordered subset of S. We now wish to demonstrate the existence of some $\hat{f} \in S$ such that $\hat{f}$ is an upper bound for T. For this purpose, consider the function $\hat{f}$ whose domain is $\bigcup_\alpha D_{\hat{f}_\alpha}$. If

$$x \in \bigcup_\alpha D_{\hat{f}_\alpha},$$

there must be some α such that

$$x \in D_{\hat{f}_\alpha}$$

and we define

$$\hat{f}(x) = \hat{f}_\alpha(x).$$

Before even showing that $\hat{f}$ is well-defined, we must verify that its domain is a subspace (remember that a linear functional can only be defined on a subspace). If $x \in \bigcup_\alpha D_{\hat{f}_\alpha}$, then, $x \in D_{\hat{f}_\alpha}$, for some α, which since $D_{\hat{f}_\alpha}$ must be a subspace, implies that, for any scalar β, $\beta x \in D_{\hat{f}_\alpha}$. Now suppose

$$x, y \in \bigcup_\alpha D_{\hat{f}_\alpha}.$$

This implies, for some α_1 and α_2,

$$x \in D_{\hat{f}_{\alpha_1}} \quad \text{and} \quad y \in D_{\hat{f}_{\alpha_2}},$$

but since T is totally ordered, either $D_{f_{\alpha_1}} \subset D_{f_{\alpha_2}}$ or $D_{f_{\alpha_2}} \subset D_{f_{\alpha_1}}$. Assuming, without loss of generality, the former relationship to hold, we have

$$x, y \in D_{f_{\alpha_2}},$$

which implies, since $D_{f_{\alpha_2}}$ is a subspace, that

$$x + y \in D_{f_{\alpha_2}}.$$

Thus the domain of $\hat{f}$, $\bigcup_\alpha D_{f_\alpha}$, is a subspace. Now we must show $\hat{f}$ is well-defined. Suppose

$$x \in D_{f_\alpha} \qquad \text{and} \qquad x \in D_{f_\beta}.$$

By the definition of $\hat{f}$, we have

$$\hat{f}(x) = \hat{f}_\alpha(x) \qquad \text{and} \qquad \hat{f}(x) = \hat{f}_\beta(x).$$

But, by the total ordering of T, either $\hat{f}_\alpha$ extends $\hat{f}_\beta$ or vice versa. In either case,

$$\hat{f}_\alpha(x) = \hat{f}_\beta(x)$$

and $\hat{f}$ is seen to be well-defined. It is clear that $\hat{f}$ is a linear mapping, that it extends f, that

$$\hat{f}(x) \le p(x)$$

when $x \in D_f$ and, also, that for any $\hat{f}_\alpha \in T$,

$$\hat{f}_\alpha < \hat{f}.$$

In summary $\hat{f}$ is an upper bound for T in S and, hence, S is inductively ordered.

By Zorn's lemma there must be some $F \in S$ such that F is a maximal element for S. Since $F \in S$, F must be a linear functional extending f with the additional property that

$$F(x) \le p(x)$$

when $x \in D_F$. To complete the proof of the theorem, it only remains to show that $D_F = X$.

Suppose not; that is, suppose there exists $x_0 \in X$ such that $x_0 \notin D_F$. Applying the preceding lemma, we see that F could be extended to another linear functional F' extending f and such that

$$F'(x) \le p(x)$$

when $x \in [D_F \cup \{x_0\}]$. Thus F' would belong to S and, since F' also extends F, would contradict the maximality of F. Therefore no such x_0 can exist and the domain of F must be the whole space. The proof is now complete.

Having obtained this result for real linear spaces, we wish to obtain a similar statement for complex spaces. In order to push it through some slight modification of the hypothesis [and of the condition $f(x) \le p(x)$ which does not make sense if f is complex-valued] is necessary. With this in mind, we make the following definition.

DEFINITION 11.2. The convex functional $p(x)$ is said to be *symmetric* if, for all scalars α, $p(\alpha x) = |\alpha| p(x)$. Symmetric convex functionals are also called *seminorms*.

THEOREM 11.2. Let M be a subspace of the complex linear space X, p a symmetric convex functional defined on X, and f a linear functional defined on M with the property that

$$|f(x)| \leq p(x)$$

for $x \in M$. In this case there exists a linear functional F defined on all of X, extending f, and satisfying the condition

$$|F(x)| \leq p(x)$$

for any $x \in X$.

REMARK. In this case the pictorial representation becomes:

$$X \xrightarrow{\ F\ } C \quad [|F(x)| \leq p(x)],$$

$$M \xrightarrow{\ f\ } C \quad [|f(x)| \leq p(x)].$$

Proof. It is clear that the complex vector space X can be viewed as a real vector space. Adopting this viewpoint and letting $f_1(x)$ denote the real part of $f(x)$ and $f_2(x)$ the imaginary part, we can write

$$f(x) = f_1(x) + if_2(x).$$

We now claim that $f_1(x)$ and $f_2(x)$ are real-valued real linear functionals, where by a *real* linear functional we mean the following: g is a *real linear functional* on the complex space V if $\alpha \in R$ implies $g(\alpha x) = \alpha g(x)$ for any $x \in V$. To prove that f_1 and f_2 have this property, let $\alpha \in R$ and consider

$$\alpha f(x) = \alpha f_1(x) + i\alpha f_2(x).$$

Since f is a linear functional, this must equal

$$f(\alpha x) = f_1(\alpha x) + if_2(\alpha x).$$

Equating real and imaginary parts, we have

$$f_1(\alpha x) = \alpha f_1(x) \quad \text{and} \quad f_2(\alpha x) = \alpha f_2(x).$$

In a similar fashion we can show that sums are also preserved, and f_1 and f_2 are indeed seen to be real linear functionals. Now consider

$$i\big(f_1(x) + if_2(x)\big) = if(x) = f(ix) = f_1(ix) + if_2(ix).$$

Equating real and imaginary parts allows us to make one of the key observations of the proof; namely

$$f_1(ix) = -f_2(x). \tag{11.3}$$

Clearly, by the hypothesis, the real-valued real linear functional f_1 satisfies

$$f_1(x) \leq p(x)$$

for any $x \in M$. By Theorem 11.1 there exists a real-valued real linear functional F_1 defined on the whole space, extending f_1 and satisfying

$$F_1(x) \le p(x)$$

for every $x \in X$. We now define

$$F(x) = F_1(x) - iF_1(ix).$$

[Note that, although we cannot immediately say $F(ix) = iF(x)$, to consider $F(ix)$ is sensible.] We now assert that F extends f. To prove this let $x \in M$ and consider

$$F(x) = F_1(x) - iF_1(ix).$$

F_1 extends f_1, so

$$F_1(x) = f_1(x) \quad \text{and} \quad F_1(ix) = f_1(ix) = -f_2(x),$$

where the last equality is obtained by invoking Eq. (11.3). Thus

$$F(x) = f_1(x) + if_2(x) = f(x),$$

and F is seen to extend f. Since F is clearly a real linear functional, to show that it is a complex linear functional it only remains to show that $F(ix) = iF(x)$. First let us compute

$$F(ix) = F_1(ix) - iF_1(-x) = F_1(ix) + iF_1(x).$$

Comparing this to

$$iF(x) = iF_1(x) + F_1(ix),$$

we see that F is indeed a complex linear functional.

Last, we must show that $|F(x)| \le p(x)$, and to do this we finally make use of the fact that $p(x)$ was assumed to be a *symmetric* convex functional.

Since the inequality is clearly satisfied if $F(x) = 0$, we assume that $F(x) \ne 0$. In this case, let us write the complex number $F(x)$ as

$$F(x) = re^{i\theta}.$$

Hence

$$F(e^{-i\theta}x) = r = |F(x)|.$$

Thus $F(e^{-i\theta}x)$ is a purely real quantity, which implies the imaginary part of $F(e^{-i\theta}x)$, $-F_1(ie^{-i\theta}x)$, must be zero, or

$$F(e^{-i\theta}x) = F_1(e^{-i\theta}x).$$

But, since $|F_1(x)| \le p(x)$ for all x, we can say

$$|F(x)| = F_1(e^{-i\theta}x) \le p(e^{-i\theta}x) = p(x).$$

The proof is now complete.

11.2 Bounded Linear Functionals

The first theorem demonstrates a rather remarkable way to ascertain whether a linear functional is continuous.

THEOREM 11.3. Let f be a linear functional on the normed linear space X. Then if f is continuous at $x_0 \in X$, it must be continuous at every point of X.

Proof. By the hypothesis, if $x_n \to x_0$, then $f(x_n) \to f(x_0)$. To prove continuity everywhere, we must show, for any $y \in X$, that if $y_n \to y$, then $f(y_n) \to f(y)$. To this end suppose $y_n \to y \in X$ and consider

$$f(y_n) = f(y_n - y + x_0 + y - x_0)$$
$$= f(y_n - y + x_0) + f(y) - f(x_0).$$

Since $y_n - y + x_0 \to x_0$,

$$f(y_n - y + x_0) \to f(x_0)$$

and

$$f(y_n) \to f(y).$$

Before proceeding, a new definition is necessary.

DEFINITION 11.3. The linear functional f on the normed linear space X is called *bounded* if there exists a real constant k, such that, for all $x \in X$,

$$|f(x)| \le k\|x\|. \tag{11.4}$$

It is immediately evident that there may be many such real numbers k satisfying (11.4) for a particular f. For example, if k_1 works, certainly anything bigger than k_1 will also work.

Our next theorem shows that the condition of boundedness is equivalent to continuity.

THEOREM 11.4. The linear functional f defined on the normed linear space X is bounded if and only if it is continuous.

Proof. First we shall show that continuity implies boundedness, and the proof will be by contradiction; thus, suppose f is continuous but not bounded. The negation of being bounded is that for any natural number n, however large, there is some point, x_n say, such that

$$|f(x_n)| > n\|x_n\|.$$

Consider now the vectors

$$y_n = \frac{x_n}{n\|x_n\|}$$

with norms

$$\|y_n\| = \frac{1}{n}.$$

Clearly, then the sequence

$$y_n \to 0.$$

Since any linear functional maps the zero vector into the zero scalar and f is continuous, this implies

$$f(y_n) \to f(0) = 0.$$

But

$$|f(y_n)| = \frac{1}{n\|x_n\|} f(x_n)$$

and

$$f(x_n) > n\|x_n\|,$$

which implies

$$|f(y_n)| > 1$$

and precludes the possibility of $f(y_n)$ approaching 0. Since we have arrived at contradictory results, the assumption (the only one we made) that f was not bounded must be false when f is continuous.

To prove the converse, that boundedness implies continuity, we need only note that boundedness certainly implies continuity at the origin and apply the preceding theorem.

11.3   The Conjugate Space

Suppose f and g are bounded linear functionals over the normed linear space X. Two such linear functionals f and g will be said to be equal, $f = g$, if and only if $f(x) = g(x)$ for all $x \in X$. By defining $f + g$ to be the linear functional whose value, at any point x, is $f(x) + g(x)$ and, with α a scalar and f as above, αf to be $\alpha f(x)$ at any point x, it is easy to verify that the class of such linear functionals is a linear space. Note that so far we have not availed ourselves of the fact that we are dealing with only the bounded linear functionals. We make use of this now, in the following definition.

DEFINITION 11.4. If f is a bounded linear functional on the normed linear space X, we define

$$\|f\| = \sup_{\|x\| \neq 0} \frac{|f(x)|}{\|x\|}.$$

Let us first verify that the least upper bound mentioned in Definition 11.4 exists. Since f was assumed to be bounded, there must be some k such that, for all x,

$$|f(x)| \le k\|x\|,$$

which implies, for $x \neq 0$,

$$\frac{|f(x)|}{\|x\|} \le k.$$

In view of this, the set of real numbers

$$\left\{ \frac{f(x)}{\|x\|} \,\middle|\, x \neq 0 \right\}$$

has an upper bound, hence a least upper bound. Further, denoting by K the set of all real numbers k satisfying Eq. (11.4), we can say that $\|f\| \in K$ for, since, for all $x \neq 0$,

$$\frac{|f(x)|}{\|x\|} \le \|f\|,$$

we have, for every x,

$$|f(x)| \le \|f\| \cdot \|x\|.$$

The normed linear space consisting of all bounded linear functionals over X is called the *conjugate space* and will be denoted by $\tilde{X}$, it being easily verified that $\|f\|$ is truly a norm.

We now wish to determine some equivalent ways of expressing $\|f\|$.

(1) $\|f\| = \inf_{k \in K} k$. First we note that, since $\|f\| \in K$,

$$\|f\| \ge \inf_{k \in K} k.$$

Next consider $x \ne 0$ and $k \in K$. In this case

$$\frac{|f(x)|}{\|x\|} \le k.$$

Since this is true for any $x \ne 0$, however, we have

$$k \ge \sup_{\|x\| \ne 0} \frac{|f(x)|}{\|x\|} = \|f\|$$

for any $k \in K$; therefore

$$\inf_{k \in K} k \ge \|f\|.$$

Hence

$$\|f\| = \inf_{k \in K} k.$$

We now claim that the following formulas for $\|f\|$ also hold. The proof of (3) involves only a modification of the proof of (2); hence only (2) will be proved here.

(2) $\|f\| = \sup_{\|x\| \le 1} |f(x)|.$

(3) $\|f\| = \sup_{\|x\| = 1} |f(x)|.$

Proof (2). Consider an $x \in X$ such that $\|x\| \le 1$. In this case

$$|f(x)| \le \|f\| \|x\| \le \|f\|,$$

which implies

$$\sup_{\|x\| \le 1} |f(x)| \le \|f\|. \tag{11.5}$$

By the definition of $\|f\|$, for any $\varepsilon > 0$, there exists an $x' \ne 0$, such that

$$|f(x')| > (\|f\| - \varepsilon)\|x'\|.$$

We now define

$$\hat{x} = \frac{x'}{\|x'\|}$$

and note that $\|\hat{x}\| = 1$, which implies

$$\sup_{\|x\| \le 1} |f(x)| \ge |f(\hat{x})| = \frac{1}{\|x'\|} |f(x')| > \|f\| - \varepsilon$$

or, for any $\varepsilon > 0$, that

$$\sup_{\|x\| \le 1} |f(x)| > \|f\| - \varepsilon.$$

Therefore,

$$\sup_{\|x\| \leq 1} |f(x)| \geq \|f\|.$$

Combining this with (11.5) yields

$$\|f\| = \sup_{\|x\| \leq 1} |f(x)|.$$

In the study of finite-dimensional linear spaces, the *conjugate* space is usually defined to be *all* linear functionals over the space and seems to conflict with the definition we gave. The following theorem shows that no conflict arises.

THEOREM 11.5. If X is a finite-dimensional linear space, all linear functionals are bounded.

Proof. Let f be a linear functional on X, and let $x_1, x_2, ..., x_n$ be a basis for X. There must exist scalars $\alpha_1, \alpha_2, ..., \alpha_n$ associated with any vector $x \in X$ such that we can write

$$x = \sum_{i=1}^{n} \alpha_i x_i.$$

Since f is a linear functional, we have

$$f(x) = \sum_{i=1}^{n} \alpha_i f(x_i),$$

$$|f(x)| \leq \sum_{i=1}^{n} |\alpha_i| \, |f(x_i)|$$

$$\leq \|x\|_0 \sum_{i=1}^{n} |f(x_i)|.$$

Letting the constant

$$k = \sum_{i=1}^{n} |f(x_i)|,$$

we see that

$$f(x) \leq k \|x\|_0$$

or that f is bounded with respect to $\| \quad \|_0$, hence with respect to any norm on X. (See Chapter 8 for a discussion of $\| \quad \|_0$ and the notion of equivalent norms.)

Having already noted that the conjugate space $\tilde{X}$ of a normed linear space X is itself a normed linear space, we wish to show, in addition, that $\tilde{X}$ is a Banach space regardless of whether X is a Banach space.

THEOREM 11.6. Let $\tilde{X}$ denote the normed linear space of all bounded linear functionals over the normed linear space X. Then $\tilde{X}$ is a Banach space.

Proof. We must show that any Cauchy sequence (in the norm on $\tilde{X}$) converges (in the norm on $\tilde{X}$) to some element of $\tilde{X}$. To this end, let $\{f_n\}$ be a Cauchy sequence of elements of $\tilde{X}$.

This means that for any $\varepsilon > 0$, there exists an N such that $n, m > N$ implies

$$\|f_n - f_m\| < \varepsilon$$

or, for any $x \in X$,

$$|f_n(x) - f_m(x)| \le \|f_n - f_m\| \, \|x\| \le \varepsilon \|x\|.$$

Thus for any particular x, $\{f_n(x)\}$ is a Cauchy sequence of scalars; since R and C are complete, the limit

$$\lim_n f_n(x) = f(x) \tag{11.6}$$

must exist. The function, defined in this way is clearly a linear functional. We wish also to show that it is bounded. For any j and an appropriate N, we can say

$$\|f_{n+j} - f_n\| < 1 \qquad (n > N),$$

which implies

$$\|f_{n+j}\| < 1 + \|f_n\|$$

or

$$|f_{n+j}(x)| < (\|f_n\| + 1)\|x\|.$$

Since this must also be true as $j \to \infty$, we have

$$|f(x)| \le (\|f_n\| + 1)\|x\|,$$

and, therefore, f is a bounded linear functional.

Last, we wish to show that $\{f_n\}$ converges to f in the norm on $\tilde{X}$. For any $\varepsilon > 0$, there must be an N such that $n, m > N$ implies

$$\|f_n - f_m\| < \varepsilon.$$

So, for any x,

$$|f_n(x) - f_m(x)| \le \varepsilon \|x\|;$$

hence

$$\lim_m |f_n(x) - f_m(x)| = |f_n(x) - \lim_m f_m(x)| \le \varepsilon \|x\|$$

or

$$|f_n(x) - f(x)| \le \varepsilon \|x\|$$

for every x, which implies

$$\|f_n - f\| \le \varepsilon,$$

and completes the proof.

EXERCISES 11

1. Let p be a sublinear functional defined on X, a real linear space. Prove that for $x_0 \in X$, there exists a linear functional F defined on X such that $F(x_0) = p(x_0)$, and $F(x) \le p(x)$ for all $x \in X$.

2. If $p(x)$ is a symmetric convex functional defined on the complex linear space X, then prove, for arbitrary $x_0 \in X$, that there exists a linear functional F defined on X such that $F(x_0) = p(x_0)$ and $|F(x)| \le p(x)$ for all $x \in X$.

3. Let X be a normed linear space and let $X \xrightarrow{f} R$. If f is additive $[f(x+y) = f(x) + f(y)]$ and if f is continuous, then show that f is homogeneous, that is, $f(\alpha x) = \alpha f(x)$ for all $\alpha \in R$.

4. Let X be a finite-dimensional linear space; let $x_1, \ldots, x_n$ be a basis. If $x = \sum_{i=1}^{n} \alpha_i x_i$ define $\|x\| = (\sum_{i=1}^{n} |\alpha_i|^2)^{\frac{1}{2}}$. If f is a bounded linear functional on X, find $\|f\|$. Do the same thing only now define $\|x\| = \max_i |\alpha_i|$.

5. Let F be the linear functional defined on $C[a, b]$ (with sup norm) by $f(x) = \int_a^b x(t)\, dt$ for $x \in C[a, b]$. Find $\|f\|$.

6. Let f be a continuous linear functional defined on the normed linear space X. Prove that, if the sequence $\{x_n\}$ is a Cauchy sequence in X, then $\{f(x_n)\}$ is a Cauchy sequence of complex numbers.

7. If p is a convex functional on X, show that, for arbitrary $x_0 \in X$ and $a > 0$, $S = \{x \in X \mid p(x - x_0) \leq a\}$ is a convex set.

8. On the space l_1 (Example 8.4) for $x = (\alpha_1, \alpha_2, \ldots)$ define $f(x) = \sum_{n=1}^{\infty} \alpha_n$ and introduce a new norm: $\|x\| = \sup_n |\alpha_n|$. Prove that f is linear but not continuous with respect to this norm.

9. Let M be a subspace of the normed linear space X. Let

$$L = \{f \in \tilde{X} \mid f(x) = 0 \quad \text{for all } x \in M\}.$$

Show that L is a closed subspace of $\tilde{X}$ and that $\tilde{M}$ is isometrically isomorphic to $\tilde{X}/L$.

10. In real euclidean n-space, show that there is only one norm-preserving extension of a bounded linear functional, defined on a subspace, to the entire space.

11. Consider the Banach space $C[a, b]$ (sup norm). For any fixed $x \in C[a, b]$, define $f_x(y)$ for any $y \in C[a, b]$ as follows:

$$f_x(y) = \int_a^b y(t) x(t)\, dt.$$

Show that f_x is a bounded linear functional and compute $\|f_x\|$.

APPENDIX

A.1 The Problem of Measure and the Hahn–Banach Theorem

In the development of Lebesgue measure on the real line R, one develops a set function defined on a class of subsets of R known as the Lebesgue-measurable sets. And indeed this is a "large" class of sets, for it can be shown to have cardinal number 2^c, where c denotes the power of the continuum.† It is so large that one wonders if the Lebesgue-measurable sets coincide with the class of all subsets of the real line. The answer to this question, however, is in the negative; there are subsets of R that are not Lebesgue-measurable‡ (that is, they do not belong to the class of Lebesgue-measurable sets).

Thus, the following problem arises:

(1) Is it possible, for each bounded subset E of the real line, to assign a non-negative real number $\mu(E)$, so that the following three conditions are satisfied:

(a) $\mu([0, 1]) = 1$.

(b) Letting two subsets of R, A and B, be called isometric if there exists an isometry of R mapping A onto B, we require that if A and B are *isometric*, $\mu(A) = \mu(B)$.

(c) If the sequence of subsets of R, $E_1, E_2, \ldots$ is pairwise disjoint, then

$$\mu\left(\bigcup_{i=1}^{\infty} E_i\right) = \sum_{i=1}^{\infty} \mu(E_i).$$

(This last requirement will be referred to as *countable additivity* of μ.)

In answer to this question, one can exhibit a disjoint sequence of isometric sets§ $E_1, E_2, \ldots$ such that

$$[-\tfrac{1}{2}, \tfrac{1}{2}] \subset \bigcup_{i=1}^{\infty} E_i \subset [-\tfrac{3}{2}, \tfrac{3}{2}].$$

The first and third requirements imply that, if $A \subset B \subset R$, $\mu(A) \leq \mu(B)$; therefore,

$$\mu([-\tfrac{1}{2}, \tfrac{1}{2}]) \leq \mu\left(\bigcup_{i=1}^{\infty} E_i\right) \leq \mu([-\tfrac{3}{2}, \tfrac{3}{2}]) < \infty.$$

Since $[-\tfrac{1}{2}, \tfrac{1}{2}]$ is isometric (since pure translations are isometries) to $[0, 1]$, we have,

$$1 = \mu([0, 1]) = \mu([-\tfrac{1}{2}, \tfrac{1}{2}]),$$

and the preceding equation becomes, using (c) again,

$$1 \leq \mu(E_1) + \mu(E_2) + \cdots < \infty.$$

Since the E_i are isometric, $\mu(E_1) = \mu(E_2) = \cdots$, and denoting their common value by a, a must satisfy

$$1 \leq a + a + \cdots < \infty,$$

a condition no real number can satisfy. Hence we have Theorem A.1.

---

† To see that this is so, recall that every subset of a set of Lebesgue measure zero is measurable and has measure zero. The Cantor set is then a set with power c, Lebesgue measure zero, and having the property that every subset is measurable. The cardinality of the class of all subsets of a set with power c is 2^c.

‡ A rather lucid example of such a set is given in I. P. Natanson, *Theory of Functions of a Real Variable*, Vol. 1, pp. 77–78.

§ Cf. Natanson, *op. cit.*, pp. 76–79.

THEOREM A.I. Problem (1) cannot be solved on R.

Knowing that the requirements of problem (1) are too stringent, let us weaken it slightly and see if it can be solved then. We state the weakened problem:

(2) Everything in (1) remains the same except condition (c), which we replace by (c').

(c') If $E_1, E_2, ..., E_n$ is a *finite*, disjoint, sequence of sets in R,

$$\mu\left(\bigcup_{i=1}^{n} E_i\right) = \sum_{i=1}^{n} \mu(E_i).$$

(This property will be referred to as *finite additivity* of μ.)

As it happens, this problem can be solved on R and can also be solved in the euclidean plane [requirement (a) being altered to involve the unit square instead of the unit interval]. It cannot, however, be solved in euclidean n-space for $n > 2$, the proof of this fact first being demonstrated by Hausdorff. Using a delightful application of the Hahn–Banach theorem (Theorem 11.1), we now present Banach's proof that (2) is solvable on R.

As will be clarified later, if the problem can be solved on just $[0, 1)$, it is easy to extend the solution to all subsets of R. To solve it on $[0, 1)$, we first get an "integral" defined on the class of all bounded, periodic functions, with period 1 and from this integral get the desired set function μ. Thus, we now wish to solve the following problem which we again refer to as (2).

(2) To every subset E of $[0, 1)$ we wish to assign a nonnegative number $\mu(E)$ such that $0 \leq \mu(E) \leq 1$:

(a) $\mu([0, 1)) = 1$;
(b) If $A, B \subset [0, 1)$ are isometric sets, then $\mu(A) = \mu(B)$;
(c) If $E_1, E_2, ..., E_n$ is a finite, disjoint collection of subsets of $[0, 1)$, then

$$\mu\left(\bigcup_{i=1}^{n} E_i\right) = \sum_{i=1}^{n} \mu(E_i).$$

Let X denote the class of all bounded, real-valued, periodic functions with period 1. It is easy to see, with respect to the usual operations of addition $[(x_1 + x_2)(t) = x_1(t) + x_2(t)]$ and multiplication by a real scalar $[(\alpha x)(t) = \alpha x(t)]$, that X is a real vector space. We now wish to exhibit a functional $\int x(t)$, defined for all $x \in X$, with the following properties:

(i) for $\alpha, \beta \in R$, $x, y \in X$,

$$\int (\alpha x(t) + \beta y(t)) = \alpha \int x(t) + \beta \int y(t);$$

(ii) for any fixed $t_0 \in R$, $\int x(t + t_0) = \int x(t)$;
(iii) if $x \in X$ is nonnegative for all t, $\int x(t) \geq 0$;
(iv) for $x \in X$, $\int x(1 - t) = \int x(t)$;
(v) letting 1 denote the function that is equal to 1 for every t,

$$\int 1 = 1.$$

The functional $\int x(t)$ is called a *Banach integral* on X.

THEOREM A.2. There exists a (nonunique) Banach integral defined on X.

Proof. In this existence proof the Banach integral will actually be constructed. To obtain it, we first obtain a sublinear functional p on X. Then, as a consequence of Exercise 11.1, it follows that there exists a (nonunique) linear functional f defined on the whole space such that $f(x) \leq p(x)$ for all $x \in X$. It will then be shown that

$$\tfrac{1}{2}[f(x) + f(\hat{x})],$$

where $\hat{x}(t) = x(1 - t)$, is a Banach integral on X.

Let $\alpha_1, \alpha_2, ..., \alpha_n \in R$ and form, for any $x \in X$,

$$p(x) = \inf_{\alpha_i}\left\{ \sup_{0 \leq t < 1} \frac{1}{n} \sum_{i=1}^{n} x(t + \alpha_i)\right\}. \tag{A.1}$$

(Note that n is not intended to be held fixed in computing the infimum.) We now show p to be a sublinear functional on X.

Since $x \in X$, there must be some $M > 0$ such that

$$|x(t)| \leq M$$

for all t. This implies that

$$-M \leq \frac{1}{n} \sum_{i=1}^{n} x(t + \alpha_i) \leq M$$

for any $\alpha_1, \alpha_2, ..., \alpha_n$ and every t; hence,

$$-M \leq \sup_t \frac{1}{n} \sum_{i=1}^{n} x(t + \alpha_i) \leq M;$$

therefore,

$$|p(x)| \leq M,$$

and p is seen to be real-valued.

If A is a subset of R, then, for any nonnegative $\alpha \in R$, $\sup \alpha A = \alpha \sup A$ and $\inf \alpha A = \alpha \inf A$. In view of this, it follows that $p(\alpha x) = \alpha p(x)$ for $\alpha \geq 0$; that is, it follows that p is positive homogeneous. To complete proving that p is a sublinear functional on X, it remains to show that it is subadditive. By the definition of p (as an infimum), for any $\varepsilon > 0$, and any $x, y \in X$, there must exist $\alpha_1, \alpha_2, ..., \alpha_n \in R$ and $\beta_1, \beta_2, ..., \beta_m \in R$ such that

$$\frac{1}{n} \sum_{i=1}^{n} x(t + \alpha_i) \leq p(x) + \varepsilon \quad \text{and} \quad \frac{1}{m} \sum_{j=1}^{m} y(t + \beta_j) \leq p(y) + \varepsilon$$

for every t. Let $\gamma_{ij} = \alpha_i + \beta_j$ ($i = 1, 2, ..., n, j = 1, 2, ..., m$) and consider

$$\frac{1}{nm} \sum_{i, j=1}^{n, m} \{x(t + \gamma_{ij}) + y(t + \gamma_{ij})\}.$$

Since only finite sums are involved, we can carry out the summation in any way we please and, in particular, can write it as

$$\frac{1}{m} \sum_{j=1}^{m} \left\{\frac{1}{n} \sum_{i=1}^{n} x(t + \alpha_i + \beta_j)\right\} + \frac{1}{n} \sum_{i=1}^{n} \left\{\frac{1}{m} \sum_{j=1}^{m} y(t + \alpha_i + \beta_j)\right\},$$

which (letting $t' = t + \beta_j$ in the first expression and $t'' = t + \alpha_i$ in the second, say) is less than or equal to $p(x) + p(y) + 2\varepsilon$; thus we have

$$\frac{1}{nm} \sum_{i,j=1}^{n,m} \{x(t + \gamma_{ij}) + y(t + \gamma_{ij})\} \leq p(x) + p(y) + 2\varepsilon$$

for any t—hence for the supremum over all t. Noting that

$$\sup_t \frac{1}{nm} \sum_{i,j=1}^{n,m} \{x(t + \gamma_{ij}) + y(t + \gamma_{ij})\}$$

is a member of the set whose greater lower bound yields $p(x + y)$, we see that we have shown

$$p(x + y) \leq p(x) + p(y) + 2\varepsilon,$$

where ε is arbitrary. The subadditivity of p is thus established. As noted previously, as a consequence of Exercise 11.1, we can now assert the existence of a linear functional f defined on X such that $f(x) \leq p(x)$ for all $x \in X$. Since $f(x) \leq p(x)$ for every x, we also have

$$f(-x) \leq p(-x),$$

which implies

$$-f(-x) = f(x) \geq -p(-x);$$

therefore,

$$-p(-x) \leq f(x) \leq p(x). \tag{A.2}$$

It will now be shown that the linear functional f satisfies conditions (i)–(iii) and (v) of the requirements that a functional must satisfy to be a Banach integral on X. The first requirement is clearly satisfied because f is linear. For any $x \in X$, we wish to show that $f(x(t + t_0)) = f(x(t))$, where $t_0 \in R$. Let

$$y(t) = x(t + t_0) - x(t).$$

In this notation, since f is linear, we must show that $f(y) = 0$. Let $\alpha_1 = 0$, $\alpha_2 = t_0$, $\alpha_3 = 2t_0$, ..., $\alpha_n = (n - 1)t_0$ and form

$$\frac{1}{n} \sum_{i=1}^{n} y(t + \alpha_i) = \frac{1}{n} \{[x(t + t_0) - x(t)] + [x(t + 2t_0) - x(t + t_0)]$$
$$+ \cdots + [x(t + nt_0) - x(t + (n - 1)t_0)]\} \tag{A.3}$$
$$= \frac{1}{n} \{x(t + nt_0) - x(t)\}.$$

Since $x \in X$, there must be some real number M such that $|x(t)| \leq M$ for all t; using this in the last expression, we see that

$$\frac{1}{n} \sum_{i=1}^{n} y(t + \alpha_i) \leq 2M/n.$$

Hence, by choosing a huge n, $(1/n) \sum_{i=1}^{n} y(t + \alpha_i)$ can be made correspondingly small, which implies for arbitrary $\varepsilon > 0$, $p(y) \leq \varepsilon$ and therefore,

$$p(y) \leq 0.$$

If we had used $-y$ instead of y, the only change that would occur is that where, in (A.3), we had $x(t + nt_0) - x(t)$, we now get $x(t) - x(t + nt_0)$. Thus we also have $p(-y) \leq 0$, which implies

$$-p(-y) \geq 0.$$

Using (A.2) we have

$$0 \leq -p(-y) \leq f(y) \leq p(y) \leq 0,$$

which implies $f(y) = 0$—the desired result. As for (iii), suppose that $x(t) \geq 0$; this implies $-x(t) \leq 0$ for every t. By the definition of p, we must have $p(-x) \leq 0$. Therefore, using the left half of (A.2), we have

$$0 \leq -p(-x) \leq f(x).$$

With regard to requirement (v) we see that if $x(t) = 1$, then, by the definition of p, $p(x) = 1$; similarly $p(-x) = -1$. An application of (A.2) demonstrates that $f(x) = 1$. We now modify f slightly and get a functional that also satisfies (iv) as well as (i)–(iii) and (v). We denote by $\int x(t)$:

$$\int x(t) = \tfrac{1}{2}[f(x) + f(\hat{x})],$$

where $\hat{x}(t) = x(1 - t)$. [The notation $\int x(t)$ will be justified by what is proved below.]

The linearity of f clearly implies (noting that $\widehat{x + y} = \hat{x} + \hat{y}$) the linearity of $\int x(t)$.

If $t_0 \in R$,

$$\int x(t + t_0) = \tfrac{1}{2}[f(x(t + t_0)) + f(x(1 - t - t_0))].$$

Since $f(x(t + t_0)) = f(x(t))$, replacing t by $(1 - t - t_0)$ we can conclude that

$$f(x(1 - t - t_0)) = f(x(1 - t)).$$

Therefore,

$$\int x(t + t_0) = \tfrac{1}{2}[f(x) + f(\hat{x})] = \int x(t).$$

Requirement (iii) is clearly satisfied. As for (iv), consider

$$\int x(1 - t) = \int \hat{x}(t)$$
$$= \tfrac{1}{2}[f(\hat{x}) + f(\hat{\hat{x}})]$$
$$= \tfrac{1}{2}[f(\hat{x}) + f(x)]$$
$$= \int x(t).$$

Last, requirement (v) is also clearly satisfied, and we have proved the existence of a Banach integral on R. We remark once again that it is by no means unique.

Having done this, we can now show that Problem (2) is solvable on $[0, 1)$ by taking, for any $E \subset [0, 1)$, $\mu(E)$ to be

$$\mu(E) = \int k_E(t),$$

where $k_E(t)$ denotes the characteristic function of E. By (iii), μ is clearly a non-negative set function on the class of all subsets of $[0, 1)$. As for the finite additivity of μ, it suffices to consider two subsets, and for this purpose we consider two disjoint subsets of $[0, 1)$, E and F. Since they are disjoint

$$k_{E \cup F} = k_E + k_F,$$

which implies

$$\mu(E \cup F) = \int [k_E(t) + k_F(t)]$$

$$= \int k_E(t) + \int k_F(t) = \mu(E) + \mu(F).$$

Before showing that isometric sets map into the same number under μ, some facts about the isometries of R will be recalled. Certainly the following three mappings are isometries of R:

(1) $\psi(x) = x + d$ (translation);
(2) $\psi(x) = -x$ (reflection in the origin);
(3) $\psi(x) = -x + d.$

It can be shown† that these exhaust all possible isometries of R. Similarly, one can show that the only possible isometries of $[0, 1)$ in $[0, 1)$ are translations and reflections about the point $\frac{1}{2}$ (or combinations thereof).

Suppose that E and F are isometric subsets of $[0, 1)$. One possibility is that the isometry is a translation—that there is some $t_0 \in R$ such that

$$F = E + t_0 = \{x + t_0 | x \in E\}.$$

In this case

$$\mu(F) = \int k_F(t) = \int k_E(t - t_0) = \int k_E(t) = \mu(E).$$

On the other hand, suppose that the isometry was a reflection about $\frac{1}{2}$; that is, suppose that

$$E = \{t \, | \, t \in E\} \quad \text{and} \quad F = \{1 - t \, | \, t \in E\}.$$

Then

$$\mu(F) = \int k_F(t) = \int k_E(1 - t) = \int k_E(t) = \mu(E).$$

In view of property (v) of the Banach integral, it also follows that $\mu([0, 1)) = 1$.

To extend μ to any bounded subset of R, first consider a subset A of R contained in $[i, i + 1)$, where i is an integer. Any subset of $[i, i + 1)$ is clearly isometric to some subset B of $[0, 1)$, and it is the value $\mu(B)$ that we shall take as $\mu(A)$. For any bounded subset of R, F there must exist an integer n such that

$$F \subset [-n, n).$$

Write F as

$$F = \bigcup_{i=-n}^{n-1} (F \cap [i, i + 1))$$

and define

$$\mu(F) = \sum_{i=-n}^{n-1} \mu(F \cap [i, i + 1)).$$

We have now completely proved the next theorem.

THEOREM A.3 (*Banach*). Problem (2) is solvable on R (but not uniquely).

† Cf., for example, I. P. Natanson, *Theory of Functions of a Real Variable*, Vol. 1, p. 72.

EXERCISES II APPENDIX

1. Show that any Banach integral $\int x(t)$ lies between the lower and upper Riemann integrals of $x(t)$.

2. Suppose μ is a solution of Problem (2). Show that

(a) $E \subset F \Rightarrow \mu(E) \leq \mu(F)$;

(b) $\mu(E) = 0$, if $E = \{x\}$;

(c) $\mu([a, b]) = b - a$.

3. Show that there exists a Banach integral that coincides with the Lebesgue integral for all bounded measurable functions on $[0, 1)$. (*Hint:* Consider a linear functional on the subspace of X, X_0, of all bounded Lebesgue measurable functions on $[0, 1)$.)

4. Show that the Banach measure μ is not countably additive.

5. Let X be the set of all bounded real-valued functions on $[0, \infty)$ with the usual operations. Show that for all $x \in X$ there exists a real number $\text{Lim}_{t \to \infty} x(t)$ such that

(a) $\displaystyle \text{Lim}_{t \to \infty} (ax(t) + by(t)) = a \, \text{Lim}_{t \to \infty} x(t) + b \, \text{Lim}_{t \to \infty} y(t)$;

(b) $\displaystyle \text{Lim}_{t \to \infty} x(t) \geq 0$, if $x(t) \geq 0$,

(c) $\displaystyle \text{Lim}_{t \to \infty} x(t + t_0) = \text{Lim}_{t \to \infty} x(t)$;

(d) $\displaystyle \text{Lim}_{t \to \infty} 1 = 1$;

(e) $\displaystyle \varliminf_{t \to \infty} x(t) \leq \text{Lim}_{t \to \infty} x(t) \leq \varlimsup_{t \to \infty} x(t)$.

Specialize these results to the case of bounded sequences $\{\alpha_n\}$ by defining $x(t) = \alpha_n$ for $n - 1 < t \leq n$ $(n = 1, 2, \ldots)$. [*Hint:* Analogous to the proof of Theorem A.2, consider the real vector space of all bounded real-valued functions defined on $[0, \infty)$. For $\alpha_1, \alpha_2, \ldots, \alpha_n \in R$ and x a member of this space show that $p(x) = \inf_{\alpha_i} \varlimsup_{t \to \infty} (1/n) \sum_{i=1}^{n} x(t + \alpha_i)$ is a sublinear functional on, this space. As in the proof of Theorem A.2, there must exist a linear functional f defined on the whole space such that $f(x) \leq p(x)$ for every x in the space. One can then take $f(x) = \text{Lim}_{t \to \infty} x(t)$.]

REFERENCES

1. A. Zaanen, "Linear Analysis."

2. S. Banach, "Théorie des Opérations Linéaires." The original discussion of the problem of measure.

CHAPTER **12**

Consequences of the Hahn–Banach Theorem

Certain important direct consequences of the Hahn–Banach theorem will be obtained in this chapter. For the most part, our concern will be with applications to *bounded* linear functionals; in particular, it will be shown that, given an arbitrary vector in the space x, there always exists a bounded linear functional which assumes the value $\|x\|$ when applied to x. This consequence is of such importance that often this, and not the result proved in the last chapter, is referred to as the Hahn–Banach theorem.

Having defined the conjugate or dual space of a normed linear space in Chapter 11, we wish to obtain some insight into the structure of this space. In particular, it is shown that whenever the conjugate space is separable, so is the original space. Proceeding in the same vein, our next concern is with what sort of functionals (assuming of course that they are bounded) exist on the conjugate space; that is, we wish to consider the dual space of the dual space. In many cases it turns out that the dual of the dual space is, under a very specific mapping mentioned here, essentially, just the original space. Such spaces are called *reflexive* spaces—a category into which every finite-dimensional space and every Hilbert space fits.

Another structural consideration of interest is that of the dual space of l_p. An extremely interesting result emerges from this investigation; namely, that the dual of l_p is (in the sense of a congruence) just l_q, where p and q are related by

$$\frac{1}{p} + \frac{1}{q} = 1.$$

In proving this we also obtain an expression for the norm of any linear functional in the conjugate space of l_p.

Finally, in the special setting of Hilbert space, we get a representation theorem (in terms of a fixed vector and the inner product) for any bounded linear functional on the space. The expression obtained is very similar to the Riesz representation theorem obtained for the finite-dimensional case, proved in Theorem 1.9

12.1 Some Consequences of the Hahn–Banach Theorem

In the last chapter it was shown that a linear functional, defined on a subspace, could be extended in such a way as to preserve the property of being less than or equal to some symmetric convex functional (or sublinear functional, as the case may be). Using this fact, we are now able to prove, for bounded linear functionals, the existence of a new linear functional, defined on the whole space, extending the original one, and having the same norm as the original.

THEOREM 12.1. Suppose f is a bounded linear functional, defined on the subspace M of the complex normed linear space X. Then there exists a bounded linear functional F, extending f, defined on the whole space, and having the same norm as f. Pictorially we have

$$F: \quad X \longrightarrow C$$

and $\qquad \|F\| = \|f\|$.

$$f: \quad M \longrightarrow C$$

Proof. Since f is assumed to be a bounded linear functional, then, for all $x \in M$,

$$|f(x)| \le \|f\| \, \|x\|.$$

For every $x \in X$ we now define

$$p(x) = \|f\| \, \|x\|$$

and claim the following assertions about $p(x)$ are valid:

(1) $p(x) \geq 0$, for all $x \in X$;
(2) for $x, y \in X$, $p(x + y) \leq p(x) + p(y)$;
(3) for $x \in X$ and $\alpha \in C$, $p(\alpha x) = |\alpha| p(x)$.

Thus, after easily verifying the validity of these assertions, it is seen that $p(x)$ is a symmetric convex functional; furthermore, for every $x \in M$,

$$|f(x)| \leq p(x).$$

Therefore, by the Hahn–Banach theorem, (Theorem 11.2), we can extend f to a new linear functional F, defined on all of X and such that

$$|F(x)| \leq p(x) = \|f\| \, \|x\| \tag{12.1}$$

for all $x \in X$. In view of this result it is clear that F is a bounded linear functional and also that

$$\|F\| \leq \|f\| \tag{12.2}$$

for, since $\|f\|$ is a bound for F in the sense of the definition of a bounded linear functional (see Sec. 11.2), it is certainly greater than or equal to the greatest lower bound of the set of all such bounds. [It was shown in Sec. 11.3, Observation (1), that $\|F\|$ can be written as $\inf\{k \, | \, |F(x)| \leq k\|x\|$ for all $x \in X\}$.] We wish to show now that the norm of F is actually equal to the norm of f. To prove this, we note that, if $x \in M$, then

$$|f(x)| = |F(x)| \leq \|F\| \, \|x\|,$$

which, by the same reasoning as used in getting (12.2), implies

$$\|f\| \leq \|F\|. \tag{12.3}$$

Combining (12.2) and (12.3), we have the desired result and the proof is complete. It is clear that, with only slight modification, the result also holds for real spaces.

Our next result also deals with the existence of a bounded linear functional having certain properties. The essential feature in the proof, as in the preceding theorem, is to fit the problem into a framework where the Hahn–Banach theorem may be applied.

THEOREM 12.2. Let x_0 be a nonzero vector in the normed linear space X. Then there exists a bounded linear functional F, defined on the whole space, such that

$$\|F\| = 1$$

and

$$F(x_0) = \|x_0\|.$$

Before proceeding to the proof we wish to note two immediate consequences of the theorem.

CONSEQUENCE 1. If X is not the trivial space (the vector space consisting solely of the zero vector), the conjugate space is not trivial either. In other words, nonzero bounded linear functionals must exist on any nontrivial normed space.

CONSEQUENCE 2. If all bounded linear functionals vanish on a given vector, the vector must be zero. This is certainly true, because one of the bounded linear functionals, when applied to the vector, must assume the norm of the vector as its value; hence the norm must be zero, which implies the vector is zero.

We now proceed to the proof.

Proof. Consider the subspace

$$M = [\{x_0\}]$$

consisting of all scalar multiples of x_0, and consider the functional f, defined on M as follows:

$$f : M \to F$$

$$\alpha x_0 \to \alpha \|x_0\|,$$

where F, as usual, denotes either the real or complex numbers. Clearly, f is a linear functional with the property that

$$f(x_0) = \|x_0\|.$$

Further, since for any $x \in M$,

$$|f(x)| = |\alpha| \, \|x_0\| = \|\alpha x_0\| = \|x\|, \qquad (12.4)$$

we see that f is a bounded linear functional. The last equation tells us that

$$\|f\| \leq 1$$

but, if there were a real constant k such that

$$k < 1 \quad \text{and} \quad |f(x)| \leq k\|x\| \quad \text{(for all } x \in M),$$

this would contradict the equality dictated by (12.4). Thus,

$$\|f\| = 1.$$

We have now established that f is a bounded linear functional, defined on the subspace M with norm 1. It now remains only to apply Theorem 12.1 to assert the existence of a bounded linear functional F, defined on the whole space, extending f, and having the same norm as f, that is, $\|F\| = 1$. The proof is, thus, seen to be complete.

In the same spirit as the above two results, our next theorem is concerned with the existence of a bounded linear functional, with still another property.

THEOREM 12.3. Let M be a subspace of the normed linear space X and suppose that $x_1 \in X$ has the property that the distance from x_1 to M, $d(x_1, M) = d$, is positive. Then there exists a bounded linear functional F, with norm 1, such that $F(x_1) = d$ and, for any $x \in M$, $F(x) = 0$.

REMARK. It is to be noted that if M is a closed proper subspace of X and $x_1 \notin M$, the conditions of the hypothesis are met. Usually, this will be the situation in which we shall have occasion to apply the theorem.

Proof. Consider

$$N = [M \cup \{x_1\}],$$

the subspace spanned by M and x_1. Any $y \in N$ can be written

$$y = x + \alpha x_1,$$

where $x \in M$ and α is a scalar. Also, since $x_1 \notin M$, the representation is unique. (Assuming y could be written in two different ways immediately implies that $x_1 \in M$.) Because of this uniqueness, we are assured that the functional f defined by taking

$$f(y) = \alpha d,$$

where y is as above and d as in the hypothesis, is a well-defined mapping from N into the scalar field. In addition f is clearly linear and has the property that

$$x \in M \Rightarrow f(x) = 0 \quad \text{and} \quad f(x_1) = d.$$

If we can now show that f is bounded (a bounded linear functional, that is) and has norm 1, we can apply Theorem 12.1 and complete the proof. To this end suppose α is a nonzero scalar and consider, with y as above,

$$\|y\| = \|x + \alpha x_1\| = \left\| -\alpha\left(-\frac{x}{\alpha} - x_1\right)\right\|$$

$$= |\alpha| \left\| -\frac{x}{\alpha} - x_1 \right\|.$$

Since $-(x/\alpha) \in M$ and $d = d(x_1, M) = \inf_{x \in M}\|x - x_1\|$, we have that

$$\left\| -\frac{x}{\alpha} - x_1 \right\| \geq d;$$

hence

$$\|y\| \geq |\alpha|d,$$

or

$$|f(y)| = |\alpha|d \leq \|y\|$$

for all $y \in N$. This implies that f is a bounded linear functional and that

$$\|f\| \leq 1. \tag{12.5}$$

For arbitrary $\varepsilon > 0$, by the definition of d, there must exist an $x \in M$ such that

$$\|x - x_1\| < d + \varepsilon. \tag{12.6}$$

For such an x consider

$$z = \frac{x - x_1}{\|x - x_1\|}.$$

Clearly, $\|z\| = 1$ and, by the definition of f and (12.6),

$$|f(z)| = \frac{d}{\|x - x_1\|} > \frac{d}{d + \varepsilon}. \tag{12.7}$$

Since, by Observation (3) of Sec. 11.3,

$$\|f\| = \sup_{\|y\| = 1} |f(y)|$$

and the fact that $\|z\| = 1$, Eq. (12.7) implies that

$$\|f\| > \frac{d}{d + \varepsilon}.$$

Since $\varepsilon > 0$ is arbitrary it follows that

$$\|f\| \geq 1. \tag{12.8}$$

Combining (12.5) and (12.8), we see that $\|f\| = 1$. By Theorem 12.1 it now follows that there is a bounded linear functional F, extending f, defined on the whole space, and such that

$$\|F\| = \|f\| = 1.$$

Our next result, also a consequence of the Hahn–Banach theorem, relates the separability of the dual space to the separability of the original space.

THEOREM 12.4. Denote the dual space of the normed linear space X by $\tilde{X}$; then if $\tilde{X}$ is separable, so is X.

REMARK. That the converse of this theorem is false is easily verified by considering the separable space $X = l_1$. In this case $\tilde{X}$ is congruent to l_∞—a nonseparable space.

Proof. Consider the set

$$S = \{f \in \tilde{X} \mid \|f\| = 1\}.$$

Since any subset of a separable metric space must be separable, S must be separable. In view of this, let

$$f_1, f_2, \ldots, f_n, \ldots$$

be a countable dense subset of S. Since each of the $f_n \in S$, then

$$\|f_n\| = 1$$

for all n. By Observation (3) of Section 11.3, it follows that, for each n, there must exist some vector x_n with norm 1 such that

$$|f_n(x_n)| > \tfrac{1}{2}.$$

(If such x_n did not exist, this would contradict the fact that $\|f_n\| = 1$.) Now let

$$M = \overline{[\{x_1, x_2, \ldots, x_n, \ldots\}]}$$

and suppose that

$$M \neq X. \tag{12.9}$$

By virtue of (12.9) there must exist some vector $x_0 \in X - M$. Consequently, the distance $d = d(x_0, M)$ must be strictly positive and Theorem 12.3 is applicable. On the basis of that theorem, there must exist some bounded linear functional F such that

$$\|F\| = 1,$$

$$F(x_0) \neq 0,$$

and

$$x \in M \Rightarrow F(x) = 0.$$

Hence, since $\|F\| = 1$, $F \in S$, and $F(x_n) = 0$ $(n = 1, 2, \ldots)$, because each $x_n \in M$. Now consider

$$f_n(x_n) = f_n(x_n) - F(x_n) + F(x_n);$$

thus

$$|f_n(x_n)| \leq |f_n(x_n) - F(x_n)| + |F(x_n)|. \tag{12.10}$$

But, since $F(x_n) = 0$ and

$$f_n(x_n) - F(x_n) = (f_n - F)(x_n),$$

we can rewrite (12.10) as

$$|f_n(x_n)| \leq |(f_n - F)(x_n)|.$$

Hence

$$\tfrac{1}{2} < |f_n(x_n)| \leq |(f_n - F)(x_n)| \leq \|f_n - F\| \, \|x_n\|$$

or, since $\|x_n\| = 1$,

$$\tfrac{1}{2} < \|f_n - F\|$$

for every n. This last statement contradicts the fact that $f_1, f_2, \ldots, f_n, \ldots$ was a dense subset of S, since, because they are dense in S and $F \in S$, there must be some f_{n_ε}, for any $\varepsilon > 0$, such that $\|f_{n_\varepsilon} - F\| < \varepsilon$. Hence our lone assumption (12.9) must be incorrect and M must indeed be the whole space, so the set of all (finite) linear combinations of the x_i is dense in X. To verify that this truly implies that X is separable we give the following "counting argument":

(1) By a *gaussian rational number* we mean a number of the form $a + bi$, where a and b are rational.

(2) Since the cartesian product of a finite number of denumerable sets is, itself, a denumerable set, the set of all gaussian rationals is a denumerable set, and we can list them as follows:

$$\alpha_1, \alpha_2, \ldots, \alpha_n, \ldots.$$

(3) Since the rationals are dense in the real numbers, the gaussian rationals are dense in the complex numbers.

(4) Since the set $\{\alpha_m x_n\}$ with m fixed and $n = 1, 2, \ldots$ is denumerable and the fact that a denumerable union of denumerable sets is denumerable, the set

$$\bigcup_{n, m = 1}^{\infty} \{\alpha_m x_n\}$$

is denumerable.

(5) Now, since the set of all finite linear combinations of this set constitutes a denumerable dense set, and since $M = X$, the separability of X follows.

12.2 The Second Conjugate Space

Having defined the conjugate space of a normed linear space to be the set consisting of all bounded linear functionals on the space, we now wish to ascertain some things about the conjugate space of this normed linear space. To avoid excessive use of the awkward "conjugate space of the conjugate space," we shall often refer to this as the *second conjugate space*. Before proceeding, we wish to introduce a convenient notation concerning linear functionals on a vector space.

Let y be a linear functional on the vector space X. We shall denote $y(x)$, where $x \in X$, by $[x, y]$. Thus whenever such a bracket appears it is understood that the vector appears in the first argument and the linear functional appears in the second. An important property of the bracket is that it is bilinear—it is linear in both arguments. To paraphrase this, if α_1 and α_2 are scalars, and x_1 and x_2 are vectors,

$$[\alpha_1 x_1 + \alpha_2 x_2, y] = \alpha_1[x_1, y] + \alpha_2[x_2, y]$$

and, similarly, if y_1 and y_2 are linear functionals,

$$[x, \alpha_1 y_1 + \alpha_2 y_2] = \alpha_1[x, y_1] + \alpha_2[x, y_2].$$

Suppose now that we adopted a slightly different viewpoint from the one we usually take; suppose instead of letting x vary and holding y fixed, we held x fixed and let y vary throughout the vector space of linear functionals on X. In this case

$$[x, y] \qquad (x \text{ fixed, } y \text{ varying})$$

constitutes, by virtue of the bilinearity, a linear functional on the space of linear functionals. Further, if X is a *normed* space and y ranges through $\tilde{X}$, the *bounded* linear functionals on X, the above process yields a bounded linear functional on $\tilde{X}$ (that is, a member of $\tilde{\tilde{X}}$). For consider the mapping

$$J : X \to \tilde{\tilde{X}}$$

$$x \to x'',$$

where the linear functional x'' is defined as follows: Let $x' \in \tilde{X}$; then

$$x''(x') = [x', x''] = [x, x'].$$

With every vector $x \in X$, we are associating a linear functional, x'' on $\tilde{X}$. The only difficulty that might arise here is one of category; that is, it is somewhat incorrect to say that x *is* the linear functional on $\tilde{X}$. Instead, one should say every vector is *associated* (uniquely) with a linear functional on $\tilde{X}$. Our next job is to verify that x'' is indeed a *bounded* linear functional on $\tilde{X}$. To do this consider, for $x' \in \tilde{X}$,

$$|x''(x')| = |x'(x)| \le \|x'\| \|x\|.$$

Thus the constant $\|x\|$ is a bound (in the sense of a bounded linear functional) for x''; furthermore, by a now standard argument,

$$\|x''\| \le \|x\|. \tag{12.11}$$

Since the identification between x and x'' is so strong, it seems reasonable to suspect that an even stronger statement can be made about the relationship between $\|x''\|$ and $\|x\|$, namely, that $\|x''\| = \|x\|$, and indeed this is so.

To verify this contention, first consider the case when $x = 0$; in this event we have

$$0 \leq \|x''\| \leq 0$$

and, consequently,

$$\|x''\| = \|x\|.$$

If x is a nonzero vector then, by Theorem 12.2, there must be some bounded linear functional x_0' with norm 1, such that $x_0'(x) = \|x\|$. But

$$\|x''\| = \sup_{\|x'\|=1} |x''(x')| = \sup_{\|x'\|=1} |x'(x)|,$$

which, since $\|x\| = |x_0'(x)| \leq \sup_{\|x'\|=1} |x'(x)|$, implies,

$$\|x''\| \geq \|x\|. \tag{12.12}$$

Combining this with (12.11), we see that the contention is verified.

Let us look at the mapping J in a little more detail now. Since, as noted previously, the bracket is bilinear, it readily follows that J is a linear mapping between the vector spaces X and $\tilde{X}$ or, to paraphrase this, J is a linear transformation. We have just noted that

$$\|x''\| = \|Jx\| = \|x\|,$$

which implies that *J is an isometry*. Since an isometry is always a $1 : 1$ *mapping*, it follows that *J is an isomorphism*. Note however that we are not saying the spaces X and $\tilde{X}$ are isomorphic; that is, J is also onto. To cover this special situation, we make the following definition.

DEFINITION 12.1. If the mapping J is onto $\tilde{X}$, then X is called *reflexive*.

For linear transformations between finite-dimensional spaces of the same dimension if such a linear transformation is $1 : 1$, it immediately follows that it is onto. We noted above that J is always $1 : 1$; therefore, since it is simple to verify that the conjugate space of a finite-dimensional space is also finite-dimensional and of the same dimension, every finite-dimensional space is reflexive. To round out the picture, we now wish to note the existence of nonreflexive spaces. In Theorem 11.6 it was shown that the conjugate space of any space was a Banach space. Thus $\tilde{X}$ must always be a complete normed space. In view of this, no incomplete space has a chance of being reflexive. (Note that for two metric spaces that are isometric, if one is complete, so is the other.) It can also be shown that every Hilbert space possesses the property of reflexivity, as will be shown in Sec. 12.5.

Looking back at the above discussion one might wonder why only bounded linear functionals were considered: Why didn't we consider the vector space of *all* linear functionals on the space? The answer is, loosely speaking, that this class of functionals is just too big to handle conveniently. Further, if we considered the mapping J above as being into this class, it can be shown that, if X is not finite-dimensional, J is never an onto map (onto the class of all linear functionals over the space of *all* linear functionals, that is) and in Exercise 12.10 the reader is asked to verify this claim.

12.3 The Conjugate Space of l_p

Thus far we have been discussing the conjugate space in the general case. We now wish to ascertain some facts about the conjugate space of the specific normed space l_p, where $1 < p < \infty$. Letting $\tilde{l}_p$ denote the conjugate space of l_p, it will be shown here that $\tilde{l}_p$ is congruent to l_q, where $(1/p) + (1/q) = 1$, but before doing this we wish to note some facts about the "signum" function, sgn α.

For a complex number α we define

$$\operatorname{sgn} \alpha = \begin{cases} 0 & (\alpha = 0), \\ \dfrac{\alpha}{|\alpha|} & (\alpha \neq 0). \end{cases}$$

The following two observations are immediately evident:

$$|\operatorname{sgn} \alpha| = \begin{cases} 0 & (\alpha = 0), \\ 1 & (\alpha \neq 0), \end{cases} \tag{12.13}$$

$$\alpha \operatorname{sgn} \bar{\alpha} = \begin{cases} 0 & (\alpha = 0), \\ \dfrac{\alpha \bar{\alpha}}{|\alpha|} = |\alpha| & (\alpha \neq 0). \end{cases} \tag{12.14}$$

Let $x = (\xi_1, \xi_2, ..., \xi_n, ...)$ be a generic element of l_p; this implies

$$\sum_{n=1}^{\infty} |\xi_n|^p < \infty. \tag{12.15}$$

Next consider the sequence of elements from l_p:

$$e_1 = (1, 0, 0, ..., 0, ... \quad)$$
$$e_2 = (0, 1, 0, ..., 0, ... \quad)$$
$$e_3 = (0, 0, 1, 0, ..., 0, ...)$$
$$\vdots \qquad\qquad \vdots$$

Since

$$x - \sum_{k=1}^{n} \xi_k e_k = (0, 0, ..., \xi_{n+1}, \xi_{n+2}, ...),$$

we see that

$$\left\| x - \sum_{k=1}^{n} \xi_k e_k \right\| = \left(\sum_{k=n+1}^{\infty} |\xi_k|^p \right)^{1/p} \to 0,$$

since it is the "tail" of the convergent series (12.15). Equivalently, we have

$$x = \sum_{k=1}^{\infty} \xi_k e_k.$$

Let $f \in l_p$ and let

$$s_n = \sum_{k=1}^{n} \xi_k e_k .$$

Since f is linear,

$$f(s_n) = \sum_{k=1}^{n} \xi_k f(e_k).$$

Since $s_n \to x$ and f must be continuous because it is a bounded linear transformation,

$$f(s_n) \to f(x)$$

or, equivalently, we arrive at the intuitively plausible result

$$f(x) = \sum_{k=1}^{\infty} \xi_k f(e_k).$$

We now proceed to our main result.

CONTENTION. Letting $\alpha_k = f(e_k)$ $(k = 1, 2, ...)$, where $f \in l_p$, $p > 1$, and $(1/p) + (1/q) = 1$, the mapping

$$T : l_p \to l_q$$

$$f \to (\alpha_1, \alpha_2, ..., \alpha_n, ...)$$

constitutes a congruence between l_p and l_q.

REMARK. In considering finite-dimensional spaces, one notes that a linear functional can be completely characterized by knowing only what it does to each vector in some fixed set—the basis. The result here is quite analogous in that bounded linear functionals will be characterized only by what values they assume on the set e_k $(k = 1, 2, ...)$.

Proof of Contention. It will first be shown that the mapping does indeed take l_p into l_q. Consider the vector from l_p

$$x = (\beta_1, \beta_2, ..., \beta_n, 0, 0, ...),$$

where

$$\beta_k = \begin{cases} |\alpha_k|^{q-1} \operatorname{sgn} \bar{\alpha}_k, & (1 \le k \le n), \\ 0 & (n > k), \end{cases}$$

and α_k is as in the statement of the contention. We note for $1 \le k \le n$, using Eq. (12.13),

$$|\beta_k| = |\alpha_k|^{q-1}.$$

Now, since $(1/p) + (1/q) = 1$, $p(q-1) = q$, and we have

$$|\beta_k|^p = |\alpha_k|^{(q-1)p} = |\alpha_k|^q.$$

Thus, by (12.14),

$$\alpha_k \beta_k = |\alpha_k|^q = |\beta_k|^p, \tag{12.16}$$

which implies

$$\|x\| = \left(\sum_{k=1}^{n} |\beta_k|^p \right)^{1/p}$$

$$= \left(\sum_{k=1}^{n} |\alpha_k|^q \right)^{1/p}. \tag{12.17}$$

Now, since we can write

$$x = \sum_{k=1}^{n} \beta_k e_k,$$

it follows that

$$f(x) = \sum_{k=1}^{n} \beta_k f(e_k) = \sum_{k=1}^{n} \alpha_k \beta_k,$$

which, by (12.16), yields

$$f(x) = \sum_{k=1}^{n} |\alpha_k|^q. \tag{12.18}$$

But for every $x \in l_p$, we have

$$|f(x)| \le \|f\| \, \|x\|,$$

which, using (12.17) and (12.18), becomes

$$|f(x)| = \sum_{k=1}^{n} |\alpha_k|^q \le \|f\| \left(\sum_{k=1}^{n} |\alpha_k|^q \right)^{1/p}.$$

Dividing through by the last term on the right, we now obtain

$$\left(\sum_{k=1}^{n} |\alpha_k|^q \right)^{1/q} \le \|f\|,$$

and the sequence of partial sums

$$s_n = \left(\sum_{k=1}^{n} |\alpha_k|^q \right)^{1/q}$$

is seen to be a bounded, monotone increasing, sequence which, therefore, converges. Hence the image of f under T, the sequence $\{\alpha_k\}$, is indeed a member of l_q.

Next we wish to show that T is an isometry, or that $\|Tf\| = \|f\|$. As an immediate consequence of the above equation, we have

$$\left(\sum_{k=1}^{\infty} |\alpha_k|^q \right)^{1/q} = \|Tf\| \le \|f\|;$$

hence it only remains to get the inequality going the other way around to show that T is an isometry. Suppose that $x \in l_p$ and write

$$x = \sum_{k=1}^{\infty} \xi_k e_k.$$

Hence

$$f(x) = \sum_{k=1}^{\infty} \xi_k f(e_k) = \sum_{k=1}^{\infty} \xi_k \alpha_k$$

and, by the triangle and Hölder inequalities,†

$$|f(x)| \le \sum_{k=1}^{\infty} |\xi_k|\, |\alpha_k| \le \left(\sum_{k=1}^{\infty} |\alpha_k|^q \right)^{1/q} \left(\sum_{k=1}^{\infty} |\xi_k|^p \right)^{1/p}.$$

But since $\left(\sum_{k=1}^{\infty} |\xi_k|^p \right)^{1/p}$ is just $\|x\|$, the last inequality is seen to be

$$|f(x)| \le \left(\sum_{k=1}^{\infty} |\alpha_k|^q \right)^{1/q} \|x\| \qquad \text{(for all } x\text{)},$$

which implies

$$\|f\| \le \left(\sum_{k=1}^{\infty} |\alpha_k|^q \right)^{1/q} = \|Tf\|;$$

therefore,

$$\|Tf\| = \|f\|,$$

and T is seen to be an isometry. Since T is clearly a linear transformation and is also $1 : 1$ because it is an isometry, it only remains to show that T is onto l_q to complete the proof. Hence for every point $\{\beta_k\}$ of l_q, we must find a linear functional $g \in \bar{l}_p$ that maps into $\{\beta_k\}$. Consider $x \in l_p$, where

$$x = \sum_{k=1}^{\infty} \xi_k e_k. \tag{12.19}$$

We now claim that such a functional is given by

$$g(x) = \sum_{k=1}^{\infty} \xi_k \beta_k.$$

To verify this, we first note that since representation (12.19) is unique, g is well-defined. Clearly g is linear on l_p and, to show that it is bounded, consider (using the triangle and Hölder inequalities)

$$|g(x)| \le \sum_{k=1}^{\infty} |\beta_k \xi_k| \le \left(\sum_{k=1}^{\infty} |\xi_k|^p \right)^{1/p} \left(\sum_{k=1}^{\infty} |\beta_k|^q \right)^{1/q}$$

or, since $\|x\| = \left(\sum_{k=1}^{\infty} |\xi_k|^p \right)^{1/p}$,

$$|g(x)| \le \left(\sum_{k=1}^{\infty} |\beta_k|^q \right)^{1/q} \|x\|;$$

therefore g is bounded. To find out what T maps g into, all we need know is what $g(e_k)$ is for $k = 1, 2, \dots$. Since $g(e_k) = \beta_k$, for any k, we see that $Tg = \{\beta_k\}$ and T is seen to be an onto map. The proof of the contention is now complete.

We now wish to draw four immediate consequences of the above result.

CONSEQUENCE 1. With x an element of l_p as above and f a bounded linear functional on l_p, we have the unique representation for f:

$$f(x) = \sum_{k=1}^{\infty} \xi_k f(e_k).$$

---

† Here, in applying the Hölder inequality, we use the fact that $p > 1$.

CONSEQUENCE 2. The norm of f is given by

$$\|f\| = \left(\sum_{k=1}^{\infty} |f(e_k)|^q \right)^{1/q}.$$

CONSEQUENCE 3. For the Hilbert space l_2 (recall that l_2 is a Hilbert space but all the other l_p, $p > 2$, are not) we can say it is, essentially, its own dual space; more precisely, l_2 is congruent to $\tilde{l}_2$.

CONSEQUENCE 4. In the special setting of l_2 suppose x and f are as above. By Consequences 1 and 2,

$$\|f\| = \left(\sum_{k=1}^{\infty} |\alpha_k|^2 \right)^{1/2}$$

and

$$f(x) = \sum_{k=1}^{\infty} \xi_k \alpha_k.$$

Alternatively, one might say, by defining y to be the sequence $\{\bar{\alpha}_k\}$, that there is a unique vector $y \in l_2$ such that $f(x) = (x, y)$ for all $x \in l_2$. Thus we have achieved an inner product representation for bounded linear functionals on l_2.

12.4 The Riesz Representation Theorem for Linear Functionals on a Hilbert Space

Thus far we have already met representation theorems for linear functionals; for example, in Chapter 1 (Theorem 1.9), we showed that any linear functional over a finite-dimensional inner product space was representable by an inner product with some unique vector y. Again, in the preceding section, Consequence 4 is a representation for a bounded linear functional on l_2. Theorems of this type allow us, essentially, to devote our full efforts to only one specific type of linear functional with no loss of generality, and it is for this reason that they are so useful in functional analysis. In addition they also have wide application in analysis since, there, they yield existence statements (about certain integrals, etc.) but, for these applications, one clearly needs theorems about representations on infinite-dimensional spaces. In this spirit we now present the following representation theorem for linear functionals on a Hilbert space.

THEOREM 12.5 (*Riesz*). f is a bounded linear functional on the Hilbert space X if and only if there exists a unique vector $y \in X$ such that $f(x) = (x, y)$ for all $x \in X$. Alternatively, denoting the conjugate space of X by $\tilde{X}$, one might say that

$$\tilde{X} = \{f_y(x) = (x, y) | y \in X\}.$$

Proof. In any case, it is clear that such a y must be unique, for if y_1 and y_2 each had the above property, this would mean

$$(x, y_1) = (x, y_2)$$

or

$$(x, y_1 - y_2) = 0$$

for every x. It remains only to choose $x = y_1 - y_2$ to prove that $y_1 = y_2$. We now show the condition to be sufficient. Suppose $y \in X$ and consider the functional f defined by

$$f(x) = (x, y).$$

Clearly, f is linear. To show it is bounded, consider (using the Cauchy–Schwarz inequality)

$$|f(x)| = |(x, y)| \leq \|x\| \, \|y\|.$$

Thus f is bounded and

$$\|f\| \leq \|y\|.$$

But for $x = y$, we have

$$|f(y)| = \|y\| \, \|y\|;$$

hence nothing smaller than $\|y\|$ will work as a bound for $f(x)$, and we have

$$\|f\| = \|y\|.$$

Proof (*necessity*). Suppose f is a bounded linear functional on X and let M denote the null space of f:

$$M = \{x \in X \mid f(x) = 0\}.$$

By merely using the continuity of f, one can easily prove that M is a closed subspace of X. In the case that $M = X$, f is trivial, and $y = 0$ will satisfy the conditions of the theorem, that is,

$$0 = f(x) = (x, 0).$$

On the other hand, suppose $M \neq X$. In this case there must be a nonzero vector w, such that $w \in M^{\perp}$. To prove this we could simply apply Theorem 10.6 or note that, if no such w existed, $M^{\perp} = \{0\}$, which, taking orthogonal complements, implies

$$M^{\perp\perp} = M = \{0\}^{\perp} = X,$$

the relation $M = M^{\perp\perp}$ being true because M is a closed subspace of a Hilbert space [Observation (5) of Sec. 10.4]. At any rate, such a w must exist. We now make the following contention.

CONTENTION. There exists a scalar α such that $y = \alpha w$ satisfies the conditions of the theorem.

To prove this we shall look at, separately, different "types" of vectors. The first type we wish to consider are those that are in M, the null space of f. Thus let $x \in M$ and compute

$$f(x) = 0 = (x, \alpha w) = \bar{\alpha}(x, w) = 0.$$

Thus, for vectors in M, *any* multiple of w works. Still having the choice of a particular α at our disposal, we now consider, as the second type, those vectors that are multiples of w or [$\{w\}$]. In this case let $x = \beta w$ (it is assumed that $\beta \neq 0$ for if it were, x would belong to M) and consider

$$\beta f(w) = f(\beta w) \qquad \text{and} \qquad (\beta w, y) = (\beta w, \alpha w) = \beta \bar{\alpha}(w, w).$$

Hence $f(\beta w) = (\beta w, y)$ if and only if

$$f(w) = \bar{\alpha}(w, w)$$

or

$$\alpha = \frac{\overline{f(w)}}{(w, w)} = \frac{\overline{f(w)}}{\|w\|^2}.$$

We now see that, with α as above, $y = \alpha w$ satisfies the conditions of the theorem for all $x \in M$ and all $x \in [\{w\}]$. Last, let x be any vector in the space and form the difference

$$x - \beta w,$$

where

$$\beta = \frac{f(x)}{f(w)},$$

it being clear that $f(w) \neq 0$ because $w \notin M$. We now have

$$f(x - \beta w) = f(x) - \beta f(w) = 0$$

and conclude that $x - \beta w \in M$. Writing

$$x = x - \beta w + \beta w,$$

we see that

$$f(x) = f(x - \beta w) + f(\beta w). \tag{12.20}$$

But the argument of the first term on the right is in M, which implies

$$f(x - \beta w) = (x - \beta w, \alpha w), \tag{12.21}$$

while the argument of the second term is a multiple of w, which implies

$$f(\beta w) = (\beta w, \alpha w) \tag{12.22}$$

Combining (12.21) and (12.22) into (12.20), we have, letting $y = \alpha w$

$$f(x) = (x - \beta w, y) + (\beta w, y)$$
$$= (x, y),$$

and the theorem is proved.

12.5 Reflexivity of Hilbert Spaces

In this section, we wish to show that every Hilbert space is reflexive. One recalls (see Definition 12.1) that this means that a certain mapping, denoted by J, from the space into the second conjugate space be onto. Before showing this, let us examine the following mapping: Let X be a Hilbert space, let $\tilde{X}$ denote the conjugate space of X, and consider the mapping T defined by

$$T: X \to \tilde{X},$$

$$y \to Ty = f, \tag{12.23}$$

where the bounded linear functional f is, for any $x \in X$, given by

$$(Ty)(x) = f(x) = (x, y). \qquad (12.24)$$

Suppose now that, under T,

$$y_1 \rightarrow f_1$$

and

$$y_2 \rightarrow f_2$$

and let

$$y_1 + y_2 \rightarrow g.$$

Thus,

$$g(x) = (x, y_1 + y_2)$$

$$= (x, y_1) + (x, y_2)$$

$$= f_1(x) + f_2(x),$$

and we conclude that

$$T(y_1 + y_2) = Ty_1 + Ty_2;$$

hence, T is additive. Now suppose,

$$T : y \rightarrow f,$$

and, for a scalar α, let $T(\alpha y) = h$. In this case,

$$h(x) = (x, \alpha y) = \bar{\alpha}(x, y) = \bar{\alpha} f(x);$$

therefore,

$$T(\alpha y) = \bar{\alpha} T y,$$

and T is seen to be *conjugate linear*.

The Riesz representation theorem for bounded linear functionals on a Hilbert space, proved in Sec. 12.4, states that, for every bounded linear functional on $\tilde{X}$, g, there exists a unique $y \in X$ such that, for any $x \in X$, $g(x) = (x, y)$ and that $\|g\| = \|y\|$. In view of this, the mapping T is seen to be onto all of $\tilde{X}$, and, further, for y and f as in the defining equation for T, Eq. (12.23),

$$\|y\| = \|f\| = \|Ty\|;$$

therefore, T is a norm-preserving mapping or isometry. Since an isometry is always a 1 : 1 mapping, we can summarize as follows: The mapping T constitutes a 1 : 1 onto, isometric, conjugate linear mapping from a Hilbert space into its conjugate space. Thus, we see that Hilbert spaces and their conjugate spaces are indistinguishable metrically and "almost" indistinguishable algebraically. Were it not for the fact that T was *conjugate* linear instead of linear, the spaces would be congruent and we could remove the word "almost" from the preceding sentence.

To arrive at the above properties of T, one notes how strongly the fact that X was a Hilbert space was used via the Riesz representation theorem. In the next result, we shall again make strong use of the fact that the space is a Hilbert space by using that same theorem.

THEOREM 12.6. If X is a Hilbert space, then X is reflexive.

Proof. Let g be a bounded linear functional on X and $x \in X$. We have, as the definition of $[x, g]$,

$$g(x) = [x, g]. \qquad (12.25)$$

We note in this equation, if x is held fixed and g is allowed to range through the conjugate space of X, $\tilde{X}$, that this defines a bounded linear functional on $\tilde{X}$. Thus, to every element x of X, we associate some member of the conjugate space of $\tilde{X}$, $\tilde{\tilde{X}}$, which we shall denote by Jx. Symbolically, we have

$$J: X \to \tilde{\tilde{X}},$$
$$x \to Jx,$$

(12.26)

where, for the defining equation for Jx we have, for any $g \in \tilde{X}$,

$$(Jx)(g) = [g, Jx] = [x, g] = g(x).$$

(12.27)

We now proceed to show that, for Hilbert spaces, the mapping J is onto all of $\tilde{\tilde{X}}$. To this end, let f be any element of $\tilde{\tilde{X}}$. We must find $z \in X$ such that $Jz = f$. For T as defined in Eq. (12.24), consider the functional g defined below:

$$g: X \to F,$$
$$x \to \overline{f(Tx)}.$$

First, it is noted that the above operation is categorically sensible; namely, $Tx \in \tilde{X}$ and $f \in \tilde{\tilde{X}}$. Let us investigate the mapping g a little further now. For $x_1, x_2 \in X$ consider

$$g(x_1 + x_2) = \overline{f\big(T(x_1 + x_2)\big)}$$
$$= \overline{f(Tx_1 + Tx_2)}$$
$$= \overline{f(Tx_1)} + \overline{f(Tx_2)}$$
$$= g(x_1) + g(x_2);$$

(12.28)

thus, g is seen to be additive. Now let $x \in X$, $\alpha \in F$, and consider

$$g(\alpha x) = \overline{f(T\alpha x)}$$
$$= \overline{f(\bar{\alpha}Tx)}$$
$$= \bar{\alpha}\overline{f(Tx)}$$
$$= \alpha g(x).$$

(12.29)

Combining (12.28) and (12.29), the linearity of g is proved. We now wish to show that g is a *bounded* linear functional. To this end, recall that T is an isometry and consider

$$|g(x)| = |\overline{f(Tx)}| = |f(Tx)| \le \|f\| \, \|Tx\| = \|f\| \cdot \|x\|$$

or

$$|g(x)| \le \|f\| \cdot \|x\|;$$

hence, $\|f\|$ is seen to serve as a bound for the linear functional g. To paraphrase this, we can say $g \in \tilde{X}$, and appealing to the Riesz representation theorem of Sec. 12.4, we can assert the existence of $z \in X$ such that, for all $x \in X$,

$$g(x) = (x, z).$$

Equivalently, we have

$$\overline{f(Tx)} = (x, z)$$

or

$$f(Tx) = (z, x). \tag{12.30}$$

By the definitions of J [Eq. (12.27)] and T [Eq. (12.24)], we see that

$$(Jz)(Tx) = (Tx)(z) = (z, x). \tag{12.31}$$

[Note that the expression $(Jz)(Tx)$ is categorically sensible because $Tx \in \tilde{X}$ and $Jz \in \tilde{\tilde{X}}$.] Combining (12.30) and (12.31) and using the fact that any bounded linear functional can be written in the form Tx (T was shown to be onto $\tilde{X}$), we have that the bounded linear functionals f and Jz agree on every member of $\tilde{X}$; hence, the linear functionals must be the same and the proof is complete, the vector z being the desired vector.

EXERCISES 12

1. Let X be a normed linear space. Show that if $f(x) = f(y)$ for all $f \in \tilde{X}$, then $x = y$.

2. Let f be a linear functional on the Hilbert space X; let N be the null space of f. Show that, if f is not continuous, then $\bar{N} = X$.

3. Prove that if f is a linear functional on the Hilbert space X, with null space N, then f is continuous if and only if N is a closed subspace.

4. Let f be a linear functional on the vector space X over F, $f \neq 0$, and N the null space of f. Show that there exists a vector y such that every $x \in X$ can be written uniquely in the form: $x = \lambda y + z$, where $z \in N$ and $\lambda \in F$.

5. Let X be a normed linear space and let $x_1, x_2, \ldots, x_n, \ldots$ be a sequence of elements of X. Show that $y \in \overline{[\{x_1, \ldots, x_n, \ldots\}]}$ if and only if $f(y) = 0$ for all bounded linear functionals f which vanish on $x_1, x_2, \ldots$.

6. Show that, if two linear functionals on a vector space X have the same null space, they must be scalar multiples of each other.

7. Let x be an element of the normed linear space X. Prove that $\|x\| = \sup\{|f(x)| \mid f \in \tilde{X}, \|f\| = 1\}$.

8. Prove that a closed subspace of a reflexive Banach space is reflexive.

9. Show that, if X is a Banach space, then X is reflexive if and only if $\tilde{X}$ is reflexive.

10. Let $\hat{X}$ be the space of *all* linear functionals on the linear space X. Define $\hat{J} : X \to \hat{\hat{X}}$ analogously to the mapping J, that is, for $x \in X$:

$$[x', \hat{J}x] = [x, x']$$

for all $x' \in \hat{X}$. Show that $\hat{J}$ is an onto mapping if and only if X is finite-dimensional.

11. If X is an inner product space and if $M^{\perp\perp} = M$ for every closed subspace M of X, show that X is a Hilbert space. [*Hint:* Use the mapping T mentioned in Eq. (12.23).]

REFERENCES

1. A. Taylor, "Introduction to Functional Analysis."
2. F. Riesz and B. Sz.-Nagy, "Functional Analysis."
3. A. Zaanen, "Linear Analysis."

CHAPTER 13

The Conjugate Space of $C[a, b]$

Results of extreme importance in functional analysis are representation theorems such as the representation theorem obtained for bounded linear functionals on l_p in the last chapter. Here we shall obtain a similar result for the bounded linear functionals on $C[a, b]$—complex-valued, continuous functions defined on the closed interval $[a, b]$; addition of functions and multiplication by a scalar is defined in the usual way (for function spaces) and the norm of $x \in C[a, b]$ is taken as

$$\|x\| = \max_{t \in [a, b]} |x(t)|. \tag{13.1}$$

It will be shown here that, to every bounded linear functional f on this space, there is some corresponding function of bounded variation $g(t)$ defined on $[a, b]$ with the property that, for every $x \in C[a, b]$, the value of $f(x)$ is given by the Riemann–Stieltjes integral

$$f(x) = \int_a^b x(t) \, dg(t). \tag{13.2}$$

In this way we shall have achieved a way to represent *any* bounded linear functional on $C[a, b]$. Needless to say, armed with the information that any bounded linear functional on $C[a, b]$ can be dressed in the above guise, when further facts about the conjugate space of $C[a, b]$ are desired, we need only examine this very particular type expression; thus we have discovered a great deal.

Just as in the preceding chapter, when we found that the conjugate space of l_p was congruent to l_q, it will be shown here that the conjugate space of $C[a, b]$ is congruent to a subspace of the (normed linear space of) complex-valued functions of finite variation on $[a, b]$. The whole chapter will be devoted to obtaining the mapping that will establish this congruence. The first stage of the proof will be Theorem 13.1 to show that to every bounded linear functional f on $C[a, b]$ there is *some* function of bounded variation $g(t)$ on $[a, b]$ satisfying (13.2). In view of this, one might think that we might be able to take this as the desired mapping. Unfortunately, however, the correspondence between bounded linear functionals f and functions of bounded variation $g(t)$ satisfying (13.2) is not $1:1$, so this possibility is out. What must be done to salvage it is to find some way of cutting down the number of functions of bounded variation that can correspond to f by satisfying (13.2). To further this end, we next introduce an equivalence relation between functions of bounded variation. Once this is done, only equivalence classes remain. Whenever an equivalence relation is introduced on a set it is always of interest to determine what the classes "look like" or to find out what a representative of the class is, and here is no exception. We show that every equivalence class can be represented by a *normalized* function of bounded variation and at last we are in business. Our main result of the chapter (Theorem 13.2) then shows that a congruence exists between the conjugate space of $C[a, b]$ and the normalized functions of bounded variation.

Some facts about functions of bounded variation will be needed throughout the course of the chapter and they will be listed separately as they are needed. As noted in Example 8.8, the space $BV[a, b]$ is a Banach space with norm of the function of bounded variation on $[a, b]$, $x(t)$, taken as

$$\|x\| = V(x) + |x(a)|, \tag{13.3}$$

where $V(x)$ denotes the total variation of $x(t)$ on $[a, b]$. The notation $BV[a, b]$ will be used throughout this chapter to denote this Banach space.

Another space that we shall have to consider is $B[a, b]$: all those complex-valued functions defined on $[a, b]$ such that, for some real number M,

$$|f(x)| \leq M.$$

(Note that we do not require the same M to work for every f.)

It is clear that this space, the space of all bounded functions on $[a, b]$, is a linear space (usual definition of addition and scalar multiplication). Further, it is a normed space (in fact, a Banach space) with norm of $x \in B[a, b]$ taken as

$$\|x\| = \sup_{t \in [a, b]} |x(t)|. \tag{13.4}$$

One can easily verify that $C[a, b]$ is a subspace of the normed linear space $B[a, b]$.

As a final word on notation, we shall denote the conjugate space of $C[a, b]$ by $\tilde{C}[a, b]$. We now proceed to establish the first link in the chain that will lead to the congruence mapping mentioned before.

13.1 A Representation Theorem for Bounded Linear Functionals on $C[a, b]$

The following theorem, an existence statement, is one in which we shall actually construct a function having the desired properties.

THEOREM 13.1. Let $f \in \tilde{C}[a, b]$. Then there exists a function $g(t) \in BV[a, b]$ such that, for all $x \in C[a, b]$,

$$f(x) = \int_a^b x(t) \, dg(t)$$

and such that

$$\|f\| = V(g),$$

where $V(g)$ denotes the total variation of $g(t)$.

Proof. By virtue of Theorem 12.1, if we view $C[a, b]$ as a subspace of $B[a, b]$, there exists a bounded linear functional F defined on all of $B[a, b]$, extending f, and such that $\|F\| = \|f\|$. Now for $a < s \leq b$, consider the characteristic function of $[a, s]$, $k_{[a, s]}(t)$:

$$k_{[a, s]}(t) = \begin{cases} 1 & (t \in [a, s]), \\ 0 & (t \notin [a, s]). \end{cases}$$

Clearly, for every such s,

$$k_{[a, s]}(t) \in B[a, b].$$

With F, the extension of f, as above, we now define (omitting the argument of the characteristic function)

$$g(s) = F(k_{[a, s]}) \qquad (a < s \leq b), \tag{13.5}$$

and

$$g(a) = 0;$$

also note that, categorically, this consideration is sensible since F is defined on $B[a, b]$. Next we partition $[a, b]$ as follows:

$$a = t_0 < t_1 < t_2 < \cdots < t_n = b,$$

and define, for $i = 1, 2, \ldots, n$,

$$\alpha_i = \overline{\operatorname{sgn}} \,[g(t_i) - g(t_{i-1})], \tag{13.6}$$

where the bar above "sgn" denotes, as usual, the complex conjugate. (The function sgn α is discussed in Sec. 12.3.)

We now define

$$y(t) = \begin{cases} \alpha_1 & (t_0 \le t \le t_1), \\ \alpha_i & (t_{i-1} < t \le t_i, \quad i = 2, \ldots, n) \end{cases} \tag{13.7}$$

and claim that

$$y(t) \in B[a, b].$$

To verify this claim we need only note that

$$|\alpha_i| = 0 \text{ or } 1 \quad (i = 1, 2, \ldots, n), \tag{13.8}$$

which implies that not only is $y(t)$ bounded but

$$\|y\| \le 1. \tag{13.9}$$

Our next contention is that

$$y(t) = \sum_{i=1}^{n} \alpha_i(y_i(t) - y_{i-1}(t)), \tag{13.10}$$

where

$$y_i = k_{[a, \, t_i]} \quad (i = 1, 2, \ldots, n), \tag{13.11}$$

and

$$y_0 = 0.$$

We shall actually verify now that (13.10) is valid in the case when $t_0 \le t \le t_1$, the verification for other t following in an exactly analogous manner. The sum on the right of (13.10) for $t_0 \le t \le t_1$ yields

$$\alpha_1(y_1(t) - y_0(t)) + \alpha_2((y_2(t) - y_1(t)) + \cdots + \alpha_n(y_n(t) - y_{n-1}(t))$$

$$= \alpha_1(1 - 0) + \alpha_2(1 - 1) + \cdots + \alpha_n(1 - 1)$$

$$= \alpha_1$$

$$= y(t).$$

We now apply F to y:

$$F(y) = \sum_{i=1}^{n} \alpha_i(F(y_i) - F(y_{i-1})),$$

which, since $F(y_i) = g(t_i)$ [using (13.5) and (13.11)], yields

$$F(y) = \sum_{i=1}^{n} \alpha_i(g(t_i) - g(t_{i-1})).$$

But since $\alpha \,\overline{\operatorname{sgn}}\, \alpha = |\alpha|$ [see Eq. (12.4)], then, by the definition of α_i [Eq. (13.6)], we can write

$$F(y) = \sum_{i=1}^{n} |g(t_i) - g(t_{i-1})|,$$

which implies, using (13.9) and the fact that F is a bounded linear functional with the same norm as f, that

$$\sum_{i=1}^{n} |g(t_i) - g(t_{i-1})| = |F(y)| \leq \|F\| \|y\| \leq \|F\| = \|f\| \tag{13.12}$$

for any n. Therefore the function $g(t)$, defined by (13.5), is of bounded variation or

$$g(t) \in BV[a, b].$$

Inequality (13.12) further implies that

$$V(g) \leq \|f\|. \tag{13.13}$$

Suppose now that $x \in C[a, b]$ and define

$$z(t) = \sum_{i=1}^{n} x(t_{i-1})(y_i(t) - y_{i-1}(t)). \tag{13.14}$$

In this case

$$F(z) = \sum_{i=1}^{n} x(t_{i-1})(g(t_i) - g(t_{i-1})), \tag{13.15}$$

where, as above, we have used the fact that $F(y_i) = g(t_i)$. It is now evident that

$$z(t) - x(t) = \begin{cases} x(t_0) - x(t) & (t_0 \leq t \leq t_1), \\ x(t_{i-1}) - x(t) & (t_{i-1} < t \leq t_i). \end{cases}$$

Letting

$$\eta = \max_{i} |t_i - t_{i-1}|,$$

we note that, since $x(t) \in C[a, b]$ is uniformly continuous (a continuous function defined on a compact set is uniformly continuous there) if $\eta \to 0$, then

$$\|z - x\| \to 0.$$

But, since F is a bounded linear functional and therefore continuous, this implies

$$\lim_{\eta \to 0} F(z) = \lim_{z \to x} F(z) = F(x).$$

By the definition of the Riemann–Stieltjes integral, though,

$$\lim_{\eta \to 0} F(z) = \int_{a}^{b} x(t) \, dg(t);$$

hence†

$$F(x) = \int_{a}^{b} x(t) \, dg(t).$$

Now, since x was an arbitrary continuous function on $[a, b]$ and F must agree with f on $C[a, b]$, we can write

$$f(x) = \int_{a}^{b} x(t) \, dg(t) \tag{13.16}$$

---

† There is no problem about the existence of this integral for we have the following theorem about the Riemann–Stieltjes integral:

(I) If $x(t)$ is continuous on $[a, b]$ and $g(t)$ is of finite variation on $[a, b]$, then $\int_{a}^{b} x(t) \, dg(t)$ exists.

for any $x \in C[a, b]$, and the first part of the theorem is proved. We now wish to prove that $\|f\| = V(g)$. Since, by (13.13), we have already shown that $V(g) \leq \|f\|$, if we can now show that $\|f\| \leq V(g)$, we shall be done. We first state the following fact:

(II) If $\varphi(t)$ is continuous on $[a, b]$ and $\psi(t)$ is of finite variation on $[a, b]$, then

$$\left| \int_a^b \varphi(t) \, d\psi(t) \right| \leq \max_{t \in [a, b]} |\varphi(t)| \cdot V(\psi).$$

[To render this intuitively plausible consider the familiar result about the Riemann integral:

$$\left| \int_a^b \varphi(t) \, dt \right| \leq \max_{t \in [a, b]} |\varphi(t)| \cdot (b - a).]$$

Applying (II) to (13.16) yields

$$|f(x)| \leq \max_{t \in [a, b]} |x(t)| \cdot V(g) = \|x\| V(g)$$

for all $x \in C[a, b]$. Thus

$$\|f\| \leq V(g), \qquad\qquad (13.17)$$

and the result is proved; that is, (13.13) and (13.17) imply

$$\|f\| = V(g).$$

An Equivalence Relation between Functions of Bounded Variation

Through the above theorem we have achieved a certain correspondence between functionals from $\tilde{C}[a, b]$ and functions of bounded variation. This correspondence is by no means well-defined, however, for if $g(t)$ satisfies (13.16), then, clearly, $g(t) + c$, where c is any constant, will also satisfy (13.16). [One can easily demonstrate this fact about the Riemann–Stieltjes integral by just examining the partial sums induced by $g(t) + c$ and comparing them to those induced by $g(t)$.]

Further, if $h(a) = g(a)$, $h(b) = g(b)$, and $h(t) = g(t)$ at all interior points of $[a, b]$, where $g(t)$ is continuous, we claim that

$$f(x) = \int_a^b x(t) \, dg(t) = \int_a^b x(t) \, dh(t) \qquad\qquad (13.18)$$

regardless of how $h(t)$ is defined at interior points of $[a, b]$, where $g(t)$ is discontinuous (provided, of course, that $h(t)$ is still of finite variation on $[a, b]$).

To prove the above equation's validity, we note that, since a function of bounded variation must be continuous on all but a denumerable set of points, we can partition the interval $[a, b]$ in such a way as to let η shrink to zero while passing through only the points of continuity of $g(t)$. [Note here that, since we are assured of the integral's existence for such an $x(t)$ and $g(t)$, we can pass to the limit in any way whatsoever and still arrive at the same value.] Clearly, if this limiting process is chosen, each and every partial sum will be identical; hence the integrals must be equal and the validity of (13.18) is apparent.

In the hope of retaining something of the above correspondence, we introduce now the following equivalence relation between elements of $BV[a, b]$: For $x_1, x_2 \in BV[a, b]$, we define

$$x_1 \sim x_2 \tag{13.19}$$

if and only if

$$\int_a^b y(t)\, dx_1(t) = \int_a^b y(t)\, dx_2(t) \tag{13.20}$$

for all $y \in C[a, b]$. It is easily verified that $\sim$ is an equivalence relation. Note also how well-suited the above equivalence relation is to our needs; if we consider only equivalence classes instead of functions themselves the correspondence induced by (13.16) seems to have a much better chance of being 1 : 1.

Before proceeding further, we wish to establish some

CRITERIA FOR $x \sim 0$. We claim that, for $x \in BV[a, b]$, $x \sim 0$ if and only if, for any c such that $a < c < b$,

$$x(a) = x(b) = x(c + 0) = x(c - 0), \tag{13.21}$$

where by $x(c + 0)$ and $x(c - 0)$ we mean, respectively, $\lim_{t \to 0} x(c + t)$, $t > 0$, and $\lim_{t \to 0} x(c - t)$, $t > 0$. Note that requirement (13.21) does not imply that $x(t)$ be continuous at each interior point, as is evidenced by the function shown in Fig. 13.1.

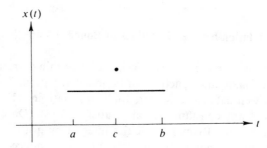

FIG. 13.1. A function $x(t)$ satisfying condition (13.21).

We now proceed to the proof of the above claim.

Proof (necessity). Suppose $x \sim 0$. Using (13.20) with $y(t) = 1$, we get

$$0 = \int_a^b dx(t) = x(b) - x(a); \tag{13.22}$$

hence,

$$x(b) = x(a).$$

Before finishing the proof, the following two facts, which can easily be demonstrated, are listed:

For $a \le c < b$ and $h > 0$,

$$\frac{1}{h} \int_c^{c+h} x(t)\, dt \to x(c + 0) \tag{13.23}$$

as $h \to 0$; similarly, for $a < c \le b$, $h > 0$,

$$\frac{1}{h} \int_{c-h}^{c} x(t) \, dt \to x(c - 0) \tag{13.24}$$

as $h \to 0$.

We now show that $x(a) = x(c + 0)$. The argument to show that this is also equal to $x(c - 0)$ is quite similar and will be omitted.

Consider the continuous function

$$y(t) = \begin{cases} 1 & (a \le t \le c), \\ 1 - \dfrac{t - c}{h} & (c < t \le c + h), \\ 0 & (c + h < t \le b) \end{cases} \tag{13.25}$$

illustrated in Fig. 13.2.

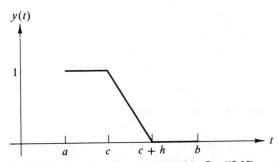

FIG. 13.2. The function y(t) defined by Eq. (13.25).

Now, with $x(t)$ as above, we consider

$$0 = \int_{a}^{b} y(t) \, dx(t) = \int_{a}^{c} dx(t) + \int_{c}^{c+h} y(t) \, dx(t) = x(c) - x(a) + \int_{c}^{c+h} y(t) \, dx(t). \tag{13.26}$$

The formula for integration by parts for the Riemann–Stieltjes integral† and Eq. (13.25) imply

$$\int_{c}^{c+h} y(t) \, dx(t) = -x(c) + \frac{1}{h} \int_{c}^{c+h} x(t) \, dt. \tag{13.27}$$

Thus (13.26) becomes

$$0 = x(c) - x(a) - x(c) + \frac{1}{h} \int_{c}^{c+h} x(t) \, dt$$

or

$$x(a) = \frac{1}{h} \int_{c}^{c+h} x(t) \, dt. \tag{13.28}$$

† (III) If the integral $\int_{a}^{b} \varphi(t) \, d\psi(t)$ exists, then $\int_{a}^{b} \psi(t) \, d\varphi(t)$ exists and $\int_{a}^{b} \varphi(t) \, d\psi(t) + \int_{a}^{b} \psi(t) \, d\varphi(t) = \varphi(b)\psi(b) - \varphi(a)\psi(a)$.

Letting $h \to 0$ and applying (13.23), we get

$$x(a) = x(c + 0). \tag{13.29}$$

In a similar fashion, we can also show that $x(b) = x(c - 0)$ and, in view of this, the necessity of the condition has been established.

Proof (sufficiency). Suppose now that $x(t)$ satisfies condition (13.21). In this case if we define $\hat{x}(t) = x\,(a)$ for all $t \in [a, b]$, then $\hat{x}(b) = x(b)$, and $\hat{x}(t) = x(t)$ at all points t interior to $[a, b]$, where $x(t)$ is continuous. Now, using the same argument used in establishing (13.18), we can say, for all $y \in C[a, b]$,

$$\int_a^b y(t)\, dx(t) = \int_a^b y(t)\, d\hat{x}(t); \tag{13.30}$$

hence, for all $y \in C[a, b]$, since $\hat{x}(t)$ is constant and, as such, has total variation equal to zero:

$$\int_a^b y(t)\, dx(t) = 0,$$

which implies

$$x \sim 0.$$

Normalized Functions of Bounded Variation

DEFINITION 13.1. The function $g(t) \in BV[a, b]$ is said to be *normalized* if $g(a) = 0$ and g is continuous from the right; that is, for all $t \in (a, b)$, $g(t + 0) = g(t)$. The collection of normalized functions of bounded variation will be denoted by $NBV[a, b]$ and it may readily be verified that $NBV[a, b]$ is a subspace of $BV[a, b]$.

We now wish to establish the following lemma.

LEMMA. Let $x_1, x_2 \in BV[a, b]$, where $x_1 \sim x_2$ and x_1 and x_2 are normalized. In this case $x_1 = x_2$.

Proof. To prove this, we first note that $x_1 - x_2 \sim 0$ and implies that the criteria for $(x_1 - x_2) \sim 0$, previously established, has been satisfied. Thus,

$$(x_1 - x_2)(a) = (x_1 - x_2)(b). \tag{13.31}$$

But, since x_1 and x_2 are normalized, $x_1(a) = x_2(a) = 0$ and (13.31) becomes

$$x_1(b) = x_2(b). \tag{13.32}$$

Further, since $(x_1 - x_2) \sim 0$, for any c such that $a < c < b$,

$$(x_1 - x_2)(a) = (x_1 - x_2)(c + 0)$$

or

$$0 = x_1(c + 0) - x_2(c + 0),$$

and, since x_1 and x_2 must be continuous from the right,

$$0 = x_1(c) - x_2(c)$$

or

$$x_1(c) = x_2(c).$$

Since c was any interior point of $[a, b]$, we have proved that

$$x_1(t) = x_2(t).$$

A Normalized Representative for Each Equivalence Class of $BV[a, b]$

We now wish to show that for each equivalence class of $BV[a, b]$ under (13.19) there is one normalized representative. Then, using the lemma just proved, we can immediately say that this representative is unique for, if any other normalized function were equivalent to it under (13.19) it would have to be equal to it. Thus, given the function $x(t) \in BV[a, b]$, we now wish to define a function $\hat{x} \in NBV[a, b]$ such that $(x - \hat{x}) \sim 0$.

Let $x(t) \in BV[a, b]$. Consider now the function $\hat{x}(t)$ defined as follows: at the left endpoint of $[a, b]$

$$\hat{x}(a) = 0,$$

whereas, for any interior point t,

$$\hat{x}(t) = x(t + 0) - x(a)$$

and at the right endpoint

$$\hat{x}(b) = x(b) - x(a).$$

We now claim that the following assertions are valid:

$$\hat{x}(t) \in NBV[a, b], \tag{13.33}$$

$$V(\hat{x}) \leq V(x), \tag{13.34}$$

$$\hat{x} \sim x. \tag{13.35}$$

The proofs of statements (13.33) and (13.35) are straightforward and will be omitted. [One need only use the definition of $\hat{x}(t)$ to get either of them.] We now prove (13.34).

Let

$$a = t_0 < t_1 < \cdots < t_n = b$$

be a partition of $[a, b]$. By the definition of $x(t + 0)$, there must exist

$$c_1, c_2, \ldots, c_{n-1}$$

"to the right" of the $t_1, t_2, \ldots, t_{n-1}$, that is, $t_i < c_i$ $(i = 1, \ldots, n - 1)$, such that, for any prescribed $\varepsilon > 0$,

$$|x(t_i + 0) - x(c_i)| < \varepsilon/2n \tag{13.36}$$

for $i = 1, \ldots, n - 1$. Now, using the definition of $\hat{x}(t)$, we can write for $i = 2, \ldots, n - 1$,

$$\hat{x}(t_i) - \hat{x}(t_{i-1}) = x(t_i + 0) - x(a) - \big(x(t_{i-1} + 0) - x(a)\big)$$

$$= x(t_i + 0) - x(t_{i-1} + 0)$$

$$= \big(x(t_i + 0) - x(c_i)\big) - \big(x(t_{i-1} + 0) - x(c_{i-1})\big) + \big(x(c_i) - x(c_{i-1})\big).$$

Taking $c_0 = a$ and $c_n = b$, we can now write

$$\sum_{i=1}^{n} |\hat{x}(t_i) - \hat{x}(t_{i-1})| \le \sum_{i=1}^{n} |x(c_i) - x(c_{i-1})| + \varepsilon \le V(x) + \varepsilon. \qquad (13.37)$$

Hence

$$V(\hat{x}) \le V(x) + \varepsilon,$$

and, since ε was arbitrary,

$$V(\hat{x}) \le V(x)$$

and (13.34) is established.

We are now in a position to get our main result.

The Conjugate Space of C[a, b]

THEOREM 13.2. The spaces $\tilde{C}[a, b]$ and $NBV[a, b]$ are congruent. In particular, the congruence is given by the following mapping:

$$T : NBV[a, b] \to \tilde{C}[a, b],$$

$$g(t) \to Tg = f(x),$$

where, for $x \in C[a, b]$,

$$f(x) = \int_a^b x(t)\, dg(t).$$

Proof. Clearly, T is a linear transformation and f is a linear functional. By Eq. (II) of this lecture, we can also say

$$|f(x)| \le V(g)\|x\|$$

for any $x \in C[a, b]$, which implies $f(x)$ is a bounded linear functional and that

$$\|f\| \le V(g) = \|g\|. \qquad (13.38)$$

[Note that, since $g \in NBV[a, b]$, $\|g\| = |g(a)| + V(g)$ is just $V(g)$.]

It will now be shown that T is an onto mapping. Let $f \in \tilde{C}[a, b]$. By Theorem 13.1 there exists some function $h(t) \in BV[a, b]$ such that

$$f(x) = \int_a^b x(t)\, dh(t).$$

By our previous results [Eqs. (13.33)–(13.35) and the lemma] we know there exists some unique normalized function $g \in NBV[a, b]$ such that

$$h \sim g.$$

Thus,

$$f(x) = \int_a^b x(t)\, dg(t)$$

and, clearly,

$$T : g \to f,$$

which proves that T is an onto mapping.

Further, with f, g, and h as above, Theorem 13.1 also states that

$$\|f\| = V(h).$$

Applying (13.34) and (13.38), we have

$$\|f\| \le V(g) \le V(h) = \|f\|,$$

which implies

$$\|f\| = V(g) = \|g\|$$

or

$$\|Tg\| = \|f\| = \|g\|.$$

Hence T is an isometry, therefore $1 : 1$, and the theorem is proved.

A feature to be noted about the above proof, and which can be said about representation theorems in general, is that they usually entail quite a bit of work. The above proof, for example, is quite simple compared to what one must go through in order to obtain the conjugate space of $L_p(E, \mu)$. That they do demand a great deal of effort is rather reasonable, however, when one takes into account the tremendous amount of information and insight they provide, that is, a bit of work should be expected. Since our function is primarily to survey, other such proofs will not be done here, and below a number of results of this type are listed. For proofs the reader is referred to, for example, Ref. 2.

13.2 A List of Some Spaces and Their Conjugate Spaces

In Table I, as usual, $\tilde{X}$ denotes the conjugate space of the normed linear space X. Instead of actually listing $\tilde{X}$, however, we list here a space to which $\tilde{X}$ is congruent.

TABLE I

Spaces and Their Conjugate Spaces

X	$\tilde{X}$
1. $C[a, b]$	$NBV[a, b]$
2. l_p	l_q $(1/p + 1/q = 1)$
3. l_1	l_∞
4. l_∞	$ba(Z, P(Z), \mu)$: bounded, additive set functions on all subsets of Z, where we take as the norm of such a set function μ, $\|\mu\| = |\mu|(Z)$, the total variation of μ on Z. Z, as usual, denotes the integers
5. $L_p(E, \mu)$	$L_q(E, \mu)$ $(1/p + 1/q = 1)$
6. $L_1(E, \mu)$	$L_\infty(E, \mu)$
7. c: the space of all sequences of complex numbers $(\xi_1, \xi_2, ..., \xi_n, ...)$ such that the $\lim_n \xi_n$ exists. Addition and scalar multiplication is the same as for l_p and as the norm of $x = (\xi_1, \xi_2, ..., \xi_n, ...) \in c$, we take $\|x\| = \sup_n |\xi_n|$	l_1
8. c_0: the subspace of c with the property that $\lim_n \xi_n = 0$	l_1

EXERCISES 13

1. Prove that c is a Banach space.

*2. Prove that $\tilde{c} = l_1$.

3. Consider the complex euclidean n-space where, for $x = (\alpha_1, ..., \alpha_n)$, we define

$$\|x\| = \left(\sum_{i=1}^{n} |\alpha_i|^p \right)^{1/p} \qquad (p \geq 1).$$

Determine the conjugate space. (*Hint:* Consider first $p > 1$ and then $p = 1$.)

4. Consider, as in Exercise 3, complex euclidean n-space but take

$$\|x\| = \sup_i |\alpha_i|.$$

Determine the conjugate space.

5. For $x \in L_p(a, b)$, $p > 1$, show that

$$f(x) = \int_a^b y(t)x(t) \, dt \qquad [y \in L_q(a, b), 1/p + 1/q = 1]$$

exists and is a bounded linear functional on $L_p(a, b)$.

6. Show that the conjugate space of $L_p(a, b)$, $1 < p < \infty$ (see Ref. 3) is $L_q(a, b)$.

7. Show that the spaces c and c_0 are not reflexive.

8. Show the conjugate space of l_1 is l_∞.

9. Without using the form of l_∞ in Table I, show that the conjugate space of l_∞ is not l_1. (*Hint:* See Theorem 12.4.)

10. Prove that the spaces l_p ($1 < p < \infty$) are reflexive.

11. With respect to the norm for $BV[a, b]$ given in Eq. (13.3), show that $BV[a, b]$ is a Banach space. Show that the subset of all absolutely continuous functions on $[a, b]$ is a closed subspace of $BV[a, b]$.

REFERENCES

1. A. Taylor, "Introduction to Functional Analysis." Discussion of the conjugate space of $C[a, b]$.

2. N. Dunford and J. Schwartz, "Linear Operators. Part I. General Theory." A very extensive table of spaces and their conjugate spaces.

3. L. Liusternik and W. Sobolev, "Elements of Functional Analysis."

CHAPTER 14

Weak Convergence and
Bounded Linear Transformations

We wish to introduce a new notion of convergence of a sequence of vectors in a normed linear space, called *weak* convergence. To distinguish this type of convergence from our old notion of convergence, we shall, when there is possible ambiguity, refer to ordinary convergence as *strong* convergence. That the words "strong" and "weak" are aptly chosen is illustrated by a theorem, which demonstrates that strong convergence implies weak convergence, and a counterexample, which shows that, in general, weak convergence does not imply strong convergence. Having made this definition, we wish to see what weak convergence looks like in the particular case of l_p; that is, in this space we obtain conditions that are both necessary and sufficient for weak convergence. In general, however, such investigations are quite involved and for this reason we limit ourselves to examining just this one special case.

By viewing the underlying field of a vector space as a one-dimensional space, it is clear that every linear functional can be viewed as a linear transformation between the space and the underlying field. Thus, after observing so many properties of linear functionals, especially bounded linear functionals, one might wonder about generalizing these results to linear transformations. In a manner exactly analogous to the way in which bounded linear functionals were defined and presented, we next introduce the notion of a *bounded linear transformation* between two normed linear spaces. The only thing which enables one to distinguish between one discussion and the other is that wherever absolute values appeared in the discussion of bounded linear functionals, norm symbols appear in the discussion of bounded linear transformations. All theorems and proofs that were valid for linear functionals also hold for bounded linear transformations, after we have made the interchange of norm and absolute value symbols mentioned earlier. One interesting fact that we shall obtain about bounded linear transformations is that bounded sets are mapped into bounded sets (and conversely) under such mappings.

Proceeding along in the same analogous spirit, just as we were able to view the class of all bounded linear functionals on a given normed space as a normed linear space, we can also view the class of all bounded linear transformations between two normed spaces as a normed linear space after introducing appropriate operations. After defining this space, certain facts about it will be deduced, most of which are analogous to statements about the now familiar conjugate space.

14.1   Weak Convergence

In this section, we wish to introduce a new kind of convergence.

DEFINITION 14.1. The sequence $\{x_n\}$ from the normed linear space X is said to *converge weakly to* $x \in X$, written $x_n \overset{w}{\to} x$, if for every $f \in \tilde{X}$ ($\tilde{X}$, as usual, denotes the conjugate space of X),

$$f(x_n) \to f(x). \tag{14.1}$$

To avoid any possible ambiguity, we shall often refer to our old notion of convergence as *strong* convergence.

The first fact about weak convergence we wish to demonstrate is that the "weak" limit is unique; more precisely, if

$$x_n \overset{w}{\to} x \quad \text{and} \quad x_n \overset{w}{\to} y, \tag{14.2}$$

which means, for any $f \in \tilde{X}$, that

$$f(x_n) \to f(x) \quad \text{and} \quad f(x_n) \to f(y),$$

then we wish to show that x must equal y. Since ordinary convergence always leads to a unique limit, the above assumptions imply that

$$f(x) = f(y).$$

By the linearity of f, this is equivalent to

$$f(x - y) = 0$$

for any $f \in \tilde{X}$. Since f can be any bounded linear functional, we can apply Theorem 12.2 (especially Consequence 2 of that theorem) to assert that $x - y = 0$ or $x = y$; hence, it is impossible for a sequence to converge weakly to two different things.

Another property that is equally simple to demonstrate is that if

$$x_n \overset{w}{\to} x \quad \text{and} \quad y_n \overset{w}{\to} y,$$

then

$$x_n + y_n \overset{w}{\to} x + y.$$

Similarly, if α is a scalar and

$$x_n \overset{w}{\to} x,$$

then

$$\alpha x_n \overset{w}{\to} \alpha x.$$

The definition of continuity of the function $f(x)$ requires that, if $x_n \to x$, then $f(x_n) \to f(x)$. Since (Theorem 11.4) every bounded linear functional is continuous, it is clear that strong convergence implies weak convergence. That the converse of this implication is not true in general is demonstrated by the following counterexample.

Weak Convergence Does Not Imply Strong Convergence

In Table I at the end of Chapter 13, concerning some spaces and the space that their conjugate space is congruent to, it is noted that (for $1 < p < \infty$) the conjugate space of $L_p(E, \mu)$ is congruent to $L_q(E, \mu)$, where p and q are related by the equation

$$\frac{1}{p} + \frac{1}{q} = 1.$$

Thus, for the special case of real $L_2(0, 2\pi)$, we see that the conjugate space is congruent to the original space. Denoting the Hilbert space real $L_2(0, 2\pi)$, by X and its conjugate space by $\tilde{X}$, and applying the Riesz representation theorem of Sec. 12.4, we see that for any $f \in \tilde{X}$ there exists some $g \in X$ such that, for any $x \in X$,

$$f(x) = (x, g) = \int_0^{2\pi} x(t)g(t)\, dt. \tag{14.3}$$

Consider the sequence of vectors $\{x_n\}$ from X now, where

$$x_n(t) = \frac{\sin nt}{\pi}$$

for $n = 1, 2, \ldots$. We now claim that

$$x_n \overset{w}{\to} 0, \tag{14.4}$$

but

$$x_n \nrightarrow 0. \tag{14.5}$$

First the validity of (14.4) will be shown. Let $f \in \tilde{X}$ and let g be as in (14.3); then

$$f(x_n) = \frac{1}{\pi} \int_0^{2\pi} (\sin nt)g(t)\, dt.$$

In view of the Riemann–Lebesgue lemma (see discussion immediately preceding Theorem 9.4), we have that

$$f(x_n) \to 0$$

for any $f \in \tilde{X}$. Therefore,

$$x_n \overset{w}{\to} 0.$$

With regard to showing (14.5), consider

$$\| x_n - 0 \|^2 = \frac{1}{\pi^2} \int_0^{2\pi} (\sin^2 nt)\, dt = \frac{1}{\pi}$$

for every n; hence,

$$x_n \nrightarrow 0.$$

and the counterexample is complete. (See Exercise 14.1 for further counterexamples.)

Thus far many notions which have given rise to genuinely different things in the infinite-dimensional case have been seen to collapse altogether when viewed in a finite-dimensional space. For example, when the notion of equivalent norms was introduced (Sec. 8.5), it was seen that all norms on a finite-dimensional space were equivalent. Similarly, when the concept of a bounded linear functional was defined, it turned out that *all* linear functionals on a finite-dimensional space were bounded. The concept of weak convergence (as opposed to strong convergence) provides no exception to this trend, and our next theorem demonstrates the indistinguishability of strong and weak convergence in finite-dimensional spaces.

THEOREM 14.1. If X is a finite-dimensional normed space, strong convergence is equivalent to weak convergence.

Proof. Since we have already shown that, in any space, strong convergence implies weak convergence, we need only show that in this situation weak convergence implies strong convergence. To this end, suppose $e_1, e_2, ..., e_k$ is a basis for X and that

$$x_n \overset{w}{\to} x,$$

where

$$x_n = \alpha_1^{(n)} e_1 + \cdots + \alpha_k^{(n)} e_k,$$

for $n = 1, 2, ...,$ and

$$x = \alpha_1 e_1 + \cdots + \alpha_k e_k.$$

We now consider the linear functionals $f_i \in \tilde{X}$ ($i = 1, 2, ..., k$) defined as follows:

$$f_i(e_j) = \delta_{ij}$$

where δ_{ij} is the Kronecker delta.† Since we are assuming that $x_n \overset{w}{\to} x$, it follows that, for $i = 1, 2, ..., k$,

$$f_i(x_n) \to f_i(x).$$

Since, by the definition of f_i, $f_i(x_n) = \alpha_i^{(n)}$, whereas $f_i(x) = \alpha_i$, we have that

$$\alpha_i^{(n)} \to \alpha_i$$

for $i = 1, 2, ..., k$. Now let $M = \max_i \|e_i\|$. For any prescribed $\varepsilon > 0$, there must exist some integer N such that, for all $n > N$, and every $i = 1, 2, ..., k$,

$$\|\alpha_i^{(n)} - \alpha_i\| < \varepsilon/Mk;$$

hence,

$$\|x_n - x\| = \left\| \sum_{i=1}^{k} (\alpha_i^{(n)} - \alpha_i) e_i \right\|$$

$$\leq \sum_{i=1}^{k} |\alpha_i^{(n)} - \alpha_i| \, \|e_i\|$$

$$< \varepsilon,$$

which is the statement that $x_n \to x$. Thus, it has been shown that weak convergence implies strong convergence, and the proof is complete.

Suppose now that X is a normed linear space and that, for the sequence $\{x_n\}$, $x_n \to x$. For any given $\varepsilon > 0$, there is some integer N such that $n > N$ implies

$$\|x_n - x\| < \varepsilon.$$

But, since

$$\|x_n\| - \|x\| \leq \|x_n - x\|,$$

this implies that

$$\|x_n\| < \varepsilon + \|x\|$$

for all $n > N$. Now let $M = \max_{i \leq N} \|x_i\|$. It is now clear that

$$\|x_n\| < \varepsilon + \|x\| + M$$

---

† The linear functionals $f_1, f_2, ..., f_k$ actually form a basis for the (k-dimensional) conjugate space as is easily verified. For a further discussion of this dual basis, see Halmos, *Finite-Dimensional Vector Spaces*, pp. 23–24.

for all n. Thus if $\{x_n\}$ is a convergent sequence, the terms of the sequence are uniformly bounded. As it happens, we have the same statement for sequences that are weakly convergent; that is, we have the next theorem.

THEOREM 14.2. If X is a normed linear space and $x_n \overset{w}{\to} x$, then there exists some positive constant M such that

$$\|x_n\| < M$$

for all n.

The proof of this fact will be postponed until Theorem 15.4 (Corollary), however.

One notes that, for a sequence to be weakly convergent to a given vector, *any* bounded linear functional, when applied to the sequence, must converge to the functional value that arises when applied to the proposed limit vector. Our next result enables us to cut down the number of bounded linear functionals that must be checked; in the situation suggested in the theorem, one need not check *all* bounded linear functionals.

THEOREM 14.3. Let $\{x_n\}$ be a sequence of vectors from the normed linear space X with the property that, for every n, there is a positive constant M such that

$$\|x_n\| \leq M.$$

Further, suppose that $f(x_n) \to f(x)$ for all bounded linear functionals f belonging to the set A, where A has the property that $[A]$ is dense in $\tilde{X}$, that is,

$$\overline{[A]} = \tilde{X}.$$

If this situation prevails, then $x_n \overset{w}{\to} x$.

REMARK. By virtue of Theorem 14.2, we immediately note that the converse of this statement also holds.

Proof. To prove this, we must show, for any $h \in \tilde{X} = \overline{[A]}$, that $h(x_n) \to h(x)$ and this will be done in two steps. First, it will be shown that this is true for all $h \in [A]$, then for $\overline{[A]}$. To prove the first contention, let $g \in [A]$. By the definition of $[A]$ this means that g must be expressible as a linear combination of elements of A or there exists scalars $\alpha_1, \alpha_2, ..., \alpha_k$ and members $f_1, f_2, ..., f_k$ of A such that

$$g = \sum_{i=1}^{k} \alpha_i f_i.$$

Now consider

$$g(x_n) = \sum_{i=1}^{k} \alpha_i f_i(x_n).$$

Since, by hypothesis, $f_i(x_n) \to f_i(x)$, we now have that

$$\sum_{i=1}^{k} \alpha_i f_i(x_n) \to \sum_{i=1}^{k} \alpha_i f_i(x) = g(x)$$

or that

$$g(x_n) \rightarrow g(x),$$

and the first contention is established.

As for the second contention, suppose $h \in \overline{[A]}$. For any given $\varepsilon > 0$, there must exist $g_j \in [A]$ $(j = 1, 2, ...)$ and an integer N such that $j > N$ implies

$$\|h - g_j\| < \varepsilon.$$

By hypothesis, there is some constant M such that

$$\|x_n\| \leq M \qquad (n = 1, 2, ...)$$

and

$$\|x\| \leq M.$$

Since we wish to show $h(x_n) \rightarrow h(x)$ to complete the proof, consider now

$$
\begin{aligned}
|h(x_n) - h(x)| &= |h(x_n) - g_j(x_n) + g_j(x_n) - g_j(x) + g_j(x) - h(x)| \\
&\leq |h(x_n) - g_j(x_n)| + |g_j(x_n) - g_j(x)| + |g_j(x) - h(x)| \\
&\leq \|h - g_j\| \|x_n\| + |g_j(x_n) - g_j(x)| + \|g_j - h\| \|x\| \\
&< \varepsilon M + |g_j(x_n) - g_j(x)| + \varepsilon M,
\end{aligned}
$$

provided $j > N$. Since $g_j \in [A]$ then, by the first part of the proof,

$$g_j(x_n) \rightarrow g_j(x),$$

and the middle term above can, therefore, also be made arbitrarily small, which implies

$$h(x_n) \rightarrow h(x),$$

and it follows that (since h was any bounded linear functional)

$$x_n \xrightarrow{w} x.$$

Our discussion so far has been quite general, and we now wish to examine what weak convergence means in a very specific situation. In particular, we wish to obtain necessary and sufficient conditions for a sequence to converge weakly in the space l_p $(1 < p < \infty)$. It so happens that this is not too difficult a task in this space; in other spaces, however, it can be quite formidable.

Weak Convergence in $l_p(1 < p < \infty)$

THEOREM 14.4. For $1 < p < \infty$, suppose

$$x_n = (\alpha_1^{(n)}, \alpha_2^{(n)}, ..., \alpha_k^{(n)}, ...) \in l_p$$

and

$$x = (\alpha_1, \alpha_2, ..., \alpha_k, ...) \in l_p.$$

Then $x_n \xrightarrow{w} x$ if and only if

(1) $\|x_n\| \leq M$ (M a positive constant) for all n;
(2) for every i, $\alpha_i^{(n)} \rightarrow \alpha_i$.

Proof (sufficiency). In this proof, the reader will be asked to recall some facts about l_p that were proved in Sec. 12.3, where it was shown that the conjugate space of l_p was congruent to l_q. Specifically, a set of bounded linear functionals A on l_p will be exhibited with the property that $\overline{[A]} = \tilde{X}$. Applying Theorem 14.3 will then enable us to confirm the theorem.

In Sec. 12.3, it was shown that any $y = (\xi_1, \xi_2, \ldots, \xi_k, \ldots) \in l_q$ could be written

$$y = \sum_{n=1}^{\infty} \xi_n e_n$$

where

$$e_1 = (1, 0, 0, \ldots, 0, \ldots)$$
$$e_2 = (0, 1, 0, \ldots, 0, \ldots)$$
$$\vdots \qquad \qquad \vdots$$

Since $\tilde{l}_p$ is congruent to l_q, it follows that there is a corresponding set of bounded linear functionals

$$f_1, f_2, \ldots, f_n, \ldots,$$

where f_i corresponds to e_i under the congruence mapping mentioned in Sec. 12.3, with the property that any $f \in \tilde{l}_p$ can be written

$$f = \sum_{n=1}^{\infty} \beta_i f_i;$$

in fact $\beta_i = f(e_i)$.

Denoting the set

$$\{f_i, f_2, \ldots, f_n, \ldots\}$$

by A, then, we see that

$$\overline{[A]} = \tilde{l}_p$$

the property required by Theorem 14.3. By that theorem, if we can show that, for every i,

$$f_i(x_n) \to f_i(x)$$

or, equivalently [since $f_i(x_n) = \alpha_i^{(n)}$ and $f_i(x) = \alpha_i$] that

$$\alpha_i^{(n)} \to \alpha_i,$$

then we are done. Since, in part (2) of the hypothesis, we have assumed this last condition to hold, Theorem 14.3 implies that

$$x_n \overset{w}{\to} x$$

and the sufficiency of the conditions has been proved.

Proof (necessity). Suppose now that $x_n \overset{w}{\to} x$. In this case, Theorem 14.2 guarantees that there is some constant M such that

$$\|x_n\| \leq M$$

for every n; part (1) is therefore satisfied. Since weak convergence implies that $f(x_n) \to f(x)$ for all $f \in \bar{l}_p$, then certainly

$$f_i(x_n) \to f_i(x),$$

where the f_i are as mentioned in the first part of the proof. But, again, since $f_i(x_n) = \alpha_i^{(n)}$ and $f_i(x) = \alpha_i$, the above condition means that

$$\alpha_i^{(n)} \to \alpha_i,$$

and part (2) is seen to hold. The proof is now complete.

Loosely speaking, we now know what weak convergence "looks like" in l_p. We now turn to generalizing the concept of a bounded linear functional.

14.2   Bounded Linear Transformations

Clearly, every linear functional is also a linear transformation between the vector space and the one-dimensional field underlying the vector space. (Any field can be thought of as a one-dimensional vector space.) In view of this, since so many interesting things have emerged from the consideration of bounded linear functionals, it seems natural to conjecture about generalizing these results to linear transformations. This brings us to our next definition.

DEFINITION 14.2. Let A be a linear transformation mapping the normed linear space X into the normed linear space Y:

$$A : X \to Y.$$

(We shall not distinguish between the symbol for norm on X and norm on Y.) A is said to be a *bounded linear transformation* if there exists some positive constant k such that, for all $x \in X$,

$$\|Ax\| \leq k\|x\|. \tag{14.6}$$

A number of observations follow immediately from the definition, the proofs of which exactly parallel the proofs of the analogous statements about bounded linear functionals in Secs. 11.2 and 11.3, the only change which need be made in the previous proofs to make them apply to this situation is to replace the absolute value symbol by a norm symbol. In view of this similarity, only one result will be proved to convince the reader of how strong an analogy prevails; the rest will only be stated. Should the reader wish to consult the proof of the corresponding statement about bounded functionals, the analogous theorem is listed, in parenthesis, after each result.

THEOREM 14.5   (THEOREM 11.3). Let X and Y be normed linear spaces, and let A be a linear transformation such that $A : X \to Y$. Then if A is continuous at x_0, it must be continuous everywhere.

THEOREM 14.6   (THEOREM 11.4). With A, X, and Y as above, A is a bounded linear transformation if and only if it is continuous.

In Sec. 11.3 the operations of addition and scalar multiplication were introduced on the set of all bounded linear functionals over a given normed linear space. Once this was done, it was noted that the algebraic structure so defined was a linear space. As one might expect when the analogous operations are introduced to the class of all bounded linear transformations between two normed linear spaces, we again obtain a new linear space. Again, just as we were able to introduce a norm on the linear space of all bounded linear functionals over a normed space, so can we introduce a norm on the class of all bounded linear transformations between two normed spaces; this we do as follows:

DEFINITION 14.3. Suppose $A : X \to Y$, where X and Y are normed linear spaces and A is a bounded linear transformation. We define

$$\|A\| = \sup_{x \neq 0} \frac{\|Ax\|}{\|x\|}.$$

The normed linear space consisting of all bounded linear transformations between X and Y will be denoted by $L(X, Y)$, it being easy to verify that $\|A\|$ does indeed possess the desired properties.

We now list the following useful, alternative ways of computing $\|A\|$.

(1) [Sec. 11.3, (1)]: Denote by K the set of all real numbers k such that $\|Ax\| \leq k\|x\|$ for all $x \in X$. Then

$$\|A\| = \inf_{k \in K} k.$$

(2) [Sec. 11.3, (2)]:

$$\|A\| = \sup_{\|x\| \leq 1} \|Ax\|$$

(3) [Sec. 11.3, (3)]:

$$\|A\| = \sup_{\|x\| = 1} \|Ax\|.$$

Since, by now, it seems rather routine for things to collapse in finite-dimensional spaces, the following theorem is not at all surprising. We prove it just to illustrate the analogy between proofs about bounded linear transformations and bounded linear functionals.

THEOREM 14.7 (THEOREM 11.5). Suppose $A : X \to Y$, where X and Y are normed linear spaces and A is a linear transformation. If X is finite-dimensional, then A is bounded.

Proof. Suppose dim $X = n$ and that a basis for X is given by

$$x_1, x_2, \ldots, x_n.$$

In view of this, for any $x \in X$, there exist scalars $\alpha_1, \alpha_2, \ldots, \alpha_n$ such that

$$x = \sum_{i=1}^{n} \alpha_i x_i.$$

Since A is linear, we have

$$Ax = \sum_{i=1}^{n} \alpha_i Ax_i.$$

Hence, letting $k = \sum_{i=1}^{n} \|Ax_i\|$, we have

$$\|Ax\| = \left\| \sum_{i=1}^{n} \alpha_i Ax_i \right\|$$

$$\leq \sum_{i=1}^{n} |\alpha_i| \, \|Ax_i\|$$

$$\leq \|x\|_0 \cdot k,$$

where it is recalled that $\max_i |\alpha_i|$ is the definition of $\|x\|_0$. Since all norms in a finite-dimensional space are equivalent and A is bounded with respect to the zeroth norm, it follows that A must be a bounded linear transformation no matter what norm is chosen for X. (See Sec. 8.5 for a discussion of the notion of equivalent norms and the zeroth norm.)

Our next result yields still another way of describing a bounded linear transformation.

THEOREM 14.8. Let A be a linear transformation mapping the normed linear space X into the normed linear space Y. Then A is bounded if and only if A maps bounded sets in X into bounded sets in Y.

Proof. One recalls that a subset S of a normed space X is bounded if and only if there exists some positive constant R, such that

$$\|x\| \leq R$$

for all $x \in S$. (In R^2 this corresponds to being able to enclose S in some circle of sufficiently large radius about the origin.)

Suppose first that S is a bounded subset of X, and A is a bounded linear transformation. In this case there is some R such that

$$\|x\| \leq R$$

for all $x \in S$. Thus for any $x \in S$ we have that

$$\|Ax\| \leq \|A\| \, \|x\| \leq \|A\| \cdot R,$$

which is the statement that the image of S under A, $A(S)$, is bounded.

Conversely, suppose that A maps bounded sets in X into bounded sets in Y. Consider the bounded set in X, $\overline{S_1(0)}$, the closed unit sphere about the origin of X. By our assumption $A\overline{(S_1(0))}$ must also be bounded, which implies the existence of a positive number k such that $\|Az\| \leq k$ for all $z \in \overline{S_1(0)}$. Consider now any nonzero vector $x \in X$; clearly,

$$\frac{x}{\|x\|} \in \overline{S_1(0)}.$$

Therefore,

$$\left\| A\left(\frac{x}{\|x\|} \right) \right\| \leq k$$

or

$$\|Ax\| \leq k\|x\|. \tag{14.7}$$

Since (14.7) clearly holds for $x = 0$ as well, it follows that A is a bounded linear transformation.

Having obtained this equivalent description of a bounded linear transformation, we now wish to make some statement about the inverse of such a transformation. Our next result provides criteria for when the inverse exists and when it, too, will be a bounded linear transformation.

THEOREM 14.9. If A is a linear transformation, mapping the normed linear space X into the normed linear space Y, the inverse function A^{-1} exists and is bounded on the range of A, $A(X)$, if and only if there exists some $k > 0$ such that

$$k\|x\| \leq \|Ax\| \quad \text{(for all } x \in X).$$

Proof (sufficiency). To show that A^{-1} exists on $A(X)$, all we need show is that A is $1 : 1$. But, to show that a linear transformation is a $1 : 1$ mapping, it suffices to show that the zero vector of the domain is the only thing mapped into the zero vector of the range. Hence, for A as above, we shall now show that A is $1 : 1$ by showing that $Ax = 0$ implies $x = 0$.

Suppose $Ax = 0$. Then

$$k\|x\| \leq 0 \Rightarrow \|x\| = 0 \Rightarrow x = 0,$$

and we see that the condition insures the existence of A^{-1}. We now show that it is also a bounded linear transformation on $A(X)$. Since it is quite straightforward to demonstrate the linearity of A^{-1}, we shall only verify that it is bounded here.

By hypothesis, for any $x \in X$, we have

$$k\|x\| \leq \|Ax\|.$$

Now, since any $x \in X$ can be written [for some $y \in A(X)$]

$$x = A^{-1}y,$$

the above condition becomes

$$k\|A^{-1}y\| \leq \|A(A^{-1}y)\| = \|y\|,$$

and we see that, for all $y \in A(X)$,

$$\|A^{-1}y\| \leq \frac{1}{k}\|y\|.$$

Therefore A^{-1} is bounded on $A(X)$.

Proof (necessity). Suppose now that A^{-1} exists and is a bounded linear transformation on $A(X)$. This implies the existence of some $M > 0$ such that, for any $x \in X$,

$$\|A^{-1}(Ax)\| \leq M\|Ax\|$$

or

$$\frac{1}{M}\|x\| \leq \|Ax\|,$$

and taking the k mentioned in the theorem to be $1/M$, we see that the theorem is proved.

In Sec. 11.3 we saw that, regardless of the metric nature of X, the conjugate space $\tilde{X}$ was a Banach space. In proving this result, we relied most heavily upon the fact that the underlying field F (either the real or the complex numbers) was a complete metric space; that is, denoting $\tilde{X}$ by $L(X, F)$, we used the completeness of F to prove that $L(X, F)$ was a complete normed linear space. We have an analogous statement for the case when F is replaced by the normed linear space Y. In this case, however, we require that Y be a Banach space, as is stated in our next theorem.

THEOREM 14.10. With X and Y assumed to be normed linear spaces, if Y is a Banach space, then $L(X, Y)$ is a Banach space, too.

Proof. To prove this, we must show that every Cauchy sequence of elements of $L(X, Y)$ converges to some point in that space. To this end, suppose that $\{A_n\}$ is a Cauchy sequence from $L(X, Y)$. This means that for any $\varepsilon > 0$ there is some N such that, for all $n, m > N$,

$$\|A_n - A_m\| < \varepsilon.$$

This implies, for any fixed vector $x \in X$, that

$$\|A_n x - A_m x\| = \|(A_n - A_m)x\| \le \|A_n - A_m\| \, \|x\| < \varepsilon \|x\|$$

or that the sequence $\{A_n x\}$ is a Cauchy sequence in Y. Since Y is a Banach space; this sequence must have a limit, and we shall call this limit Ax; we now have

$$Ax \equiv \lim_n A_n x,$$

being assured that the function Ax is defined for every $x \in X$. This transformation is clearly linear; we now show that it is a bounded linear transformation and that $A_n \to A$. As for the boundedness of A, since $\{A_n\}$ is a Cauchy sequence, the terms must be uniformly bounded; there must be some real number M such that, for every n,

$$\|A_n\| \le M.$$

This implies that, for all n and any $x \in X$,

$$\|A_n x\| \le \|A_n\| \, \|x\| \le M\|x\|,$$

but, since this is true for every n, we must also have

$$\lim_n \|A_n x\| \le M\|x\|.$$

Since the norm is a continuous mapping, we can bring the limit inside to obtain

$$\|Ax\| \le M\|x\|,$$

which proves that A is a bounded linear transformation or that $A \in L(X, Y)$. We now show that $A_n \to A$. With N and ε as above, we have

$$\|A_n x - A_m x\| < \varepsilon\|x\|$$

for all $n, m > N$ and any vector x in X. Hence, taking the limit of this expression as $n \to \infty$ and using the continuity of the norm, we have

$$\lim_n \| A_n x - A_m x \| = \| Ax - A_m x \| \leq \varepsilon \| x \|,$$

which, for all $m > N$, implies that

$$\| A - A_m \| \leq \varepsilon$$

and proves that $A_n \to A$. We have now shown that any Cauchy sequence of elements from $L(X, Y)$ converges to some other element of this space and have proved that this space is a Banach space.

EXERCISES 14

1. Prove the following:

   (a) If X is an inner product space and if $x_n \overset{w}{\to} x$, then $(x_n, y) \to (x, y)$ for each $y \in X$.

   (b) If X is a Hilbert space, $x_n \overset{w}{\to} x$ if and only if $(x_n, y) \to (x, y)$ for each $y \in X$.

   (c) If $\{x_n\}$ is an orthonormal sequence in a Hilbert space X, then $x_n \overset{w}{\to} 0$ but $x_n \nrightarrow 0$.

2. Let X be a normed linear space. Show that if $x_n \overset{w}{\to} x$, there exists a sequence of linear combinations $\{\sum_{i=1}^{k_n} \alpha_i^{(n)} x_i^{(n)}\}$ which converges to x where $x_i^{(n)}$ are terms of the original sequence.

3. Let X be a normed linear space. If $x_n \overset{w}{\to} x$, then show that

$$\| x \| \leq \overline{\lim_n} \| x_n \| \qquad \left(\overline{\lim_n} \| x_n \| = \lim_k \sup_{n \geq k} \| x_n \| \right).$$

4. If X and Y are normed linear spaces and if $A : X \to Y$ is an isomorphism onto Y such that both A and A^{-1} are continuous, then show that $\| A^{-1} \| \geq \| A \|^{-1}$.

5. If X is a normed linear space, if Y is a B-space, and if $A : X \to Y$ is an onto isomorphism such that both A and A^{-1} are continuous, show that X is also a B-space.

6. Show that if X and Y are normed linear spaces such that $X \neq \{0\}$, then if $L(X, Y)$ is complete, Y must be complete.

7. Show that a mapping $A : X \to Y$, where X and Y are linear spaces, is a linear transformation if and only if its graph $G = \{\langle x, Ax \rangle \mid x \in X\}$, is a subspace of $X \times Y$.

8. Let X and Y be normed linear spaces and let $A : X \to Y$ be a bounded linear transformation. Show that $N(A)$, the *null space of A*, given by $\{x \mid Ax = 0\}$, is a closed subspace of X.

9. If X is an inner product space and $x_n \overset{w}{\to} x$ in X and $\|x_n\| \to \|x\|$, then $x_n \to x$.

10. Prove that the sequence $\{x_n\}$ in $L_p(a, b)$, $1 < p < \infty$ converges weakly to $x \in L_p(a, b)$ if and only if (a) the sequence $\{\|x_n\|\}$ is bounded and (b)

$$\int_a^y x_n(t) \, dt \to \int_a^y x(t) \, dt \qquad \text{for all } y \in [a, b].$$

[*Hint*: Use the representation of bounded linear functionals on $L_p(a, b)$ given in Exercises 13.5 and 13.6.]

11. Let X and Y be normed linear spaces and $A : X \to Y$ be a bounded linear transformation. Show that A can be extended uniquely to $B : X^* \to Y^*$, where B is a bounded linear transformation such that $\|A\| = \|B\|$ and where X^* and Y^* designate the completions of X and Y, respectively.

12. Show that in l_1 weak and strong convergence coincide.

13. Let X be a normed linear space. Show that $x_n \overset{w}{\to} x$ if $\text{Re}(f(x_n)) \to \text{Re}(f(x))$ for all $f \in \tilde{X}$.

14. Let X be a finite-dimensional space with basis $x_1, \ldots, x_n$. For $x \in X$, write $x = \sum_{i=1}^n \alpha_i x_i$ and define $\|x\|_0 = \max_i |\alpha_i|$. Let A be the linear transformation defined on the basis elements by

$$Ax_i = \sum_{j=1}^n \alpha_{ij} x_j \qquad (\alpha_{ij} \in F).$$

Compute $\|A\|$ with respect to the above norm.

15. Let X be a normed linear space and $A : X \to Y$ be a bounded linear transformation. Suppose there exists a vector $x \neq 0$ and a scalar λ such that $Ax = \lambda x$. Prove that $\|A\| \geq |\lambda|$.

16. Let X be a Hilbert space and let M be a closed subspace of X. For each $z \in X$, write $z = x + y$, where $x \in M$ and $y \in M^\perp$. Define $Ez = x$. Show that E is a bounded linear transformation such that $\|E\| = 1$.

REFERENCES

1. A. Zaanen, "Linear Analysis."
2. N. Dunford and J. Schwartz, "Linear Operators, Part I: General Theory."

CHAPTER 15

Convergence in $L(X, Y)$ and the Principle of Uniform Boundedness

In Chapter 14 the notion of a sequence of vectors from a normed linear space converging "weakly" to some other vector in the space was introduced. In making this definition, there was a certain amount of interplay between the space and its dual space. In this chapter, another kind of convergence will be introduced, defined on the normed linear space that was first encountered in Sec. 14.2—the normed linear space of all bounded linear transformations between two normed linear spaces. The two new concepts we wish to introduce, namely, that of *strong and weak convergence of a sequence of bounded linear transformations*, are quite peculiar to this type of space; that is, it is quite meaningless to speak of *strong convergence* as defined here of a sequence of vectors from an arbitrary normed linear space and equally meaningless to apply the term weak convergence to the general space.

As stated above, there are but two new concepts of convergence that will be introduced here. There is, however, some additional new terminology. The terminology is introduced only for convenience in distinguishing between the various types of convergence that are present in the space $L(X, Y)$. [The space $L(X, Y)$ is defined and discussed in Sec. 14.2]. Alas, the new terminology does not stop here though. When one considers the special case of $L(X, F)$—F, as usual, denotes the underlying scalar field of the real or complex numbers—some of the notions that are genuinely different in the general case of $L(X, Y)$ collapse to the same thing in this special setting, and to avoid triviality we again do some relabeling. In the hope of bringing all these things into their proper perspective, the various notions are summarized in Table II with the feeling that, when viewed side by side, the true similarities and differences will become apparent.

One of the more remarkable results proved in this lecture is the *Banach–Steinhaus theorem* also referred to as the *principle of uniform boundedness*—a result which allows statements to be made about when an arbitrary collection of bounded linear transformations will have a finite least upper bound.

15.1   Convergence in $L(X, Y)$

Suppose that X and Y are each normed linear spaces and that $L(X, Y)$ is the normed linear space of all bounded linear transformations between X and Y as described in Sec. 14.2. Since this is a normed linear space, it certainly is sensible to say that the sequence $\{A_n\}$, of elements of $L(X, Y)$ converges to A. By this, as usual, we mean that for any $\varepsilon > 0$ there is some integer N such that, for all $n > N$,

$$\|A_n - A\| < \varepsilon.$$

Ordinarily, one would say only that the sequence $\{A_n\}$ converges to A. In this special vector space, though, if this situation prevails, we shall denote it by saying that the sequence $\{A_n\}$ *converges uniformly to* A. It is to be emphasized that only in this special normed space $L(X, Y)$ will this terminology be used. The main reason for introducing the term at all is because of the following definition in which a new kind of convergence is defined and in light of Theorem 15.1.

DEFINITION 15.1. The sequence $\{A_n\}$ from $L(X, Y)$ is said to *converge strongly to* A, denoted by

$$A_n \overset{s}{\to} A,$$

if for any $x \in X$,

$$A_n x \to Ax.$$

(This last convergence is, of course, with respect to the norm on Y.)

In Sec. 14.1 the notion of weak convergence was defined for the general normed linear space. To review briefly, the sequence $\{x_n\}$ was said to converge weakly to x if, for all bounded linear functionals f, $f(x_n) \to f(x)$. In the special case of $L(X, Y)$ the term weak convergence will carry the following connotation: The sequence of bounded linear transformations $\{A_n\}$ is said to *converge weakly to* A, denoted by $A_n \overset{w}{\to} A$, if for any $x \in X$, the sequence of elements of Y, $\{A_n x\}$, converges weakly to Ax; in other words, we say that $A_n \overset{w}{\to} A$ if, for any $x \in X$ and for $f \in \tilde{Y}$, the dual space of Y,

$$f(A_n x) \to f(Ax).$$

As a final paraphrase, we say $A_n \overset{w}{\to} A$ if, for any $x \in X$, $A_n x \overset{w}{\to} Ax$.

Even though the terms weak and strong convergence have been used previously to mean other things (in a different setting), when either term is encountered it should be clear from the context which notion is meant. In the following discussion the symbols $\{A_n\}$, A, etc., are as above.

Thus we now have three kinds of convergence on $L(X, Y)$—uniform, strong, and weak. We should like to see what relation, if any, exists between the three. If one compares the definitions of strong and weak convergence, it is seen that each entails only the usual notions of strong and weak convergence of the vectors $A_n x$, a sequence of vectors from Y. Since this is so and we have already seen that strong convergence of vectors always implies weak convergence (of these vectors), we see that, in the newly defined sense of the terms, strong convergence implies weak convergence. (The discussion about strong convergence implying weak convergence in the general case is found in Sec. 14.1.) We shall now show that uniform convergence implies strong convergence. To this end, suppose uniform convergence prevails between the sequence $\{A_n\}$ and A. This means that

$$\|A_n - A\| \to 0.$$

Therefore, for any $x \in X$,

$$\|A_n x - Ax\| \le \|A_n - A\|\,\|x\| \to 0,$$

and the implication is proved. Hence, uniform $\Rightarrow$ strong $\Rightarrow$ weak.

We would now like to focus attention on an even more special situation, namely, when the normed linear space Y is just the scalar field F. As above, then, we have the three kinds of convergence defined on $L(X, F)$, or at least there seem to be three. As we shall see, there are really only two.

We show now that the notions of strong and weak convergence are equivalent in $L(X, F)$. If $\{f_n\}$ is a sequence of elements from $L(X, F)$, we recall that $\{f_n\}$ will converge to f strongly if, for any $x \in X$,

$$f_n(x) \to f(x). \tag{15.1}$$

Similarly, we say that $\{f_n\}$ converges weakly to f if for any $x \in X$

$$f_n(x) \overset{w}{\to} f(x). \tag{15.2}$$

The strong and weak convergences, (15.1) and (15.2), take place in the one-dimensional space F, however, and in the last chapter (Theorem 14.1) it was shown that strong and weak convergence are equivalent in a finite-dimensional space. Thus we cannot distinguish between the convergence (15.1) and (15.2) in this case, and the desired equivalence has been demonstrated. In view of this we shall drop the term "strong" convergence for this space, and shall refer to this type of convergence as *weak* convergence.

Just as in the general case [the case of $L(X, Y)$ where Y is any normed linear space, that is] we still have our usual notion of what it means for the sequence $\{f_n\}$ to converge to f with respect to the norm on $L(X, F)$. To conform to the more general situation, one would wish to call this "uniform" convergence. Unfortunately, this is not standard practice, however, and this sort of convergence for linear functionals is called *strong* convergence; we shall say that bounded linear *transformations* converge *uniformly* and that bounded linear *functionals* converge *strongly* if $\|A_n - A\| \to 0$ or $\|f_n - f\| \to 0$, where $A_n, A \in L(X, Y)$ and $f_n, f \in L(X, F)$, respectively.

We beg the reader's indulgence as one last notion of convergence is defined.

Consider the normed linear space X and its conjugate space $\tilde{X} = L(X, F)$. Clearly, since $\tilde{X}$ is a normed linear space, the usual notions of strong and weak convergence apply. We wish now to modify the notion of weak convergence for this special normed space. First, let us suppose that the sequence of elements from $\tilde{X}$, $\{f_n\}$, converge weakly to f. This means that, for every $g \in \tilde{\tilde{X}}$,

$$g(f_n) \to g(f). \tag{15.3}$$

Since requiring all bounded linear functionals on $\tilde{X}$ to have this property is such a stringent requirement (this being such a large class of functionals, in general), we shall not use this; we wish now to modify it by requiring only a subset of $\tilde{\tilde{X}}$ to have property (15.3).

Adhering to the notation of Sec. 12.2, suppose we required only all bounded linear functionals of the form Jx, where $x \in X$ and J is as in Sec. 12.2, to have property (15.3).† In this case the sequence $\{f_n\}$ is said to be *w*-convergent* (*weak-**) to f, written $f_n \overset{w*}{\to} f$. Thus for *w**-convergence of $\{f_n\}$ to f, we require, for all $x \in X$,

$$(Jx)(f_n) \to (Jx)(f)$$

or, equivalently, that

$$f_n(x) \to f(x).$$

To consolidate these notions, it is hoped that Table II will be helpful.

---

† The mapping J is as follows: Suppose $y \in \tilde{X}$. Thus we have $J : X \to \tilde{\tilde{X}}$; $x \to Jx$, where the bounded linear functional Jx is defined by $[y, Jx] = [x, y] = y(x)$.

TABLE II

The Various Notions of Convergence

$L(X, Y)$,	$\{A_n\}$			$L(X, F)$,	$\{f_n\}$	
Type of convergence	Notation	Definition	Type of convergence	Notation	Definition	
Uniform	$A_n \to A$	$\|A_n - A\| \to 0$	Strong	$f_n \xrightarrow{w} f$	$\|f_n - f\| \to 0$	
Strong	$A_n \xrightarrow{s} A$	$A_n x \to Ax$ for all $x \in X$	Weak	$f_n \to f$	$f_n(x) \to f(x)$ for all $x \in X$	
					$\updownarrow_{w}$	
Weak	$A_n \xrightarrow{w} A$	$A_n x \xrightarrow{w} Ax$ for all $x \in X$	Weak	$f_n \xrightarrow{w} f$	$f_n(x) \xrightarrow{w} f(x)$ for all $x \in X$	
					$\updownarrow$	
			w^*	$f_n \xrightarrow{w*} f$	$(Jx)(f_n) \to (Jx)(f)$ for all $x \in X$	
					$\updownarrow$	
					$f_n(x) \to f(x)$ for all $x \in X$	
	X, $\{x_n\}$					
Strong	$x_n \to x$	$\|x_n - x\| \to 0$				
Weak	$x_n \xrightarrow{w} x$	$f(x_n) \to f(x)$ for all $f \in \tilde{X}$				

Our next theorem provides a strong motivation for using the word "uniform" in the sense we did in $L(X, Y)$.

THEOREM 15.1. The sequence $\{A_n\}$ from $L(X, Y)$ converges uniformly to A if and only if $A_n x \to Ax$ uniformly for all $x \in X$ such that $\|x\| \le 1$.

Proof. First it will be shown $A_n \to A$ does indeed imply that the condition holds. Hence, suppose for a given $\varepsilon > 0$, there is some N such that $n > N$ implies

$$\|A_n - A\| < \varepsilon.$$

Thus, for x such that $\|x\| \le 1$,

$$\|A_n x - Ax\| \le \|A_n - A\| \, \|x\| \le \|A_n - A\| < \varepsilon.$$

Clearly, the integer N mentioned above does not depend on any particular choice of x in the closed unit sphere, and the uniform convergence mentioned in the condition is seen to be demonstrated.

Conversely, suppose that, with ε as above, there is an integer N whose choice depends only on ε such that $n > N$ implies

$$\|A_n x - Ax\| < \varepsilon \tag{15.4}$$

for all x such that $\|x\| \le 1$. Now choose any nonzero vector $x \in X$ and define

$$y = \frac{x}{\|x\|}.$$

Clearly, (15.4) implies that, for $n > N$,

$$\|A_n y - Ay\| < \varepsilon$$

or

$$\|A_n x - Ax\| < \varepsilon \|x\|, \tag{15.5}$$

for any nonzero x. Since it is obvious that (15.5) holds when x is zero, it follows that (15.5) holds for every x and that

$$\|A_n - A\| < \varepsilon$$

for $n > N$. It is to be noted again how strongly we needed the fact that N depended only on ε and not on x, the proof now being complete.

In the special case of $L(X, X)$ we would now like to introduce a new operation—that of *multiplication* of bounded linear transformations. As one might expect for any $A, B \in L(X, X)$, we define, for any $x \in X$,

$$(AB)(x) = A(B(x)); \tag{15.6}$$

that is, we shall take the composite of A and B as the product. One notes that this could not be done in the more general situation of $L(X, Y)$. We now wish to demonstrate some of the algebraic properties of this new operation. We first wish to demonstrate that the space $L(X, X)$ is closed under multiplication, by virtue of the following observation: With A, B, and x as above,

$$\|ABx\| \le \|A\| \, \|Bx\| \le \|A\| \, \|B\| \, \|x\|$$

for any x, which implies that AB is bounded and that

$$\|AB\| \le \|A\| \, \|B\| \tag{15.7}$$

or that the product of bounded linear transformations as defined above is also a bounded linear transformation, it being simple to verify the linearity of AB. That the operation is also associative follows directly, and as another note we see that the product of the identity transformation with any other leaves it alone; thus the identity transformation constitutes an identity element for this product. Last, it is easily verified that

$$A(B + C) = AB + AC \tag{15.8}$$

and that, for any scalar α,

$$\alpha(AB) = (\alpha A)B = A(\alpha B). \tag{15.9}$$

We shall very shortly have occasion to return to this space, from a somewhat enlightened viewpoint (see Chapter 19).

15.2 The Principle of Uniform Boundedness

The following theorem is the most important theorem of the chapter. It enables us to determine whether the norms of a given collection of bounded linear transformations [members of $L(X, Y)$, that is] $\{A_\alpha\}$ have a finite least upper bound or, equivalently, if there is some uniform bound for the set $\{\|A_\alpha\|\}$. Clearly, since the norm was defined to be a real-valued (not an extended real-valued) mapping, the norm of each A_α must be finite, but there is no guarantee that they might not form an increasing sequence. The following, pervasive, theorem provides a criterion for determining when such an increasing sequence is not formed.

THEOREM 15.2 (BANACH–STEINHAUS) (*principle of uniform boundedness*). Let Λ be an index set of arbitrary cardinality and suppose $\{A_\alpha\}$, where α runs through Λ, is a collection of members of $L(X, Y)$ [which has, as usual, the connotation of Sec. 14.2] where X is a Banach space. Now if $\sup_\alpha \|A_\alpha x\| < \infty$ for all $x \in X$, then $\sup_\alpha \|A_\alpha\| < \infty$.

REMARK. Clearly, if $\sup_\alpha \|A_\alpha\| < \infty$, there exists some positive number M such that $\|A_\alpha\| < M$ for all α, or the set of real numbers $\{\|A_\alpha\|\}$ has a uniform bound—M.

Proof. First we suppose that Λ is a denumerable set, that is, we suppose that the denumerable set $\{A_\alpha\}$, $\alpha \in \Lambda$, satisfies the condition of the theorem. After proving it for this case, we easily extend it to the desired result encompassing sets of arbitrary cardinality.

Hence suppose $A_1, A_2, \ldots, A_n, \ldots$ satisfies the condition of the hypothesis. We now claim that it suffices to prove that the following condition holds:

For some $x_0 \in X$ and some positive number δ there exists a $k > 0$ such that

$$\|A_n x\| < k \tag{15.10}$$

for all n and all $x \in \overline{S_\delta(x_0)}$.

We will now show that (15.10) really does imply the theorem and then show that (15.10) holds. Hence suppose it is satisfied and let x be a vector such that $\|x\| \le \delta$. In this case, we have

$$\|A_n x\| = \|A_n(x + x_0) - A_n x_0\|$$
$$\le \|A_n(x + x_0)\| + \|A_n x_0\|. \tag{15.11}$$

But, since

$$\|(x + x_0) - x_0\| = \|x\| \le \delta,$$

we see that

$$x + x_0 \in \overline{S_\delta(x_0)},$$

which implies, by assumption (15.10),

$$\|A_n(x + x_0)\| < k$$

for every n. Since $\|A_n x_0\| < k$, too, (15.11) implies that, for all n,

$$\|A_n x\| < 2k$$

for all x such that $\|x\| \le \delta$. Now let x be any nonzero vector and consider

$$\frac{\delta x}{\|x\|}.$$

In this case,

$$\|A_n x\| = \left\| \frac{\|x\|}{\delta} A_n \left(\frac{\delta x}{\|x\|} \right) \right\|$$

$$\leq \frac{\|x\|}{\delta} 2k$$

for all n. Since the above statement must also hold for $x = 0$, we see that we must have

$$\|A_n\| \leq 2k/\delta$$

for all n; therefore,

$$\sup_n \|A_n\| \leq \frac{2k}{\delta},$$

and we have indeed shown that (15.10) implies the uniform boundedness of $\{\|A_n\|\}$. Now we shall show (15.10) holds. We do this by contradiction.

Suppose no such $k > 0$ as mentioned in (15.10) exists; that is, suppose for any spherical neighborhood S of any point and any real number a, there exists an integer n_0 and an $x_0 \in S$ such that

$$\|A_{n_0} x_0\| > a.$$

Since the function $f(x) = \|A_{n_0} x\|$ is a real-valued continuous function of x that is greater than a at x_0, it follows from the continuity that there must be a whole neighborhood of x_0 in which $f(x) > a$; furthermore, we can assume that this neighborhood of x_0 is wholly contained in S. Symbolically, we are asserting the existence of a neighborhood of x_0, S_a, such that $\bar{S}_a \subset S$ and

$$\|A_{n_0} x\| > a$$

for all $x \in \bar{S}_a$. We are now in a position to pull the strings together and complete the proof. In the above discussion S was any neighborhood and a was any (positive) number. We continue to let S be any neighborhood, but we now take $a = 1$. By virtue of the above discussion, we know there exists some *closed* set S_1 such that

$$S_1 \subset S$$

and an integer n_1 such that

$$\|A_{n_1} x\| > 1$$

for all $x \in S_1$. Further, we can also assume, without loss of generality, that the diameter of S_1, $d(S_1)$, is less than 1.

Now let us go through the above procedure again but with S replaced by S_1, and $a = 2$. In this event, we get some closed set $S_2 \subset S_1$ such that $d(S_2) < \frac{1}{2}$, and such that for some n_2,

$$\|A_{n_2} x\| > 2$$

for all $x \in S_2$. Continuing on in this fashion, we generate a nested sequence of closed sets with diameters shrinking to zero, which we denote by $\{S_k\}$. Since X is a Banach space, we can appeal to Theorem 5.1 to assert the existence of some

$$y \in \bigcap S_k,$$

which implies

$$\|A_{n_k} y\| > k \qquad (15.12)$$

for every k. But we were assuming

$$\sup_n \|A_n x\| < \infty,$$

for all x, which is incompatible with (15.12) and the assumption that (15.10) does not hold has led to a contradiction. The proof of the theorem for the case when Λ is denumerable is now seen to be complete. We now proceed to the general case.

Suppose $\{A_\alpha\}$, $\alpha \in \Lambda$, Λ arbitrary, satisfies the hypothesis (certainly if this is so then any denumerable subset must also satisfy the hypothesis) and suppose at the same time, that

$$\sup \|A_\alpha\| = \infty. \qquad (15.13)$$

Assumption (15.13) implies that there exist $A_1, A_2, ..., A_n, ...$ such that

$$\|A_1\| > 1$$
$$\|A_2\| > 2 \qquad (15.14)$$
$$\vdots \quad \vdots$$

But since the set $\{A_1, A_2, ..., A_n, ...\}$ satisfies the hypothesis, by what we have just proved,

$$\sup_n \|A_n\| < \infty,$$

a fact which is certainly discordant with (15.14). The theorem now follows.

15.3 Some Consequences of the Principle of Uniform Boundedness

As we already know, all the terms of an ordinary Cauchy sequence must be bounded. We now obtain a somewhat similar result in the space $L(X, Y)$.

THEOREM 15.3. Let $\{A_n\}$ be a sequence of terms from $L(X, Y)$, where X is a Banach space, converging strongly to A. Then there exists a positive constant M such that $\|A_n\| < M$ for all n.

Proof. Since $A_n \xrightarrow{s} A$, then

$$\lim_n A_n x = Ax$$

for all $x \in X$. This in turn, implies that

$$\sup_n \|A_n x\| < \infty$$

for all x and, now, by the principle of uniform boundedness, we see that we must have

$$\sup_n \|A_n\| < \infty,$$

which is the desired result.

DEFINITION 15.2. The sequence $\{A_n\}$ from $L(X, Y)$ is said to be a *strong Cauchy sequence*, if the sequence $\{A_n x\}$ is a Cauchy sequence for all $x \in X$.

Having made this definition, it now becomes meaningful to consider a space $L(X, Y)$ to be *complete in the strong sense*; by this we shall mean that every strong Cauchy sequence converges strongly to some member of the space. We now have the following corollary to the above theorem.

COROLLARY. If the spaces X and Y are each Banach spaces, then $L(X, Y)$ is complete in the strong sense.

Proof. Let $\{A_n\}$ be a strong Cauchy sequence in $L(X, Y)$. We must show that there is some element A of $L(X, Y)$ to which $\{A_n\}$ converges strongly. Since we have assumed that $\{A_n\}$ is a strong Cauchy sequence, it follows by definition that, for any $x \in X$, $\{A_n x\}$ is a Cauchy sequence of elements of Y. Since Y is a Banach space, though, the limit of this sequence must exist in Y. Thus, for every $x \in X$, the function

$$Ax = \lim_n A_n x \tag{15.15}$$

is defined. Clearly, A is a linear transformation and (15.15) is equivalent to saying that

$$A_n \overset{s}{\to} A.$$

We would now like to show that A is a bounded linear transformation. By Theorem 15.3, since X is a Banach space,

$$\|A_n\| < M$$

for all n and some positive constant M. Since, for any $x \in X$, we can say

$$\|A_n x\| \le \|A_n\| \|x\|,$$

this implies that

$$\|A_n x\| \le M \|x\|$$

for any x, and every n. Since it is true for every n, it must also be true in the limit or

$$\lim_n \|A_n x\| \le M \|x\|.$$

By the continuity of the norm, we can bring the limit inside the norm symbol to assert that

$$\|Ax\| \le M \|x\|$$

for every x, which proves that A is bounded and the corollary follows.

In a way analogous to the way in which we just defined what was meant by a strong Cauchy sequence of elements of $L(X, Y)$, we now define what is meant by a *weak Cauchy sequence* of elements of the normed linear space X.

DEFINITION 15.3. The sequence of elements $\{x_n\}$ of the normed linear space X is said to be a *weak Cauchy sequence* if $\{f(x_n)\}$ is a Cauchy sequence of elements for all $f \in \tilde{X}$, the conjugate space of X.

We now show that a weak Cauchy sequence must be bounded.

THEOREM 15.4. In the normed linear space X every weak Cauchy sequence is bounded.

Proof. Let $\{x_n\}$ be a weak Cauchy sequence of elements of X; that is, $\{f(x_n)\}$ is a Cauchy sequence for all $f \in \tilde{X}$. We recall the mapping J of Sec. 12.2 now:

$$J : X \to \tilde{\tilde{X}},$$

$$x \to Jx,$$

where the bounded linear functional Jx is defined for any $y \in \tilde{X}$ by

$$[y, Jx] = [x, y].$$

Clearly, an equivalent way of representing $\tilde{\tilde{X}}$ is $L(\tilde{X}, F)$; that is, we view J as

$$J : X \to L(\tilde{X}, F).$$

In addition, we recall, for any x, that $\|Jx\| = \|x\|$.

Since $\{f(x_n)\}$ is a Cauchy sequence of complex numbers, we must have, for any $f \in \tilde{X}$,

$$\sup_n |Jx_n(f)| = \sup_n |f(x_n)| < \infty. \tag{15.16}$$

Recalling that $\tilde{X}$ must be a Banach space (Theorem 11.6), we see that the principle of uniform boundedness (Theorem 15.2) and Eq. (15.16) imply that

$$\sup_n \|Jx_n\| < \infty,$$

which, since $\|Jx_n\| = \|x_n\|$, yields

$$\sup_n \|x_n\| < \infty,$$

which is the desired result.

As a simple corollary we can now prove the one theorem that was not proved in the preceding lecture—Theorem 14.2.

COROLLARY. In the normed linear space X, if the sequence $\{x_n\}$ converges weakly to x, $x_n \overset{w}{\to} x$, there exists some positive constant M such that $\|x_n\| \leq M$ for all n.

To prove this we need only note that, if

$$x_n \overset{w}{\to} x,$$

certainly $\{x_n\}$ is a weak Cauchy sequence, and the theorem can be applied.

DEFINITION 15.4. A normed linear space X is said to be *weakly complete* if every weak Cauchy sequence of elements of X converges weakly to some other member of X.

Our next theorem shows that any reflexive space has this property.

THEOREM 15.5. If the normed space X is reflexive, it is also weakly complete.

REMARK. In Theorem 12.6 it was shown that all Hilbert spaces are reflexive. This theorem then implies that every Hilbert space is weakly complete—an intuitively plausible property one might tend to expect of a Hilbert space.

Proof. Suppose $\{x_n\}$ is a weak Cauchy sequence of elements of X. This means that $\{f(x_n)\}$ is a Cauchy sequence for all $f \in \tilde{X}$. Now, with J as in the preceding theorem, we note that, since

$$Jx_n(f = f(x_n),$$

$\{Jx_n(f)\}$ is a Cauchy sequence of scalars for all $f \in \tilde{X}$. Since the underlying scalar field is either the real or complex numbers (each of which is a complete metric space) this implies that the functional y defined (on $\tilde{X}$) by

$$y(f) = \lim_n Jx_n(f)$$

exists for every $f \in \tilde{X}$. That y is a *linear* functional is easily verified, and we shall now show that y is a *bounded* linear functional.

Since $\|Jx_n\| = \|x_n\|$ and $\{x_n\}$ is a weak Cauchy sequence, it follows, by the preceding theorem, that there is some positive number M such that

$$\|x_n\| \le M$$

for all n. This implies that

$$|Jx_n(f)| \le M\|f\|$$

for any $f \in \tilde{X}$ and all n; hence it is true in the limit. Since we can interchange the operations of taking absolute value and limit by the continuity of the absolute value operation, it follows that

$$|y(f)| \le M\|f\|$$

for all $f \in \tilde{X}$. This, however, implies that y is a *bounded* linear functional, or that $y \in \tilde{\tilde{X}}$. By the reflexivity of X, there must be some $x \in X$ that we can "identify" with y; more precisely, there must exist some $x \in X$ such that

$$y = Jx.$$

Hence for any $f \in \tilde{X}$ we can say

$$\lim_n f(x_n) = \lim_n Jx_n(f) = y(f) = Jx(f) = f(x);$$

therefore, since this is so for any $f \in \tilde{X}$, we have

$$x_n \overset{w}{\to} x$$

and the theorem is proved.

EXERCISES 15

1. Show that if $A_n \xrightarrow{s} A$ and $B_n \xrightarrow{s} B$, then $A_n + B_n \xrightarrow{s} A + B$. Show also that, if $A_n \xrightarrow{s} C$, then $A = C$. Similarly, for weak convergence.

2. Let X be a normed linear space and Y a Hilbert space, and let $\{A_n\}$ be a sequence in $L(X, Y)$. Show that $A_n \xrightarrow{w} A$, where $A \in L(X, Y)$ if and only if $(A_n x, y) \to (Ax, y)$ for all $x \in X$ and all $y \in Y$.

3. Let S be a subset of the normed linear space X. Show that S is bounded if and only if $f(S)$ is bounded for each $f \in \tilde{X}$.

4. If $\{A_n\}$ is a sequence in $L(X, Y)$, where X is a Banach space such that $A_n \xrightarrow{w} A$, then show that there exists an M such that $\|A_n\| \le M$ for all n.

5. Prove that the strong limit of continuous linear transformations is continuous; that is, if $A_n \in L(X, Y)$, where X is a Banach space, and $A_n \xrightarrow{s} A$, then $A \in L(X, Y)$.

6. Let X be a normed linear space and let $\{f_n\}$ be a sequence in $\tilde{X}$. If $f_n \xrightarrow{w} f$ [in the sense of $L(X, F)$ of Table II] and if X is a Banach space, the sequence $\{f_n\}$ is bounded.

7. Let X be a Banach space and $\{f_n\}$ be a sequence in $\tilde{X}$. Show that $f_n \xrightarrow{w} f$ [in the sense of $L(X, F)$ as in Table II] where $f \in \tilde{X}$ if and only if the sequence $\{\|f_n\|\}$ is bounded and $f_n(x) \to f(x)$ for all $x \in A$, where $\overline{[A]} = X$.

8. Show that the Banach–Steinhaus theorem ceases to be true if X is not complete.

9. Show that, if X is a reflexive space, every bounded infinite set $A \subset X$ contains a weak Cauchy sequence. (*Hint:* Use Exercise 12.8.)

10. Show that, if X is a separable space, an arbitrary bounded sequence of elements of $\tilde{X}$ contains a weakly convergent [in the sense of $L(X, F)$] subsequence.

11. Let $\{A_n\}$ be a sequence in $L(X, Y)$ where Y is a Banach space such that $\|A_n\| \le M$ for all n. Suppose that $\lim_n A_n(x)$ exists for all x in a set W, dense in a sphere S; then prove that $\lim_n A_n(x)$ exists for all x and that the limit is a bounded linear transformation with norm less than or equal to $\underline{\lim}_n \|A_n\|$.

12. Show that, if X and Y are finite-dimensional spaces, then the three notions of convergence in $L(X, Y)$ introduced in this chapter (namely, uniform, strong, and weak) coincide.

REFERENCES

1. A. Zaanen, "Linear Analysis."
2. S. Banach, "Théorie des Opérations Linéaires."
3. N. Dunford and J. Schwartz, "Linear Operators, Part I: General Theory."
4. A. Taylor, "Introduction to Functional Analysis."

CHAPTER **16**

Closed Transformations and the
Closed Graph Theorem

By now, the notion of a bounded linear transformation is quite a familiar one. In view of the large amount of machinery we have already developed about such transformations, if one, in practice, could demonstrate a given operator to be a bounded linear transformation, this would certainly be a step toward the solution of whatever the problem at hand. As one's past experience in analysis might indicate, other transformations which do not have all the desirable properties that continuous transformations have might also arise quite frequently and, as it happens, many operators defined in terms of differentiation (ordinary or partial) are discontinuous; thus, the powerful theorems we have developed could not be thrown at them. Many important operators do have another property, though, which to some extent makes up for the fact that they are not continuous, and it is to this type of linear transformation—the *closed* linear transformation—that we devote this chapter.

After defining such transformations, we are immediately concerned with what linkage there may be between closed linear transformations and bounded linear transformations. Although it is shown that closed linear transformations are not generally continuous, there are certain special settings when the property of closure implies continuity and one such result along these lines is the *closed-graph theorem*, Theorem 16.7.

This theorem however is but a consequence of the *bounded inverse theorem*, Theorem 16.5. This theorem is of a stature in functional analysis commensurate with the Hahn–Banach theorem and the principle of uniform boundedness. It tells us that, under certain conditions, when the inverse of a closed transformation exists, it is a bounded, linear transformation—a striking result. In addition to the closed-graph theorem, another well-known result called the *interior mapping principle* also follows as an immediate consequence of the bounded inverse theorem.

16.1 The Graph of a Mapping

Certainly the concept of the graph of a real-valued function of a single real variable is a familiar one; it is just the pictorial tabulation of the set of ordered pairs $\langle x, f(x) \rangle$, where x is allowed to assume all values in the domain of f. Indeed, in modern terminology it is the collection of these ordered pairs that is referred to as the function. At any rate, we now wish to generalize the notion of the " graph " of a real-valued function of a real variable. To this end suppose X and Y are sets and A is a mapping from some subset D of X, into Y. Then the collection of ordered pairs

$$G_A = \{\langle x, A(x) \rangle | x \in D\}$$

is called the *graph of A*. [We use the notation $\langle x, A(x) \rangle$ to avoid any possible confusion with inner products.] Our main concern will be with the case when X and Y are normed linear spaces and A is a linear transformation; in this case the subset D of X will have to be a subspace of X.

In Sec. 9.2 it was pointed out that the cartesian product of the vector spaces X and Y could easily be viewed as a linear space by taking addition of elements of $X \times Y$ as

$$\langle x_1, y_1 \rangle + \langle x_2, y_2 \rangle = \langle x_1 + x_2, y_1 + y_2 \rangle$$

and, for scalar multiplication,

$$\alpha\langle x, y\rangle = \langle \alpha x, \alpha y\rangle.$$

Viewing $X \times Y$ in this way is sometimes denoted by $X \oplus Y$, the resulting vector space being called the *external direct sum* of X and Y. In this case it is easily verified that A is indeed a *linear* transformation if and only if G_A is a subspace of $X \times Y$.

In Sec. 9.2 it was also pointed out that, if X and Y were each normed spaces, a norm for $X \times Y$ was immediately available [Eqs. (9.1)–(9.3)]; namely, using the same symbol for the norm on X and the norm on Y, one possibility was to take

$$\|\langle x, y\rangle\| = \|x\| + \|y\|. \tag{16.1}$$

Two other norms we might equally well have taken are

$$\|\langle x, y\rangle\| = (\|x\|^p + \|y\|^p)^{1/p}, \qquad (p \geq 1) \tag{16.2}$$

and

$$\|\langle x, y\rangle\| = \max(\|x\|, \|y\|), \tag{16.3}$$

it being noted that all three norms are equivalent on $X \times Y$.

As a final observation, we note that if X and Y are Banach spaces, so is $X \times Y$.

16.2   Closed Linear Transformations and the Bounded Inverse Theorem

DEFINITION 16.1. Let X and Y be normed linear spaces and let D be a subspace of X. Then the linear transformation.

$$A : D \to Y$$

is called *closed* if for every convergent sequence $\{x_n\}$ of points of D, where $x_n \to x \in X$ say, such that $\{Ax_n\}$ is a convergent sequence of points of Y, where $Ax_n \to y \in Y$ say, the following two conditions are satisfied:

$$x \in D \quad \text{and} \quad y = Ax.$$

As motivation for the word "closed" in describing such transformations, we now show that a linear transformation A is closed if and only if its graph G_A is a closed† subspace of $X \times Y$.

First suppose that $X \supset D \xrightarrow{A} Y$ is a closed linear transformation. To show that G_A is closed, we must show that any limit point of G_A is actually a member G_A. In view of this let $\langle x, y\rangle$ be a limit point of G_A. Since this is so, there must be a sequence of points of G_A, $\langle x_n, Ax_n\rangle$, where $x_n \in D$, converging to $\langle x, y\rangle$. This is equivalent to requiring that $\|\langle x_n - x, Ax_n - y\rangle\| \to 0$ or $\|x_n - x\| + \|Ax_n - y\| \to 0$, which implies that

$$x_n \to x \quad \text{and} \quad Ax_n \to y.$$

Since A is closed, we can say that this implies

$$x \in D \quad \text{and} \quad y = Ax$$

---

† As norm on $X \times Y$, we take that given in Eq. (16.1), namely, $\|\langle x, y\rangle\| = \|x\| + \|y\|$.

and, in view of this, we can write

$$\langle x, y \rangle = \langle x, Ax \rangle \in G_A \, ;$$

therefore, G_A is closed.

Conversely, suppose G_A is closed and that

$$x_n \rightarrow x, \qquad x_n \in D$$

for all n, and $Ax_n \rightarrow y$. We must show $x \in D$ and $y = Ax$. The condition implies that

$$\langle x_n, Ax_n \rangle \rightarrow \langle x, y \rangle \in \bar{G}_A \, .$$

Since G_A is closed, $G_A = \bar{G}_A$, and we have that

$$\langle x, y \rangle \in G_A \, .$$

By the definition of G_A this means that

$$x \in D \qquad \text{and} \qquad y = Ax;$$

A, therefore, is closed.

We have now proved the next theorem.

THEOREM 16.1. A is a closed linear transformation if and only if its graph is a closed subspace.

In view of the equivalence stated by the above theorem, it is now only a matter of personal preference as to what to take for the defining property of a closed linear transformation; in some cases using the closed graph requirement will yield an answer much more quickly, while in others, the original definition will be the judicious choice.

We now wish to exhibit another transformation.

Linear Transformation that is Closed But Not Bounded

Suppose

$$X \supset D \overset{A}{\rightarrow} Y,$$

where $X = Y = C[0, 1]$ with sup norm (see Example 6.3 for a discussion of $C[0, 1]$) and D consists of all those functions that have a continuous derivative; that is,

$$D = \{x \in X \mid x' \in C[0, 1]\}.$$

We take the linear transformation A to be differentiation, $Ax = x'$, for any $x \in D$. We first note that A is clearly a linear transformation. To show that it is not bounded, consider the sequence of points of D,

$$x_n(t) = t^n$$

for $n = 1, 2, \dots$. In this case $Ax_n = nt^{n-1}$. We now observe that, with the sup norm, $\|x_n\| = 1$ and $\|Ax_n\| = n$ for all n; hence,

$$\|Ax_n\| = n = n\|x_n\|.$$

Since

$$\sup_{\substack{x \neq 0 \\ x \in D}} \frac{\|Ax\|}{\|x\|} \geq \sup_n \frac{\|Ax_n\|}{\|x_n\|} = \sup_n n = \infty,$$

it follows that A is not a bounded linear transformation. To show that it is closed, suppose $\{x_n\}$ is a sequence of points of D such that

$$x_n \to x \quad \text{and} \quad Ax_n \to y.$$

The second requirement means that

$$x_n' \to y.$$

Translating this last statement into function-theoretic language, it says that the derived sequence of a given convergent sequence is also a uniformly convergent sequence; consequently, the derivative of the original limit function must exist and be equal to the limit function of the derived sequence. Restating this once more in its original setting, we have

$$x \in D \quad \text{and} \quad Ax = y,$$

the fact that x is continuous following because it is the limit of a uniformly convergent sequence of continuous functions. This, however, is just the statement of the fact that A is closed.

We now wish to investigate when a bounded transformation is closed. Clearly, if A is a bounded linear transformation such that

$$X \supset D \overset{A}{\to} Y,$$

where D is a *closed* subspace of X, then if $\{x_n\}$ is a convergent sequence of points of D such that

$$Ax_n \to y,$$

the limit of the sequence, it says, must belong D while the continuity of A implies that

$$Ax_n \to Ax.$$

Thus, under these conditions, A is easily seen to be closed and we have proved Theorem 16.2.

THEOREM 16.2. Let A be a bounded linear transformation such that

$$X \supset D \overset{A}{\to} Y,$$

where D is a closed subspace of X. Then A is closed.

Since, in any case, the entire space is always a closed subspace, we have the following corollary.

COROLLARY. Suppose A is a linear transformation such that

$$A : X \to Y.$$

Then if A is continuous, A is closed.

In many situations one is much more concerned with the inverse of a given operator than the operator itself [for example, suppose one is trying to solve the differential equation $y' + ay = f$ with $y(0) = 0$. The operator whose inverse we wish is then $A : y \rightarrow f$.], and it is with the properties of the inverse operator that our next theorem is concerned.

THEOREM 16.3. Suppose A is a closed linear transformation such that

$$X \supset D \overset{A}{\rightarrow} Y.$$

Then if A^{-1} exists, it is also a closed linear transformation.

Proof. By Theorem 16.1, A being closed implies that

$$G_A = \{\langle x, Ax \rangle | x \in D\}$$

is closed. Consider now the range of A, $A(D)$. Since A^{-1} exists, for any $y \in A(D)$, there is a unique $x \in D$ such that $y = Ax$ or $x = A^{-1}y$. In view of this we can describe the graph of A as

$$G_A = \{\langle A^{-1}y, y \rangle | y \in A(D)\}.$$

Consider now the mapping

$$X \times Y \rightarrow Y \times X,$$

$$\langle x, y \rangle \rightarrow \langle y, x \rangle.$$

It is trivial to verify that this mapping is an isometry. Since isometries must send closed sets into closed sets, we have that

$$\{\langle A^{-1}y, y \rangle \, | \, y \in A(D)\}$$

being closed implies that

$$\{\langle y, A^{-1}y \rangle \, | \, y \in A(D)\}$$

is also closed. This last set, however, is just the graph of A^{-1} and, since it is closed, it being obvious that A^{-1} is linear if it exists, we see that A^{-1} is a closed linear transformation.

THEOREM 16.4. Suppose A is a linear transformation such that

$$X \supset D \overset{A}{\rightarrow} Y,$$

where Y is a Banach space. If, in addition, A is closed and bounded, then D is a closed subspace of X.

Proof. To prove that D is closed, we must show that any limit point of D is also a member of D. Hence suppose x is a limit point of D. This means that there must be some sequence of points of D, $\{x_n\}$, such that

$$x_n \rightarrow x$$

Consider now

$$\| Ax_n - Ax_m \| \leq \| A \| \| x_n - x_m \|.$$

Since $\|x_n - x_m\| \to 0$ as $n, m \to \infty$, we see that the sequence of points $\{Ax_n\}$ of Y is a Cauchy sequence. Y being a Banach space implies the existence of some $y \in Y$ such that

$$Ax_n \to y.$$

Thus, having verified that the sequence $\{Ax_n\}$ is a convergent sequence, we can now utilize the fact that A is closed to assert that $x \in D$. Hence D contains all its limit points, and the proof is, therefore, complete.

Our next result is an extremely important one. It ranks in importance with the Hahn–Banach theorem and the principle of uniform boundedness. The theorem's essential concern is with the boundedness of the inverse transformation. The theorem tells us that, under the conditions of the theorem, when the inverse of a closed transformation exists, it is a bounded, linear transformation—a remarkable result. Some immediate consequences of this theorem are Theorems 16.6 and 16.7 (*closed graph theorem*) and the interior mapping principle (see Exercise 16.9). These consequences are so immediate that they might more properly be called corollaries than separate theorems; the proof of each involves nothing more than fitting things into the setting of Theorem 16.5 and just applying that theorem.

The proof of Theorem 16.5 involves several simple statements. In order not to lose the flavor or over-all viewpoint of the proof, the key steps are numbered; in the same spirit, not all the proofs of these simple statements are supplied in the body of the proof—instead they are given in the appendix to this lecture. Those that have been supplied in the body of the proof are those, it is felt, that constitute the essence of the over-all demonstration, and it is in these that the entire hypothesis is made use of. We now proceed to the theorem.

THEOREM 16.5 (*bounded inverse theorem*). Suppose between the normed linear spaces, X and Y, we have

$$X \supset D \overset{A}{\to} Y,$$

where X is a Banach space, A is a closed linear transformation (which means that D must be a *subspace* of X), and the range of A, $A(D)$, is of category II. Then:

(a) A is an onto mapping; that is, $A(D) = Y$.

(b) There exists an $m > 0$ such that, for any $y \in Y$, there is some $x \in D$ such that $Ax = y$ and $\|x\| \le m\|y\|$.

(c) If A^{-1} exists, it is a bounded linear transformation.

REMARK. As a result of Theorem 14.9, we can make the following statement: Let A be a linear transformation mapping the normed linear space D into the normed linear space Y. Then, if there exists some $k > 0$ such that

$$k\|x\| \le \|Ax\|$$

for *all* $x \in D$, then A^{-1} exists and is a bounded linear transformation on the range of A, $A(D)$.

In view of this result we see that if (b) can be established, (c) follows.

Proof. The proof proceeds by stages, the key steps being numbered 1, 2, etc. Before proceeding, we wish to introduce the following notation:

For closures of neighborhoods of the origin of X, we shall use

$$S_\alpha = \overline{S_\alpha(0)} = \{x \in X \mid \|x - 0\| \leq \alpha\}.$$

We now define

$$D_\alpha = D \cap S_\alpha$$

and note that, since A is defined on all of D, it is certainly defined on D_α for any α. We now proceed to the proof and remind the reader again that, if any step seems troublesome, the proof can probably be found in the appendix.

In proving part (a) [that $A(D) = Y$], we first show that $A(D)$ contains some neighborhood of 0, and the first six steps in the proof are devoted to proving the existence of such a neighborhood. Once we have done this, the fact that $A(D) = Y$ follows rather readily as is shown in Step 7.

Step 1.

$$D = \bigcup_{n=1}^{\infty} D_n. \tag{16.4}$$

Applying A to both sides, we have

$$A(D) = \bigcup_{n=1}^{\infty} A(D_n). \tag{16.5}$$

To avoid confusion, we shall denote a δ-neighborhood of the point $z \in Y$ by $B_\delta(z)$.

Step 2. Using this notation, it then follows that, since $A(D)$ is of category II and has been written as a countable union of sets, not *all* of these sets $A(D_n)$ can be nowhere dense. (See Sec. 6.2 for a discussion of nowhere dense sets and category.) Hence suppose $A(D_{n_0})$ is not nowhere dense. Not being nowhere dense, one recalls, means that the interior of $\overline{A(D_{n_0})}$ is not empty; thus, there must exist some

$$z \in \overline{A(D_{n_0})} \tag{16.6}$$

and some $\delta > 0$ such that

$$B_\delta(z) \subset \overline{A(D_{n_0})}. \tag{16.7}$$

Step 3 (*scaling down*). For $\alpha, \beta, \varepsilon > 0$,

$$B_\varepsilon(w) \subset \overline{A(D_\alpha)} \Leftrightarrow B_{\beta\varepsilon/\alpha}\left(\frac{\beta w}{\alpha}\right) \subset \overline{A(D_\beta)}. \tag{16.8}$$

Loosely speaking, the above equivalence shows that if one has an inclusion like the one on the left of (16.8), one can "multiply through" by the scalar factor β/α without affecting the validity of the inclusion.

Step 4. Thus, in the inclusion (16.7), we can "multiply through" by the scalar factor $1/n_0$ to get

$$B_{\delta/n_0}\left(\frac{z}{n_0}\right) \subset \overline{A(D_1)}. \tag{16.9}$$

Thus, so far, we have shown [using only the linearity of A and the fact that $A(D)$ is of category II] the existence of a point

$$z_1 \in A(D_1)$$

such that a neighborhood of this point is wholly contained in $\overline{A(D_1)}$. [We are guaranteed such a z_1 exists because, since $z/n_0 \in \overline{A(D_1)}$, there must be points z_1 from $A(D_1)$ as "close" to z/n_0 as we wish.] In particular, we can now say that there must be some $z_1 \in A(D_1)$ such that for

$$\delta_1 = \frac{\delta}{2n_0},$$

we have

$$B_{\delta_1}(z_1) \subset \overline{A(D_1)}. \tag{16.10}$$

By the definition of $A(D_1)$, since $z_1 \in A(D_1)$, it follows that there must be some $\hat{x} \in D_1$ such that

$$A\hat{x} = z_1.$$

Consider now the set

$$N = \{x - \hat{x} \mid x \in D_1\}.$$

For any $x - \hat{x} \in N$ (since $x, \hat{x} \in D_1 \subset S_1$), we have

$$\|x - \hat{x}\| \le \|x\| + \|\hat{x}\| \le 2,$$

which implies, since $x - \hat{x} \in D$,

$$x - \hat{x} \in S_2 \cap D = D_2$$

or

$$N \subset D_2. \tag{16.11}$$

Applying A to both sides of (16.11), we get

$$A(N) \subset A(D_2),$$

which immediately implies

$$\overline{A(N)} \subset \overline{A(D_2)}. \tag{16.12}$$

But, for any $x - \hat{x} \in N$, we have

$$A(x - \hat{x}) = Ax - A\hat{x} = Ax - z_1,$$

which implies

$$A(N) = A(D_1) - z_1 = \{Ay - z_1 \mid y \in D_1\};$$

hence,

$$\overline{A(N)} = \overline{A(D_1) - z_1} = \overline{A(D_1)} - z_1. \tag{16.13}$$

Using this, we are now able to establish the next step.

Step 5.

$$B_{\alpha\delta_1/2}(0) \subset \overline{A(D_\alpha)}. \tag{16.14}$$

First, it is claimed that

$$B_{\delta_1}(0) \subset \overline{A(N)},$$

since, for any $w \in B_{\delta_1}(0)$, we must have

$$\|w\| < \delta_1,$$

which implies

$$\|w + z_1 - z_1\| < \delta_1;$$

therefore,

$$w + z_1 \in B_{\delta_1}(z_1) \subset \overline{A(D_1)}$$

or

$$w \in \overline{A(D_1)} - z_1.$$

Using (16.13) and (16.12) we see that

$$w \in \overline{A(N)} \subset \overline{A(D_2)}.$$

Thus any element w of $B_{\delta_1}(0)$ is seen to also be a member of $\overline{A(D_2)}$, and we have

$$B_{\delta_1}(0) \subset \overline{A(D_2)}.$$

By virtue of Step 3 it now follows that

$$B_{\alpha\delta_1/2}(0) \subset \overline{A(D_\alpha)}. \tag{16.15}$$

We now define

$$P_\alpha = B_{\alpha\delta_1/2}(0),$$

where $\alpha = 2^{-k}$ ($k = 0, 1, 2, \ldots$). We now get to the meatiest part of the proof—the part where we shall use the facts that X is a Banach space and A is closed—in establishing that $A(D)$ contains some neighborhood of the origin. Symbolically, we show the next step.

Step 6.

$$P_1 \subset A(D_2) \subset A(D).$$

Essentially, all we wish to do is to remove the bar over $A(D_\alpha)$ in (16.14); yet, in this seemingly simple maneuver, we shall need almost the entire strength of the hypothesis.

We first note that, for any $k = 0, 1, 2, \ldots$,

$$P_{2^{-k}} \subset \overline{A(D_{2^{-k}})} \tag{16.16}$$

and the radius of $P_{2^{-k}}$, rad $P_{2^{-k}}$, is just

$$\text{rad } P_{2^{-k}} = \frac{\delta_1}{2^{k+1}} \tag{16.17}$$

Suppose now that $y \in P_1$ (the case when $k = 0$); by (16.16) we have

$$y \in \overline{A(\bar{D}_1)}.$$

Since this is so, there must be some

$$v \in A(D_1)$$

such that

$$\|y - v\| < \frac{\delta_1}{2^2}$$

But, since $v \in A(D_1)$, there must be some $x_1 \in D_1$ such that $Ax_1 = v$; consequently, we can assert the existence of $x_1 \in D_1$ such that

$$\|y - Ax_1\| < \frac{\delta_1}{2^2}.$$

This now implies that

$$y - Ax_1 \in P_{2^{-1}} \subset \overline{A(D_{2^{-1}})}.$$

Reasoning as above, we now assert the existence of $x_2 \in D_{2^{-1}}$ such that

$$\|(y - Ax_1) - Ax_2\| < \frac{\delta_1}{2^3}.$$

Proceeding in this fashion we obtain the sequence $\{x_k\}$, where

$$x_k \in D_{2^{-k+1}}$$

and

$$\|y - Ax_1 - Ax_2 - \cdots - Ax_n\| < \frac{\delta_1}{2^{n+1}} \tag{16.18}$$

Now we let

$$s_n = \sum_{k=1}^{n} x_k$$

and note that

$$\|s_n\| \leq \sum_{k=1}^{n} \|x_k\|.$$

We can get a specific bound on $\|s_n\|$, however, for, since $x_k \in D_{2^{-k+1}}$ $(k = 0, 1, ...)$, it follows that

$$\|x_1\| \leq 1$$

$$\|x_2\| \leq \tfrac{1}{2}$$

$$\vdots \qquad \vdots$$

and this implies that

$$\|s_n\| \leq 2.$$

This last statement is equivalent to saying

$$s_n \in S_2.$$

Since s_n is a sum of terms from the subspace D, it follows that

$$s_n \in D;$$

therefore,

$$s_n \in S_2 \cap D = D_2$$

for any n. Clearly, $\{s_n\}$ is a Cauchy sequence and, since X is a Banach space, there must be some $x \in X$ such that

$$s_n \to x.$$

Since $\|s_n\| \leq 2$ for every n, it follows, using the continuity of the norm, that

$$\|x\| \leq 2$$

and, therefore,

$$x \in S_2. \tag{16.19}$$

In view of the fact that [by Eq. (16.18)]

$$\|y - As_n\| < \frac{\delta_1}{2^{n+1}},$$

we see that

$$As_n \to y.$$

Using the closure of A, since we now have

$$s_n \to x \quad \text{and} \quad As_n \to y,$$

we can say

$$x \in D \quad \text{and} \quad y = Ax.$$

Combining this with (16.19), we see that

$$x \in D \cap S_2 = D_2.$$

Since $y \in P_1$, we have shown

$$P_1 \subset A(D_2) \subset A(D). \tag{16.20}$$

Now that we have shown this neighborhood of 0 to be contained in $A(D)$ it remains only to turn the crank to prove the next step.

Step 7.

$$A(D) = Y.$$

Suppose y is any vector in Y. For a suitable real number β we can say

$$\beta y \in P_1;$$

just take, for $y \neq 0$,

$$\beta = \frac{\delta_1}{4\|y\|}, \tag{16.21}$$

for example. In this case

$$\|\beta y\| = \frac{\delta_1}{4},$$

and since, by (16.17),

$$\text{rad } P_1 = \frac{\delta_1}{2},$$

we see that

$$\beta y \in P_1.$$

But $P_1 \subset A(D)$, so there must be some $x \in D$ such that

$$\beta y = Ax,$$

which implies

$$y = A\left(\frac{x}{\beta}\right).$$

Since y was any vector in Y (the result being trivially true if $y = 0$) the proof of (a) is now complete.

Proof (b). We now wish to show that, for any $y \in Y$, there is some $x \in D$ such that $y = Ax$ and

$$\|x\| \le m\|y\|. \tag{16.22}$$

Note that we do not exclude from consideration the possibility that many elements x may map into y nor, if this is the case, do we say that every such x will satisfy (16.22); we say only that there is *some* x satisfying (16.22) and such that $y = Ax$. It is for this reason the expression, "If A^{-1} exists" and not "A^{-1} must exist" is used in part (c), for if (16.22) held for every such x, it would follow that

$$Ax = 0 \Leftrightarrow x = 0$$

and A^{-1} would have to exist. Bearing this and the Remark after the statement of the theorem in mind, we realize the validity of (c) once (b) has been shown. We now proceed to the proof of (b).

Let y be any vector in Y. As noted in proving (a), there must be some scalar β and some vector $x_0 \in D_2$ (see step 7) such that

$$y = A\left(\frac{x_0}{\beta}\right).$$

A direct computation yields, for β as in (16.21),

$$\left\| \frac{x_0}{\beta} \right\| = \frac{\|x_0\|}{\beta} \le \frac{2}{\beta} = \frac{8\|y\|}{\delta_1};$$

thus, letting the x and m of the theorem be x_0/β and $8/\delta_1$, respectively, we see that (b) is proved.

16.3 Some Consequences of the Bounded Inverse Theorem

THEOREM 16.6. Suppose $A : X \to Y$, where X and Y are Banach spaces, A is a bounded linear transformation, and $A(X) = Y$ (A is an onto map). Then if A^{-1} exists, it is a bounded linear transformation.

Proof. To prove this result, all we do is show that everything fits into the setting described in the hypothesis of the bounded inverse theorem. Hence we must show that

(1) X is a Banach space;
(2) A is a closed linear transformation;
(3) the range of A, $A(X)$, is of category II.

If we can do this it remains only to appeal to part (c) of the bounded inverse theorem to complete the proof.

We see that (1) is satisfied by hypothesis and (2) is satisfied because A is bounded and defined on the whole space. (3) is seen to be satisfied because we have assumed that $A(X) = Y$ and further assumed Y to be a Banach space. Since by the Baire category theorem (Theorem 6.1), any complete metric space must be of category II, we see that the hypothesis of Theorem 16.5 is completely satisfied and the desired result now follows.

THEOREM 16.7 (*closed graph theorem*). Suppose A is a linear transformation such that $A : X \to Y$, where X and Y are Banach spaces. Then if A is closed, A is bounded.

REMARK. Compare this result to the example given in the preceding section to demonstrate that closed transformations need not be bounded. Since $C[0, 1]$ with sup norm is a Banach space and closed subspaces of a Banach space must also be Banach spaces, we can conclude from the theorem and the example that the subspace D mentioned in the example is not closed.

Proof. In this theorem, we shall just rearrange things so that the previous theorem may be applied.

Since X and Y are each Banach spaces, it follows that $X \times Y$ is a Banach space. A being a closed linear transformation implies that the graph of A, G_A is a closed subspace of $X \times Y$. Since closed subsets of complete metric spaces must also be complete, it follows that G_A is a Banach space, too. Now consider the mapping B:

$$B : G_A \longrightarrow X,$$

$$\langle x, Ax \rangle \to x.$$

Clearly, B is a linear transformation, and we claim, in addition, that B is bounded. To prove this, consider

$$\|B\langle x, Ax \rangle\| = \|x\| \leq \|x\| + \|Ax\| = \|\langle x, Ax \rangle\|.$$

Thus, B is seen to be a bounded linear transformation. Clearly, B is onto, that is, $B(G_A) = X$, and we shall now show that B is 1:1. Since B is linear, all we need do to prove this is show that $\langle 0, 0 \rangle$ is the only element B maps into 0. Hence, suppose

$$B\langle x, Ax \rangle = x = 0.$$

But $x = 0$ implies $Ax = 0$, so

$$\langle x, Ax \rangle = \langle 0, 0 \rangle$$

and B is seen to be 1:1. We have now shown that B is a 1:1, onto mapping between G_A and X, which means that the inverse function B^{-1} must exist between X and G_A. To summarize further, the situation we are now presented with is the following:

(1) G_A and X are Banach spaces;
(2) B is a bounded linear transformation;
(3) B^{-1} exists.

It now follows by the preceding theorem that B^{-1} is a bounded linear transformation. To complete the proof, we must show that

$$x_n \to x \Rightarrow Ax_n \to Ax.$$

Hence, suppose

$$x_n \to x.$$

Since B^{-1} is continuous, we can say

$$B^{-1}x_n \to B^{-1}x$$

or

$$\langle x_n, Ax_n \rangle \to \langle x, Ax \rangle$$

or

$$\langle x_n - x, Ax_n - Ax \rangle \to \langle 0, 0 \rangle;$$

therefore,

$$Ax_n \to Ax,$$

and the continuity of A has been shown.

APPENDIX

A.1 Supplement to Theorem 16.5

Proof (1). Clearly,

$$D \supset \bigcup_{n=1}^{\infty} D_n.$$

Since, for any $x \in D$, there must be some n such that

$$\|x\| \le n,$$

it follows that

$$D \subset \bigcup_{n=1}^{\infty} D_n.$$

Proof (*Step* 3). We wish to show that, if

$$X \supset D \overset{A}{\to} Y,$$

where A is linear, then, for $\alpha, \beta, \varepsilon > 0$,

$$B_\varepsilon(w) \subset \overline{A(D_\alpha)} \Leftrightarrow B_{\beta\varepsilon/\alpha}\left(\frac{\beta w}{\alpha}\right) \subset \overline{A(D_\beta)}. \tag{A.1}$$

We shall prove only that the implication from left to right holds, the other way around following immediately from this; hence, suppose

$$B_\varepsilon(w) \subset \overline{A(D_\alpha)} \tag{A.2}$$

and let

$$z \in B_{\beta\varepsilon/\alpha}\left(\frac{\beta w}{\alpha}\right).$$

This implies

$$\left\| z - \frac{\beta w}{\alpha} \right\| < \frac{\beta\varepsilon}{\alpha};$$

thus,

$$\left\| \frac{\alpha}{\beta} z - w \right\| < \varepsilon,$$

and we see that, using (A.2),

$$\frac{\alpha}{\beta} z \in \overline{A(D_\alpha)}.$$

Since this is so, there must be a sequence of points of $A(D_\alpha)$, $\{z_k\}$, such that

$$z_k \to \frac{\alpha}{\beta} z$$

or

$$\frac{\beta}{\alpha} z_k \to z.$$

Since, for any k,

$$z_k \in A(D_\alpha),$$

we can write

$$z_k = A x_k,$$

where $x_k \in D_\alpha = D \cap S_\alpha$. Since

$$\left\| \frac{\beta}{\alpha} x_k \right\| \le \frac{\beta}{\alpha} \cdot \alpha = \beta,$$

it follows that

$$\frac{\beta}{\alpha} x_k \in D \cap S_\beta = D_\beta;$$

applying A to both sides of this last expression, we get

$$\frac{\beta}{\alpha} A x_k \in A(D_\beta)$$

or

$$\frac{\beta}{\alpha} z_k \in A(D_\beta).$$

Thus,

$$\frac{\beta}{\alpha} z_k \to z \in \overline{A(D_\beta)}$$

and the inclusion is demonstrated.

EXERCISES 16

1. Let X be a normed linear space and A a linear transformation such that $X \supset D \overset{A}{\to} X$. Show that, if A is closed, then $A - \lambda = A - \lambda 1$ is also closed.

2. Let X be a normed linear space and let A and B be linear transformations such that

$$X \supset D_A \overset{A}{\to} X \quad \text{and} \quad X \supset D_B \overset{B}{\to} X.$$

If A and B are closed, does it necessarily follow that

$$A + B : D_A \cap D_B \to X$$

is closed?

3. Let X and Y be Banach spaces; then prove that $X \times Y$ is a Banach space with respect to any of the norms in Sec. 16.1.

4. Let X and Y be normed linear spaces. Suppose $A : X \to Y$ is closed. Let F be a compact subset of X. Show that $A(F)$ is a closed subset of Y.

5. Suppose A is a closed linear transformation defined on the normed linear space X. Does it necessarily follow that A maps closed sets into closed sets?

6. Let X be a Hilbert space and $A : X \to X$ a linear transformation such that $(Ax, y) = (x, Ay)$ for all $x, y \in X$. Then prove that A is bounded.

7. Let X be a Banach space with respect to the norms $\| \ \|_1$ and $\| \ \|_2$. Suppose there exists a constant a such that $\|x\|_1 \leq a\|x\|_2$ for all $x \in X$; then show that the two norms are equivalent.

8. Let X be a Banach space and A a closed linear transformation such that $X \supset D \overset{A}{\to} X$. Show that, if A^{-1} exists and is continuous on $A(D)$, then $A(D)$ is closed.

9. Under the same assumptions as Theorem 16.5, prove the *interior mapping principle*: If E is a relatively open subset of D (that is, $E = D \cap O$, where O is open in X), then $A(E)$ is an open set in Y.

10. Let A be a closed linear transformation mapping the normed linear space X into the normed linear space Y. If K is a compact subset of Y, show that $A^{-1}(K)$ (the inverse image) is a closed subset of X.

REFERENCES

1. M. Naimark, "Normed Rings."
2. A. Zaanen, "Linear Analysis."
3. A. Taylor, "Introduction to Functional Analysis."

CHAPTER **17**

Closures, Conjugate Transformations, and Complete Continuity

In this chapter the concept of a linear transformation *permitting (admitting)* *closure* will be introduced. In a sense it is a linear transformation that can be extended to a closed linear transformation. It is further shown that bounded linear transformations, which map a subspace of a normed linear space into a Banach space, are a class of linear transformations that always admit closure.

One of the most beautiful concepts one sees in linear algebra is that of the *adjoint* of a linear transformation.† The subtle intertwining of linear functional and linear transformation blended with the ever-present challenge of keeping the categories straight makes it a most intriguing issue. Here, we shall look at the adjoint in a slightly more special setting—the framework of *normed* linear spaces, *bounded* linear functionals, and *bounded* linear transformations. Later on, we shall specialize even further, when we examine the adjoint in Hilbert space. In that setting we shall even define it slightly differently, and in view of this latter interest we shall reserve the term " adjoint " for what we shall later define for Hilbert spaces; for the sake of what is meant in the present discussion, the term *conjugate map* will be used.

Another old friend from linear algebra that shall be revisited in the special setting of normed linear spaces is the *annihilator* of a subset of a linear space—all *bounded* linear functionals that vanish on that subset. Several facts about annihilators will be listed here.

After defining bounded linear transformations, we went to a weaker type, the closed linear transformation. We now wish to go to a type that is stronger than the original, continuous, linear transformation—the *completely continuous* linear transformation. To those interested in integral or integro-differential equations, this is a frequently encountered operator, and in this chapter we begin to develop some machinery about it.

17.1   The Closure of a Linear Transformation

Throughout the rest of the chapter, unless explicitly stated otherwise, X and Y will denote normed linear spaces, D a subspace of X, and A a linear transformation. With this convention in mind, suppose the following situation prevails:

$$X \supset D \xrightarrow{A} Y. \tag{17.1}$$

As noted in the last chapter the linear transformation A is closed if and only if its graph G_A is closed. Suppose in the above case that A is not closed—that G_A is not a closed subset of $X \times Y$. Suppose, in addition, that there is some linear transformation $\bar{A}$ such that the graph of $\bar{A}$, $G_{\bar{A}}$ is the closure of the graph of A, $\overline{G_A}$. If so, one says that A *permits (admits, has) a closure*; furthermore one calls the linear transformation $\bar{A}$ *the closure of A.*

The question that springs immediately to mind when one views this rather unrestrictive definition is: Does the closure of A extend A? The answer to this question is yes, as we shall now demonstrate.

Consider the sequence $\{x_n\}$ of points from D; suppose that

$$x_n \to x \quad \text{and} \quad Ax_n \to y. \tag{17.2}$$

† For a supplementary discussion of the adjoint, one might refer to Halmos, *Finite-Dimensional Vector Spaces*, pp. 78–80.

Under these assumptions then, for the sequence of points from $G_A, \langle x_n, Ax_n \rangle$, we must have

$$\langle x_n, Ax_n \rangle \to \langle x, y \rangle \in \bar{G}_A = G_{\bar{A}},$$

where the same norm for $X \times Y$ as was used in the last lecture is assumed (namely, $\|\langle x, y \rangle\| = \|x\| + \|y\|$). Since

$$\langle x, y \rangle \in G_{\bar{A}},$$

we must have, denoting the domain of $\bar{A}$ by $D_{\bar{A}}$,

$$x \in D_{\bar{A}} \quad \text{and} \quad y = \bar{A}x.$$

But since $y = \lim_n Ax_n$, this means that

$$\bar{A}x = \lim_n Ax_n \tag{17.3}$$

is a condition that $\bar{A}$ must satisfy for all x satisfying (17.2). If we now consider the special case when $x \in D$, the domain of A, and the sequences

$$x, x, \ldots, x, \ldots = \{x_n\} \tag{17.4}$$

and

$$Ax, Ax, \ldots, Ax, \ldots = \{Ax_n\}, \tag{17.5}$$

then we see that condition (17.3) does indeed imply that $\bar{A}$ extends A. We might now refer to $\bar{A}$ as the *minimal closed extension of A*. We now wish to formulate a condition to assure the existence of a closure without referring to the graph.

(I) The linear transformation A, described in (17.1), admits a closure $\bar{A}$, if and only if the conditions

$$x_n \in D, \quad x_n' \in D, \quad x_n \to x, \quad Ax_n \to y, \quad x_n' \to x, \quad Ax_n' \to y'$$

imply that

$$y = y'.$$

The proof of this equivalence is straightforward, and in Exercise 17.1 the reader is invited to demonstrate its validity.

It is clear that, if the above condition holds, the domain of $\bar{A}$ is as follows.

(II). $$D_{\bar{A}} = \{x \in X \mid x_n \in D, x_n \to x, \lim_n Ax_n \text{ exists}\},$$

and from this it is clear that, in general,

$$D \subset D_{\bar{A}} \subset \bar{D}.$$

The inclusion does not usually go the other way around, however. This is to be expected though because, a priori, there is no reason to expect that, for every convergent sequence of points of D, $\{x_n\}$, the limit, $\lim_n Ax_n$ will exist. In the class of linear transformations that we discuss in the next section, a class that always permits a closure, the inclusion does go the other way around as we shall presently see.

17.2   A Class of Linear Transformations that Admit a Closure

Once again we suppose that

$$X \supset D \xrightarrow{A} Y \qquad (17.6)$$

with A, X, D, and Y as in (17.1). This time suppose also that A is a bounded linear transformation and Y is a Banach space. In this case we shall show that A has a closure, and we shall actually construct this new linear transformation. We obtain it by a process known as *extension by continuity*, something of interest for its own sake. Hence, we first establish the following contention.

CONTENTION. There exists a unique bounded linear transformation $\hat{A}$ defined on all of $\bar{D}$, extending A and such that $\|A\| = \|\hat{A}\|$.

Before proving that $\hat{A}$ possesses any of the properties we claim it has, we must first prescribe how it operates on elements of $\bar{D}$. To this end let $x_0 \in \bar{D}$. There must be a sequence of points of D, $\{x_n\}$ such that

$$x_n \to x_0. \qquad (17.7)$$

Consider now the sequence $\{Ax_n\}$. We wish to show that this is a Cauchy sequence in Y. Since A is a bounded linear transformation, it follows that

$$\|Ax_n - Ax_m\| \le \|A\| \cdot \|x_n - x_m\|,$$

and it is clear that $\{Ax_n\}$ is a Cauchy sequence. Since Y is a Banach space, the sequence must have a limit, and we shall denote this limit in the following way:

$$\hat{A}x_0 = \lim_n Ax_n.$$

The mapping $\hat{A}$ is now defined for every point in $\bar{D}$. We now show that it is well-defined.

Suppose that, for x_n, $x_n' \in D$,

$$x_n \to x_0 \quad \text{and} \quad x_n' \to x_0. \qquad (17.8)$$

In this case it is clear that

$$(x_n - x_n') \to 0,$$

which implies, by the continuity of A, that

$$A(x_n - x_n') \to A0 = 0$$

and, therefore,

$$\lim_n Ax_n = \lim_n Ax_n'. \qquad (17.9)$$

Thus, $\hat{A}$ is seen to be well-defined. The linearity of $\hat{A}$ follows immediately from the linearity of A and some elementary facts about limits. To show that

$$Ax = \hat{A}x$$

when $x \in D$, all we need do is consider the sequence $x, x, \ldots, x, \ldots$ and apply the definition of $\hat{A}$. As for the boundedness of the linear transformation $\hat{A}$ we note that [with x_n and x_0 as in (17.7)]

$$\|Ax_n\| \leq \|A\| \|x_n\|$$

holds for every n—hence in the limit. Since the norm is a continuous operation, we are justified in interchanging the operations of limit and norm to get

$$\|\hat{A}x_0\| \leq \|A\| \|x_0\|,$$

and the boundedness of $\hat{A}$ is evident; further,

$$\|\hat{A}\| \leq \|A\|.$$

Since $\hat{A}$ extends A, however, we must have† $\|A\| \leq \|\hat{A}\|$, and we see that $\|\hat{A}\| = \|A\|$. We now show that this extension by continuity of A is unique. Suppose that B is a bounded linear transformation defined on all of $\bar{D}$, extending A. Then, with x_n and x_0 as in (17.7), since B is continuous, we must have

$$Bx_0 = \lim_n Bx_n.$$

Since B extends A, we also have, for every n, the fact that

$$Bx_n = Ax_n,$$

and therefore

$$Bx_0 = \lim_n Bx_n = \lim_n Ax_n = \hat{A}x_0;$$

which proves the uniqueness of $\hat{A}$. The contention has now been completely established.

We next wish to show that A admits a closure $\bar{A}$ and that this closure is just $\hat{A}$. In Sec. 17.1 we stated a result [see (I), Sec. 17.1] which guaranteed that a linear transformation would admit a closure. When we proved that $\hat{A}$ was well-defined [Eqs. (17.8) and (17.9)], we showed that the conditions of that result were satisfied; hence, by that result, A permits a closure. To show that the linear transformations $\bar{A}$ and $\hat{A}$ are equal, we must show two things: that their domains are the same and that they agree on every vector in the domain. First we show the domains to be equal.

By the results of the last section we know that, in general,

$$D_{\bar{A}} \subset \bar{D}.$$

Now let x be any element of $\bar{D}$ and let $\{x_n\}$ be a sequence of points of D converging to x. Since Y is a Banach space and we showed before that $\{Ax_n\}$ is a Cauchy sequence, the limit $\lim_n Ax_n$ must exist. By (II) of the preceding section, we see that $x \in D_{\bar{A}}$, and, therefore,

$$\bar{D} \subset D_{\bar{A}},$$

---

† To see the validity of the above reasoning, note that, if C and D are any two sets of real numbers such that $C \subset D$, we must have $\sup C \leq \sup D$; hence,

$$D \subset \bar{D} \Rightarrow \sup_{\substack{x \in D \\ x \neq 0}} \frac{\|Ax\|}{\|x\|} = \sup_{\substack{x \in D \\ x \neq 0}} \frac{\|\hat{A}x\|}{\|x\|} \leq \sup_{\substack{x \in \bar{D} \\ x \neq 0}} \frac{\|\hat{A}x\|}{\|x\|}.$$

and the domains are seen to be equal. Since we must also have

$$\bar{A}x = \lim_n Ax_n = \hat{A}x,$$

we see that $\hat{A} = \bar{A}$. In summary we have established:

If A is a bounded linear transformation into a Banach space, then A permits a closure; furthermore, the closure of A is just the extension of A by continuity to the closure of the domain of A.

17.3 The Conjugate Map of a Bounded Linear Transformation

Let X and Y be normed linear spaces, denote the conjugate spaces of X and Y by $\tilde{X}$ and $\tilde{Y}$, respectively, and let $L(X, Y)$ denote the class of all bounded linear transformations mapping X into Y. Now let $y' \in \tilde{Y}$ and consider, for $x \in X$, and $A \in L(X, Y)$,

$$y'(Ax) = [Ax, y']. \tag{17.10}$$

If in (17.10) we let x range through all of X, it is clear that we would generate a set of scalar values; thus, a functional on the normed linear space X is defined. We denote this functional by x' [that is, for any $x \in X$, $x'(x) = y'(Ax)$], and we now contend that (1) x' is a *linear* functional and further (2) that x' is a *bounded* linear functional.

The proof of (1) follows immediately from the linearity of A and the linearity of y'. We now prove (2). To this end consider

$$|x'(x)| = |y'(Ax)| \le \|y'\| \cdot \|Ax\| \le \|y'\| \cdot \|A\| \cdot \|x\|,$$

which proves not only that x' is a bounded linear functional but also that

$$\|x'\| \le \|y'\| \|A\|. \tag{17.11}$$

Let us look at what we have done: we started with a member of $\tilde{Y}$ and then, using A, we obtained a member of $\tilde{X}$. Thus we have some kind of a mapping which, since it depends on A, we shall denote by A' such that

$$A' : \tilde{Y} \to \tilde{X}, \tag{17.12}$$

$$y' \to A'y' = x';$$

that is, for any $x \in X$,

$$(A'y')(x) = x'(x), \tag{17.13}$$

or

$$[x, A'y'] = [x, x'] = [Ax, y'].$$

DEFINITION 17.1. The mapping A' is called the *conjugate map (or adjoint)* of the bounded linear transformation A. Most often we shall use the term conjugate map, because we wish to reserve the term "adjoint" for something to be introduced later.

We now have a mapping between two normed linear spaces and the fact that this mapping A' is a linear transformation follows directly from the definitions of the quantities involved. We claim more, though: We claim that A' is actually a *bounded* linear transformation. From (17.11) and (17.12) it follows that

$$\|A'y'\| \leq \|A\| \|y'\|,$$

which implies that A' is bounded and that

$$\|A'\| \leq \|A\|. \tag{17.14}$$

We now wish to make a slightly stronger statement and prove that equality prevails in (17.14). Before doing this, however, we shall prove another fact (which, in itself, is quite interesting) that will greatly facilitate the proof that $\|A'\| = \|A\|$. It is by now very familiar that, for a bounded linear functional x', we can compute the norm of x' by taking

$$\|x'\| = \sup \{|x'(x)| \mid x \in X \text{ and } \|x\| = 1\}.$$

We now wish to prove that a dual statement of the above also prevails—that we can compute the norm of a *vector* x via

$$\|x\| = \sup\{|x'(x)| \mid x' \in \tilde{X} \text{ and } \|x'\| = 1\}. \tag{17.15}$$

For any $x' \in \tilde{X}$ such that $\|x'\| = 1$, we see that

$$|x'(x)| \leq \|x'\| \|x\| = \|x\|,$$

which implies

$$\sup_{\|x'\|=1} |x'(x)| \leq \|x\|. \tag{17.16}$$

On the other hand, by a consequence of the Hahn–Banach theorem (Theorem 12.2), there must be some $x' \in X$ such that

$$\|x'\| = 1$$

and

$$x'(x) = \|x\|,$$

which implies

$$\|x\| \leq \sup_{\|x'\|=1} |x'(x)|. \tag{17.17}$$

Combining (17.16) and (17.17), Eq. (17.15) now follows.

In view of (17.15) we can write

$$\|Ax\| = \sup\{|y'(Ax)| \mid y' \in \tilde{Y} \text{ and } \|y'\| = 1\}.$$

But for any $y' \in \tilde{Y}$ such that $\|y'\| = 1$, we have

$$|y'(Ax)| = |x'(x)| \leq \|x'\| \|x\| = \|A'y'\| \|x\| \leq \|A'\| \|y'\| \|x\| = \|A'\| \|x\|,$$

which implies

$$\|Ax\| = \sup_{\|y'\|=1} |y'(Ax)| \leq \|A'\| \|x\|.$$

Therefore,

$$\|A\| \leq \|A'\|, \tag{17.18}$$

and combining this with (17.14), we see that

$$\|A\| = \|A'\|. \tag{17.19}$$

Thus if A is a bounded linear transformation mapping X into Y, the conjugate map of A, A' is also a bounded linear transformation mapping $\tilde{Y}$ into $\tilde{X}$ and has the same norm as A.

The following facts about the congujate map are straightforward to demonstrate (the proofs involve only using definitions) and, for this reason, will only be listed here; the only one that might present any difficulty at all is (17.23).
Suppose A, $B \in L(X, Y)$. Then

$$(A + B)' = A' + B'. \tag{17.20}$$

For α a scalar and $A \in L(X, Y)$,

$$(\alpha A) = \alpha A'. \tag{17.21}$$

If

$$X \xrightarrow{B} Y \xrightarrow{A} Z$$

and $B \in L(X, Y)$, $A \in L(Y, Z)$, then

$$(AB)' = B'A'. \tag{17.22}$$

If $A \in L(X, Y)$, A^{-1} exists and $A^{-1} \in L(Y, X)$ then $(A')^{-1}$ exists,

$$(A')^{-1} \in L(\tilde{X}, \tilde{Y}) \text{ and } (A')^{-1} = (A^{-1})'. \tag{17.23}$$

17.4 Annihilators

DEFINITION 17.2. Suppose S is an arbitrary nonempty subset of the normed linear space X. We define the *annihilator of S*, S^a to be those bounded linear functionals on X that take every vector in S into zero: symbolically,

$$S^a = \{x' \in \tilde{X} \mid x'(x) = 0 \text{ for all } x \in S\}.$$

If $S = \emptyset$, we define $S^a = \tilde{X}$.

DEFINITION 17.3. Suppose, with X as above, that E is an arbitrary nonempty subset of $\tilde{X}$. We define the *annihilator of E*, denoted by ^aE, as follows:

$$^aE = \{x \in X \mid x'(x) = 0 \text{ for all } x' \in E\}.$$

We take $^aE = X$ if E is empty.

The sets S^a and ^aE can easily be shown to be *subspaces* regardless of whether the subsets S and E were subspaces of X and $\tilde{X}$, respectively.
Let us look at the annihilator of a subset in a more special situation. In Theorem 12.5 it was shown that, if X was a Hilbert space, every bounded linear functional on X could be represented by an inner product. Thus, in view of the correspondence between bounded linear functionals and vectors of the Hilbert space [see the mapping T mentioned in Eq. (12.23)], we see that, if S is a subset of a Hilbert space, then, identifying vectors and corresponding linear functionals, we have

$$S^a = \{y \in X \mid (x, y) = 0 \text{ for all } x \in S\}$$

or, to paraphrase this result, we see that the annihilator of a subset and the orthogonal complement of a subset coincide in a Hilbert space.

Properties of Annihilators

In the following discussion all symbols will have the same connotation as in the two definitions just given.

(1) The subspaces S^a and ^aE are *closed* subspaces.

We noted above that S^a and ^aE were subspaces; thus, it remains only to demonstrate the proposed closure. We shall prove it here only for S^a, the proof that ^aE is closed following in an exactly analogous manner.
Let

$$x' \in \overline{S^a}$$

and let $\{x_n'\}$ be a sequence of points of S^a such that

$$x_n' \to x'.$$

This implies that, for any $x \in X$,

$$x_n'(x) \to x'(x).$$

Since this is true for all vectors in X, it is certainly true for just those vectors in S, but, for $x \in S$,

$$x_n'(x) = 0$$

for every n. This implies that

$$x'(x) = 0$$

if $x \in S$ and, therefore,

$$x' \in S^a,$$

and S^a is seen to be closed.

(2) For $S \subset X$ form

$$[\overline{S}] = M.$$

Then

$$^a(S^a) = M.$$

REMARK. If S is a closed subspace of X, then

$$^a(S^a) = S.$$

Proof (2). Clearly,

$$S \subset M,$$

and it is easy to verify that this implies

$$S^a \supset M^a.$$

Now suppose that x' is an arbitrary element of S^a. This implies that $x'(x) = 0$ for all $x \in S$. Clearly, we also have

$$x'(x) = 0$$

for all $x \in [S]$. By the continuity of x', the last statement can be strengthened to include any $x \in \overline{[S]} = M$; therefore

$$x' \in M^a,$$

and we have the fact that

$$M^a = S^a.$$

Thus for any $x' \in S^a = M^a$, we must have

$$x'(x) = 0$$

for all $x \in M$. For a vector to be in $^a(S^a)$, all linear functionals in S^a must vanish on it, and in view of the preceding statement, we see that

$$M \subset {}^a(S^a).$$

Suppose this inclusion were proper; that is, suppose there exists

$$x_1 \in {}^a(S^a) - M. \qquad (17.24)$$

Since M is a closed subspace and $x_1 \in M$, it follows the distance $d(x_1, M)$ is positive. Now, by a consequence of the Hahn–Banach theorem (Theorem 12.3), it follows that there exists some bounded linear functional x' such that

$$x'(x_1) = d(x_1, M) > 0$$

and which vanishes on every vector in M. But since x' takes every vector in M into zero, it follows that

$$x' \in M^a = S^a,$$

which implies $x'(x_1) = 0$—a contradiction. Therefore assumption (17.24) must be wrong, and we have proved that

$$M = {}^a(S^a).$$

A reflection of this fact in the dual space is mentioned in Exercise 17.5.

The following facts are very straightforward things to prove; one uses only definitions to prove any of them. To illustrate how these proofs go, the first will be proved here; the rest will only be stated. In these statements $R(A)$ will denote the range of A, $A(X)$ if you wish, and $N(A)$ will denote the null space of A: those $x \in X$ such that $Ax = 0$. A' denotes the adjoint of A.

(3) $\left(\overline{R(A)}\right)^a = N(A')$.

We have just seen that, for an arbitrary set S, $\left(\overline{[S]}\right)^a = S^a$. In particular, if S is a subspace we have $\bar{S}^a = S^a$. Thus the bar over $R(A)$ in statement (3) is completely optional. To prove the statement, consider:

$$y' \in R(A)^a$$

$$\Leftrightarrow \quad 0 = [Ax, y'] = [x, A'y'] \quad \text{(for any } x \in X)$$

$$\Leftrightarrow A'y' = 0$$

$$\Leftrightarrow y' \in N(A').$$

(4) $\overline{R(A)} = {}^a N(A')$.

(5) $^a\overline{(R(A'))} = N(A)$.

(6) $\overline{R(A')} \subset N(A)^a$.

For a case when equality holds in statement (6), see Exercise 17.6.

17.5 Completely Continuous Operators; Finite-Dimensional Operators

For the following discussion, a handy reference is Chapter 5, where all the topological notions used in this section are introduced and many theorems about them are proved, the first such notion we wish to recall being the definition of a *relatively compact* set, a set whose closure is compact.

DEFINITION 17.4. The linear transformation A mapping the normed linear space X into the normed linear space Y is called *completely continuous* or *compact* if it maps bounded sets (of X) into relatively compact sets (of Y).

In Theorem 14.8, we showed that a bounded linear transformation mapped bounded sets into bounded sets and conversely. Using this we now prove Theorem 17.1.

THEOREM 17.1. Every completely continuous linear transformation is a continuous linear transformation.

Proof. Let A be a completely continuous linear transformation. We shall show that A maps bounded sets into bounded sets. To this end, let S be a bounded set in the domain. Since A is completely continuous, the image of S under A, $A(S)$ must be relatively compact set. Therefore $\overline{A(S)}$ is compact which, by Theorem 5.2, implies that it is bounded. Hence A is continuous.

We shall presently show that the converse of this theorem is false. We now wish to consider completely continuous transformations on a finite-dimensional space. As one might suspect, the usual collapse takes place: Every linear transformation is completely continuous on a finite-dimensional space.

THEOREM 17.2. Let A be a linear transformation on the finite-dimensional space X. Then A is completely continuous.

Proof. We first note that the image of the entire space $A(X)$ must also be a finite-dimensional vector space by virtue of the linearity of A. By Theorem 14.7, we know that all linear transformations on a finite-dimensional space are bounded. Since A must, therefore, be bounded, it follows, for any bounded set S in X, that $A(S)$ must be bounded, too. We now have

$$A(S) \subset A(X);$$

that is, we now have a bounded set in a finite-dimensional space. One can easily verify that in a finite-dimensional space, (using Exercise 8.15) that a set is compact if and only if it is closed and bounded. Thus if K is bounded, its closure $\overline{K}$ is also bounded; that is, $\overline{K}$ is closed and bounded, therefore compact. We have now shown K, any bounded set, to be relatively compact, which implies that the bounded set $A(S)$ must be relatively compact, and the complete continuity of A is established.

We now wish to show that not every continuous linear transformation is completely continuous and, thus, justify our use of a different term to describe such linear transformations. Our example of a transformation that is bounded but not completely continuous is a quite well-known transformation—the identity transformation.

A Transformation That Is Bounded But Not Completely Continuous

In view of Theorem 17.2, to find a linear transformation that is not completely continuous, it is futile to look at finite-dimensional spaces. Hence, suppose X is an infinite-dimensional normed space and consider the identity mapping:

$$1 : X \to X.$$

That this is a bounded linear transformation is indeed clear. We now show that it is not completely continuous.

Since X was assumed to be infinite-dimensional, it follows that an infinite number of linearly independent vectors $x_1, x_2, ..., x_n, ...$ can be selected from X. We now claim that there exist y_n ($n = 1, 2, ...$) such that

$$\|y_n\| = 1 \quad \text{(for all } n\text{)},$$

$$y_n \in M_n = [\{x_1, x_2, ..., x_n\}],$$

$$\|y_{n+1} - x\| \geq \tfrac{1}{2} \quad \text{(for all } x \in M_n\text{)}.$$

To prove this claim, we start by exhibiting the first such vector:

$$y_1 = \frac{x_1}{\|x_1\|} \in M_1.$$

We now note that since M_n is a finite-dimensional subspace it must be closed (see Theorem 8.7, Corollary 2). Since the vectors x_1 and x_2 are assumed to be linearly independent, it follows that the inclusion

$$M_1 \subset M_2$$

is proper. Now we can apply the theorem due to Riesz (Theorem 8.8), which asserts the existence of a vector $y_2 \in M_2$ with norm one and such that

$$\|y_2 - x\| \geq \tfrac{1}{2}$$

for all $x \in M_1$. When we now consider $M_2 \subset M_3$, we see that precisely the same situation has occurred again, and the way to obtain the $y_1, y_2, ..., y_n, ...$ satisfying the conditions specified before is apparent. Now consider the set

$$S = \{y_1, y_2, ..., y_n, ...\}.$$

It is clear that S is a bounded set. Now consider the image of S under the identity transformation which is just S again. To prove that the transformation 1 is not completely continuous, all we need do now is to show that S is not relatively compact.

Theorem 5.6 states that a set A is relatively compact if and only if a convergent subsequence can be selected from every sequence of points in A. Since, for any n and m, $n \neq m$,

$$\|y_n - y_m\| \geq \tfrac{1}{2},$$

it is clear that one cannot extract a convergent subsequence from the sequence $\{y_n\}$, and it follows that S is not relatively compact.

A Type of Transformation That Is Always Completely Continuous

Let X be a normed linear space, z a fixed vector in X, and f a bounded linear functional on X. We now define

$$A : X \to X$$

by the following prescription: Let $x \in X$. Then

$$Ax = f(x)z.$$

The linearity of the above transformation follows directly from the linearity of f. We now show that A is completely continuous. To do this we must show that A maps bounded sets into relatively compact sets; hence, let S be a bounded set in X. To show that $A(S)$ is relatively compact we will show (Theorem 5.6) that, from every sequence of points in $A(S)$, a convergent subsequence can be selected. Let $\{Ax_n\}$ be a sequence of points from $A(S)$. Since S was assumed to be bounded, there must be some $M > 0$ such that

$$\|x_n\| \leq M \qquad \text{(for every } n\text{)},$$

which implies, since f is bounded, that

$$|f(x_n)| \leq \|f\| \cdot \|x_n\| \leq \|f\| \cdot M \qquad \text{(for every } n\text{)},$$

and we see that $\{f(x_n)\}$ is a (uniformly) bounded sequence of complex numbers. By the Bolzano–Weierstrass theorem for the plane, it is clear that a convergent subsequence of $\{f(x_n)\}$, $\{f(x_{n_k})\}$ can be selected, and we assume that the complex number λ is the limit of this convergent subsequence.

We now have

$$\{Ax_{n_k}\} = \{f(x_{n_k})z\};$$

consequently,

$$Ax_{n_k} \to \lambda z,$$

and the desired convergent subsequence has been exhibited.

In retrospect, one might enquire as to whether something of a more general nature about the structure of completely continuous operators might be gleaned from the above construction. And, as it happens, there is. One notes that the range of A, $A(X)$, where A is as above, is one-dimensional. One further notes that A is a bounded linear transformation. It seems reasonable to suspect that, if some additional conditions are imposed upon a bounded linear transformation, one can guarantee that it is completely continuous, and in view of the above discussion, it seems likely that requiring the range of the transformation to be finite-dimensional might do the trick. Indeed this is the case and will be proved in Theorem 17.3.

DEFINITION 17.5. The linear transformation $A : X \to Y$ is called *finite-dimensional* if $A(X)$ is finite-dimensional.

THEOREM 17.3. Let A be a linear transformation mapping the normed linear space X into the normed linear space Y. Then, if A is continuous and finite-dimensional, it is completely continuous.

Proof. We must show that such an A maps bounded sets into relatively compact sets. Thus, let S be a bounded set in X. Since A is a bounded linear-transformation, $A(S)$ must be a bounded set; further $A(S)$ is a bounded subset of the finite-dimensional space $A(X)$, due to the finite dimensionality of A. In finite-dimensional spaces, however, as noted in the proof of Theorem 17.2, bounded sets must be relatively compact which completes the proof.

17.6 Further Properties of Completely Continuous Transformations

Some additional facts about completely continuous operators will be demonstrated here, among which will be the fact that the set of completely continuous operators comprises a two-sided ideal in the ring $L(X, X)$.

We first show a linear combination of completely continuous operators to be a completely continuous operator. Throughout this section, it will be assumed that X and Y are normed linear spaces.

THEOREM 17.4. Suppose A and B are completely continuous operators mapping X into Y. Then $A \pm B$ is a completely continuous. Further, if α is a scalar, then αA is completely continuous.

Proof. We first show that $A \pm B$ is completely continuous. Hence suppose $S \subset X$ is a bounded set and that $\{x_n\}$ is a sequence of points of S. Since S is bounded, there must be some $M > 0$ such that

$$\|x_n\| \leq M \qquad \text{(for all } n\text{)}.$$

Since A is completely continuous, then from the sequence $\{Ax_n\}$, we must be able to select a convergent subsequence $\{Ax_{n_j}\}$, and suppose

$$Ax_{n_j} \to y.$$

Consider now the sequence $\{Bx_{n_j}\}$. Since B is completely continuous, we must be able to select a convergent subsequence from the above sequence, which we shall denote by $Bx_{n_{j_k}}$. Since $\{Ax_{n_{j_k}}\}$ is a subsequence of the original convergent sequence, it follows that this, too, is a convergent sequence and, moreover, must converge to the same limit as the original sequence y. Thus, if

$$Bx_{n_{j_k}} \to z,$$

we must have

$$(A \pm B)x_{n_{j_k}} \to y \pm z,$$

which, since we have been able to select a convergent subsequence from any sequence of points in $(A \pm B)(S)$, implies that $(A \pm B)(S)$ is relatively compact. The proof that αA, where $\alpha \in F$, is completely continuous if A is completely continuous, is quite straightforward and will be omitted.

A familiar result from analysis is that the limit of a sequence of continuous functions is also continuous if the original sequence converges uniformly to the limit function. We have a similar result for uniformly convergent sequences of completely continuous linear transformations, which we state in the next theorem.

THEOREM 17.5. Let $\{A_n\}$ be a sequence of completely continuous operators mapping X into Y, where Y is a Banach space, such that

$$\lim_n \|A_n - A\| = 0.$$

Then A is a completely continuous operator.

Proof. We shall show that, whenever S is a bounded subset of X, $A(S)$ is totally bounded. After doing this it will follow that $A(S)$ is relatively compact because Y is a complete metric space [Theorem 5.7(2)].

Let $S \subset X$ be bounded and let $\varepsilon > 0$ be given. Since the transformations $\{A_n\}$ are completely continuous, $A_n(S)$ is relatively compact—hence totally bounded [Theorem 5.7(1)] for every n. Moreover (Theorem 5.6; proof of Theorem 5.3), a finite ε-net which is *wholly contained in* $A_n(S)$, $\{A_nx_i\}$ can be selected in each case. Since S is bounded, there exists an $M > 0$ such that $\|x\| < M$ for all $x \in S$, and we can choose n so large that $\|A - A_n\| \leq \varepsilon/3M$. For this n let $\{A_nx_i\}$, $i = 1, \ldots, k$ be an $(\varepsilon/3)$-net for $A_n(S)$. Let $z \in A(S)$; as such there exists $y \in S$ such that $z = Ay$. By appropriately choosing x_i and considering

$$\|z - Ax_i\| = \|Ay - Ax_i\| \leq \|Ay - A_ny\|$$
$$+ \|A_ny - A_nx_i\|$$
$$+ \|A_nx_i - Ax_i\|,$$

the total boundedness of $A(S)$ is seen to be established. This, in conjunction with our initial remarks, proves the relative compactness of $A(S)$. Since the linearity of A is evident, it follows that A is a completely continuous operator.

THEOREM 17.6. If A and B are linear transformations mapping the normed linear space X into X, where A is completely continuous and B is bounded, then AB and BA are completely continuous operators.

REMARK. Combining this result with Theorem 17.4, we see that the class of all completely continuous operators is a two-sided ideal in the ring $L(X, X)$.

Proof. Suppose S is a bounded set in X and consider

$$(AB)(S) = A(B(S)).$$

Since B is bounded, it follows that $B(S)$ is a bounded set. A being completely continuous implies that $A(B(S))$ is relatively compact. Therefore AB is a completely continuous operator.

Consider now $(BA)(S) = B(A(S))$. By the complete continuity of A, $A(S)$ is relatively compact. Since the continuous image of a relatively compact set is, again, a relatively compact set and B is continuous, it follows that $B(A(S))$ is relatively compact. The complete continuity of BA now follows.

COROLLARY. Suppose X is an infinite-dimensional space and that A is a completely continuous operator mapping X into X. Then A^{-1}, if it exists, is not a bounded linear transformation, that is, $A^{-1} \notin L(X, X)$. (It is possible that A^{-1} could be bounded but it cannot be defined on all of X.)

Proof. Suppose $A^{-1} \in L(X, X)$. Then, since A is completely continuous, we can apply the theorem to assert the complete continuity of $1 = AA^{-1}$, which is false as was shown in the preceding section. (The identity transformation, on an infinite-dimensional space, is not completely continuous.) In view of this contradiction, the corollary is proved.

THEOREM 17.7. Suppose $A : X \to Y$. If A is completely continuous, the range of A, $R(A)$ is separable.

Proof. Before proving this result, it will be shown that a totally bounded metric space X is separable.

Under these conditions we have, for each positive integer j, a finite $(1/j)$-net N_j. We now claim that the countable set

$$N = \bigcup_j N_j$$

is dense in X. To prove this, let $\varepsilon > 0$ and $y \varepsilon X$ be given. An integer k exists such that $(1/k) < \varepsilon$ and an element $x \in N_k$ therefore exists whose distance from y is less than $1/k$ which is less than ε.

We now suppose that A is a completely continuous transformation mapping the normed linear space X into the normed linear space Y. Letting n run through the positive integers, we can write X as the union of all n-neighborhoods of 0:

$$X = \bigcup_n S_n(0),$$

because the norm of any vector in X must be less than some n. Since the direct image of a union is the union of the direct images (this is not so for intersections), it follows that

$$A(X) = \bigcup_n A(S_n(0)).$$

The complete continuity of A implies that $A\big(S_n(0)\big)$ is relatively compact for each n. By Theorem 5.7(1), relatively compact sets are totally bounded; by what was mentioned at the beginning of the proof, total boundedness implies separability. In summary we have written $A(X)$, the range of A, as a countable union of separable spaces. $A(X)$, therefore, is separable.

THEOREM 17.8. If A is completely continuous, its conjugate map A' is completely continuous.

Proof. We suppose that X and Y are normed linear spaces and that $A : X \to Y$ is completely continuous. Recalling that its conjugate map A' maps $\tilde{Y}$ into $\tilde{X}$, to show that A' is completely continuous, therefore, we must show that, if $f_1, f_2, \ldots, f_n, \ldots$ is a uniformly bounded sequence from $\tilde{Y}$, $\|f_n\| \le M$ for all n say, then we can extract a convergent subsequence from $A'f_1, A'f_2, \ldots, A'f_n, \ldots$.

Since A is completely continuous its range $R(A)$ must be separable (Theorem 17.7); let $w_1, w_2, \ldots, w_n, \ldots$ be a dense subset of $R(A)$. Consider now the sequence of complex numbers

$$f_1(w_1), f_2(w_1), \ldots, f_n(w_1), \ldots.$$

Since

$$|f_n(w_1)| \le \|f_n\| \, \|w_1\| \le M\|w_1\|,$$

$\{f_n(w_1)\}$ is a bounded sequence of complex numbers and, by the Bolzano–Weierstrass theorem, therefore must have a convergent subsequence, $f_{11}(w_1), f_{12}(w_1), \ldots, f_{1n}(w_1), \ldots$, say. (This notation is used instead of the more standard $\{f_{n_k}(w_1)\}$ because so many more subsequences will be selected.) We now consider the values of the functionals $\{f_{1n}\}$ at the vector w_2 to get a new bounded sequence of complex numbers

$$f_{11}(w_2), f_{12}(w_2), \ldots, f_{1n}(w_2), \ldots.$$

By the same reasoning, $\{f_{1n}(w_2)\}$ has a convergent subsequence which we shall denote by $\{f_{2n}(w_2)\}$. Similarly, we apply the sequence $\{f_{2n}\}$ to w_3 and get the convergent subsequence $\{f_{3n}(w_3)\}$, and so on. Let the diagonal sequence $\{f_{kk}\}$ be denoted by $\{g_k\}$. It is clear that, for any j, the limit

$$\lim_k g_k(w_j)$$

exists, because for $k \ge j$, $\{g_k(w_j)\}$ is a subsequence of the convergent sequence

$$f_{j1}(w_j), f_{j2}(w_j), \ldots, f_{jn}(w_j), \ldots.$$

We now claim that

$$\lim_k g_k(y) \tag{17.25}$$

exists for all $y \in \overline{R(A)}$. To prove this, consider

$$|g_k(y) - g_j(y)| \le |g_k(y) - g_k(w_n)| + |g_k(w_n) - g_j(w_n)| + |g_j(w_n) - g_j(y)|$$

$$\le \|g_k\| \, \|y - w_n\| + |g_k(w_n) - g_j(w_n)| + \|g_j\| \, \|w_n - y\|$$

$$\le 2M\|y - w_n\| + |g_k(w_n) - g_j(w_n)|.$$

Since $w_1, w_2, \ldots, w_n, \ldots$ is a dense subset of $R(A)$, hence of $\overline{R(A)}$, the term $\|y - w_n\|$ can be made arbitrarily small by appropriate choice of w_n; since $\lim_k g_k(w_n)$ exists by the preceding argument, $\{g_k(w_n)\}$ is a Cauchy sequence, and therefore $|g_k(w_n) - g_j(w_n)|$ can be made arbitrarily small by choosing k and j large enough. It now follows that $\{g_k(y)\}$ is a Cauchy sequence of complex numbers and, therefore, must converge. Consider now

$$A'g_k = v_k \in \tilde{X}.$$

Since

$$[x, v_k] = [x, A'g_k] = [Ax, g_k] = g_k(Ax)$$

and $Ax \in R(A)$, then, by (17.25),

$$\lim_k v_k(x) = \lim_k g_k(Ax) \bar{=} v(x) \qquad \text{(for any } x \in X)$$

exists. Clearly, v is a linear functional on X; in addition, since

$$|v(x)| = \lim_k |v_k(x)| = \lim_k |g_k(Ax)|$$

$$\leq M\|A\| \|x\|,$$

v is a *bounded* linear functional. To paraphrase these results, we can say

$$v_k \xrightarrow{w} v;$$

we contend that more is true, though; we contend that

$$v_k \to v.$$

This is the same as

$$A'g_k \to v. \tag{17.26}$$

Since $\{g_k\}$ is a subsequence of $\{f_n\}$, the theorem follows if (17.26) can be shown. To prove (17.26) we shall assume it to be false and get a contradiction; hence suppose there is $\varepsilon > 0$ such that there exists a subsequence of $\{g_k\}$, $\{g_{k_j}\}$ for which

$$\|A'g_{k_j} - v\| > \varepsilon > 0 \qquad \text{(for all } j). \tag{17.27}$$

Let

$$h_j = g_{k_j}.$$

One way of computing the norm of $(A'h_j - v)$ is via

$$\|A'h_j - v\| = \sup_{\|x\|=1} \|(A'h_j - v)(x)\|.$$

Thus there must exist vectors x_j $(j = 1, 2, \ldots)$ with norm one such that

$$|A'h_j(x_j) - v(x_j)| > \tfrac{1}{2}\|A'h_j - v\| > \frac{\varepsilon}{2} \qquad \text{(for each } j). \tag{17.28}$$

The vectors $\{x_j\}$ are clearly uniformly bounded so, since A is completely continuous, one must be able to extract a convergent subsequence $\{Ax_{j_k}\}$ from $\{Ax_j\}$. Let

$$x_{j_k} = z_k \qquad (k = 1, 2, \ldots),$$

and realize that to each z_k there corresponds h_{j_k}; h_{j_k} will be denoted by $\{\psi_k\}$. (The important thing to remember about $\{\psi_k\}$ is that it is a subsequence of $\{g_k\}$.) Let the limit of Az_k be denoted by z; that is, suppose

$$\lim_k Az_k = z. \tag{17.29}$$

Since $z \in \overline{R(A)}$, the limit

$$\lim_k g_k(z) = \alpha \tag{17.30}$$

exists by (17.25). The subsequence of $\{g_k(z)\}$, $\{\psi_k(z)\}$ must therefore be convergent and converge to α, too. Next, note that

$$v(z_k) = \lim_j v_j(z_k)$$

$$= \lim_j g_j(Az_k) = \lim_j \psi_j(Az_k).$$

Now we can write

$$|\psi_j(z) - \psi_j(Az_k)| \le \|\psi_j\| \, \|z - Az_k\|$$

$$\le M\|z - Az_k\|$$

for every j, which implies

$$\lim_j |\psi_j(z) - \psi_j(Az_k)| = |\lim_j \psi_j(z) - \lim_j \psi_j(Az_k)|$$

$$\le M\|z - Az_k\|$$

or

$$|\alpha - v(z_k)| \le M\|z - Az_k\|.$$

This implies that

$$|\psi_k(Az_k) - v(z_k)| \le |\psi_k(Az_k) - \psi_k(z)| + |\psi_k(z) - \alpha| + |\alpha - v(z_k)|$$

$$\le M\|Az_k - z\| + |\psi_k(z) - \alpha| + M\|z - Az_k\|$$

$$= 2M\|Az_k - z\| + |\psi_k(z) - \alpha|.$$

It remains now only to utilize (17.29) and (17.30) to see that

$$|\psi_k(Az_k) - v(z_k)| \to 0,$$

which contradicts our lone assumption—namely (17.27). In spite of all the subscripts, this completes the proof

EXERCISES 17

1. Show that a linear transformation A defined on $D \subset X$, a normed linear space, admits closure if and only if all sequences $\{x_n\}$ from D such that $x_n \to 0$ and $Ax_n \to y$ imply that $y = 0$.

2. Consider the following diagram:

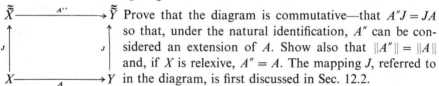 Prove that the diagram is commutative—that $A''J = JA$ so that, under the natural identification, A'' can be considered an extension of A. Show also that $\|A''\| = \|A\|$ and, if X is relexive, $A'' = A$. The mapping J, referred to in the diagram, is first discussed in Sec. 12.2.

3. If the linear transformation A has a closed extension B, prove that A permits a closure.

4. Prove that the mapping of $L(X, Y)$ into $L(\widetilde{Y}, \widetilde{X})$ given by $A \to A'$ is a bounded linear transformation.

5. Show that if X is reflexive and N is a closed subspace of $\widetilde{X}$, then $(^aN)^a = N$.

6. Show that if A is a bounded operator on a B-space, and if $R(A)$ is a B-space, then $\overline{R(A')} = N(A)^a$.

7. Suppose $A : X \to X$ is bounded and X is a B-space. Show that, if $R(A') = \widetilde{X}$, then A has a bounded inverse.

8. Show that a completely continuous transformation maps a weakly convergent sequence into a strongly convergent one.

9. Let X and Y be normed linear spaces and $A : X \to Y$ be a linear transformation where X is reflexive and such that A maps every weak Cauchy sequence (in particular every weakly converging sequence) into a strongly converging sequence. Then prove that A is completely continuous.

10. Show that Exercise 9 is no longer true if X is not reflexive, that is, $A : X \to Y$ such that A maps weakly convergent sequences into strongly convergent sequences does not imply that A is completely continuous. (*Hint:* Take $X = Y = l_1$ and $A = 1$; see also Exercise 14.12.)

11. Let $A : X \to Y$. If Y is a Banach space and A' is completely continuous, show that A is completely continuous.

12. Let X and Y be normed linear spaces and $A : X \to Y$ be a continuous linear transformation. Show that if $x_n \overset{w}{\to} x$ in X, then $Ax_n \overset{w}{\to} Ax$ in Y.

13. If X is infinite-dimensional and if A is an isometric linear transformation defined on X, prove that A is not completely continuous.

REFERENCES

1. M. Naimark, "Normed Rings." Further discussion of the closure.

For further properties of completely continuous transformations, see the following references.
2. A. Zaanen, "Linear Analysis."
3. A. Taylor, "Introduction to Functional Analysis."
4. F. Riesz and B. Sz.-Nagy, "Functional Analysis."

CHAPTER 18

Spectral Notions

Very often when an operator is encountered its inverse is of more interest than the operator itself, and it is the inverses of certain operators that are the chief concern in this chapter. In particular, if $A \in L(X, X)$, where X is, of course, a normed linear space, and λ is a scalar, we shall be concerned with the inverse of the operator $(\lambda - A)$.† When such considerations are made in finite-dimensional spaces, the situation is exceedingly clear-cut: either $(\lambda - A)^{-1}$ exists or it does not. If, for some given λ, $(\lambda - A)^{-1}$ did not exist, then λ was called an eigenvalue of A, and it was easily shown that if dim $X = n$, there were at most n (distinct) eigenvalues. Things are not so simple if X is infinite-dimensional. Aside from the fact that there may be infinitely many, indeed even a whole continuum, of scalars λ, such that $(\lambda - A)^{-1}$ does not exist, the fundamental issue of the existence of this inverse operator needs more qualification. Certainly a point of interest, assuming $(\lambda - A)^{-1}$ exists, is whether $(\lambda - A)^{-1}$ is a *bounded* linear transformation. (It will always be linear if it exists.) Another point, for reasons that will become apparent later, is whether the range of $(\lambda - A)$, $R(\lambda - A)$, is dense in X; that is, does $\overline{R(\lambda - A)} = X$? None of these issues had to be dealt with if X was finite-dimensional.

The treatment of the above questions leads us into what is known as *spectral theory* or *spectral analysis* of linear transformations. As to what direction such investigations will ultimately take, let us return to the finite-dimensional case once again. Knowing the eigenvalues of A, when A was a normal transformation on an inner product space, for example, enabled us to write A as a linear combination of projections and, further, to decompose X into an orthogonal direct sum decomposition (see Theorem 2.6, the spectral decomposition theorem for normal transformations). Things will usually be much more complicated in the infinite-dimensional case but, nevertheless, the trend is indicated by the above result.

18.1 Spectra and the Resolvent Set

Throughout this section, X will be assumed to be a complex normed linear space and A a linear transformation on X such that

$$X \supset D \overset{A}{\to} X. \tag{18.1}$$

Our concern will be with introducing several notions pertaining to the inverse operator of $(\lambda - A)$, where $\lambda \in F$ (F, as usual, denotes the scalar field which, in this case, is the complex numbers). In addition, we shall also prove a theorem concerning the character of the point spectrum of completely continuous transformations. Since we shall be so concerned with inverses, before proceeding any further, we state precisely what we mean by the inverse of the operator A above (there is no special fact about A used in the definition below, and it applies to any function mapping a given set into another).

DEFINITION 18.1. Let A be as in (18.1). The *inverse* of A, A^{-1}, is said to exist if there exists a function A^{-1} such that for any $x \in D$, $A^{-1}(Ax) = x$. The domain of A^{-1} is just the range of A, $R(A)$, and its range is D. Thus we have the following picture:

$$A^{-1} : R(A) \to X. \tag{18.2}$$

† The notation $(\lambda - A)$ is just a shorthand notation for the operator $\lambda 1 - A$, where 1 denotes the identity transformation.

In addition, we note that A must be $1:1$ and

$$A^{-1}(R(A)) = D. \tag{18.3}$$

We note that the inverse of A^{-1} is just A.

With A assumed to be only a linear transformation as in (18.1) and using the abbreviation $\lambda - A$, λ a scalar, to denote $\lambda 1 - A$ where 1 represents the identity transformation, we now make the following definitions.

DEFINITION 18.2. If the range of $\lambda - A$ is dense in X and if $\lambda - A$ has a bounded inverse on $R(\lambda - A)$, then λ is said to belong to the *resolvent set of A, $\rho(A)$*.

DEFINITION 18.3. If the range of $\lambda - A$, $R(\lambda - A)$, is dense in X and if $\lambda - A$ has an unbounded inverse (that is, we assume $(\lambda - A)^{-1}$ exists but is not bounded), then λ is said to belong to the *continuous spectrum of A, $C\sigma(A)$*.

DEFINITION 18.4. If the range of $\lambda - A$ is not dense in X but $\lambda - A$ has an inverse, bounded or unbounded, then λ is said to belong to the *residual spectrum of A, $R\sigma(A)$*.

DEFINITION 18.5. If $(\lambda - A)^{-1}$ does not exist, then λ is said to belong to the *point (discrete) spectrum of A, $P\sigma(A)$*. Note that this set consists of just the eigenvalues of A.

TABLE III

A Summary of Definitions 18.2 through 18.5

	$(\lambda - A)^{-1}$	Boundedness of $(\lambda - A)^{-1}$	$R(\lambda - A)$	Set to which λ belongs
18.2.	exists	bounded	dense in X	resolvent set, $\rho(A)$
18.3.	exists	unbounded	dense in X	continuous spectrum, $C\sigma(A)$
18.4.	exists	bounded or unbounded	not dense in X	residual spectrum, $R\sigma(A)$
18.5.	does not exist		dense or not dense in X	point spectrum, $P\sigma(A)$

The above definitions are restated in Table III. We note immediately that the sets in Table III are disjoint and that an element $\lambda \in F$ belongs to one and only one of the abovementioned sets. Thus we have the following disjoint union for F:

$$F = \rho(A) \cup C\sigma(A) \cup R\sigma(A) \cup P\sigma(A).$$

DEFINITION 18.6. The set

$$C\sigma(A) \cup R\sigma(A) \cup P\sigma(A),$$

denoted by $\sigma(A)$, is called the *spectrum of* A. Thus the union of the continuous, the residual, and the point spectra of A comprise the spectrum of A.

Suppose that X is finite-dimensional now and that A is defined on all of X. Then, if $(\lambda - A)^{-1}$ exists, it must be bounded, because all linear transformations on a finite-dimensional space are bounded (Theorem 14.7). Further, since $(\lambda - A)^{-1}$ exists, $\lambda - A$ must be $1:1$, which, since $(\lambda - A)$ is a linear transformation from a finite-dimensional space into itself, implies that $\lambda - A$ is onto X or that $R(\lambda - A) = X$. Hence if the inverse exists at all, it must be bounded and the range of $(\lambda - A)$ must be dense in X and Definitions 18.2–18.4 all collapse into the situation covered by Definition 18.2. We have now established the next theorem.

THEOREM 18.1. In a finite-dimensional space X, with A defined on all of X, $R\sigma(A) = \emptyset$ and $C\sigma(A) = \emptyset$.

A collection of linear transformations whose behavior is very similar to linear transformations from a finite-dimensional space into itself is the set of completely continuous operators introduced in the last chapter. In their case it turns out that 0 is the only possible member of the residual or continuous spectra. In addition, there are at most a countable number of elements in their point spectra, and we shall prove this in our next theorem.

THEOREM 18.2. Suppose A is a completely continuous linear transformation such that $X \supset D \overset{A}{\to} X$. Then $P\sigma(A)$ is at most countable (it could be empty), and 0 is its only possible limit point.

Proof. We contend that, for any $\varepsilon > 0$, there are at most finitely many points in

$$P_\varepsilon = \{\lambda \in P\sigma(A)|\,|\lambda| \geq \varepsilon\}.$$

Since, with the exception of 0, any $\lambda \in P\sigma(A)$ can be found in $\bigcup_{n=1}^\infty (P_{1/n})$, if the contention can be established, the theorem will follow. The contention will be proved by contradiction. Thus we now suppose the contention to be false, which means that, for some $\varepsilon > 0$, there are infinitely many $\lambda \in P_\varepsilon$. Since there are infinitely many we now can select.

$$\lambda_1, \lambda_2, \ldots, \lambda_n, \ldots \in P_\varepsilon, \tag{18.4}$$

where the λ_i are all distinct. We now select corresponding eigenvectors

$$x_1, x_2, \ldots, x_n, \ldots$$

and note that, since eigenvectors corresponding to distinct eigenvalues must be linearly independent, the above vectors constitute a linearly independent set. Form now the sets

$$M_n = [\{x_1, \ldots, x_n\}].$$

For any n, M_{n-1} is finite-dimensional and therefore closed (Theorem 8.7, Corollary 2). By virtue of the linear independence of $x_1, x_2, \ldots$, the inclusion

$$M_{n-1} \subset M_n$$

must be proper. We are now in a position to bring the Riesz theorem, Theorem 8.8, to bear against the situation and assert the existence of a $y_n \in M_n$ such that

$$\|y_n\| = 1 \quad \text{and} \quad \|y_n - x\| \geq \tfrac{1}{2} \quad \text{for all } x \in M_{n-1}. \quad (18.5)$$

Consider now the vector from M_n

$$x = \sum_{i=1}^{n} \alpha_i x_i,$$

where the α_i are scalars, and compute

$$Ax = \sum_{i=1}^{n} \alpha_i \lambda_i x_i.$$

Consequently,

$$\lambda_n x - Ax = (\lambda_n - A)x = \sum_{i=1}^{n-1} \alpha_i(\lambda_n - \lambda_i)x_i,$$

and we see that

$$(\lambda_n - A)(M_n) \subset M_{n-1}.$$

In view of this, for any m such that $1 \leq m < n$, we can say

$$w = (\lambda_n - A)y_n + Ay_m \in M_{n-1}. \quad (18.6)$$

Since M_{n-1} is a subspace, it is clear that $\lambda_n^{-1}w \in M_{n-1}$, too. We now form

$$Ay_n - Ay_m = \lambda_n y_n - (\lambda_n y_n - Ay_n + Ay_m)$$

$$= \lambda_n y_n - w$$

$$= \lambda_n(y_n - \lambda_n^{-1}w).$$

Using (18.4)–(18.6), we have

$$\|Ay_n - Ay_m\| = |\lambda_n| \, \|y_n - \lambda_n^{-1}w\| \geq \frac{|\lambda_n|}{2} \geq \frac{\varepsilon}{2}. \quad (18.7)$$

Now we shall make use of our assumption that A is completely continuous. In view of (18.5), it is clear that the sequence y_n is a (uniformly) bounded sequence which, since A is completely continuous, implies that it should be possible to extract a convergent subsequence from $\{Ay_n\}$. In view of (18.7), we see that such a convergent subsequence cannot exist, and our lone assumption about the falsehood of the contention has led to a contradiction, thus proving the validity of the contention, from which the theorem immediately follows.

We now apply the definitions stated above to some concrete examples.

18.2   The Spectra of Two Particular Transformations

In this first example the results of Chapter 9 will be relied upon extensively.

EXAMPLE 18.1. Suppose X is a separable, infinite-dimensional, Hilbert space. For such a space (which happens to be congruent to l_2), there must be a countable, complete orthonormal set

$$x_1, x_2, \ldots, x_n, \ldots$$

by virtue of Theorems 9.4 and 9.7. It further follows that any $x \in X$ has the representation

$$x = \sum_{n=1}^{\infty} \alpha_n x_n,$$ (18.8)

where $\alpha_n = (x, x_n)$. Consider a sequence of scalars $\{\lambda_n\}$, now such that

$$\lambda_n \to 1 \quad \text{with no } \lambda_n = 1.$$

We are now going to show that the linear transformation mapping X into X defined by† [with x as in (18.8)]

$$Ax = \sum_{n=1}^{\infty} \alpha_n \lambda_n x_n$$ (18.9)

has the following properties:

$$\lambda_n \in P\sigma(A) \quad (n = 1, 2, \ldots),$$ (18.10)

$$1 \in C\sigma(A),$$ (18.11)

$$\lambda \neq \lambda_n \quad \text{for any } n \text{ and } \lambda \neq 1 \Rightarrow \lambda \in \rho(A),$$ (18.12)

$$R\sigma(A) = \emptyset.$$ (18.13)

Since $Ax_n = \lambda_n x_n$ for any n, (18.10) follows immediately. We now show (18.11). To do so, we must show that the transformation $1 - A$ has an unbounded inverse and that the range of $(1 - A)$ is dense in X. Denoting $(1 - A)$ by A_1, we first show that A_1^{-1} exists [on $R(A_1)$, the range of A_1] by showing that A_1 is $1 : 1$. Hence, suppose

$$A_1 x = 0.$$

This implies that

$$x = Ax$$

or that

$$\sum_{n=1}^{\infty} \alpha_n(1 - \lambda_n)x_n = 0,$$

which by the orthonormality of the x_n implies

$$\sum_{n=1}^{\infty} |\alpha_n|^2 |1 - \lambda_n|^2 = 0.$$

Since $1 - \lambda_n \neq 0$ for every n, we must have

$$\alpha_n = 0 \quad \text{(for all } n\text{)};$$

therefore $x = 0$, and we have shown A_1 to be $1 : 1$. It now follows that A_1^{-1} exists. Consider now

$$A_1 x_n = (1 - A)x_n = x_n - Ax_n = (1 - \lambda_n)x_n;$$

consequently,

$$\|A_1 x_n\| = |1 - \lambda_n| \to 0.$$ (18.14)

† Clearly, $Ax \in X$ because, since $\{\lambda_n\}$ is convergent and therefore bounded by M say, we can write $\|Ax\|^2 = \sum_{n=1}^{\infty} |\alpha_n|^2 |\lambda_n|^2 \leq M \sum_{n=1}^{\infty} |\alpha_n|^2 = M \cdot \|x\|^2$. Thus not only do we see that the coefficients in the series (18.9) are square summable [see Theorem 9.5(1)] but also that A is a *bounded* linear transformation.

A necessary condition for A_1^{-1} to be bounded, however, is the existence of some $k > 0$ such that

$$k\|x\| \leq \|A_1 x\| \quad \text{(for all } x \in X).$$

In view of (18.14), no such k could exist, and it follows that A_1^{-1} is not bounded. To show that $R(A_1)$ is dense in X we note, for any n, that

$$A_1\left(\frac{x_n}{1 - \lambda_n}\right) = x_n;$$

therefore $x_n \in R(A_1)$. This implies that

$$[\{x_1, x_2, \ldots, x_n, \ldots\}] \subset R(A_1),$$

and since we already know that [Theorem 9.12 (1), (4)]

$$\overline{[\{x_1, x_2, \ldots, x_n, \ldots\}]} = X$$

it follows that $\overline{R(A_1)} = X$, and we have proved that

$$1 \in C\sigma(A).$$

Now suppose that λ is a scalar such that $\lambda \neq \lambda_n$ for any n and that $\lambda \neq 1$. For any such λ there must be a real number $a_\lambda > 0$ such that

$$|\lambda - \lambda_n| > a_\lambda > 0 \quad \text{(for all } n),$$

for if such an a_λ did not exist, this would contradict the fact that 1 is the only limit point of the set $\{\lambda_1, \lambda_2, \ldots\}$.† Our goal now is to show that for such λ, $\lambda \in \rho(A)$; hence we must show that the transformation $A_\lambda = \lambda - A$ has a bounded inverse and that the range of A_λ, $R(A_\lambda)$, is dense in X. With x as in (18.8), we now consider

$$A_\lambda x = \lambda x - A x = \sum_{n=1}^{\infty} \alpha_n(\lambda - \lambda_n)x_n$$

which, by the Pythagorean theorem (see Exercise 9.16), implies

$$\|A_\lambda x\|^2 = \sum_{n=1}^{\infty} |\alpha_n|^2 |\lambda - \lambda_n|^2 \geq a_\lambda^2 \|x\|^2.$$

Therefore,

$$\|A_\lambda x\| \geq a_\lambda \|x\| \quad \text{(for all } x \in X),$$

and it follows by Theorem 14.9 that A_λ has a bounded inverse. We now show A_λ is onto X and, thus, certainly has a range that is dense in X. To show that it is onto, consider any $y \in X$, where

$$y = \sum_{n=1}^{\infty} \beta_n x_n.$$

We wish to find an $x \in X$ such that $A_\lambda x = y$ or, equivalently, to find an $x = \sum_{n=1}^{\infty} \alpha_n x_n$ such that

$$\sum_{n=1}^{\infty} \alpha_n(\lambda - \lambda_n)x_n = \sum_{n=1}^{\infty} \beta_n x_n.$$

† Note that $\lambda \notin \overline{\{\lambda_1, \lambda_2, \ldots\}}$ implies that the distance between λ and the closure of $\{\lambda_1, \lambda_2, \ldots\}$ must be positive.

Taking

$$\alpha_n = \frac{\beta_n}{\lambda - \lambda_n}$$

certainly seems to do the trick, but we must verify for such α_n that $\sum_{n=1}^{\infty} \alpha_n x_n$ is indeed a member of the space; that is, we must show that $\sum_{n=1}^{\infty} |\alpha_n|^2 < \infty$. The nod in the affirmative about the above issues is given by noting that

$$\sum_{n=1}^{\infty} |\alpha_n|^2 = \sum_{n=1}^{\infty} \frac{|\beta_n|^2}{|\lambda - \lambda_n|^2} < \frac{1}{a_\lambda^2} \sum_{n=1}^{\infty} |\beta_n|^2 < \infty$$

and, therefore, $R(A_\lambda) = X$. Consequently, for all scalars λ, such that $\lambda \neq \lambda_n$ for any n and $\lambda \neq 1$, $\lambda \in \rho(A)$. In view of what we have shown it also follows, now, that $R\sigma(A) = \emptyset$.

EXAMPLE 18.2. Again, we suppose X to be an infinite-dimensional, separable Hilbert space and that

$$x_1, x_2, \ldots, x_n, \ldots$$

is a complete orthonormal set in X. For

$$x = \sum_{n=1}^{\infty} \alpha_n x_n,$$

as in (18.8), we now define a so-called *destruction operator* (mapping X into X) A by taking

$$Ax = \sum_{n=1}^{\infty} \frac{\alpha_n}{n+1} x_{n+1}.$$

Thus,

$$Ax_1 = \frac{x_2}{2}, \quad Ax_2 = \frac{x_3}{3}, \ldots.$$

Suppose that, with x as above,

$$Ax = 0.$$

This implies

$$\sum_{n=1}^{\infty} \frac{\alpha_n}{n+1} x_{n+1} = 0;$$

consequently, (by Exercise 9.16, for example) $\alpha_n = 0$ ($n = 1, 2, \ldots$), which means

$$x = 0,$$

and A^{-1} is seen to exist [on $R(A)$]. Clearly, $x_1 \notin R(A)$ and, further, it is easy to show that

$$\overline{R(A)} = [x_2, x_3, \ldots] \neq X;$$

that is, the range of A is not dense in X. Thus it follows that

$$0 \in R\sigma(A),$$

the residual spectrum of A. Last, although we shall not do so here, it can be shown that $\lambda \in \rho(A)$ for any $\lambda \neq 0$.

18.3   Approximate Proper Values

As in (18.1), suppose that $X \supset D \xrightarrow{A} X$.

DEFINITION 18.7. The scalar λ is called an *approximate proper value* of A if, for any $\varepsilon > 0$, there exists an $x \in D$ such that $\|x\| = 1$ and

$$\|(\lambda - A)x\| < \varepsilon.$$

Clearly we can modify the norm requirement on x and restate the last part of the above definition to be: For any $\varepsilon > 0$ there exists a nonzero vector $x \in D$ such that

$$\|(\lambda - A)x\| < \varepsilon\|x\|.$$

We denote by $\pi(A)$ the set of all approximate proper values and call this set the *approximate spectrum* of A.

THEOREM 18.3. λ is an approximate proper value of A if and only if $\lambda - A$ does not have a bounded inverse.

Proof. Suppose $\lambda \in \pi(A)$. If so, there must exist $x_n \in D$, $\|x_n\| = 1$, such that

$$\|(\lambda - A)x_n\| < \frac{1}{n}$$

for $n = 1, 2 \ldots$. This clearly precludes the possibility of finding a $k > 0$ such that

$$k\|x\| \leq \|(\lambda - A)x\| \tag{18.15}$$

for all $x \in D$ and, thus, implies that $\lambda - A$ does not have a bounded inverse.

Conversely, if $\lambda - A$ does not have a bounded inverse, a $k > 0$ satisfying (18.15) for all $x \in D$ cannot exist. Such a k not existing, however, means that for any $\varepsilon > 0$, an $x \in D$ with norm one can be found such that

$$\|(\lambda - A)x\| < \varepsilon,$$

which proves that λ is an approximate proper value of A.

COROLLARY. The above theorem immediately implies

$$\pi(A) \subset \sigma(A).$$

EXERCISES 18

1. Let X be a Banach space and let $A \in L(X, X)$. Show that:

(a) If $\lambda \in \rho(A)$, then $\lambda \in \rho(A')$.
(b) If $\lambda \in R\sigma(A)$, then $\lambda \in P\sigma(A')$.
(c) If $\lambda \in P\sigma(A)$, then either $\lambda \in P\sigma(A')$ or $\lambda \in R\sigma(A')$.

    (d) If $\lambda \in C\sigma(A')$, then $\lambda \in C\sigma(A)$.

    (e) If $\lambda \in \rho(A')$, then $\lambda \in \rho(A)$.

2. Show that $P\sigma(A) \cup C\sigma(A) \subset \pi(A)$.

3. Suppose $A : X \to X$ is bounded linear and X is a Banach space. Show that, if $\lambda \in \rho(A)$, then $R(\lambda - A) = X$.

4. Show that, if $\lambda \in \pi(A)$, where A is a bounded linear transformation, then $|\lambda| \leq \|A\|$.

5. Show that the transformation defined in Example 18.2 is completely continuous.

6. Let X be a Hilbert space and let $A \in L(X, X)$. Show that the following statements are equivalent:

    (a) $\lambda \in \pi(A)$.

    (b) There exists a sequence $\{B_n\}$ in $L(X, X)$ such that $\|B_n\| = 1$ for all n and $\|(\lambda - A)B_n\| \to 0$ as $n \to \infty$.

7. Let X be an infinite-dimensional separable Hilbert space and let $\{x_n\}$ be a complete orthonormal set. For $x = \sum_{n=1}^{\infty} \alpha_n x_n$, define $Ax = \sum_{n=1}^{\infty} \alpha_n x_{n+1}$.

    (a) Prove that A is bounded with $\|A\| = 1$,

    (b) Show also that $P\sigma(A) = \emptyset$;

8. If $A : X \to X$ is completely continuous and X is infinite-dimensional, prove that $0 \notin \rho(A)$.

9. Let $A : X \to X$ be completely continuous. Show that if $\lambda \neq 0$, either $\lambda \in \rho(A)$ or $\lambda \in P\sigma(A)$.

10. Let $X = C[0, 1]$ (with sup norm) and define $A : X \to X$ by

$$(Af)(x) = \int_0^x f(t)\, dt,$$

where $f \in X$ and $0 \leq x \leq 1$. Show that $\{0\} = R\sigma(A)$ and $\{\lambda \in C | \lambda \neq 0\} = \rho(A)$.

11. Let $X = l_1$ and define for $x = (\alpha_1, \alpha_2, \ldots, \alpha_n, \ldots) \in l_1$,

$$Ax = \left(\alpha_1, \tfrac{1}{2}\alpha_2, \ldots, \frac{1}{n}\alpha_n, \ldots \right).$$

Show that A is completely continuous and $0 \in C\sigma(A)$.

12. Let X be a separable infinite-dimensional Hilbert space with $x_1, x_2, \ldots, x_n, \ldots$ as a complete orthonormal set. For $x = \sum_{n=1}^{\infty} \alpha_n x_n$, define

$$Ax = \sum_{n=1}^{\infty} \frac{\alpha_n}{n} x_{n-1} \qquad (x_0 = 0).$$

Show that $P\sigma(A) = \{0\}$.

13. In the separable Hilbert space X, let $\{x_1, x_2, ..., x_n, ...\}$ be a complete ortho-normal set and let the linear transformation A behave as follows on these elements:

$$Ax_n = \lambda x_n - x_{n+1} \qquad (\lambda \in F, n = 1, 2, ...).$$

Compute $\sigma(A)$.

14. The following is an example of an *unbounded operator with a bounded spectrum*. (In this particular case the spectrum is empty, so is it certainly bounded.) Consider the linear space of complex-valued functions defined on $[0, 1]$, that are infinitely differentiable on $[0, 1]$, and such that all derivatives vanish at the origin, that is, $f(0) = 0, f'(0) = 0, f''(0) = 0, ...$ (Note that this class does not consist solely of the zero function; for example, the function defined to be $\exp -(1/t^2)$ for $t \neq 0$ and taken to be 0 when $t = 0$ is a nonzero member of this collection.) Calling this linear space with supremum norm X, we now consider

$$A : X \to X,$$

$$f \to f'.$$

This operator is linear and not bounded (verify); moreover, for any scalar α, $A - \alpha 1$ is onto all of X, for consider any $g \in X$ and consider

(1)
$$f(t) = e^{\alpha t} \int_0^t e^{-\alpha s} g(s) \, ds.$$

The reader should verify that $f \in X$. Further

$$f'(t) = e^{\alpha t} e^{-\alpha t} g(t) + \alpha e^{\alpha t} \int_0^t e^{-\alpha s} g(s) \, ds,$$

so that $Af - \alpha f = g$ and proves the ontoness of $A - \alpha 1$. It is also $1 : 1$ for, if $(A - \alpha 1)f = 0$, then $f'(t) - \alpha f(t) = 0$ and this differential equation with boundary condition $f(0) = 0$ can only be satisfied by the zero function. By making some estimates in (1), it follows that there is some constant K such that [for $(A - \alpha 1)^{-1} g = f$]

$$\|(A - \alpha 1)^{-1} g\| = \|f\| \leq K \|g\|.$$

REFERENCES

1. A. Zaanen, "Linear Analysis."
2. A. Taylor, "Introduction to Functional Analysis."
3. P. Halmos, "Introduction to Hilbert Space."
4. E. Lorch, "Spectral Theory."

CHAPTER **19**

Introduction to Banach Algebras

Among the most important results in analytic function theory one must certainly include Liouville's theorem, Cauchy's integral theorem, and Cauchy's integral formula. Indeed, these results constitute the core of the richness of analytic function theory.

In this chapter the concept of a vector-valued analytic function is defined, and it is with surprising ease that the theorems mentioned above can be extended to this new category of analytic function. The astounding pervasiveness of the Hahn–Banach theorem asserts itself once again, as one of its consequences brings the proofs of Liouville's theorem and Cauchy's integral theorem to swift and graceful endings. Although we shall not have immediate occasion to use these results, they will be called upon heavily to establish some results in the ensuing chapters. Regardless of this consideration, however, their beauty alone justifies their inclusion.

We next introduce a new algebraic structure—an *algebra*. This is a vector space with a multiplication defined between vectors in such a way that the set of vectors, together with the operations of vector addition and this multiplication, constitutes a ring, and such that scalar multiplication and vector multiplication are linked in a certain prescribed way. The notion of an algebra is then extended to the concepts of *normed algebra* and *Banach algebra*.

After obtaining some results about Banach algebras, we then apply these results to the special Banach algebra $L(X, X)$, where X is a Banach space, and in so doing we shall realize many pertinent facts about the bounded linear transformations defined on X.

19.1   Analytic Vector-Valued Functions

In this section the notion of an analytic vector-valued function will be introduced. It will then be shown that a theorem, identical in form to the ordinary Liouville theorem of analytic function theory, holds for such functions. Another extension of a familiar result that will be proved here is an extended version of Cauchy's integral theorem for such functions. Our strongest tool in proving each of these results is a consequence of the Hahn–Banach theorem.

Suppose D is a region in the complex plane C, suppose X is a normed linear space and x is a function such that

$$x : D \to X,$$

$$\lambda \to x(\lambda).$$

In this context we now make the following definition.

DEFINITION 19.1. The function $x(\lambda)$ is said to be *analytic in D* if for all $\lambda_0 \in D$ the limit

$$x'(\lambda_0) = \lim_{\lambda \to \lambda_0} \frac{x(\lambda) - x(\lambda_0)}{\lambda - \lambda_0}$$

exists. This limit, when it exists, will be referred to as the *derivative of $x(\lambda)$ at λ_0*. Note, finally, that the above limit is taken in the sense of the norm of X.

Suppose now that $x(\lambda)$ is analytic in D and let $f \in \tilde{X}$. In this case we see that the composite function fx is just a complex-valued function of a complex variable.

We shall show now that if $x(\lambda)$ is analytic in D, then $f(x(\lambda))$ is an analytic function in D as well. To show this, we must show that f possesses a derivative at each point of D. Consider now

$$\lim_{\lambda \to \lambda_0} \frac{f(x(\lambda)) - f(x(\lambda_0))}{\lambda - \lambda_0} = f\left(\lim_{\lambda \to \lambda_0} \frac{x(\lambda) - x(\lambda_0)}{\lambda - \lambda_0}\right) = f(x'(\lambda_0)),$$

this manipulation being justified by the linearity and continuity of f. The analyticity of f is now evident.

Using mainly just a consequence of the Hahn–Banach theorem, we now state and prove *Liouville's theorem for analytic vector-valued functions.*

THEOREM 19.1. Suppose X is a complex normed space and let $x(\lambda)$ be as follows:

$$x : C \to X,$$

$$\lambda \to x(\lambda).$$

If $x(\lambda)$ is analytic throughout C (that is, *entire*) and there is some $M > 0$ such that $\|x(\lambda)\| < M$ for all $\lambda \in C$, then $x(\lambda)$ must be constant.

Proof. If f is any bounded linear functional on X, then, by the previous discussion, it follows that $f(x(\lambda))$ is an entire function. It also follows that

$$|f(x(\lambda))| \le \|f\| \cdot \|x(\lambda)\| \le \|f\| \cdot M,$$

which means that f is a bounded, entire function. By the ordinary Liouville theorem, it follows that $f(x(\lambda))$ must be constant. Thus, for any two complex numbers, α and β, we must have

$$f(x(\alpha)) = f(x(\beta))$$

or, by linearity,

$$f(x(\alpha) - x(\beta)) = 0.$$

Since f was *any* bounded linear functional, we see that all bounded linear functionals must vanish on the vector $x(\alpha) - x(\beta)$ and it follows (see Consequence 2, Theorem 12.2) that the vector must be zero; hence

$$x(\alpha) = x(\beta)$$

for any α, β. This is equivalent to saying

$$x(\lambda) = \text{constant}$$

and the theorem is proved.

Before proceeding to our next result, we wish to define a notion of integrating vector-valued functions along a contour in the complex plane. Let X be a Banach space and let $x(\lambda)$ be a continuous function whose domain is a region of C, D say, and whose range is in X. Let Γ be some rectifiable Jordan arc in D connecting the points z, $w \in D$ and consider a partition of $\Gamma : z = \lambda_1, \lambda_2, ..., \lambda_n = w$; that is, $\lambda_1, \lambda_2, ..., \lambda_n \in \Gamma$. Now let $\lambda_i{}^*$ be a point on Γ "between" λ_i and λ_{i+1}. (What is meant by "between" here is, aside from its clear intuitive connotation, easily realized by considering a parametrization of Γ.) Now, letting $\eta = \max|\lambda_{i+1} - \lambda_i|$, we define

$$\lim_{\substack{n \to \infty \\ \eta \to 0}} \sum_{i=1}^{n-1} x(\lambda_i^*) \cdot (\lambda_{i+1} - \lambda_i) = \int_\Gamma x(\lambda)\, d\lambda.$$

We point out that:

(1) the limit here is a limit in the norm on X;
(2) the limit must exist by virtue of the continuity,

(actually uniform continuity) of $x(\lambda)$ on Γ and the completeness of the space. [The uniform continuity of $x(\lambda)$ on Γ follows because Γ can be regarded as the continuous image of the compact set, the closed interval $[0, 1]$, which means that $x(\lambda)$ is a continuous function on a compact set and must, therefore, be uniformly continuous there.]

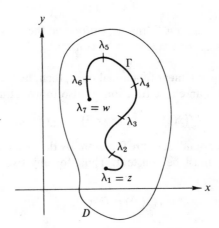

FIG. 19.1. A typical curve Γ and a partition of Γ for $n = 7$.

We now proceed to Cauchy's integral theorem for vector-valued functions.

THEOREM 19.2. Let X be a Banach space. If $x(\lambda)$ is analytic in a region bounded by a rectifiable Jordan arc Γ and continuous on Γ, then

$$\int_\Gamma x(\lambda)\, d\lambda = 0.$$

Proof. Let f be any bounded linear functional on X and let

$$y = \int_\Gamma x(\lambda)\, d\lambda.$$

From the continuity and linearity of f, it follows that

$$f(y) = \int_\Gamma f(x(\lambda))\, d\lambda.$$

Since $f(x(\lambda))$ is a complex-valued analytic function in the region bounded by Γ and continuous on Γ, then, by the ordinary Cauchy integral theorem,

$$f(y) = 0.$$

Now, as in the preceding theorem, since every bounded linear functional vanishes on y, y must be zero or

$$\int_\Gamma x(\lambda)\, d\lambda = 0.$$

In analytic function theory, a result which follows immediately from Cauchy's integral theorem is Cauchy's integral formula, and in an identical manner we can establish, for $x(\lambda)$ as in the above theorem, the following formula for nth derivatives of $x(\lambda)$:

$$x^{(n)}(\lambda_0) = \frac{n!}{2\pi i} \int_\Gamma \frac{x(\lambda)}{(\lambda - \lambda_0)^{n+1}}\, d\lambda, \tag{19.1}$$

where λ_0 is any interior point of the region bounded by Γ.

Here, as in the more familiar situation, the existence of the nth derivative for any n is implied by the existence of the first derivative. An immediate consequence of the Cauchy integral formula is the Taylor series expansion of $x(\lambda)$ with (positive) radius of convergence given by only slightly modifying the usual Cauchy–Hadamard formula; that is, we have for R, the radius of convergence of the power series for $x(\lambda)$ about λ_0,

$$R = \frac{1}{\lim_n \sup \|x_n\|^{1/n}}, \tag{19.2}$$

where

$$x_n = \frac{x^{(n)}(\lambda_0)}{n!} \quad \text{and} \quad x(\lambda) = \sum_{n=0}^\infty x_n(\lambda - \lambda_0)^n.$$

19.2 Normed and Banach Algebras

A set X admitting three operations, scalar multiplication, vector addition, and vector multiplication, is called a complex *normed algebra* if (denoting scalar and vector multiplication by juxtaposition):

(1) X is a complex normed linear space with respect to scalar multiplication and vector addition;

(2) X is a ring with respect to vector addition and vector multiplication;

(3) for a scalar α and any two vectors x and y,

$$\alpha(xy) = (\alpha x)y = x(\alpha y);$$

(4) for any two vectors x and y,

$$\|xy\| \leq \|x\| \cdot \|y\|.$$

Note that in (3) the operations of internal (vector) and external (scalar) multiplication have been linked. Also note how very similar (4) is, formally, to the triangle inequality. If, in addition, X is a Banach space, then X is called a *Banach algebra*. Several examples of Banach algebras will now be presented.

EXAMPLE 19.1. In Theorem 14.10 we showed that, if Y was complete, then $L(X, Y)$ was complete too. In Eq. (15.6) we showed that a multiplication could be defined for members of $L(X, X)$. Thus, in view of Eqs. (15.7)–(15.9), $L(X, X)$ is seen to be a Banach algebra if X is a Banach space—a fact we shall have occasion to use many times.

EXAMPLE 19.2. Let X be the linear space $C[a, b]$ with sup norm. We define the product of $f(x)$, $g(x) \in C[a, b]$ as follows:

$$(fg)(x) = f(x)g(x).$$

EXAMPLE 19.3. Let W be the set of all absolutely convergent trigonometric series,

$$\sum_{n=-\infty}^{\infty} c_n e^{int},$$

and take, as the norm of any $x(t)$ in W,

$$\|x(t)\| = \sum_{n=-\infty}^{\infty} |c_n|.$$

As the product of two such series, we shall take the Cauchy product, being assured that this multiplication is a closed operation by the fact that the Cauchy product of absolutely convergent series is, again, an absolutely convergent series. The completeness of this space is evident once one realizes that, as a normed linear space, it is congruent to l_1.

EXAMPLE 19.4 Let A consist of all functions analytic in the open unit disc in the complex plane and continuous in the closed unit disc. We shall take as the norm of a function $f \in A$,

$$\|f\| = \sup_{|z| \le 1} |f(z)| = \sup_{|z| = 1} |f(z)|,$$

and as the product of $f, g \in A$, we define

$$(fg)(z) = f(z)g(z).$$

EXAMPLE 19.5. Let P_{n+1} denote the vector space of all polynomials (with complex coefficients) of degree less than or equal to n. Since this is a finite-dimensional space it must be complete (Theorem 8.7, Corollary 1). As the norm of

$$x(t) = a_0 + a_1 t + \cdots + a_n t^n,$$

we shall take

$$\|x(t)\| = \sum_{i=0}^{n} |a_i|,$$

and as the product of $x(t) = a_0 + a_1 t + \cdots + a_n t^n$ and $y(t) = b_0 + b_1 t + \cdots + b_n t^n$, we define

$$(xy)(t) = \sum_{k=0}^{n} c_k t^k,$$

where

$$c_k = \sum_{j+l=k} a_j b_l.$$

EXAMPLE 19.6. Suppose $L_1(-\infty, \infty)$ is the normed linear space and the product of $f, g \in L_1(-\infty, \infty)$ is defined as the *convolution of f and g*,

$$(f * g)(t) = \int_{-\infty}^{\infty} f(t - x)g(x)\, dx.$$

EXAMPLE 19.7. Suppose $G = \{\sigma_1, \sigma_2, \ldots, \sigma_n\}$ is a finite group. Consider the class of all complex-valued functions defined on G; that is, all f such that $f: G \to C$. Clearly, this collection of functions constitutes a linear space with scalar multiplication and vector addition defined in the same way as for $C[a, b]$. We now introduce the following norm on this space: Let $f: G \to C$. We define

$$\|f\| = \sum_{i=1}^{n} |f(\sigma_i)|.$$

As the product of two functions f and g mapping G into C, we take

$$(f * g)(\sigma_k) = \sum_{j=1}^{n} f(\sigma_k \sigma_j^{-1})g(\sigma_j),$$

and we now verify that $\|f * g\| \le \|f\|\|g\|$:

$$\|f * g\| = \sum_{k=1}^{n} |(f * g)(\sigma_k)|$$

$$= \sum_{k=1}^{n} \left| \sum_{j=1}^{n} f(\sigma_k \sigma_j^{-1})g(\sigma_j) \right|$$

$$\le \sum_{k=1}^{n} \sum_{j=1}^{n} |f(\sigma_k \sigma_j^{-1})| \cdot |g(\sigma_j)|.$$

Interchanging the order of summation in the last term, we see that it becomes

$$\sum_{j=1}^{n} |g(\sigma_j)| \sum_{k=1}^{n} |f(\sigma_k \sigma_j^{-1})| = \|g\|\,\|f\|;$$

thus,

$$\|f * g\| \le \|f\|\,\|g\|.$$

EXAMPLE 19.8. Let Z denote the set of all integers, and take $L_1(Z)$ to be all complex-valued functions f on Z such that

$$\sum_{n=-\infty}^{\infty} |f(n)| < \infty.$$

We define the product of two functions f and g, from $L_1(Z)$, $f * g$, as

$$(f*g)(n) = \sum_{m=-\infty}^{\infty} f(n-m)g(m),$$

and take as the norm of $f \in L_1(Z)$,

$$\|f\| = \sum_{n=-\infty}^{\infty} |f(n)|.$$

Many similarities between Examples 19.6, 19.7, and 19.8 are apparent: In each case multiplication was a "convolution-type" multiplication and the norms in each case were also very similar. Actually, through the use of an abstract integral, we could also have defined multiplication in 19.7 and 19.8 as convolution by appropriately assigning a measure to the set of all subsets of the space. The following example subsumes all three.

EXAMPLE 19.9. Let G be a locally compact topological group and let μ be a right Haar measure defined on G. (It can be shown, although it is by no means easy, that a right Haar measure can be defined on any locally compact topological group.) Denote by $L_1(G)$: all equivalence classes of complex-valued Borel measurable functions, f, such that $\int_G |f| \, d\mu < \infty$, and let scalar multiplication and addition of such classes be defined in formally the same way as it was for $L_1(a, b)$. As norm of $f \in L_1(G)$, we take

$$\|f\| = \int_G |f| \, d\mu$$

and as the product of two functions f and g, from $L_1(G)$, $f*g$, we define

$$(f*g)(x) = \int_G f(xy^{-1})g(y) \, d\mu(y).$$

In this case, aside from verifying that $\|f*g\| \leq \|f\| \|g\|$, we must also verify that the definition of multiplication is a sensible one; that is, it is necessary to show that the integrand above is measurable and that the integral exists for almost every x. The reader is referred to other sources (cf. Ref. 8) for the verification of these facts, however.

We now wish to show that in any normed algebra, ring multiplication is a continuous operation; that is, we wish to prove that the mapping

$$f : X \times X \to X,$$

$$\langle x, y \rangle \to xy,$$

is continuous. To do this we must show that, if $N(xy)$ is a neighborhood of xy in X (the range), there is a neighborhood of the pair $\langle x, y \rangle$ in $X \times X$ (the domain), W say, such that

$$f(W) \subset N(xy).$$

A suitable choice for W might be of the form $S_{\varepsilon_1}(x) \times S_{\varepsilon_2}(y)$, where ε_1 and ε_2 are chosen to satisfy the following requirement:

$$f(W) = f\big(S_{\varepsilon_1}(x) \times S_{\varepsilon_2}(y)\big) = S_{\varepsilon_1}(x)S_{\varepsilon_2}(y) \subset N(xy),$$

where
$$S_{\varepsilon_1}(x)S_{\varepsilon_2}(y) = \{zw \mid z \in S_{\varepsilon_1}(x), w \in S_{\varepsilon_2}(y)\}.$$

With these notions in mind we now proceed to the theorem.

THEOREM 19.3. In any normed algebra X, ring multiplication is continuous.

Proof. Let x_0 and y_0 be fixed points in X. Consider now

$$\|xy - x_0 y_0\| = \|(x - x_0)(y - y_0) + x_0(y - y_0) + (x - x_0)y_0\|$$
$$\leq \|x - x_0\| \|y - y_0\| + \|x_0\| \|y - y_0\| + \|y_0\| \|x - x_0\|.$$

The continuity of multiplication is now evident.

19.3 Banach Algebras with Identity

In this section we shall obtain a few facts about Banach algebras X with an *identity element*, that is, an element $e \in X$ such that $ex = xe = x$ for all $x \in X$; we shall also assume that $\|e\| = 1$. Once an identity is present, it makes sense to consider whether certain elements have inverses (multiplicative), and so on. Among other things, it will be shown here that the subset of elements with two-sided inverses (the *units* or *regular* elements) is an open subset of X; elements of X which do not possess two-sided inverses are called *singular*. We also apply some of our results to the special Banach algebra $L(X, X)$, where X is a Banach space.

It has been our algebraic experience that, in any structure, the identity element has an inverse, and our first theorem shows that in Banach algebras even things close to the identity have inverses.

THEOREM 19.4. Let X be a Banach algebra with identity. Then if $\|e - x\| < 1$, x^{-1} exists and is given by the formula

$$x^{-1} = e + \sum_{n=1}^{\infty} (e - x)^n.$$

REMARK. The above series is seen to be absolutely convergent by virtue of our assumption that $\|e - x\| < 1$. By Exercise 9.2, we see that, since we are in a Banach space, the series must converge.

Proof. Write
$$x = e - (e - x),$$
and consider the product

$$(e - (e - x))(e + (e - x) + (e - x)^2 + \cdots)$$
$$= e + (e - x) + (e - x)^2 + \cdots - (e - x) - (e - x)^2 - \cdots.$$

Since the series is absolutely convergent, if follows that the terms can be grouped in any way whatsoever, and the same sum will still be arrived at. Thus the above series is seen to converge to the identity element e, which proves the theorem.

COROLLARY. Let $x \in X$ and let λ be a scalar. Then if $|\lambda| > \|x\|$, $\lambda e - x$ is a unit (any element $z \in X$ such that $z^{-1} \in X$ will be called a *unit of X*) and $(\lambda e - x)^{-1} = \sum_{n=1}^{\infty} \lambda^{-n} x^{n-1}$.

Proof. We first note that, if w is a unit of X and α is any scalar not zero, then αw is a unit too, and $(\alpha w)^{-1} = \alpha^{-1} w^{-1}$. In view of this, since

$$\lambda e - x = \lambda\left(e - \frac{x}{\lambda}\right),$$

we lose nothing by showing that $[e - (x/\lambda)]$ is a unit. Taking the norm of $e - [e - (x/\lambda)]$, we have

$$\left\| e - \left(e - \frac{x}{\lambda}\right) \right\| = \frac{\|x\|}{|\lambda|},$$

which is less than 1 by hypothesis. Using the theorem, it follows that $[e - (x/\lambda)]$ is a unit. Using the formula given in the theorem, we have

$$\lambda e - x = \lambda(e - \lambda^{-1} x)$$

and

$$(\lambda e - x)^{-1} = \lambda^{-1}(e - \lambda^{-1} x)^{-1}$$

$$= \lambda^{-1}\left(e + \sum_{n=1}^{\infty} [e - (e - \lambda^{-1} x)]^n\right)$$

$$= \lambda^{-1}\left(e + \sum_{n=1}^{\infty} (\lambda^{-1} x)^n\right)$$

or, defining $x^0 = e$,

$$(\lambda e - x)^{-1} = \sum_{n=1}^{\infty} \lambda^{-n} x^{n-1}.$$

The proof is now complete.

Let X be a Banach space and consider the special Banach algebra $L(X, X)$ mentioned in the preceding section. As a direct application of the corollary to this special setting, we see that, if $A \in L(X, X)$ and λ is a scalar such that $|\lambda| > \|A\|$, then $(\lambda - A)^{-1}$ exists and is given by

$$(\lambda - A)^{-1} = \sum_{n=1}^{\infty} \lambda^{-n} A^{n-1}. \tag{19.3}$$

Further, since $(\lambda - A)^{-1} \in L(X, X)$, we see that $(\lambda - A)^{-1}$ is defined on all of X which implies the range of $(\lambda - A)$ is all of X, hence, certainly dense in X, and we conclude that λ belongs to the resolvent set of A, $\rho(A)$. It is to be emphasized that when we just say $(\lambda - A)^{-1}$ exists, we are saying it in the sense of Definition 18.1 and not in the sense that the inverse of $(\lambda e - x)$ existed in the corollary to Theorem 19.4; that is, the assertion that the inverse function exists is much weaker than saying that the algebraic inverse element in $L(X, X)$ exists.

We now show that the units of a Banach algebra X, with identity, form an open set. To do this we must show that, for any unit x, there is some ε-neighborhood of x containing only units.

Let U denote the units of X:

$$U = \{x \in X \mid x^{-1} \in X\}.$$

Clearly, $e \in U$ and, by virtue of Theorem 19.4, we see that

$$S_1(e) \subset U.$$

If x is any unit, then

$$xx^{-1} = e \in S_1(e).$$

Since ring multiplication is continuous (Theorem 19.3), there must be an $\varepsilon > 0$ such that

$$S_\varepsilon(x)x^{-1} \subset S_1(e) \subset U,$$

where

$$S_\varepsilon(x)x^{-1} = \{yx^{-1} \mid y \in S_\varepsilon(x)\}.$$

For any $y \in S_\varepsilon(x)$, we have $yx^{-1} \in U$, which implies yx^{-1} is a unit. Thus there must be some $z \in X$ such that

$$(yx^{-1})z = z(yx^{-1}) = e,$$

which implies that y is a unit whose inverse is given by $x^{-1}z$.

This implies that $y \in U$. But y was any point in $S_\varepsilon(x)$, and x was any element of U, so that every point of U is contained in a neighborhood that lies within U; therefore U is an open set. *Thus the set of units U is an open subset of the Banach algebra X.*

Two new definitions are now necessary. In each it is assumed that X is a normed algebra with identity and U is the set of units of X.

DEFINITION 19.2. Let λ be a scalar. If $x - \lambda e$ is a unit, then λ is called a *regular point of x.*

DEFINITION 19.3. The set of nonregular points of x is called the *spectrum of x* and will be denoted by $\sigma(x)$. Thus if $\lambda \in \sigma(x)$, then $(x - \lambda e) \notin U$.

Suppose X is a Banach space and consider the element A from the Banach algebra $L(X, X)$. We wish to show now that no ambiguity arises when one considers the spectrum of A using Definition 19.3 or Definition 18.6. To prove this, we show that the complements of these two sets are equal; that is, we show that the set of regular points of A is just the resolvent set of A.

First, suppose that λ is a regular point of A. This is equivalent to saying $(\lambda - A)^{-1} \in L(X, X)$ which, in turn, means that (1) $(\lambda - A)^{-1}$ exists, and (2) $(\lambda - A)^{-1}$ is a bounded linear transformation, and last, since $(\lambda - A)^{-1}$ is defined on all of X, (3) $R(\lambda - A) = X$. Thus $\lambda \in \rho(A)$.

Conversely, suppose $\lambda \in \rho(A)$. Then $(\lambda - A)^{-1}$ must be bounded, and the range of $\lambda - A$, $R(\lambda - A)$, must be dense in X. To conclude that λ is a regular point, however, we must show that $(\lambda - A)^{-1} \in L(X, X)$, or equivalently that $R(\lambda - A) = X$. Since we already know $R(\lambda - A)$ to be dense in X, it suffices now to prove that the range of $\lambda - A$ is closed. Clearly, $\lambda - A$ is a bounded linear transformation and, since it is defined on all of X, is therefore closed (Corollary to Theorem 16.2).

Since it is closed, its inverse must also be closed (Theorem 16.3). Since, by hypothesis, $(\lambda - A)^{-1}$ is bounded we now have

$$(\lambda - A)^{-1} : R(\lambda - A) \to X,$$

where $(\lambda - A)^{-1}$ is a closed and bounded linear transformation. Using the fact that X is a Banach space and Theorem 16.4, we conclude that $R(\lambda - A)$ is closed or that

$$R(\lambda - A) = \overline{R(\lambda - A)} = X.$$

Therefore, λ is a regular point of A, and it follows that the set of regular points of A coincides with the resolvent set of A. Moreover, it is now evident, by taking complements, that the spectrum of A in the sense of Definition 19.3 agrees with the notion of spectrum defined in Definition 18.6.

Compactness of the Spectrum

Suppose now that X is a Banach algebra with identity, and denote the units of X by U. For any given element $x \in X$, we shall now show that the spectrum of x is a compact set. First, it will be shown that the regular points of x are an open set.

Let λ_0 be a regular point of x. This means that

$$x_{\lambda_0} = x - \lambda_0 e \in U.$$

Since U is an open set, there exists a neighborhood of x_{λ_0}, $S_\varepsilon(x_{\lambda_0})$, such that

$$S_\varepsilon(x_{\lambda_0}) \subset U.$$

Consider the function $x_\lambda = x - \lambda e$; that is,

$$x_\lambda : C \to X,$$

$$\lambda \to x - \lambda e.$$

This function is continuous for certainly $\lambda \to \lambda_0$ implies $x - \lambda e \to x - \lambda_0 e$. Since it is continuous, given a neighborhood of x_{λ_0},

$$S_\varepsilon(x_{\lambda_0}) \subset U,$$

there must exist a neighborhood of λ_0, $S_\delta(\lambda_0) \subset C$ such that

$$x_\lambda \in S_\varepsilon(x_{\lambda_0}) \subset U \qquad \text{for all } \lambda \in S_\delta(\lambda_0).$$

Thus, for all such λ, $x_\lambda \in U$. In view of this, all points in $S_\delta(\lambda_0)$ are regular points; hence the set of regular points is an open set, because every point is contained in a neighborhood lying wholly within the set of regular points. Summarizing, we have:

The regular points of $x \in X$, where X is a Banach algebra with identity, are an open set in the complex plane.

This implies that the $\sigma(x)$ is a closed set. We know by Theorem 19.4 [see also Eq. (19.3)] that if $|\lambda| > \|x\|$, then $(x - \lambda e)^{-1} \in X$ (that is, λ is regular) which implies that $\sigma(x)$ must be wholly contained in the closed disk of radius $\|x\|$ or, symbolically, that

$$\sigma(x) \subset \overline{S_{\|x\|}(0)}.$$

We have now established that $\sigma(x)$ is a closed and bounded set in the complex plane; it follows by the Heine–Borel theorem that $\sigma(x)$ is compact.

19.4 An Analytic Function—the Resolvent Operator

As in the preceding section, X will denote a Banach algebra with identity and we shall denote the (open) set of regular points of a given element $x \in X$ by C_R. In this case the mapping

$$x : C_R \to X,$$

$$\lambda \to (x - \lambda e)^{-1}$$

clearly makes sense, and the vector-valued function $x(\lambda) = (x - \lambda e)^{-1}$ is called the *resolvent function associated with* x, or, more simply, the *resolvent of* x. It is our goal in this section to show that this function is an analytic function of λ (with domain C_R) in the sense of analyticity defined in Definition 19.1.

First we show that taking inverses is a continuous operation.

THEOREM 19.5. With U denoting the units of X, the mapping

$$f : U \to U$$

$$x \to x^{-1}$$

is continuous.

Proof. It suffices to show that $x_n \to x$ implies $x_n^{-1} \to x^{-1}$. To this end let $\{x_n\}$ be a sequence of points from U converging to $x \in U$. Thus,

$$x_n \to x,$$

which implies

$$x^{-1} x_n \to e,$$

so that for any $\varepsilon > 0$ there must be an N such that

$$\|x^{-1} x_n - e\| < \varepsilon \qquad \text{(for } n > N\text{)}.$$

Choose N_1 now such that $\|x^{-1}x_n - e\| < 1$ for $n > N_1$ and consider the series

$$e + \sum_{k=1}^{\infty} (e - x^{-1}x_n)^k \qquad \text{(for } n > N_1\text{)}.$$

By Theorem 19.4, since $\|e - x^{-1}x_n\| < 1$, the series must converge absolutely to $(x^{-1}x_n)^{-1} = x_n^{-1}x$. We now have

$$\|e - x_n^{-1}x\| \le \sum_{k=1}^{\infty} \|(e - x^{-1}x_n)^k\| = \sum_{k=1}^{\infty} \|x^{-k}(x - x_n)^k\|$$

$$\le \sum_{k=1}^{\infty} \|x^{-1}\|^k \|x - x_n\|^k = (*).$$

Since $x_n \to x$, we can choose N_2 such that $n > N_2$ will make (*) as small as desired; hence

$$\|e - x_n^{-1}x\| \to 0$$

or

$$x_n^{-1}x \to e$$

or

$$x_n^{-1} \to x^{-1},$$

and the continuity of taking inverses has been proved.

We now establish a certain equality and then, using the equality and the above theorem, will prove the analyticity of the resolvent operator.

Suppose $x(\lambda)$ is the resolvent of the element $x \in X$. If λ_1 and λ_2 are any two regular points, then

$$x(\lambda_2) - x(\lambda_1) = (\lambda_2 - \lambda_1)x(\lambda_1)x(\lambda_2). \tag{19.4}$$

To prove this, consider

$$x(\lambda_1)^{-1}x(\lambda_2) = (x - \lambda_1 e)x(\lambda_2)$$

$$= ((x - \lambda_2 e) + (\lambda_2 e - \lambda_1 e))x(\lambda_2)$$

$$= e + (\lambda_2 - \lambda_1)x(\lambda_2).$$

This implies

$$x(\lambda_2) = x(\lambda_1) + (\lambda_2 - \lambda_1)x(\lambda_1)x(\lambda_2),$$

and (19.4) follows.

We now proceed to the theorem.

THEOREM 19.6. The resolvent of $x \in X$, $x(\lambda)$, is analytic in the set of regular points.

Proof. Let λ and λ_0 be regular points. By the preceding equality [(19.4)], it follows that

$$\lim_{\lambda \to \lambda_0} \frac{x(\lambda) - x(\lambda_0)}{\lambda - \lambda_0} = \lim_{\lambda \to \lambda_0} x(\lambda_0)x(\lambda) = x(\lambda_0) \lim_{\lambda \to \lambda_0} x(\lambda). \tag{19.5}$$

Since, as $\lambda \to \lambda_0$, $x - \lambda e \to x - \lambda_0 e$ then, using the continuity mentioned in Theorem 19.5, it follows that

$$\lim_{\lambda \to \lambda_0} (x - \lambda e)^{-1} = (x - \lambda_0 e)^{-1} = x(\lambda_0);$$

therefore the limit in (19.5) exists and

$$x'(\lambda_0) = x(\lambda_0)^2;$$

the analyticity of $x(\lambda)$ now follows.

We now show that the spectrum of any element in X is nonempty; thus, for some λ, the element x can be "translated" to the nonunit $x - \lambda e$.

THEOREM 19.7. For any $x \in X$, $\sigma(x) \neq \emptyset$.

Proof. We will show that the assumption $\sigma(x) = \emptyset$ leads to contradictory results.

By Theorem 19.6, we know that the function $x(\lambda) = (x - \lambda e)^{-1}$ is an analytic function over the set of regular points of X. If $\sigma(x) = \emptyset$, then $x(\lambda)$ is analytic everywhere, hence entire. Clearly,

$$e - \frac{x}{\lambda} \to e$$

as $|\lambda| \to \infty$, which implies, by Theorem 19.5 that

$$\left(e - \frac{x}{\lambda}\right)^{-1} \to e^{-1} = e.$$

Consequently,

$$\|(x - \lambda e)^{-1}\| = |\lambda^{-1}| \, \|\left(e - \frac{x}{\lambda}\right)^{-1}\| \to \frac{\|e\|}{|\lambda|} \to 0 \qquad (19.6)$$

as $|\lambda| \to \infty$. Therefore, for any given $M_1 > 0$, there is some $R > 0$ such that

$$\|(x - \lambda e)^{-1}\| < M_1$$

for $|\lambda| > R$. Letting

$$M_2 = \max_{|\lambda| \le R} \|x(\lambda)\|,$$

we see that $x(\lambda) = (x - \lambda e)^{-1}$ is bounded everywhere by the larger of M_1 and M_2. By Liouville's theorem (Theorem 19.1), we see that $x(\lambda)$, since it is a bounded, entire function, must be a constant, and Eq. (19.6) implies that this constant must be zero. This is absurd, however, for if $(x - \lambda e)^{-1} = 0$, how can

$$(x - \lambda e)(x - \lambda e)^{-1} = e?$$

Thus our assumption that $\sigma(x) = \emptyset$ has led to contradictory results, and it now follows that, for any $x \in X$, $\sigma(x) \neq \emptyset$.

As a special case, if $A \in L(X, X)$, where X is a complex Banach space, we see that $\sigma(A) \neq \emptyset$.

Our next theorem gives us some insight into the structure of certain types of Banach algebras. It asserts that if X is a complex Banach algebra with identity such that an inverse exists for every nonzero element, then X is isomorphic to the complex numbers C. Such algebras (that is, algebras such that the nonzero elements form a group with respect to vector multiplication) are called *division algebras*.

THEOREM 19.8. (Gelfand). Let X be a complex Banach algebra with identity. If X is a division algebra, then X is isomorphic to the complex numbers.

Proof. Let $x \in X$. The preceding theorem states that $\sigma(x) \neq \emptyset$, or that there must be some $\lambda \in C$ such that $x - \lambda e$ is not invertible. But we are in a division algebra. This means that all elements except zero have inverses; hence,

$$x - \lambda e = 0 \qquad \text{or} \qquad x = \lambda e.$$

We see now that every $x \in X$ can be written as a unique scalar multiple of the identity. The following mapping now suggests itself as the isomorphism mentioned in the theorem:

$$f : X \to C,$$

$$x = \lambda e \to \lambda.$$

The mapping is easily seen to be 1 : 1, linear, and multiplicative. For any scalar λ, the vector λe maps into λ, which proves the ontoness of f and, therefore, establishes the desired isomorphism.

We state without proof that if a X is a *real* Banach algebra that is also a division algebra then X is isomorphic either to the real numbers, to the complex numbers, or to the quaternions. Thus, the real case is much more complicated indeed.

19.5 Spectral Radius and the Spectral Mapping Theorem for Polynomials

Suppose X is a complex Banach algebra with identity. In Sec. 19.3 we saw that the spectrum of $x \in X$ was wholly contained in the closed sphere $\overline{S_{\|x\|}(0)}$. It is possible that $\sigma(x)$ might be contained in some smaller circle about the origin, however, and the radius of the "smallest" circle containing $\sigma(x)$ is referred to as the *spectral radius of x*. More precisely, the *spectral radius of x*, denoted by $r_\sigma(x)$, is defined to be

$$r_\sigma(x) = \sup_{\lambda \in \sigma(x)} |\lambda|. \tag{19.7}$$

Presently, we shall show that an alternate way of computing $r_\sigma(x)$ is given by

$$r_\sigma(x) = \lim_n \|x^n\|^{1/n}. \tag{19.8}$$

An important result that will aid us in proving (19.8) is the following theorem, known as the *spectral mapping theorem for polynomials*.

THEOREM 19.9. Let X be a complex normed algebra with identity, let $x \in X$, and let $p(t)$ be a polynomial with complex coefficients. Then $\sigma(p(x)) = p(\sigma(x))$.

Proof. The proof will proceed by stages. First we shall treat constant polynomials. To this end let $p(t) = \alpha_0 = \alpha_0 t^0$, and compute

$$\sigma(p(x)) = \sigma(\alpha_0 e) = \{\lambda \mid (\alpha_0 e - \lambda e)^{-1} \notin X\}$$

$$= \{\alpha_0\}.$$

Now consider

$$p(\sigma(x)) = \{p(\lambda) \mid \lambda \in \sigma(x)\}$$

$$= \{\alpha_0 \lambda^0 \mid \lambda \in \sigma(x)\}$$

$$= \{\alpha_0\}.$$

Therefore, in this case,

$$\sigma(p(x)) = p(\sigma(x)).$$

Before proceeding to the case of nonconstant polynomials, we wish to show that for any vector z and any scalar α,

$$\sigma(\alpha z) = \alpha\sigma(z). \tag{19.9}$$

The result is clear when $\alpha = 0$. Suppose now that $\alpha \neq 0$. Then

$$\lambda \in \sigma(\alpha z)$$

$$\Leftrightarrow \quad \alpha z - \lambda e \quad \text{is not invertible,}$$

$$\Leftrightarrow \quad z - \frac{\lambda}{\alpha} e \quad \text{is not invertible,}$$

$$\Leftrightarrow \quad \frac{\lambda}{\alpha} \in \sigma(z),$$

$$\Leftrightarrow \quad \lambda \in \alpha\sigma(z).$$

Seeing that (19.9) holds, we now need only consider polynomials with leading coefficient equal to 1. Suppose that

$$p(t) = t^n + \alpha_{n-1}t^{n-1} + \cdots + \alpha_1 t + \alpha_0,$$

where $n \geq 1$, and consider, $p(t) - \lambda$ for any $\lambda \in C$. Since C is an algebraically closed field, $p(t) - \lambda$ must factor completely into linear terms:

$$p(t) - \lambda = (t - \beta_1)(t - \beta_2) \cdots (t - \beta_n). \tag{19.10}$$

Substituting x for t we get†

$$p(x) - \lambda e = (x - \beta_1 e)(x - \beta_2 e) \cdots (x - \beta_n e). \tag{19.11}$$

If $\lambda \in \sigma(p(x))$, then one of the $x - \beta_j e$ must be noninvertible or, equivalently, there must be some j such that

$$\beta_j \in \sigma(x),$$

which implies

$$p(\beta_j) \in p(\sigma(x)) = \{p(\lambda) \mid \lambda \in \sigma(x)\}. \tag{19.12}$$

Substituting β_j for t in (19.10), we see the $p(\beta_j) = \lambda$ and (19.12) becomes

$$\lambda \in p(\sigma(x)),$$

and we have shown that

$$\sigma(p(x)) \subset p(\sigma(x)).$$

---

† The reason for the preservation of form between Eqs. (19.10) and (19.11) lies in observing that the substitution mapping

$$C[t] \to C[x],$$
$$p(t) \to p(x),$$

is a ring homomorphism, where

$$C[t] = \left\{ p(t) = \sum_{i=0}^{n} \alpha_i t^i \mid \alpha_i \in C, n = 0, 1, 2, \ldots \right\}$$

$$\text{and} \quad C[x] = \left\{ p(x) = \alpha_0 e + \sum_{i=1}^{n} \alpha_i x^i \mid \alpha_i \in C, n = 1, 2, \ldots \right\}$$

To complete the proof, we must get the above inclusion going the other way around. To this end, suppose $\lambda \in p(\sigma(x))$. By the definition of $p(\sigma(x))$ [Eq. (19.12)], there must be some $\gamma_j \in \sigma(x)$ such that $\lambda = p(\gamma_j)$. Now form

$$p(t) - \lambda = (t - \gamma_1) \cdots (t - \gamma_j) \cdots (t - \gamma_n),$$

it being clear that γ_j is a root of $p(t) - \lambda$. Substituting x for t, we get

$$p(x) - \lambda e = (x - \gamma_1 e) \cdots (x - \gamma_j e) \cdots (x - \gamma_n e). \tag{19.13}$$

Note now that in any complex normed algebra, commutative or not, elements of the form $x - \alpha e$, $x - \beta e$, where α and β are scalars, must always commute. In view of this, if $\lambda \notin \sigma(p(x))$, that is, if $p(x) - \lambda e$ were invertible, we could multiply both sides of (19.13) on the left by $(p(x) - \lambda e)^{-1}$ and, by the commutativity just mentioned, move $(x - \gamma_j e)$ all the way to the right to get

$$e = (p(x) - \lambda e)^{-1}[(x - \gamma_1 e) \cdots (x - \gamma_n e)](x - \gamma_j e) \tag{19.14}$$

to conclude that $(x - \gamma_j e)$ has a left inverse. Similarly, we could show that $(x - \gamma_j e)$ has right inverse (hence an inverse) which contradicts the fact that $\gamma_j \in \sigma(x)$. We therefore conclude that $\lambda \in \sigma(p(x))$, which implies that

$$p(\sigma(x)) \subset \sigma(p(x)),$$

and completes the proof.

COROLLARY. Let X be a complex normed linear space and consider the complex normed algebra with identity $L(X, X)$. If $A \in L(X, X)$ then, letting $p(t)$ be any polynomial with complex coefficients,

$$\sigma(p(A)) = p(\sigma(A)).$$

Suppose now that X is a complex Banach algebra with identity and that $x \in X$. Clearly, by the above theorem, $\sigma(x^n) = \sigma(x)^n$. Using this, we now show that $r_\sigma(x^n) = r_\sigma(x)^n$. Consider

$$r_\sigma(x^n) = \sup_{\lambda \in \sigma(x^n)} |\lambda| = \sup_{\lambda \in \sigma(x)^n} |\lambda|$$

$$= \sup_{\mu \in \sigma(x)} |\mu|^n$$

$$= \left(\sup_{\mu \in \sigma(x)} |\mu| \right)^n$$

$$= r_\sigma(x)^n. \tag{19.15}$$

By the corollary to Theorem 19.4, it follows that

$$r_\sigma(x^n) \leq \|x^n\|.$$

Now we can write

$$r_\sigma(x)^n \leq \|x^n\|$$

or

$$r_\sigma(x) \leq \|x^n\|^{1/n} \quad \text{(for any } n\text{)}.$$

Since this is true for any n, taking the inferior limit of both sides yields

$$r_\sigma(x) \leq \underline{\lim}_n \|x^n\|^{1/n}. \tag{19.16}$$

At present we must content ourselves with this statement, that is, taking the inferior limit instead of the limit, because we have no assurance that the limit, $\lim_n \|x^n\|^{1/n}$ exists, whereas the inferior limit always exists. Since, in general, the inferior limit of a sequence is always less than or equal to the superior limit, if we can show that

$$r_\sigma(x) = \overline{\lim_n} \|x^n\|^{1/n} \tag{19.17}$$

we shall have that

$$\overline{\lim_n} \|x^n\|^{1/n} \leq \underline{\lim}_n \|x^n\|^{1/n},$$

which implies that $\lim_n \|x^n\|^{1/n}$ exists and $r_\sigma(x) = \lim_n \|x^n\|^{1/n}$. Our next theorem, among other things, shows Eq. (19.17) to hold.

THEOREM 19.10. Let X be a complex Banach algebra with identity and let $x \in X$. Then

(1) for $|\lambda| > r_\sigma(x)$, $(\lambda e - x)^{-1}$ exists and is given by

$$\sum_{n=1}^{\infty} \lambda^{-n} x^{n-1}; \tag{19.18}$$

(2) if the series (19.18) converges for $|\lambda| = r_\sigma(x)$, then it represents $(\lambda e - x)^{-1}$;
(3) the series (19.18) diverges for $|\lambda| < r_\sigma(x)$.

Proof. A well-known result from analytic function theory is the following: If both power series

$$\sum_{n=0}^{\infty} a_n(z - z_0)^n \quad \text{and} \quad \sum_{n=0}^{\infty} b_n(z - z_0)^n$$

have a positive radius of convergence, and if their sums coincide for all points of a neighborhood of z_0, or only for an infinite number of such points (distinct from one another and from z_0) with the limit point z_0, then they are identical. This result is known as the *identity theorem for power series*,† and it is a simple matter to extend it to the case of power series with vector-valued coefficients such as the series of (19.18). With this result in mind, we observe that if λ is such that $|\lambda| > r_\sigma(x)$, then λ must be a regular point of x [if not, if there was some λ outside the closed circle of radius $r_\sigma(x)$ that was a spectral point of x, this would contradict the way in which $r_\sigma(x)$ was obtained]. Thus, by Theorem 19.6, the function $(\lambda e - x)^{-1}$ is analytic everywhere outside the closed disk of radius $r_\sigma(x)$, and as such can be expanded in a Taylor series, $\sum_{n=0}^{\infty} y_n \lambda^{-n}$, in powers of $1/\lambda$ about the point ∞. On the other hand, by the corollary to Theorem 19.4, $(\lambda e - x)^{-1}$ is given by (19.18) for $|\lambda| > \|x\|$. By the identity theorem mentioned above, it follows that the Taylor expansion of $(\lambda e - x)^{-1}$ is identical with (19.18); moreover, since the radius of convergence can be characterized as the distance to the nearest singular point, this series in $1/\lambda$ must converge at least for all $|\lambda| > r_\sigma(x)$. This proves part (1).

---

† See, for example, Knopp, *Theory of Functions, Part 1*. The above theorem is given as Theorem 2, p. 81.

Proof (2). Suppose the series (19.18) converges to y. We will now show that $(\lambda e - x)y = e$. To this end, we compute

$$(\lambda e - x)y = \sum_{n=1}^{\infty} \lambda^{-n+1} x^{n-1} - \sum_{n=1}^{\infty} \lambda^{-n} x^n$$

$$= e + \sum_{n=2}^{\infty} \lambda^{-(n-1)} x^{n-1} - \sum_{n=1}^{\infty} \lambda^{-n} x^n$$

$$= e + \sum_{n=1}^{\infty} \lambda^{-n} x^n - \sum_{n=1}^{\infty} \lambda^{-n} x^n$$

$$= e.$$

It is equally simple to verify that $y(\lambda e - x) = e$. Therefore part (2) is proved.

Proof (3). Compare the statement of part (2) to the proof of part (2). Note that nowhere in the proof of part (2) did we make use of the fact that $|\lambda| = r_\sigma(x)$. What we have really obtained in the above proof is the fact that anywhere the series (19.18) converges, it represents the inverse of $(\lambda e - x)$. Now if there were a λ_0, $|\lambda_0| < r_\sigma(x)$, such that (19.18) converged, it follows that it converges for all λ, such that $|\lambda| > |\lambda_0|$, by virtue of the fact that (19.18) is a *power series*. (Note that, if the power series, $\sum_{n=0}^{\infty} a_n z^n$ converges for $z = z_0$, it must converge for all z such that $|z| < |z_0|$.) Using the proof of part (2), this means that all points outside the circle of radius $|\lambda_0|$ are regular and contradicts the way in which $r_\sigma(x)$ was chosen. Therefore, the series diverges for all λ whose absolute value is less than $r_\sigma(x)$. The proof of part (3) is now complete.

Combining parts (1) and (3) of this theorem, we see that $r_\sigma(x)$ is the radius of convergence of the series (19.18). We can compute the radius of convergence of any power series, however, by the Cauchy–Hadamard formula. This yields

$$\text{rad. of conv. of (19.18)} = \overline{\lim_n} \|x^{n-1}\|^{1/n} = \overline{\lim_n} \|x^n\|^{1/n}.$$

Hence,

$$r_\sigma(x) = \overline{\lim_n} \|x^n\|^{1/n}. \tag{19.19}$$

Combining this with (19.16), we see that

$$\lim_n \|x^n\|^{1/n} \tag{19.20}$$

exists (we have shown $\underline{\lim}_n \|x^n\|^{1/n} = \overline{\lim}_n \|x^n\|^{1/n}$) and

$$r_\sigma(x) = \lim_n \|x^n\|^{1/n}. \tag{19.21}$$

We now observe that, if X is a Banach space and $A \in L(X, X)$, then

$$r_\sigma(A) = \lim_n \|A^n\|^{1/n}. \tag{19.22}$$

19.6 The Gelfand Theory

We assume throughout this entire section,† unless otherwise stated, that X is a complex commutative Banach algebra with an identity e such that $\|e\| = 1$.

DEFINITION 19.4. A nonempty subset I of X is called an *ideal* if

(a) I is a subspace of X;
(b) for any $x \in X$ and $y \in I$, $xy \in I$.

We observe that, in defining an ideal, it would suffice to demand that for $x, y \in I$, $x + y \in I$ and for $x \in X$, $y \in I$, $xy \in I$. It then follows that, for $\lambda \in C$ and $x \in X$, $\lambda e \cdot x = \lambda x \in I$ and so I must indeed be a subspace.

Any algebra X certainly contains the two ideals $\{0\}$ and X itself. An ideal $I \neq X$ is called a *proper* ideal.

THEOREM 19.11. Let I be a proper ideal of X; then $\bar{I}$ is also a proper ideal.

Proof. We first show that $\bar{I}$ is an ideal. Let $x, y \in \bar{I}$. Then there exist sequences $\{x_n\}$ and $\{y_n\}$ such that

$$x_n \to x \qquad \text{and} \qquad y_n \to y \qquad (x_n, y_n \in I).$$

By the continuity of vector addition, $x_n + y_n \to x + y$, which implies that $x + y \in \bar{I}$. Similarly, one sees that $x \in \bar{I}$ and $z \in X$ implies that $zx \in \bar{I}$, so $\bar{I}$ is an ideal.

Next we show that $\bar{I}$ is a proper ideal. If $x \in I$, certainly x is not a unit, for otherwise $x^{-1} \in X$; whence $e = x^{-1}x \in I$, which clearly implies that the ideal $I = X$—a contradiction. Thus $x \in I$ implies that x is singular, so $I \subset CU = S$, the set of singular elements of X, but this is a closed proper subset of X (Sec. 19.3). Hence $\bar{I} \subset S$ and $\bar{I}$ is also a proper ideal.

We now wish to consider certain special ideals which will play a fundamental role in what is to follow.

DEFINITION 19.5. A proper ideal M of X is called *maximal* if it is not properly contained in any proper ideal of X.

It is clear that any maximal ideal is closed, for, if M is maximal, it is, in particular, proper and therefore, by Theorem 19.11 $\bar{M}$ is a proper ideal. If $M \subset \bar{M}$ properly, this would contradict the fact that M is a maximal ideal; hence, $M = \bar{M}$, so M is closed.

THEOREM 19.12. Let I be a proper ideal of X (a commutative, complex, Banach algebra, as usual, with identity). Then there exists a maximal ideal M such that $I \subset M$.

---

† The results of this section may be omitted on a first reading.

Proof. Let S be the set of all proper ideals of X which contain I. S is, of course, partially ordered with respect to set inclusion. It is, moreover, inductively ordered, for if $T = \{I_\alpha\}_{\alpha \in \Lambda}$ is any totally ordered subset of S, then $U_\alpha I_\alpha \supset I$ is an upper bound for T; it also is in S, since it is an ideal by the total ordering and is proper, for, if $U_\alpha I_\alpha = X$, then $e \in I_\alpha$ for some α which contradicts the fact that I_α is proper. Thus S is inductively ordered and by Zorn's lemma (Appendix to Chapter 9) has a maximal element $M \supset I$ and M is clearly a maximal ideal.

COROLLARY. If x is a singular element of X, there exists a maximal ideal that contains x.

Proof. We apply the preceding theorem to the proper ideal $I = (x) = \{yx | y \in X\}$. We note that this is clearly an ideal and is proper, for if $(x) = X$ for some $y \in X$, $e = yx$, contradicting the fact that x is singular.

In Secs. 8.6 and 8.7 we showed that, if X was a normed linear space and M was a closed subspace of X, then X/M could also be viewed as a normed linear space, by defining addition and scalar multiplication in X/M as:

$$(x + M) + (y + M) = (x + y) + M; \qquad \alpha(x + M) = \alpha x + M.$$

In the event that X is an *algebra* and M is an ideal in X, a further operation can be introduced—that of multiplication:

$$(x + M)(y + M) = xy + M.$$

The above operation is well-defined (independent of the particular choice of representatives x and y), because M is an ideal and X/M with respect to the three operations noted above is actually an algebra (indeed, a normed algebra with respect to the norm mentioned in Sec. 8.6)—a fact we leave for the reader to verify. Moreover, as shown in Sec. 8.7, X/M is complete when X is.

Suppose for the moment we forget about scalar multiplication in X/M and just view it as a ring. In this case we claim that *if M is a maximal ideal, then X/M is a field.*

Proof. Assuming the reader has already verified that X/M is a ring, it only remains to verify that the nonzero elements of X/M form a multiplicative group, that is, for $x + M \neq 0 + M$ (equivalently, $x \notin M$) we must show that $x + M$ has a multiplicative inverse. To this end, let I be the ideal generated by $M \cup \{x\}$ (the smallest ideal containing M and x, that is). Since $e \in X$, it is clear that

$$I = \{y + zx \mid y \in M, z \in X\}.$$

Now I contains M properly (equivalently: $x \notin M$):, since $x \in I$ but $x \notin M$. Therefore, since M is maximal, $I = X$ and, since $e \in X$, there must exist elements $y_0 \in M$ and $z_0 \in X$ such that

$$e = y_0 + z_0 x,$$

which implies that

$$e + M = (y_0 + z_0 x) + M$$
$$= (y_0 + M) + (z_0 x + M) = M + (z_0 x + M)$$
$$= z_0 x + M = (z_0 + M)(x + M),$$

and $z_0 + M$ is the inverse of $x + M$, which completes the proof.

We must now pause to prove the following elementary fact: Let $f: X \to Y$ be a homomorphism of the complex algebra X onto the complex algebra Y (that is, f is linear and f is multiplicative in the sense that $f(xy) = f(x)f(y)$ for all $x, y \in X$). Then Y is isomorphic to X/K, denoted by $Y \simeq X/K$, where K is the kernel of the homomorphism f. More precisely, *we can decompose the homomorphism f into the product of two mappings $f = gh$, where $h : X \to X/K$ is a homomorphism onto and $g : X/K \to Y$ is an isomorphism onto.*

Proof. Consider the kernel of f, K:

$$K = \{x \in X \mid f(x) = 0\}.$$

K is clearly an ideal, and X/K is therefore an algebra. Consider the mapping

$$X \overset{h}{\to} X/K \overset{g}{\to} Y,$$

$$x \to x + K \to f(x).$$

The mapping h is clearly a well-defined homomorphism of the algebra X onto the algebra X/K. The map g is well-defined, for, if $z \in K$, then

$$g(x + z + K) = f(x + z) = f(x) + f(z) = f(x).$$

Moreover, g is clearly a homomorphism onto Y. We contend that it is an isomorphism, for suppose $g(x + K) = 0$, that is, $f(x) = 0$, which implies $x \in K$ and therefore $x + K = 0 + K$; it follows that g is indeed an isomorphism, so $Y \simeq X/K$. Finally, it is observed that

$$g(h(x)) = g(x + K) = f(x),$$

so that $f = gh$.

Returning once more to consideration of X/M as an algebra instead of just a ring, we see that X/M is certainly a division algebra if M is a maximal ideal, and so by the Gelfand theorem (Theorem 19.8), it must differ from the complex numbers in name only; more precisely, we have the following isomorphism $x + M \to x(M)$ [$x(M)$ is described below]:

$$X \to X/M \longrightarrow C,$$
$$x \to x + M = x(M)(e + M) \to x(M). \qquad (19.23)$$

As for $x(M)$, it was shown in the Gelfand theorem that in such a division algebra every element was a scalar multiple of the identity. In our case the identity element is $(e + M)$ and the scalar multiple λ of $e + M$ such that $x + M = \lambda(e + M)$ we denote by $x(M)$; we shall continue to use this notation throughout the rest of the discussion.

In addition to being an isomorphism, the mapping $x + M \to x(M)$ is also an isometry. To see this, we first prove that

$$\|e + M\| = 1. \tag{19.24}$$

Clearly, by the definition of the norm in X/M,

$$\|e + M\| \le \|e\| = 1. \tag{19.25}$$

Now let $y \in e + M$, so $y = e + x$, where $x \in M$, or $y = e - (-x)$. If $\|y\| < 1$, we have

$$\|e - (-x)\| < 1,$$

which implies, using Theorem 19.4 that $-x$ and, consequently, x is invertible; but this is impossible since $x \in M$ and M is a proper ideal because it was assumed to be maximal. In summary, $y \in e + M$ implies $\|y\| \ge 1$, so $\|e + M\| = 1$, and, combining this with (19.25), Eq. (19.24) is established.

With regard to isometry, we now note that

$$x + M = x(M)(e + M)$$

so

$$\|x + M\| = \|x(M)(e + M)\| = |x(M)| \|e + M\| = |x(M)|.$$

The set of all maximal ideals of X will henceforth be denoted by $\mathcal{M}$. The above considerations show that each $M \in \mathcal{M}$ determines a homomorphism of X onto C. Conversely, each nontrivial homomorphism of X onto C (nontrivial meaning not identically equal to 0) determines a maximal ideal M, namely, its kernel: the set of elements mapped into zero by the homomorphism. Moreover, X/M is isomorphic to C. (This is a special case of the fundamental theorem of homomorphisms for groups with operators, usually considered in an algebra course, but again not difficult to prove.) Under the isomorphism between X/M and C, we have $e + M$ corresponding to 1 so that $x(M)(e + M) \leftrightarrow x(M)$, which shows, together with our previous considerations that there is a $1:1$ correspondence between mappings of the type (19.23) and homomorphisms of X onto C.

Consider now an arbitrary, but fixed, element of X, x. The element x determines a complex-valued function $\hat{x}$ on $\mathcal{M}$ given by

$$\hat{x} : \mathcal{M} \to C, \tag{19.26}$$

$$M \to x(M).$$

Thus we obtain a mapping $x \to \hat{x}$ of X into $F(\mathcal{M})$, the set of all complex-valued functions defined on $\mathcal{M}$. We list several properties of this mapping, many of which we have already established, in the following theorem.

THEOREM 19.13. The mapping $x \to \hat{x}$ given by (19.26) has the following properties:

(a) if $x = x_1 + x_2$, then $\hat{x} = \hat{x}_1 + \hat{x}_2$, that is, $x(M) = x_1(M) + x_2(M)$ for all $M \in \mathcal{M}$;

(b) if $x = \alpha x_1$, then $\hat{x} = \alpha \hat{x}_1$, that is, $x(M) = \alpha x_1(M)$ for all $M \in \mathcal{M}$;

(c) if $x = x_1 x_2$, then $\hat{x} = \hat{x}_1 \hat{x}_2$, that is, $x(M) = x_1(M) x_2(M)$ for all $M \in \mathcal{M}$;

(d) $\hat{e} = 1$, that is, $e(M) = 1$ for all $M \in \mathcal{M}$;

(e) $\hat{x}(M) = 0$ if and only if $x \in M$, that is, $x(M) = 0$ if and only if $x \in M$;

(f) if $M_1, M_2 \in \mathcal{M}$ and $M_1 \neq M_2$, there exists an $x \in X$ such that $\hat{x}(M_1) \neq \hat{x}(M_2)$, that is, $x(M_1) \neq x(M_2)$;

(g) $|\hat{x}(M)| = |x(M)| \leq \|x\|$ for all $M \in \mathcal{M}$.

Proof. Properties (a) to (d) follow immediately from the fact that the mapping (19.23): $X \to X/M \to C$ given by $x \to x + M = x(M)e + M \to x(M)$ is a homomorphism of X onto C. Thus, for example,

$$x_1 + x_2 \to (x_1 + x_2)(M) = x_1(M) + x_2(M)$$

directly because the mapping (19.23) is a homomorphism.

Property (e) follows from the fact that the mapping $X/M \leftrightarrow C$ given by $x + M = x(M)e + M \leftrightarrow x(M)$ is an isomorphism, so $x(M) = 0$ if and only if $x + M = 0$, that is, if and only if $x \in M$.

(f) follows now directly from (e). For if $M_1 \neq M_2$, there exists an element $x \in X$ such that $x \in M_1$ but $x \notin M_2$ so $x(M_1) = 0$ but $x(M_2) \neq 0$; hence, $x(M_1) \neq x(M_2)$.

Finally, we prove (g). Write

$$x + M = x(M)e + M = x(M)(e + M),$$

so

$$\|x + M\| = |x(M)| \, \|e + M\| = |x(M)|,$$

recalling Eq. (19.24) of this section. However, $\|x + M\| \leq \|x\|$ so $|x(M)| \leq \|x\|$ and this completes the proof of the theorem.

In terms of notations we have introduced in this section, we shall next characterize some of the concepts introduced earlier in this chapter.

THEOREM 19.14. $\sigma(x) = \{x(M) | M \in \mathcal{M}\}$, where $x \in X$ and where $\mathcal{M}$ is the set of maximal ideals of X.

Proof. Let $x(M) = \lambda$. Then

$$(x - \lambda e)(M) = x(M) - \lambda e(M) = \lambda - \lambda = 0.$$

Hence, $x - \lambda e \in M$, which implies (as we know) that $x - \lambda e$ is not a unit or, equivalently, $\lambda \in \sigma(x)$.

Conversely, if $\lambda \in \sigma(x)$, then $x - \lambda e$ is not invertible. By the Corollary to Theorem 19.12 then,

$$x - \lambda e \in M$$

for some maximal ideal M. Hence,

$$0 = (x - \lambda e)(M) = x(M) - \lambda e(M) = x(M) - \lambda,$$

which completes the proof.

Next we give an alternate characterization of the spectral radius.

THEOREM 19.15. If $x \in X$, the spectral radius,

$$r_\sigma(x) = \sup_{M \in \mathcal{M}} |x(M)|$$

where X, as usual, denotes a commutative Banach algebra with identity.

Proof. To prove this it suffices to apply the previous theorem:

$$r_\sigma(x) = \sup_{\lambda \in \sigma(x)} |\lambda| = \sup_{M \in \mathcal{M}} |x(M)|.$$

DEFINITION 19.6. The *radical* of X is the intersection $\bigcap_{M \in \mathcal{M}} M$ of all the maximal ideals in X.

Clearly the radical always contains the 0 vector, since each ideal does. If, however, the radical consists of only the 0 vector, then X is called *semisimple*.

We return now to the set $F(\mathcal{M})$ of all complex-valued functions on the set $\mathcal{M}$. $F(\mathcal{M})$ is an algebra if we define operations in the pointwise fashion, that is,

$$(f + g)(M) = f(M) + g(M); \quad (\alpha f)(M) = \alpha f(M); \quad (fg)(M) = f(M)g(M),$$

where $f, g \in F(\mathcal{M})$, $\alpha \in C$ and $M \in \mathcal{M}$. Consider again the mapping

$$X \to F(\mathcal{M}),$$
$$x \to \hat{x}. \tag{19.27}$$

By Theorem 19.13, we know that

$$\widehat{x_1 + x_2} = \hat{x}_1 + \hat{x}_2, \quad \widehat{\alpha x} = \alpha \hat{x}, \quad \widehat{x_1 x_2} = \hat{x}_1 \hat{x}_2,$$

so that the mapping (19.27) is an algebra homomorphism. The kernel of this homomorphism (the elements mapped into the zero element of $F(\mathcal{M})$, that is) is precisely the radical of X, for if $\hat{x} = 0$, then $\hat{x}(M) = x(M) = 0$ for all M, which is equivalent to $x \in M$ for all $M \in \mathcal{M}$. If, conversely, $x \in \bigcap_{M \in \mathcal{M}} M$, then clearly $\hat{x} = 0$. Hence we have proved the next theorem.

THEOREM 19.16. The mapping (19.27), $x \to \hat{x}$, of X into $F(\mathcal{M})$, the algebra of all complex-valued functions defined on $\mathcal{M}$, is a homomorphism. It is an isomorphism into $F(\mathcal{M})$ if and only if X is semisimple.

A particular example of a semisimple Banach algebra is the algebra W considered in Example 19.3. We shall see this by actually determining the set $\mathcal{M}$ of all maximal ideals of W. Let $t \in [0, 2\pi)$ and consider the mapping $W \to C$ given by $x \to x(t_0)$— a substitution mapping. This is clearly a homomorphism of W onto C and therefore determines a maximal ideal

$$M_{t_0} = \{x \in W \mid x(t_0) = 0\}.$$

We now desire to show that *every* maximal ideal is of this form. We must show that, given a maximal ideal M of W, there exists a $t_0 \in [0, 2\pi)$ such that $M = M_{t_0}$. For this purpose let M be an arbitrary maximal ideal of W. M, as we know, determines a homomorphism of W:

$$W \to W/M \leftrightarrow C. \tag{19.28}$$

Under this homomorphism, let the element $x_0 = e^{it} \in W$ map into the complex number $\alpha = x_0(M)$ as follows:

$$x_0 \to \alpha \quad \text{[under (19.28)]}.$$

Then

$$x_0^{-1} = e^{-it} \to \alpha^{-1} \quad \text{[under (19.28)]}.$$

Recalling the norm on W, it is clear that $\|x_0\| = 1$; therefore,

$$|\alpha| = |x_0(M)| \le \|x_0\| = 1.$$

Similarly,

$$|\alpha^{-1}| \le \|x_0^{-1}\| = 1.$$

Combining these two results, it is evident that $|\alpha| = 1$, so there exists a $t_0 \in [0, 2\pi)$ such that $\alpha = e^{it_0}$ and under the homomorphism (19.28),

$$e^{int} \to e^{int_0}$$

which implies that

$$S_N = \sum_{n=-N}^{N} c_n e^{int} \to \sum_{n=-N}^{N} c_n e^{int_0}$$

under (19.28). But property (g) of Theorem 19.13 shows that the mapping (19.26), $x \to \hat{x}$, is norm-decreasing, so if $\{x_n\}$ converges to x, then $\{x_n(M)\}$ converges to $x(M)$. In our case, supposing that S_N converges to $\sum_{n=-\infty}^{\infty} c_n e^{int}$, $S_N(M)$ converges to $\sum_{n=-\infty}^{\infty} c_n e^{int_0}$; that is,

$$\sum_{n=-\infty}^{\infty} c_n e^{int} \to \sum_{n=-\infty}^{\infty} c_n e^{int_0}$$

under (19.28). In view of this, (19.28) is seen to be just a substitution mapping too and $M = M_{t_0}$. From this, in turn, it is clear that if $x \in \bigcap_{M \in \mathcal{M}} M$, where $\mathcal{M}$ is the set of all maximal ideals of W, then $x = 0$.

To rephrase this a little, we have seen that the set $\mathcal{M}$ of maximal ideals of W can be identified with the interval $[0, 2\pi)$. Furthermore, if $x(t) = \sum_{n=-\infty}^{\infty} c_n e^{int} \in W$ and $x(t)$ does not vanish at any t, then

$$0 \notin \sigma(x) = \{x(M) \mid M \in \mathcal{M}\},$$

which means that $x(t)$ is invertible, that is, $1/x(t) \in W$, a rather short proof of a fact once proved using classical methods by Wiener (next theorem).

THEOREM 19.17. (Wiener). If the absolutely convergent trigonometric series $x(t) = \sum_{n=-\infty}^{\infty} c_n e^{int}$ vanishes for no value of t, the function

$$\frac{1}{x(t)} = 1 / \sum_{n=-\infty}^{\infty} c_n e^{int}$$

can be expanded in an absolutely convergent trigonometric series.

Before going on to some topological considerations, we want to consider one more application of the general theory. Before doing this, however, we must establish a few preliminary results, among which is the following remarkable theorem. It states: If a homomorphism is nontrivial (an algebraic fact), it must be continuous (an analytic fact). This is something like saying: f is a homomorphism, therefore f is green.

THEOREM 19.18. Let X be a complex Banach algebra with identity (not necessarily commutative) and $f: X \to C$ a nontrivial homomorphism. Then f is continuous; in fact $\|f\| = 1$.

Proof. Let $x \in X$ and suppose $f(x) = \lambda$. Since f is a homomorphism

$$f(x - \lambda e) = f(x) - \lambda f(e) = \lambda - \lambda = 0.$$

The element $x - \lambda e$ is thus seen to be a member of the kernel of f, a maximal ideal, and is therefore not invertible $[\lambda \in \sigma(x)]$. By the Corollary to Theorem 19.4 then $|\lambda| \leq \|x\|$; in other words,

$$|f(x)| \leq \|x\|$$

for any x, which means that f is bounded and that $\|f\| \leq 1$. However,

$$1 = |f(e)| \leq \|f\| \|e\| = \|f\|,$$

so $\|f\| = 1$, which completes the proof.

We now present a mapping with properties very similar to those of the ordinary derivative.

DEFINITION 19.7. Letting X be an arbitrary complex Banach algebra (not necessarily commutative), a mapping $D: X \to X$ is called a *derivation* if (a) D is a linear transformation; (b) $D(xy) = Dx \cdot y + x Dy$, for all $x, y \in X$ (note that the order here is important).

The following result about derivations was first established by Singer and Wermer [Ref. 6].

THEOREM 19.19. Let X be a commutative complex Banach algebra with an identity and let $D: X \to X$ be a *bounded* (in the sense of bounded linear transformation) derivation; then D must map X into its radical.

Proof. With D as above, let $\alpha \in C$ and consider the series

$$\sum_{n=0}^{\infty} \frac{\alpha^n D^n}{n!}.$$

This converges absolutely in $L(X, X)$ and, therefore, converges to some operator in $L(X, X)$ which will be denoted by $e^{\alpha D}$. Next let f be an arbitrary homomorphism of X into C and let

$$g_\alpha(x) = f(e^{\alpha D}(x)).$$

First, it is clear that g_α is a linear functional. We now contend that it is also multiplicative and is consequently a homomorphism. To see this, we observe that since D is a derivation,

$$\frac{D^n(xy)}{n!} = \sum_{i+j=n} \frac{D^i(x)}{i!} \frac{D^j(y)}{j!} \tag{19.29}$$

Thus

$$g_\alpha(xy) = f\left(e^{\alpha D}(xy)\right)$$

$$= f\left(\sum_{n=0}^{\infty} \frac{\alpha^n D^n(xy)}{n!}\right)$$

$$= \sum_{n=0}^{\infty} \alpha^n \frac{f(D^n(xy))}{n!} \quad \text{[by Theorem (19.18)]} \tag{19.30}$$

$$= \sum_{n=0}^{\infty} \alpha^n \sum_{i+j=n} \frac{f(D^i(x))f(D^j(y))}{i!j!}$$

by (19.29) and since f is a homomorphism. Also

$$g_\alpha(x)g_\alpha(y) = \left[\sum_{i=0}^{\infty} \frac{\alpha^i f(D^i(x))}{i!}\right]\left[\sum_{j=0}^{\infty} \frac{\alpha^j f(D^j(x))}{j!}\right]. \tag{19.31}$$

But the infinite series of complex numbers occurring in (19.31) converges absolutely and so we can, in multiplying the two sums, form the Cauchy product. Comparing (19.30) and (19.31) under these circumstances, we see that

$$g_\alpha(xy) = g_\alpha(x)g_\alpha(y).$$

Thus g_α is a homomorphism and, as such, Theorem 19.18 implies that

$$\|g_\alpha(x)\| \le \|x\| \quad \text{(for all x).} \tag{19.32}$$

However, since

$$|g_\alpha(x)| \le \sum_{n=0}^{\infty} |\alpha|^n \frac{\|D^n(x)\|}{n!} \le \sum_{n=0}^{\infty} \frac{|\alpha|^n \|D\|^n \|x\|}{n!},$$

$g_\alpha(x)$ is seen to be an entire function of α, holding x fixed but arbitrary. (Note that at this point it is vital that D be a *bounded* derivation.) By (19.32), this entire function is bounded in the entire complex plane; by the ordinary Liouville theorem for complex-valued analytic functions proved in Theorem 19.1, it must therefore be constant. The only way for $g_\alpha(x)$ to be constant for every α, however, is for

$$f(D^n(x)) = 0 \quad (n = 1, 2, \ldots)$$

and, in particular,

$$f(D(x)) = 0.$$

But f was *any* homomorphism! Thus by the general correspondence between maximal ideals and homomorphisms, $D(x)$ belongs to every maximal ideal; that is, $D(x)$ belongs to the radical, which is what we wanted to prove.

The theorem shows, in particular, that if D is a bounded derivation and X is a *semisimple*, commutative, complex Banach algebra with identity, then D must be zero. It has been conjectured that this last result is true even when D is not bounded (that is, the only derivation on a semisimple, commutative, complex Banach algebra is the trivial one), but it has only been proved for certain special types of Banach algebras.

As another application of Theorem 19.18 we have the following corollary.

COROLLARY. Let $A, B \in L(X, X)$, where X is a Banach space. Suppose $AB - BA$ belongs to the closed algebra generated by 1 (the identity transformation) and A; then $\sigma(AB - BA) = \{0\}$. In particular, $AB - BA \neq 1$.

Proof. We let Y designate the closed subalgebra generated by 1 and A. [Note that this is a commutative subalgebra of $L(X, X)$.] Define

$$D(C) = CB - BC.$$

It is readily verified that D is a derivation on $L(X, X)$. Moreover, D is bounded since

$$\|D(C)\| = \|CB - BC\|$$
$$\leq \|CB\| + \|BC\|$$
$$\leq \|C\| \|B\| + \|B\| \|C\| = (2\|B\|)\|C\|.$$

We contend finally that D maps the subalgebra Y, which is a commutative Banach algebra with identity, into itself. For if $p(A)$ is a polynomial in A, then (denoting the formal derivative of $p(A)$ by $p'(A)$),

$$D(p(A)) = p'(A)D(A),$$

since D is a derivation; also $p'(A)D(A) \in Y$, for $D(A) = AB - BA \in Y$, by hypothesis. Now the general element C of Y is a limit of polynomials in A and, since D is bounded, it follows immediately that $D(C) \in Y$ for $C \in Y$. Applying Theorem 19.19 to the algebra Y, we conclude that D maps Y into its radical. Thus $D(A) = AB - BA$ belongs to the radical of Y. By Theorem 19.14, however, any element in the radical must have a spectrum consisting of just 0, so that $\sigma(AB - BA) = \{0\}$. Hence certainly $AB - BA \neq 1$, which completes the proof.

The fact that $AB - BA \neq 1$ for linear transformations on a finite-dimensional space X can be established simply by considering the associated matrices in a fixed basis and then by computing traces. The above corollary extends this result to bounded operators on a Banach space. The result is no longer valid if the operators are unbounded (counterexample?). $AB - BA$ is frequently called the *commutator* of the operators A and B, and commutators are of particular importance in mathematics and certain areas of physics, but we shall not pursue these matters any further, since it would take us too far afield.

19.7   Weak Topologies and the Gelfand Topology†

In Sec. 7.3 we discussed what is meant by putting the *weak* topology on an arbitrary set X associated with a family $\{f_\alpha\}_{\alpha \in \Lambda}$ of mappings, $f_\alpha : X \to X_\alpha$, where each

† The results of this section will not be used again in this volume, and may therefore be omitted on a first reading.

X_α was a topological space. We then specialized a bit when we considered the particular situation where all the X_α were the same, either C or R. (Throughout the following discussion, we assume that all $X_\alpha = C$.) We then saw that a basis for the weak topology on X consisted of all sets of the form

$$V(x_0; f_1, f_2, \ldots, f_n, \varepsilon) = \{x \mid |f_i(x) - f_i(x_0)| < \varepsilon, i = 1, \ldots, n\} \qquad (19.33)$$

where $x_0 \in X$, $\varepsilon > 0$, and the $f_i \in \{f_\alpha\}_{\alpha \in \Lambda}$.

Suppose now that X is a complex normed space and $\{f_\alpha\}_{\alpha \in \Lambda} = \tilde{X}$, the set of all bounded linear functionals on X. The collection of open sets of the resulting weak topology on X, generated by the family $\{f_\alpha\}_{\alpha \in \Lambda} = \tilde{X}$ will be denoted by $\mathcal{O}_w$ to distinguish it from the family $\mathcal{O}$ of open sets of the metric topology (with respect to the norm) of X. From our general considerations of weak topologies (see, in particular, Sec. 7.3), the topological space $(X, \mathcal{O}_w)$ is a Hausdorff space—if $x_0 \neq y_0$, that is, $x_0 - y_0 \neq 0$, there exists an $f \in \tilde{X}$ (Theorem 12.2) such that $f(x_0 - y_0) \neq 0$ or $f(x_0) \neq f(y_0)$.

Since each f_α is continuous in the metric topology $\mathcal{O}$, each

$$V(x_0; f_1, \ldots, f_n, \varepsilon) \in \mathcal{O};$$

hence $\mathcal{O}_w \subset \mathcal{O}$; that is, the metric topology is stronger than the weak topology.

All notions of a topological nature pertaining to the topological space $(X, \mathcal{O}_w)$ will be preceded by the word "weak" to distinguish it from the corresponding situation in the space $(X, \mathcal{O})$; for example, we speak of weak compactness, weak convergence, weak closure, etc. We must be careful, however, in just one case of this terminology: namely, we have already introduced (see Definition 14.1) a notion of weak convergence: $x_n \overset{w}{\to} x_0$, in a normed linear space X. Let us show now that this coincides with the notion of weak convergence just introduced with respect to the topology $(X, \mathcal{O}_w)$. To this end, suppose that $x_n \to x_0$ in the topology $\mathcal{O}_w$, which means that every $O \in \mathcal{O}_w$ and containing x_0 contains almost all (all but a finite number, that is) of the x_n. Thus almost all $x_n \in V(x_0 : f, \varepsilon)$ or

$$|f(x_n) - f(x_0)| < \varepsilon \qquad (f \in \tilde{X}, \varepsilon > 0),$$

for all n sufficiently large. This, however, just means that $x_n \overset{w}{\to} x_0$ in our old sense.

On the other hand, suppose that $x_n \overset{w}{\to} x_0$ and let $O \in \mathcal{O}_w$ be an arbitrary weakly open set containing x_0. Since the sets of the form (19.33) constitute a basis for $\mathcal{O}_w$, there exist $f_1, f_2, \ldots, f_m \in \tilde{X}$ such that

$$V(x_0; f_1, \ldots, f_m, \varepsilon) \subset O.$$

But $x_n \overset{w}{\to} x_0$ implies that

$$|f_i(x_n) - f_i(x_0)| < \varepsilon \qquad (i = 1, 2, \ldots, m),$$

for n sufficiently large, that is,

$$x_n \in V(x_0; f_1, \ldots, f_m, \varepsilon)$$

for all n sufficiently large; whence the sequence $\{x_n\}$ converges to x_0 with respect to the topology $\mathcal{O}_w$ and shows consistent usage of the word "weak" with regard to convergence in X.

To emphasize at times that we are dealing with the metric topology $\mathcal{O}$ of X, we shall sometimes preface topological properties here with the word "strong"; for example, we say strong closure, strongly convergent sequence, etc.

If E is a subset of X, we denote by $\bar{E}^s$ its strong closure and by $\bar{E}^w$ its weak closure. Clearly, $\bar{E}^s \subset \bar{E}^w$, since, as we have already noted, $\mathcal{O}_w \subset \mathcal{O}$. Thus if $\bar{E}^w = E$, then $\bar{E}^s = E$—every weakly closed subset is strongly closed. The converse is in general not true; it is, however, true in the case of subspaces as the following theorem shows.

THEOREM 19.20. A subspace M of a normed linear space X (over C or R) is weakly closed if and only if it is strongly closed.

Proof. We have already noted, in general, that every weakly closed subset is strongly closed, so it remains only to show that, if M is strongly closed, it is also weakly closed. Hence suppose $\bar{M}^s = M$, M is strongly closed. If we can show that $\bar{M}^w \subset \bar{M}^s = M$, we are done. Equivalent to this statement is

$$CM \subset C\bar{M}^w,$$

so suppose that $x \notin M = \bar{M}^s$. For such an x, $d(x, M) = d > 0$ [$d(x, M)$ as usual denotes the distance from x to M with respect to the norm-generated metric]; hence by Theorem 12.3 there exists an $f \in \tilde{X}$ such that $f(M) = 0, f(x) \neq 0$. The set

$$O_w = \{y \in X \mid f(y) \neq 0\}$$

must be a member of $\mathcal{O}_w$, because it is the inverse image of the open set $C - \{0\}$ in the complex plane and $\mathcal{O}_w$ was taken as the weakest topology with respect to which all functionals of $\tilde{X}$ are continuous. We now have $O_w \in \mathcal{O}_w$ and $x \in O_w$. But $M \cap O_w = \emptyset$, since f vanishes on M; therefore, $x \notin \bar{M}^w$, which proves that $CM \subset C\bar{M}^w$ and completes the proof.

Let us now return to the general situation of weak topologies generated by a family of functions mapping a given set into C; only now instead of taking the underlying set X to be an arbitrary normed linear space over C, let us take it to be $\tilde{X}$; as family of functions $\{f_\alpha\}_{\alpha \in \Lambda}$, mapping X into C, take $\tilde{\tilde{X}}$, the second conjugate space of X (discussed in Sec. 12.2). The resulting weak topology for $\tilde{X}$ will not interest us as much as something else known as the *weak-* topology*, which will be discussed now.

Instead of taking $\{f_\alpha\}_{\alpha \in \Lambda}$ to be all of $\tilde{\tilde{X}}$ we shall consider only the subset $J(X)$, where $J : X \to \tilde{\tilde{X}}$ is the natural correspondence first mentioned in Sec. 12.2. If the normed linear space X happens to be reflexive, then $J(X) = \tilde{\tilde{X}}$ and the weak topology for $\tilde{X}$ and the weak-* topology for $\tilde{X}$ will, of course, coincide. Generally, however, the weak-* topology is weaker than the weak topology, a fact which is easy to see and is left as an exercise for the reader.

We observe that a typical basis set for the weak-* topology is of the form

$$V(f_0; Jx_1, \ldots, Jx_n, \varepsilon) = \{f \in \tilde{X} \mid |Jx_i(f) - Jx_i(f_0)| < \varepsilon, i = 1, \ldots, n\}$$
$$= \{f \in \tilde{X} \mid |f(x_i) - f_0(x_i)| < \varepsilon, i = 1 \quad ., n\}, \qquad (19.34)$$

since $Jx_i(f) = f(x_i)$ by the definition of J. It is customary to abbreviate and to write, instead of $V(f_0; Jx_1, \ldots, Jx_n, \varepsilon)$,

$$V(f_0; x_1, \ldots, x_n, \varepsilon), \tag{19.35}$$

and we shall adhere to this convention.

The set $\tilde{X}$ now has three topologies associated with it: $\mathcal{O}$, the norm-metric topology; $\mathcal{O}_w$, the weak topology; and $\mathcal{O}_{w*}$, the weak-$*$ topology. As before, to emphasize which topology we are dealing with, we shall preface topological notions with the words strong, weak, and weak-$*$. At first sight there seems to be a danger here because we have already introduced weak-$*$ convergence in Sec. 15.1, but just as in the case of weak convergence, discussed previously, it is easily demonstrated that we are employing a consistent usage of weak-$*$ convergence.

We next note that all three topological spaces are Hausdorff. For example, $(\tilde{X}, \mathcal{O}_{w*})$ is Hausdorff for if $f_1, f_2 \in \tilde{X}$ and if $f_1 \neq f_2$, there exists an $x \in X$ such that $f_1(x) \neq f_2(x)$ or $Jx(f_1) \neq Jx(f_2)$, and we now appeal to the general considerations of Sec. 7.3 and 7.4 to see that $(\tilde{X}, \mathcal{O}_{w*})$ is Hausdorff.

The following important result pertaining to weak-$*$ topologies will subsequently be applied to the set of maximal ideals of a Banach algebra.

THEOREM 19.21. The solid unit sphere,

$$S = \{ f \in \tilde{X} \mid \|f\| \leq 1 \}$$

of $\tilde{X}$ is compact in the weak-$*$ topology.

Proof. For any $f \in S$,

$$|f(x)| \leq \|f\| \, \|x\| \leq \|x\|.$$

If the underlying space X is a real normed linear space, take $I_x = [-\|x\|, \|x\|]$ with the induced topology of the real axis, while if it is a complex space take $I_x = \{ \alpha \in C \mid |\alpha| \leq \|x\| \}$ with the induced topology of the complex plane. In either case, I_x is compact by the Heine–Borel theorem so, by Tychonoff's Theorem (Theorem 7.16) $E = \prod_{x \in X} I_x$ is compact (with respect to the Tychonoff topology). E, we recall, consists of all functions $g : X \to \bigcup_{x \in X} I_x$ such that $g(x) \in I_x$; but we have just observed that $|f(x)| \leq \|x\|$ for $f \in S$ so $f(x) \in I_x$. Thus $S \subset E$.

Recalling the initial discussion (Sec. 7.3) of the product (Tychonoff) topology, it is remembered that this topology for E was the weakest topology for which all the projection mappings were continuous, and a typical basis set in this particular situation is of the form

$$V(g_0; x_1, \ldots, x_n, \varepsilon) = \{ g \in E \mid |g(x_i) - g_0(x_i)| < \varepsilon, \, i = 1, \ldots, n \},$$

where $g_0 \in E$, $\varepsilon > 0$, for here $pr_x(g) = g(x)$.

S as a subset of E has an induced topology, namely, that induced by the product topology of E, and a typical basis set in this topology is

$$V(g_0; x_1, \ldots, x_n, \varepsilon) \cap S = \{ g \in S \mid |g(x_i) - g_0(x_i)| < \varepsilon, \, i = 1, \ldots, n \}. \tag{19.36}$$

However, S as a subset of $\tilde{X}$ inherits a topology, namely, the one induced on S by the weak-∗ topology of $\tilde{X}$. Examining the basis open sets of the weak-∗ topology given by (19.34) or (19.35) and comparing these with (19.36), it is apparent that these two topologies for S coincide.

Since E is compact in the product topology it will follow (Theorem 17.11) that, if S is a closed subspace of E, then S is also compact. This, coupled with our previous observation concerning the coincidence of the weak-∗ topology for S and the induced product topology for S, will complete the proof. We proceed therefore to show that S is closed. Let $g_0 \in \bar{S}$. In this case every open set containing g_0 contains elements of S. In particular, the basis open set $V(g_0; x, y, x + y, \varepsilon)$, $\varepsilon > 0$, $x, y \in X$, must contain points of S. Suppose that

$$f \in V(g_0; x, y, x + y, \varepsilon) \cap S.$$

Since $f \in S \subset \tilde{X}$, f is linear; in particular, $f(x + y) = f(x) + f(y)$. Also since $f \in V(g_0; x, y, x + y, \varepsilon)$, we have

$$|g_0(x) - f(x)| < \varepsilon, \qquad |g_0(y) - f(y)| < \varepsilon,$$

and

$$|g_0(x + y) - f(x + y)| < \varepsilon.$$

Thus,

$$|g_0(x + y) - (g_0(x) + g_0(y))| \le |g_0(x + y) - f(x + y)| + |f(x + y) - (g_0(x) + g_0(y))|$$

$$= |g_0(x + y) - f(x + y)| + |f(x) + f(y) - (g_0(x) + g_0(y))|$$

$$\le |g_0(x + y) - f(x + y)| + |f(x) - g_0(x)| + |f(y) - g_0(y)|$$

$$< 3\varepsilon,$$

and since ε is an arbitrary positive real number, it follows that $g_0(x + y) = g_0(x) + g_0(y)$. Similarly,

$$g_0(\alpha x) = \alpha g_0(x) \qquad (\alpha \in C \text{ or } R, \text{ as the case may be}).$$

For any $x \in X$, there exists an $f_x \in V(g_0; x, \varepsilon) \cap S$ so

$$|g_0(x) - f_x(x)| < \varepsilon,$$

and

$$|g_0(x)| \le |f_x(x)| + \varepsilon$$

$$\le \|f_x\| \, \|x\| + \varepsilon$$

$$\le \|x\| + \varepsilon,$$

which implies that

$$|g_0(x)| \le \|x\|.$$

Thus g_0 is a bounded linear functional such that $\|g_0\| \le 1$; in other words, $g_0 \in S$ and S is therefore closed. The proof is now complete.

Some of the results we have just established will now be applied to the Banach algebra situation.

The Gelfand Topology

Let X be a complex commutative Banach algebra with an identity, and let $\mathcal{M}$ be the set of maximal ideals of X. Each $x \in X$ determines a function $\hat{x}$ on $\mathcal{M}$ with the defining property

$$\hat{x}(M) = x(M) \qquad (\text{for all } M \in \mathcal{M}),$$

where $x(M)$ is as in Eq. (19.23). We now topologize $\mathcal{M}$ by taking the weak topology determined by the family $\{\hat{x}\}_{x \in X}$, that is, the weakest topology with respect to which all $\hat{x}$ are continuous. We know by the general considerations of Sec. 7.3 that a typical basis open set in $\mathcal{M}$ is of the form

$$V(M_0; \hat{x}_1, ..., \hat{x}_n, \varepsilon) = \{M \in \mathcal{M} \,|\, |\hat{x}_i(M) - \hat{x}_i(M_0)| < \varepsilon, \, i = 1, ..., n\}$$
$$= \{M \in \mathcal{M} \,|\, |x_i(M) - x_i(M_0)| < \varepsilon, \, i = 1, ..., n\}. \tag{19.37}$$

Rather than write $V(M_0; \hat{x}_1, ..., \hat{x}_n, \varepsilon)$ for a typical basis set, it is customary to write instead just

$$V(M_0; x_1, ..., x_n, \varepsilon), \tag{19.38}$$

and we shall do so.

The topology just introduced on $\mathcal{M}$ is sometimes called the *Gelfand topology*. We note that $\mathcal{M}$ with respect to this topology is a Hausdorff space, for, if M_1, $M_2 \in \mathcal{M}$ and $M_1 \neq M_2$, then, by Theorem 19.13, there exists an $x \in X$ such that $x(M_1) \neq x(M_2)$, which implies by the considerations of Sec. 7.3 and 7.4 that the space is Hausdorff.

We observed in Sec. 19.6 that there is a 1 : 1 correspondence between maximal ideals (elements of $\mathcal{M}$) and nontrivial homomorphisms of X onto C. In Theorem 19.18, we noted that every nontrivial homomorphism of X onto C is bounded and has norm equal to one. Thus we can view $\mathcal{M}$ as a subset of S, the solid unit sphere of $\tilde{X}$, by simply identifying M (a fixed maximal ideal) and the homomorphism f, where $f(x) = x(M)$ for all $x \in X$. $\mathcal{M}$ so viewed, as a subset of $\tilde{X}$, has a topology induced by the weak-$*$ topology of $\tilde{X}$, the basis sets being of the type given by (19.34) or (19.35); comparing this with the basis sets (19.37) or (19.38) of the Gelfand topology for $\mathcal{M}$, we see easily that the two coincide.

THEOREM 19.22. The space $\mathcal{M}$ is compact with respect to the Gelfand topology.

Proof. Since, as just noted above, the Gelfand topology coincides with the topology induced by the weak-$*$ topology of $\tilde{X}$ and since $\mathcal{M}$ is actually a subset of S, the solid unit sphere of $\tilde{X}$ which is compact in the weak-$*$ topology by Theorem 19.21, it suffices to show that $\mathcal{M}$ is a closed subspace of S. To this end, let $g_0 \in \bar{\mathcal{M}} \cap S$. Then $g_0 \in S$, so g_0 is a bounded linear functional such that $\|g_0\| \leq 1$. But $g_0 \in \bar{\mathcal{M}}$, so that

$$V(g_0; x, y, xy, \varepsilon) \cap \mathcal{M} \neq \emptyset,$$

where ε is an arbitrary positive number and x and y are arbitrary elements of X.

Since the intersection is not empty, we have an $f \in V(g_0; x, y, xy, \varepsilon) \cap M$. Since $f \in M, f(xy) = f(x)f(y)$, and since $f \in V(g_0; x, y, xy, \varepsilon)$,

$$|f(x) - g_0(x)| < \varepsilon, \qquad |f(y) - g_0(y)| < \varepsilon, \qquad |f(xy) - g_0(xy)| < \varepsilon. \quad (19.39)$$

Now,

$$g_0(xy) - g_0(x)g_0(y) = \big(g_0(xy) - f(xy)\big) + \big(f(xy) - f(x)f(y)\big)$$
$$+ f(y)\big(f(x) - g_0(x)\big) + g_0(x)\big(f(y) - g_0(y)\big).$$

Taking absolute values, using (19.39), noting that $f(xy) = f(x)f(y)$ and that $\|f\| \le 1$ and $\|g_0\| \le 1$, we get

$$|g_0(xy) - g_0(x)g_0(y)| \le \varepsilon(1 + \|y\| + \|x\|),$$

which implies that $g_0(xy) = g_0(x)g_0(y)$ and shows that g_0 is a homomorphism, so $g_0 \in M$. Therefore, M is a closed subspace of S which, as we have observed, completes the proof.

19.8 Topological Vector Spaces and Operator Topologies

In this final section,† we wish to introduce the concepts of a topological vector space and a topological algebra, and also to give a variety of operator topologies. The discussion will be just introductory and is given only to expose the reader rather early to these notions. Considerable attention will be given to these matters in the second volume. Relevant references are also given at the end of this chapter.

First, a particular example will be considered. Let X be just a real or complex linear space, say complex for definiteness, and let $\mathscr{F}$ be a subset of the set of all linear functionals on X. We consider the weak topology on X generated by $\mathscr{F}$. A typical basis open set is, as we know, of the form

$$V(x_0; f_1, \ldots, f_n, \varepsilon) = \{x \in X \,|\, |f_i(x) - f_i(x_0)| < \varepsilon, \, i = 1, \ldots, n\}, \quad (19.40)$$

where $\varepsilon > 0$, $x_0 \in X$, and the $f_i \in \mathscr{F}$.

Consider the mapping

$$\psi : X \times X \to X,$$

where $\psi\langle x, y \rangle = x + y$. Clearly, $\psi\langle 0, 0 \rangle = 0$. It is now contended that ψ is continuous at $\langle 0, 0 \rangle$ (see Exercise 7.5). Let $V(0; f_1, \ldots, f_n, \varepsilon)$ be an arbitrary basis open set containing the vector 0. Let

$$\langle x, y \rangle \in V(0; f_1, \ldots, f_n, \varepsilon/2) \times V(0; f_1, \ldots, f_n, \varepsilon/2)$$

be a basis open set in the product topology. Then

$$|f_i(x + y)| = |f_i(x) + f_i(y)|$$
$$\le |f_i(x)| + |f_i(y)|$$
$$< \frac{\varepsilon}{2} + \frac{\varepsilon}{2} = \varepsilon \qquad (\text{for} \quad i = 1, 2, \ldots, n);$$

---

† This section will not be relied upon throughout the rest of this volume, and may be omitted on a first reading.

that is, we have shown that

$$\psi(V(0; f_1, ..., f_n, \varepsilon) \times V(0; f_1, ..., f_n, \varepsilon/2)) \subset V(0; f_1, ..., f_n, \varepsilon) \qquad (19.41)$$

which proves that ψ is continuous at $\langle 0, 0 \rangle$.

Since the functions involved are all linear, it follows immediately that (see Exercise 19.24)

$$V(x_0; f_1, ..., f_n, \varepsilon) = V(0; f_1, ..., f_n, \varepsilon) + x_0. \qquad (19.42)$$

From this and the fact that ψ is continuous at $\langle 0, 0 \rangle$, it follows that ψ is continuous on all of $X \times X$. To elaborate on this briefly, let $\langle x_0, y_0 \rangle$ be an arbitrary point of $X \times X$ and let $V(x_0 + y_0; f_1, ..., f_n, \varepsilon)$ be an arbitrary basis open set containing $x_0 + y_0$. Using (19.42), it follows that

$$V(0; f_1, ..., f_n, \varepsilon) + (x_0 + y_0) = V(x_0 + y_0; f_1, ..., f_n, \varepsilon).$$

Using (19.41) and (19.42) together with this result, we now have

$$\psi(V(x_0; f_1, ..., f_n, \varepsilon) \times V(y_0; f_1, ..., f_n, \varepsilon)$$
$$= \psi(V(0; f_1, ..., f_n, \varepsilon/2) + x_0 \times V(0; f_1, ..., f_n, \varepsilon/2) + y_0)$$
$$\subset V(0; f_1, ..., f_n, \varepsilon) + x_0 + y_0$$
$$= V(x_0 + y_0; f_1, ..., f_n, \varepsilon).$$

Since $\langle x_0, y_0 \rangle$ was an arbitrary point of $X \times X$ and we have shown ψ to be continuous at $\langle x_0, y_0 \rangle$, it follows that ψ is continuous everywhere on $X \times X$.

Next consider the mapping

$$\varphi : C \times X \to X,$$

$$\langle \alpha, x \rangle \to \alpha x.$$

In similar fashion it can be shown that φ is continuous on $C \times X$.

By the considerations of Sec. 7.5, we know that the weak topology on X generated by $\mathscr{F}$ will be Hausdorff if $\mathscr{F}$ contains enough functionals to separate points of X; that is, if for $x_1, x_2 \in X$, $x_1 \neq x_2$, there exists an $f \in \mathscr{F}$ such that $f(x_1) \neq f(x_2)$. This is certainly the case if $\mathscr{F}$ consists of the set of *all* linear functionals on X.

DEFINITION 19.8. A *topological vector space* is a real or complex vector space X plus a topology for X such that the mappings

(a) $\psi : X \times X \to X$,
   $\langle x, y \rangle \to x + y$,
(b) $\varphi : C \times X \to X$,
   $\langle \alpha, x \rangle \to \alpha x$

are continuous.

Note in (b) that we have assumed a complex vector space. If X is real, we must replace C by R in (b). It is possible in certain cases to extend this definition to a vector space over an arbitrary field, but we shall not deal with this here.

A topological vector space is called *locally convex* if there exists a basis of open sets all of which are convex sets.

We note that the above example of a topological vector space is an example of a locally convex space, for consider the basis-open sets $V(x_0; f_1, ..., f_n, \varepsilon)$. Suppose that $x, y \in V(x_0; f_1, ..., f_n, \varepsilon)$; then [see (19.40)]

$$|f_i(x) - f_i(x_0)| < \varepsilon \qquad (i = 1, ..., n)$$

and

$$|f_i(y) - f_i(x_0)| < \varepsilon \qquad (i = 1, ..., n).$$

Let α, β be nonnegative real numbers such that $\alpha + \beta = 1$. Then for any i ($1 \le i \le n$),

$$f_i(\alpha x + \beta y) - f_i(x_0) = \alpha f_i(x) + \beta f_i(y) - f_i(x_0)$$
$$= \alpha f_i(x) + \beta f_i(y) - (\alpha + \beta) f_i(x_0)$$
$$= \alpha f_i(x) - \alpha f_i(x_0) + \beta f_i(y) - \beta f_i(x_0).$$

Therefore,

$$|f_i(\alpha x + \beta y) - f_i(x_0)| \le \alpha |f_i(x) - f_i(x_0)| + \beta |f_i(y) - f_i(x_0)|$$
$$< (\alpha + \beta)\varepsilon = \varepsilon \qquad (i = 1, ..., n).$$

Hence, $\alpha x + \beta y \in V(x_0; f_1, ..., f_n, \varepsilon)$ and this set is, consequently, a convex set.

Clearly, another example of a locally convex topological vector space is given by a complex (or real) normed linear space with its norm-metric topology.

We now wish to discuss briefly the issue of various operator topologies. This will enable us to put the discussion of Chapter 15 in a more appropriate framework and also affords us an opportunity to introduce the reader to these topologies.

Let X be a complex (or real) normed linear space and consider $L(X, X)$, the algebra of all bounded linear transformations on X. The *uniform topology* for $L(X, X)$ is just the metric topology, that is, $L(X, X)$ is a normed linear space, too, and we take as a basis all sets of the form

$$S_\varepsilon(A_0) = \{A \in L(X, X) \mid \|A - A_0\| < \varepsilon\} \qquad [A_0 \in L(X, X)].$$

There is nothing particularly new that happens when $L(X, X)$ is considered with this topology, even though we indicate its consideration by prefacing all topological notions about $L(X, X)$ pertaining to this topology by the word *uniform*. In particular, the notion of convergence in this topology, uniform convergence, coincides with the notion of uniform convergence introduced in Chapter 15—a fact which, because of its simplicity, is left for the reader to verify. The open sets of the uniform topology will be denoted by $\mathscr{O}_u$ and the closure of a set $S \subset L(X, X)$ in the uniform topology is denoted by $\bar{S}^u$.

With X as above, let $\tilde{X}$ designate the conjugate space of X and let $y' \in \tilde{X}$. Consider the mapping

$$f_{xy'} : L(X, X) \to C,$$
$$A \longrightarrow y'(Ax).$$

The weakest topology for $L(X, X)$ such that each member of the family $\{f_{xy'} \mid x \in X, y' \in \tilde{X}\}$ is continuous is called the *weak topology* for $L(X, X)$. In the special case when X is a Hilbert space by the Riesz representation theorem

(Theorem 12.5), we can consider the family $\{f_{xy}|x, y \in X\}$ which has as a typical member $f_{xy} \in L(X, X)$,

$$f_{xy}(A) = (Ax, y).$$

Returning once more to the general case, a typical basis set in the weak topology is of the form

$$V(A_0; f_{x_i y_i'}, \; i = 1, \ldots, n, \varepsilon),$$

which we write simply as

$$V(A_0; x_1, \ldots, x_n, y_1', \ldots, y_n', \varepsilon)$$
$$= \{A \in L(X, X) \mid |y_i'(Ax_i) - y_i'(A_0 x_i)| < \varepsilon, \; i = 1, \ldots, n\} \qquad (19.43)$$

whereas in the Hilbert space situation, we write

$$V(A_0; x_1, \ldots, x_n, y_1, \ldots, y_n, \varepsilon)$$
$$= \{A \in L(X, X) \mid |((A - A_0)x_i, y_i)| < \varepsilon, \; i = 1, \ldots, n\}. \qquad (19.44)$$

We adopt the convention of prefacing all topological notions about $L(X, X)$ with respect to the weak topology with the word "weak," it being easy to verify that in the special case of "weak convergence" there is no ambiguity with regard to weak convergence in the sense of the weak topology or as introduced in Chapter 15. The class of weakly open sets will be denoted by $\mathcal{O}_w$ and the closure of a set $S \subset L(X, X)$ with respect to the weak topology will be denoted by $\bar{S}^w$. Clearly, $\mathcal{O}_w \subset \mathcal{O}_u$, which implies $\bar{S}^u \subset \bar{S}^w$.

Both of the preceding examples are special cases of a more general structure known as a *topological algebra*, defined below.

DEFINITION 19.9. A complex (or real) algebra X which is also a locally convex topological vector space (with respect to the same additive operation and scalar multiplication) is called a *topological algebra* if, in addition, the mapping

$$\varphi : X \times X \to X,$$
$$\langle x, y \rangle \to xy,$$

is a continuous function of each argument separately.

It is possible to give a more general definition of a topological algebra, but this will suffice for our short introductory purposes. At any rate, it is not difficult to see that $L(X, X)$ with both the uniform and weak topologies is a topological algebra.

Next consider the mapping

$$f_x : L(X, X) \to X,$$
$$A \to Ax.$$

The *strong topology* for $L(X, X)$ is the topology generated by the family $\{f_x\}_{x \in X}$. (Note that X here is taken with its metric topology, where the metric is that generated by the norm on X.) The basis open sets are all sets of the form $V(A_0; f_{x_1}, \ldots, f_{x_n}, \varepsilon)$, which we write simply as

$$V(A_0; x_1, \ldots, x_n, \varepsilon) = \{A \in L(X, X) \mid \|f_{x_i}(A) - f_{x_i}(A_0)\| < \varepsilon, \; i = 1, \ldots, n\}$$
$$= \{A \in L(X, X) \mid \|Ax_i - A_0 x_i\| < \varepsilon, \; i = 1, \ldots, n\}. \qquad (19.45)$$

In this situation we preface topological notions about $L(X, X)$ with respect to the strong topology with the word "strong"; the topology for $L(X, X)$ is denoted by $\mathcal{O}_s$ and closures of subsets S of $L(X, X)$ by $\bar{S}^s$. The notion of strong convergence in this latter sense is identical to that defined in Chapter 15, and we also have $\mathcal{O}_s \subset \mathcal{O}_u$ and $\mathcal{O}_w \subset \mathcal{O}_s$. This last fact will be verified.

Let

$$V(A_0; x_1, ..., x_n, y_1', ..., y_n', \varepsilon)$$

given by (19.43) be an arbitrary weak basis set. We contend that it contains the strong basis set

$$V(A_0; x_1, ..., x_n, \delta),$$

where $\delta = \varepsilon/\max \|y_i'\|$, given by (19.45). To see this, let $A \in V(A_0; x_1, x_2, ..., x_n, \varepsilon)$. Then

$$\|(A - A_0)x_i\| < \delta \qquad (i = 1, 2, ..., n).$$

Thus

$$|y_i'(Ax_i - A_0 x_i)| \leq \|y_i'\| \, \|Ax_i - A_0 x_i\|$$

$$< \delta \max_i \|y_i'\| = \varepsilon,$$

that is, $A_0 \in V(A_0; x_1, ..., x_n, y_1', ..., y_n', \varepsilon)$. It follows at once from this that $\mathcal{O}_w \subset \mathcal{O}_s$.

We wish to introduce one further topology on $L(X, X)$, but before doing so, it is necessary to introduce certain concepts closely related to those of Sec. 24.1. Let $X_1, X_2, ..., X_n, ...$ be normed linear spaces all over C or R. The (external) direct sum $Y = X_1 \oplus X_2 \oplus ...$ of these spaces is defined to be the set of all vectors

$$y = \langle x_1, x_2, ... \rangle \qquad (x_i \in X_i),$$

such that $\sum_{i=1}^{\infty} \|x_i\|^2 < \infty$. Y can be made into a vector space by defining addition and scalar multiplication component-wise and can be made into a normed linear space by defining

$$\|y\| = \left(\sum_{i=1}^{\infty} \|x_i\|^2 \right)^{1/2}. \tag{19.46}$$

If each X is also a Hilbert space, Y can be made into a Hilbert space by defining the inner product for y and z, where

$$z = \langle x_1', x_2', ... \rangle$$

by

$$(y, z) = \sum_{i=1}^{\infty} (x_i, x_i').$$

It is quite straightforward to verify that all the necessary axioms are satisfied, and thus it will not be done here.

Now, again, let X be a normed linear space and consider the mapping, for $y = \langle x_1, x_2, ... \rangle \in X \oplus X \oplus \cdots$,

$$f_y : L(X, X) \to X \oplus X \oplus \cdots,$$

$$A \to \langle Ax_1, Ax_2, ... \rangle.$$

It is noted that

$$\sum_{i=1}^{\infty} \|Ax_i\|^2 \le \|A\|^2 \sum_{i=1}^{\infty} \|x_i\|^2 < \infty,$$

so that, indeed, $f_y(A) \in X \oplus X \oplus \cdots$. We define the *strongest topology* for $L(X, X)$ to be the weak topology on $L(X, X)$ generated by the family $\{f_y\}_{y \in Y}$, where Y is taken with respect to its norm-derived metric topology, the norm being as in (19.46). A typical subbasis set in this topology is of the form $V(A_0; f_y, \varepsilon)$, where $y = \langle x_1, x_2, \ldots \rangle$, which we write as

$$V(A_0; x_1, x_2, \ldots, \varepsilon) = \left\{ A \,\middle|\, \sum_{i=1}^{\infty} \|(A - A_0)x_i\|^2 < \varepsilon^2 \right\}. \tag{19.47}$$

A typical *basis* set is thus seen to be of the form

$$\bigcap_{i=1}^{k} V(A_0; x_1^{(i)}, x_2^{(i)}, \ldots, \varepsilon),$$

where the $\langle x_1^{(i)}, x_2^{(i)}, \ldots \rangle \in Y$.
However, considering the sequences

$$x_1^{(1)}, x_2^{(1)}, \ldots$$
$$x_1^{(2)}, x_2^{(2)}, \ldots$$
$$\vdots \qquad \vdots$$
$$x_1^{(k)}, x_2^{(k)}, \ldots$$

we can, by the usual method, write all these terms in a single sequence, $z_1, z_2, \ldots,$ $z_n, \ldots,$ which will be square summable, and clearly

$$\bigcap_{i=1}^{k} V(A_0; x_1^{(i)}, x_2^{(i)}, \ldots, \varepsilon) \supset V(A_0; z_1, z_2, \ldots, \varepsilon).$$

Thus the sets of the form (19.47) actually constitute a basis for the strongest topology. As with the preceding notions, we prefix considerations in $L(X, X)$ pertaining to the strongest topology with "strongest," write $\mathcal{O}_{st}$ for the topology and $\bar{S}^{st}$ for the closure of a subset S of $L(X, X)$. If $\{A_n\}$ converges to A_0 in this topology, we write $A_n \overset{st}{\to} A_0$, which means that

$$\sum_{i=1}^{\infty} \|(A_n - A_0)x_i\|^2 < \varepsilon^2$$

for n sufficiently large and $\sum_{i=1}^{\infty} \|x_i\|^2 < \infty$. In particular, taking all $x_i = 0$ but x_j, we get

$$\|A_n x_j - A_0 x_j\|^2 < \varepsilon^2,$$

for n sufficiently large and $x_j \in X$. Thus $A_n \overset{st}{\to} A_0$ implies $A_n \overset{s}{\to} A_0$. We also have $\mathcal{O}_s \subset \mathcal{O}_{st}$ as can be seen by letting $V(A_0; x_1, \ldots, x_n, \varepsilon)$, given by (19.45), be an arbitrary strong basis set. Then, clearly,

$$V(A_0; x_1, \ldots, x_n, 0, 0, \ldots, \varepsilon) \subset V(A_0; x_1, \ldots, x_n, \varepsilon)$$

for, if A belongs to the set on the left side of this inclusion, then $\sum_{i=1}^{\infty} \|(A - A_0)x_i\|^2$

$< \varepsilon^2$, and it follows that $\|(A - A_0)x_i\| < \varepsilon$ $(i = 1, 2, ..., n)$. It is now clear that $\mathcal{O}_s \subset \mathcal{O}_{st}$.

Last, it is noted that $\mathcal{O}_{st} \subset \mathcal{O}_u$, for if $\|A - A_0\| < \delta$, then for $\sum_{i=1}^{\infty} \|x_i\|^2 = M < \infty$,

$$\sum_{i=1}^{\infty} \|(A - A_0)x_i\|^2 \leq \|A - A_0\|^2 \sum_{i=1}^{\infty} \|x_i\|^2$$

$$= \|A - A_0\|^2 M$$

$$< \delta^2 M,$$

which can be made less than any prescribed ε^2 by taking δ sufficiently small, that is, for δ sufficiently small,

$$S_\delta(A_0) \subset V(A_0; x_1, ..., x_n, ..., \varepsilon).$$

We note that $L(X, X)$ with either the strong or strongest topologies is also a topological algebra. In summary, we have that

$$\mathcal{O}_w \subset \mathcal{O}_s \subset \mathcal{O}_{st} \subset \mathcal{O}_u;$$

hence,

$$\bar{S}^u \subset \bar{S}^{st} \subset \bar{S}^s \subset \bar{S}^w$$

for an arbitrary subset S of $L(X, X)$. It is remarked too that, in general, the inclusions cannot be replaced by equalities.

EXERCISES 19

In the exercises below, X is always assumed to have an identity e; when subalgebras Y of X are considered, it is assumed that $e \in Y$.

1. Let X be a Banach algebra and u a unit of X. Show that if $x \in X$ is such that $\|x - u\| < 1/\|u^{-1}\|$, then x is also a unit.

2. Suppose that X is just a normed linear space and that $A \in L(X, X)$. Then show that for $|\lambda| > \|A\|$, $(\lambda - A)^{-1}$ exists and is bounded and that

$$(\lambda - A)^{-1}y = \sum_{1}^{\infty} \lambda^{-n}A^{n-1}y \qquad \text{for each } y \in R(\lambda - A).$$

3. Let $A \in L(X, X)$, where X is a complex B-space. Show that, for $\lambda \in \rho(A)$,

$$\frac{d^n}{d\lambda^n} R_\lambda = (-1)^n n! \, R_\lambda^{n+1},$$

where R_λ is the resolvent operator of A, that is, $R_\lambda = (\lambda - A)^{-1}$.

4. Let X be a Banach algebra and let $x \in X$. x is called a *topological divisor of zero* if there exists a sequence $\{x_n\}$ in X such that $\|x_n\| = 1$ and either $xx_n \to 0$ or $x_n x \to 0$. If x is a topological divisor of zero, show that x is not a unit. Show also that the boundary of the set of nonunits is a subset of the set of topological divisors of zero. (*Note*: the boundary of a set S is $\bar{S} \cap \overline{CS}$.)

5. Let Y be a closed subalgebra of the Banach algebra X. Let $x \in Y$, and denote by $\sigma_X(x)$ and $\sigma_Y(x)$ the spectrum of x in X and Y, respectively. Show that

   (a) $\sigma_X(x) \subset \sigma_Y(x)$.
   (b) Each boundary point of $\sigma_Y(x)$ is also a boundary point of $\sigma_X(x)$.

6. Let X be a complex normed algebra which is also a division algebra. Prove that X is isomorphic to C.

7. Let X be a complex Banach algebra such that 0 is the only topological divisor of zero in X; then show that X is isomorphic to C.

8. Let X be a complex Banach algebra. If there exists a constant $m > 0$ such that $\|xy\| \geq m\|x\|\|y\|$ for all $x, y \in X$, show that X is isomorphic to C.

9. If X is a commutative Banach algebra and if $x, y \in X$, show that $r_\sigma(xy) \leq r_\sigma(x)r_\sigma(y)$.

10. Show that Theorem 19.4 is false if X is just a normed algebra with an identity, that is, not necessarily a Banach algebra.

The following exercises pertain to Secs. 19.6–19.8.

11. Let h be a nontrivial homomorphism of the commutative Banach algebra with identity X onto C. If $x \in X$, show that $h(x) \in \sigma(x)$.

12. Let X be a vector space and f a nontrivial linear functional defined on X. Show that f maps X *onto* the underlying scalar field.

13. Let X be a Banach space and let $A, B \in L(X, X)$. Suppose that $C = AB - BA$ and that $AC = CA$ and $BC = CB$. Then prove that $\sigma(C) = \{0\}$.

14. Let X be a commutative Banach algebra with identity and let $\mathscr{R}$ be the radical of x. Show that $X/\mathscr{R}$ is semisimple.

15. Determine $\mathscr{M}$ for the Banach algebra A of all functions analytic in $|z| < 1$ and continuous in $|z| \leq 1$.

16. Same as Exercise 15, only consider the algebra $C(X)$ of all continuous complex-valued functions on a compact Hausdorff space X with the usual point-wise operations and sup norm.

17. Let X be a commutative Banach algebra with identity. Show that $x \in X$ belongs to the radical if and only if for every $\lambda \in C$, the sequence $\{(\lambda x)^n\}$ converges to 0.

18. Let X be an algebra with an identity e and h a nontrivial linear functional on X. Show that h is multiplicative $[h(xy) = h(x)h(y)$ for all $x, y \in X]$ if and only if its kernel is a subalgebra of X and $h(e) = 1$.

19. Let X be a reflexive space. Show that the solid unit sphere is closed in the weak topology.

20. Let X be a normed l   ˙˙ˢ Show that the only linear functionals on X
    which are continuou⸳   ˙˙ᵉ the functionals of $\tilde{X}$.

21. Let X and Y be Banach spaces and $A : X \to Y$ a linear transformation. Show that A is continuous with respect to the norm-metric topologies in X and Y if only if it is continuous with respect to the weak topologies.

22. Prove that the weak-$*$ topology for $\tilde{X}$ is weaker than the weak topology.

23. Consider $\tilde{X}$ and the topologies $\mathcal{O}_w$ and $\mathcal{O}_{w*}$, respectively. Show that convergence with respect to these topologies coincides with weak convergence and weak-$*$ convergence, respectively, as introduced in Chapter 15 for normed linear spaces.

24. Verify Eq. (19.42).

25. Show that there exist complex algebras which cannot be normed (in the sense of a normed algebra).

REFERENCES

1. M. Naimark, "Normed Rings."
2. E. Lorch, "Spectral Theory."
3. C. Rickart, "The General Theory of Banach Algebras."
4. E. Hille and R. Phillips, "Functional Analysis and Semigroups."
5. K. Knopp, "Theory of Functions of a Complex Variable," Vol. 1.
6. I. Singer and J. Wermer, "Derivations on Commutative Normed Algebras," *Math. Ann.* 129, 260–264 (1955).
7. C. Putnam, "On the Spectra of Commutators," Proc. Am. Math. Soc., 5 (6), 929–931 (1954). Another approach to Exercise 19.13 is given here.
8. G. Bachman, "Elements of Abstract Harmonic Analysis."

CHAPTER **20**

Adjoints and Sesquilinear Functionals

In the previous chapters we have noted some of the conveniences available in Hilbert spaces. For example we noted such properties as being able to identify the annihilator of a subset with its orthogonal complement (Sec. 17.4) and being able to identify vectors of the space with bounded linear functionals (Sec. 12.5). Both these qualities are consequences of the Riesz representation theorem (Sec. 12.4), which enabled us to make a certain identification between brackets (bounded linear functionals) and parentheses (inner products). In view of this identification it seems reasonable that a restatement or paraphrasing of the notion of conjugate map, defined in terms of brackets (Sec. 17.3), in terms of parentheses might be possible in such spaces. Since the identification between space and dual space for Hilbert spaces is a conjugate isometry (Sec. 12.5) and not a full-fledged congruence, we might suspect that the analog might also invlove such complex conjugate differences, and indeed this happens.

What we shall actually do is abandon the conjugate map in Hilbert space setting and use the notion of *adjoint*. Where we had

$$A : X \to Y$$

and

$$A' : \tilde{Y} \to \tilde{X}$$

via

$$[Ax, y] = [x, A'y]$$

for normed spaces, we shall now use A^*, the *adjoint* of $A \in L(X, Y)$, defined as follows for the case when X and Y are Hilbert spaces:

$$A^* : Y \to X,$$

via

$$(Ax, y) = (x, A^*y).$$

Needless to say, certain differences arise; for example, where we had, for α a scalar, $(\alpha A)' = \alpha A'$, we now get $(\alpha A)^* = \bar{\alpha} A^*$ (that is, a complex conjugate pops up).

Along with some other things, we now proceed to carry out the above program.

20.1 The Adjoint Operator

In Chapter 1 a representation theorem for linear functionals on finite-dimensional, inner product spaces was proved. Using this theorem, the concept of the adjoint of a linear transformation mapping such a space into itself was defined [see Eq. (1.11)]. We now wish to extend this notion to the infinite-dimensional case. For the sake of the present discussion, we assume that X and Y are inner product spaces and that A is an arbitrary function (A does not have to be linear or continuous or anything else) defined as follows:

$$X \supset D_A \overset{A}{\to} Y; \tag{20.1}$$

it is further assumed (and this is a vital assumption) that

$$\bar{D}_A = X. \tag{20.2}$$

Suppose that y is some vector in Y. It is possible that there is some vector $z \in X$ such that

$$(Ax, y) = (x, z) \tag{20.3}$$

for all $x \in D_A$. Denote by D^* the set of all $y \in Y$ for which such a $z \in X$ exists; that is,

$$D^* = \left\{ y \in Y \,\middle|\, \begin{array}{l} \text{there exists } z \in X \text{ such that} \\ (Ax, y) = (x, z) \text{ for all } x \in D_A \end{array} \right\}. \tag{20.4}$$

We now define

$$Y \supset D^* = D_{A^*} \xrightarrow{A^*} X, \tag{20.5}$$

$$y \to z,$$

the mapping A^* being called the *adjoint of A*. In summary if, for $y \in Y$, there exists $z \in Y$ such that

$$(Ax, y) = (x, z) \tag{20.6}$$

for all $x \in D_A$, we say that

$$y \in D_{A^*} \quad \text{and} \quad A^*y = z. \tag{20.7}$$

Once any mapping has been proposed, one must immediately ascertain whether it is a bona fide function, that is, whether it is well-defined. We now show A^* to be well-defined, and it is in this endeavor that the assumption that $\bar{D}_A = X$ is used.

Suppose y and z are as in (20.6) and suppose also that

$$(Ax, y) = (x, w)$$

for all $x \in D_A$. If so,

$$(x, z) = (x, w)$$

for all $x \in D_A$, which implies $(x, z - w) = 0$ for all $x \in D_A$. Equivalently, one might say

$$z - w \perp D_A.$$

From this it readily follows that

$$z - w \perp \bar{D}_A = X,$$

which implies that $z - w$ is orthogonal to itself, and it follows that $z - w = 0$. Therefore A^* is well-defined.

Note that no special assumptions about A were made; A is just an arbitrary mapping from D_A into Y. In spite of this weak hypothesis about A, we shall show A^* to be a linear transformation. To prove this conjecture, we must show that D_{A^*} is a subspace and that A^* preserves linear combinations. Hence suppose

$$y_1 \in D_{A^*}, \quad y_2 \in D_{A^*},$$

and

$$A^*y_1 = z_1, \quad A^*y_2 = z_2.$$

Now, for scalars α and β, consider the sum of

$$(Ax, \alpha y_1) = \bar{\alpha}(Ax, y_1) = \bar{\alpha}(x, A^*y_1) = (x, \alpha z_1)$$

and

$$(Ax, \beta y_2) = \bar{\beta}(Ax, y_2) = \bar{\beta}(x, A^*y_2) = (x, \beta z_2);$$

this yields

$$(Ax, \alpha y_1 + \beta y_2) = (x, \alpha z_1 + \beta z_2) \quad \text{(for all } x \in D_A).$$

Therefore,

$$\alpha y_1 + \beta y_2 \in D_{A^*}$$

and

$$A^*(\alpha y_1 + \beta y_2) = \alpha z_1 + \beta z_2 = \alpha A^* y_1 + \beta A^* y_2,$$

and the linearity of A^* is apparent.

In Theorem 9.2, it was shown that the inner product was a continuous mapping. Using this, we now wish to show that A^* is a closed linear transformation. Let $\{y_n\}$ be a sequence of vectors from D_{A^*} such that

$$y_n \to y \quad \text{and} \quad z_n = A^* y_n \to z.$$

We must show that $y \in D_{A^*}$ and $A^* y = z$. For every n we have

$$(Ax, y_n) = (x, A^* y_n) = (x, z_n).$$

Hence,

$$\lim_n (Ax, y_n) = \lim_n (x, z_n).$$

Using the continuity of the inner product, we have

$$(Ax, y) = (x, z) \quad \text{(for all } x \in D_A\text{)}.$$

Consequently,

$$y \in D_{A^*} \quad \text{and} \quad A^* y = z,$$

which proves A^* is closed. We have now established the following theorem.

THEOREM 20.1. The adjoint A^* of an arbitrary function, $A, X \supset D_A \xrightarrow{A} Y$, where X and Y are inner product spaces and D_A is dense in X, is a closed linear transformation.

Our next theorem appeals to one's love of symmetry.

THEOREM 20.2. Let A, X, and Y be as in the preceding theorem. In addition, suppose A^{-1} exists and that $D_{A^{-1}}$ is dense in Y. Then

$$(A^{-1})^* = (A^*)^{-1}.$$

Proof. To show the equality of these two functions, we must show that their domains are the same and that they agree on every vector in the domain. First, suppose $x \in D_A$ and that $y \in D_{(A^{-1})^*}$. We have

$$(x, y) = (A^{-1} Ax, y) = (Ax, (A^{-1})^* y).$$

Therefore [see Eqs. (20.6) and (20.7)], since x is any vector in D_A,

$$(A^{-1})^* y \in D_{A^*} \tag{20.8}$$

and

$$A^*(A^{-1})^* y = y. \tag{20.9}$$

On the other hand, suppose $x \in D_{A^{-1}}$ and $y \in D_{A^*}$. In this case we have

$$(x, y) = (AA^{-1} x, y) = (A^{-1} x, A^* y),$$

which implies

$$A^* y \in D_{(A^{-1})^*} \tag{20.10}$$

and

$$(A^{-1})^* A^* y = y. \tag{20.11}$$

Equation (20.11) shows $(A^*)^{-1}$ to exist and that the restriction of $(A^{-1})^*$ to $R(A^*)$ is just $(A^*)^{-1}$, that is,

$$D_{(A^*)^{-1}} \subset D_{(A^{-1})^*}.$$

Since, if $y \in D_{(A^{-1})^*}$, y admits the representation (20.9), that is, $y \in R(A^*) = D_{(A^{*-1})}$, we also have

$$D_{(A^{-1})^*} \subset D_{(A^*)^{-1}},$$

and the theorem now follows.

We now list several additional properties of adjoints. In the statements below it is assumed that A and B are mappings from X into Y and that A and B, as well as any other transformation occurring below, have domains which are dense in X. The only difficult feature in proving any of the statements is in showing the equality of the domains of the operators.

(1) For $\alpha \in C$, $(\alpha A)^* = \bar{\alpha} A^*$.
(2) If $A < B$ [read "B extends A", that is, $D_A \subset D_B$ and $B|_{D_A} = A$], then $B^* < A^*$.
(3) $A^* + B^* < (A + B)^*$. (Recall that $D_{A+B} = D_A \cap D_B$.)
(4) $B^* A^* < (AB)^*$.
(5) For $\alpha \in C$, $(A + \alpha)^* = A^* + \bar{\alpha}$.

To illustrate how proving such statements goes, we prove (3) here.

Proof (3). Suppose $x \in D_{A+B} = D_A \cap D_B$, and suppose $y \in D_{A^*+B^*} = D_{A^*} \cap D_{B^*}$. Now consider

$$((A + B)x, y) = (Ax, y) + (Bx, y)$$
$$= (x, A^*y) + (x, B^*y)$$
$$= (x, (A^* + B^*)y).$$

Thus

$$y \in D_{(A+B)^*}$$

and

$$(A + B)^* y = (A^* + B^*)y.$$

(3) now follows.

20.2 Adjoints and Closures

In the discussion below we assume X and Y to be complex Hilbert spaces. We shall consider certain effects of assuming a linear transformation A from X into Y to admit a closure. Whenever adjoints are spoken of in this section, it is assumed of course that the necessary domains are dense. We then consider the very special types of operators known as Hermitian, symmetric, and self-adjoint operators.

If X and Y are Hilbert spaces, an inner product on the cartesian product $X \times Y$ is defined by taking

$$(\langle x_1, x_2 \rangle, \langle y_1, y_2 \rangle) = (x_1, y_1) + (x_2, y_2),$$

where $\langle x_1, x_2 \rangle$, $\langle y_1, y_2 \rangle \in X \times Y$ (see Sec. 9.2). In an identical fashion we can also introduce an inner product on $Y \times X$. With this inner product in mind, it is easy to establish that the mappings U_1 and U_2

$$U_1 : X \times Y \to Y \times X,$$
$$\langle x, y \rangle \to \langle iy, -ix \rangle;$$
$$U_2 : Y \times X \to X \times Y,$$
$$\langle y, x \rangle \to \langle ix, -iy \rangle$$

establish congruences between $X \times Y$ and $Y \times X$. Moreover, it is easy to verify that

$$U_1 U_2 = U_2 U_1 = 1$$

and, using Exercise 1.3, for $B \subset X \times Y$ or $C \subset Y \times X$,

$$U_1(B^\perp) = U_1(B)^\perp \quad \text{and} \quad U_2(C^\perp) = U_2(C)^\perp. \tag{20.12}$$

Suppose that A is an arbitrary mapping such that

$$X \supset D_A \overset{A}{\to} Y$$

and $\bar{D}_A = X$. Consider the graph of A:

$$G_A = \{ \langle x, Ax \rangle \mid x \in D_A \}$$

and apply U_1 to this subset of $X \times Y$. We have

$$U_1(G_A) = \{ \langle iAx, -ix \rangle \mid x \in D_A \}$$

and claim that

$$U_1(G_A)^\perp = G_{A^*}, \tag{20.13}$$

where A^* denotes the adjoint of A. The proof is as follows:

$$\langle y, z \rangle \in U_1(G_A)^\perp$$
$$\Leftrightarrow (\langle y, z \rangle, \langle iAx, -ix \rangle) = 0 \quad \text{(for all } x \in D_A),$$
$$\Leftrightarrow (y, iAx) + (z, -ix) = 0 \quad \text{(for all } x \in D_A)$$
$$\Leftrightarrow i(Ax, y) - i(x, z) = 0 \quad \text{(for all } x \in D_A)$$
$$\Leftrightarrow (Ax, y) = (x, z) \quad \text{(for all } x \in D_A)$$
$$\Leftrightarrow y \in D_{A^*} \quad \text{and} \quad A^* y = z$$
$$\Leftrightarrow \langle y, z \rangle \in G_{A^*}.$$

In the event that A is a linear transformation *admitting a closure*, we can show the domain of A^* to be dense in Y; then, similar to the way (20.13) was demonstrated, it can be shown that

$$U_2(G_{A^*})^\perp = G_{A^{**}}. \tag{20.14}$$

To prove that the domain of A^* is dense in Y (as we must in order to consider A^{**}), we suppose to the contrary and arrive at a contradiction. Hence suppose $\bar{D}_{A^*} \neq Y$. Since Y is a Hilbert space, by the projection theorem, Theorem 10.8,

$$Y = \bar{D}_{A^*} \oplus (\bar{D}_{A^*})^\perp.$$

Consequently there must be some nonzero vector w such that $w \perp D_{A^*}$, that is, $w \perp y$ for all $y \in D_{A^*}$. This implies

$$\langle 0, w \rangle \perp \langle iA^* y, -iy \rangle \qquad \text{(for all } y \in D_{A^*} \text{)},$$

or, equivalently,

$$\langle 0, w \rangle \in U_2(G_{A^*})^\perp. \qquad (20.15)$$

But Eq. (20.13) implies that

$$G_{A^*}{}^\perp = U_1(G_A)^{\perp\perp} = \overline{U_1(G_A)},$$

where the right equality holds by observing that $X \times Y$ is a Hilbert space and invoking (7) of Sec. 10.4. Recalling that isometries must preserve closure, we now apply the isometry U_2 to the above equation to get

$$U_2(G_{A^*}{}^\perp) = U_2\overline{\left(U_1(G_A) \right)} = \overline{U_2 \left(U_1(G_A) \right)},$$

which, since $U_2 U_1 = 1$, implies

$$U_2(G_{A^*}{}^\perp) = \overline{G}_A = G_{\bar{A}}. \qquad (20.16)$$

Therefore, using (20.15), and (20.12), we have

$$\langle 0, w \rangle \in G_{\bar{A}},$$

which means that there must be some $x \in D_{\bar{A}}$ such that

$$\langle 0, w \rangle = \langle x, \bar{A}x \rangle.$$

Hence $x = 0$ and $w = \bar{A}x = 0$—a contradiction. The denseness of the domain of A^* is now proved.

Another result which follows immediately from (20.13) is the fact that A^* is a closed transformation (a fact we have already noted in Theorem 20.1). To prove this from (20.13), we need only recall that the orthogonal complement of a set is always closed, which implies that the graph of A^* is closed.

In Sec. 17.1 we first discussed the notion of a linear transformation A admitting a closure $\bar{A}$. In our next theorem we suppose A to be a linear transformation admitting a closure, and investigate the relationships that ensue between A, $\bar{A}$, $\bar{A}^*$, A^*, and A^{**}.

THEOREM 20.3. Let X and Y be complex Hilbert spaces, D_A a subspace of X which is also dense in X, and A a linear transformation

$$X \supset D_A \xrightarrow{A} Y.$$

Suppose that A permits the closure $\bar{A}$. Then

(1) $\bar{A}^* = A^*$;
(2) $\bar{A} = A^{**}$;
(3) if A is a closed transformation, then $A = A^{**}$.

REMARK. Part (2) implies that A^{**} is an extension of A, since $\bar{A}$ is an extension of A.

Proof. We first prove part (1), but before doing this note that if the graphs of any two mappings are equal, the mappings must be identical. With this in mind we will now show that $G_{\bar{A}^*} = G_{A*}$ to prove the equality of $\bar{A}^*$ and A^*. Since $\bar{A}$ is the closure of A, we must have

$$G_{\bar{A}} = \bar{G}_A,$$

which implies

$$U_1(G_{\bar{A}}) = U_1(\bar{G}_A). \tag{20.17}$$

From Eq. (20.13), we see that

$$U_1(G_{\bar{A}})^{\perp} = G_{\bar{A}^*}. \tag{20.18}$$

Since U_1 is an isometry, we have

$$U_1(\bar{G}_A) = \overline{U_1(G_A)},$$

and it follows, using (20.13) again, that

$$U_1(G_{\bar{A}})^{\perp} = \overline{U_1(G_A)}^{\perp} = U_1(G_A)^{\perp} = G_{A*};$$

hence,

$$G_{\bar{A}^*} = G_{A*},$$

which implies

$$\bar{A}^* = A^*$$

and proves part (1).

To prove (2) we note that since U_2 maps $Y \times X$ isometrically onto $X \times Y$ (see Exercise 1.3),

$$U_2(G_{A*}{}^{\perp}) = U_2(G_{A*})^{\perp}. \tag{20.19}$$

Using (20.14), (20.16), and (20.19), we now have

$$G_{\bar{A}} = U_2(G_{A*})^{\perp} = G_{A**},$$

which implies $\bar{A} = A^{**}$ and proves (2). Part (3) now follows as a consequence of (2).

We now wish to consider linear transformations of the following type: With X a complex inner product space,

$$X \supset D_A \xrightarrow{A} X. \tag{20.20}$$

In the event that

$$(Ax, y) = (x, Ay) \qquad \text{for all } x, y \in D_A, \tag{20.21}$$

then A is said to be *Hermitian*. Examples of such transformations are 1 and 0. If, in addition, $\bar{D}_A = X$, then A is said to be *symmetric*.

If A is symmetric, then

$$(Ax, y) = (x, Ay) \qquad (\text{for all } x, y \in D_A)$$

$$\Leftrightarrow y \in D_{A*}, \, A^*y = Ay \qquad (\text{for all } y \in D_A)$$

$$\Leftrightarrow A < A^*. \tag{20.22}$$

In view of this we could have used the requirement that A^* extend A to define a symmetric operator. If instead of requiring that A^* extend A, we make the stronger requirement that $A^* = A$, then A is called *self-adjoint*. Since A^* is always a closed transformation, we see that all self-adjoint transformations are closed.

Note that these three definitions have been presented in order of increasing strength: that is

$$\text{self-adjoint} \Rightarrow \text{symmetric} \Rightarrow \text{Hermitian}.$$

If $D_A = X$ and A is Hermitian, however, everything collapses—A must also be symmetric and self-adjoint.

Using the definition of a self-adjoint operator and results (1) and (4) of the preceding section, it is easy to establish the following result.

THEOREM 20.4. Suppose $A = A^*$. Then for any real numbers α and β, where $\alpha \neq 0$, the operator $\alpha A + \beta$ is self-adjoint.

Our next theorem provides a criterion for determining when a symmetric operator is self-adjoint.

THEOREM 20.5. Suppose A is a symmetric operator and suppose that $R(A) = X$, that is, A is an onto mapping, then A is self-adjoint.

Proof. In view of (20.22), we know that $A < A^*$. If we can show that $D_{A^*} \subset D_A$, the theorem will follow. We now proceed to establish this.

Let $y \in D_{A^*}$ and suppose $A^*y = z \in X$. Since $R(A) = X$, there must be some $w \in D_A$ such that ·

$$z = Aw.$$

For any $x \in D_A$, we now have

$$(Ax, y) = (x, A^*y)$$

$$= (x, Aw).$$

Since $w \in D_A$ and $A < A^*$, we must have

$$Aw = A^*w;$$

therefore,

$$(Ax, y - w) = 0 \qquad \text{(for all } x \in D_A\text{)}.$$

Since A is an onto mapping, it follows that $y - w = 0$ or

$$y = w \in D_A;$$

consequently,

$$D_{A^*} \subset D_A,$$

which establishes the theorem.

The following important theorem will be used when the spectral theorem for unbounded, self-adjoint operators is discussed in Chapter 29.

THEOREM 20.6. Let X be a complex Hilbert space and suppose $X \supset D_A \xrightarrow{A} X$, where $\bar{D}_A = X$ and A is a closed linear transformation. Then the linear transformation $(1 + A^*A)^{-1} = B$ exists; so does $C = A(1 + A^*A)^{-1}$. Moreover, B and C

are each defined on the whole space, each is bounded and such that $\|B\| \leq 1$ and $\|C\| \leq 1$, and B is such that† $(Bx, x) \geq 0$ for all $x \in X$.

Proof. Since A is closed then, by definition, its graph G_A, is a closed subspace of the Hilbert space $X \times X$. Thus, by the projection theorem (Theorem 10.8), we can write

$$X \times X = G_A \oplus G_A^{\perp}.$$

Another property that A possesses because it is closed is that

$$A^{**} = A,$$

which implies that

$$G_A = G_{A^{**}}.$$

Letting U_2 be the mapping mentioned earlier in this section, Eqs. (20.14), (20.16), and (20.19) state that

$$U_2(G_{A^*})^{\perp} = G_A = G_{A^{**}}.$$

Since U_2 is an onto isometry, the fact that G_{A^*} is closed implies that $U_2(G_{A^*})$ is closed too. Therefore

$$U_2(G_{A^*})^{\perp\perp} = U_2(G_{A^*}).$$

Substituting this in the previous equation yields $U_2(G_{A^*}) = G_A^{\perp}$, and it now follows that the above orthogonal direct sum decomposition of $X \times X$ can be rewritten as

$$X \times X = G_A \oplus U_2(G_{A^*}). \tag{20.23}$$

We now have the following *unique* representation of any $\langle z, 0 \rangle \in X \times X$: There exist unique x and w belonging to X such that

$$\langle z, 0 \rangle = \langle x, Ax \rangle + \langle iA^*w, -iw \rangle,$$

or, letting $y = iw$,

$$\langle z, 0 \rangle = \langle x, Ax \rangle + \langle A^*y, -y \rangle. \tag{20.24}$$

To paraphrase this, we have found unique x and y belonging to X satisfying the two equations

$$z = x + A^*y \quad \text{and} \quad 0 = Ax - y.$$

Hence to each $z \in X$ there corresponds a unique x and a unique y, and we shall denote this correspondence by the mappings B and C; that is,

$$Bz = x \quad \text{and} \quad Cz = y.$$

It will now be shown that these two mappings can be taken as the B and C mentioned in the statement of the theorem.

Since all the equations were linear, the mappings B and C are linear. In this notation, we have

$$z = Bz + A^*Cz \quad \text{and} \quad 0 = ABz - Cz.$$

---

† Transformations with this property are called positive (see Sec. 22.3).

Since the above relations are satisfied by every $z \in X$, it follows that

$$1 = B + A^*C \quad \text{and} \quad AB - C = 0. \tag{20.25}$$

Substituting $C = AB$ in the former equation yields

$$1 = (1 + A^*A)B. \tag{20.26}$$

For any $z \in X$ we clearly have $\|z\|^2 = \|\langle z, 0 \rangle\|^2$ and since the decomposition in (20.23) is *orthogonal*, we can write, using (20.24) and the Pythagorean theorem,

$$\|z\|^2 = \|\langle z, 0 \rangle\|^2 = \|\langle x, Ax \rangle\|^2 + \|\langle A^*y, -y \rangle\|^2$$
$$= \|x\|^2 + \|Ax\|^2 + \|A^*y\|^2 + \|y\|^2,$$

which implies that

$$\|x\|^2 + \|y\|^2 = \|Bz\|^2 + \|Cz\|^2 \le \|z\|^2, \tag{20.27}$$

which, in turn, implies

$$\|Bz\| \le \|z\| \quad \text{and} \quad \|Cz\| \le \|z\|$$

for all z. Therefore B and C are bounded and

$$\|B\| \le 1 \quad \text{and} \quad \|C\| \le 1.$$

To complete the proof it only remains to show that $(1 + A^*A)^{-1}$ exists, that it is defined on the whole space, that $B = (1 + A^*A)^{-1}$, and that $(Bx, x) \ge 0$ for all x. We shall show that $(1 + A^*A)^{-1}$ exists and is defined on the whole space by showing that $(1 + A^*A)$ is a 1:1 mapping of X onto X. First the 1:1-ness is considered. Let w be any vector in D_A such that $Aw \in D_{A^*}$, and consider

$$((1 + A^*A)w, w) = (w, w) + (A^*Aw, w).$$

Since $A^{**} = A$, this is equal to

$$(w, w) + (Aw, Aw) \ge (w, w).$$

So, if $(1 + A^*A)w = 0$, we have

$$0 \ge (w, w) \ge 0,$$

which implies that $w = 0$. Combining this with the linearity of A, the 1:1-ness follows.

By virtue of Eq. (20.26), any $z \in X$ can be written

$$z = (1 + A^*A)(Bz);$$

that is, any vector $z \in X$ is the image of another vector Bz under $(1 + A^*A)$, and $(1 + A^*A)$ is seen to be onto X.

Since $(1 + A^*A)$ is 1:1 and onto, its inverse exists and must, by virtue of (20.26), be given by B. Last, for any $x \in X$, consider

$$(Bx, x) = (Bx, (1 + A^*A)Bx) = (Bx, Bx) + (ABx, ABx) \ge 0.$$

The proof is now complete.

20.3 Adjoints of Bounded Linear Transformations in Hilbert Spaces

Suppose that X and Y are Hilbert spaces (complex) and that A is a bounded linear transformation such that

$$X \supset D_A \xrightarrow{A} Y \tag{20.28}$$

and whose domain D_A is dense in X. We now wish to establish that, when this situation prevails, there is no loss in generality in assuming A to be defined on the whole space.

In Sec. 17.2 it was noted that transformations such as A, bounded linear transformations whose range is in a Banach space, always permit a closure $\bar{A}$, actually bounded with the same norm as A, defined on $\bar{D}_A$, which, in this case, is all of X. In Theorem 20.3 it was shown that $\bar{A}^* = A^*$. Hence, every result valid for the adjoint of $\bar{A}$ will also hold for A^*. In view of this, any time a transformation such as A is encountered, it can always be extended to its closure, $\bar{A}$, which is defined on the whole space; for this reason nothing is lost by assuming such transformations to be defined on the whole space initially. We shall make this assumption throughout the ensuing discussion, except in the last theorem.

With X and Y as above, we now suppose $A \in L(X, Y)$. Letting y be an arbitrary fixed element of Y, we define

$$f_y : X \to C,$$
$$x \to (Ax, y).$$

f_y is clearly a linear functional and the fact that it is bounded follows from the Cauchy–Schwarz inequality (Sec. 1.2); applying this inequality also yields

$$\|f_y\| \leq \|A\| \, \|y\|. \tag{20.29}$$

Appealing to the Riesz representation theorem of Sec. 12.4, we now assert the existence of a unique vector $z \in X$ such that

$$f_y(x) = (x, z) \quad \text{(for all } x\text{)} \tag{20.30}$$

and

$$\|f_y\| = \|z\|. \tag{20.31}$$

Rewriting Eq. (20.30) as $(Ax, y) = (x, z)$, we see that

$$y \in D_{A^*} \quad \text{and} \quad A^*y = z.$$

Since y was any vector in Y, we see that $D_{A^*} = Y$; moreover,

$$\|A^*y\| = \|z\| = \|f_y\| \leq \|A\| \, \|y\|$$

for any $y \in D_{A^*} = Y$, which implies A^* is a *bounded* linear transformation and that

$$\|A^*\| \leq \|A\|. \tag{20.32}$$

By exactly the same reasoning, it can also be shown that A^{**} is a bounded linear transformation and that

$$\|A^{**}\| \leq \|A^*\|. \tag{20.33}$$

Since $A \in L(X, Y)$, A is closed (Corollary to Theorem 16.2); therefore, by Theorem 20.3 (3), $A = A^{**}$. Substituting A for A^{**} in (20.33) and comparing the resulting inequality to (20.32), we conclude that

$$\|A\| = \|A^*\|. \tag{20.34}$$

We have now proved the next theorem.

THEOREM 20.7. If A is a bounded linear transformation mapping the complex Hilbert space X into the complex Hilbert space Y, the adjoint of A, A^*, always exists and is a bounded linear transformation defined everywhere on Y; moreover, $\|A\| = \|A^*\|$.

For the case when $A, B \in L(X, Y)$, since $D_{A^*} = D_{B^*} = Y$, results (3) and (4) of Sec. 20.1 can be strengthened to equalities; that is,

$$(A + B)^* = A^* + B^* \tag{20.35}$$

and

$$(AB)^* = B^*A^*. \tag{20.36}$$

A number of results obtained for the conjugate map of a bounded linear transformation in Sec. 17.4 still look just about the same for the adjoint of A. In particular, with A, X, and Y as in Theorem 20.7, results (3)–(6) of Section 17.4 become

$$\overline{R(A)}^{\perp} = R(A)^{\perp} = N(A^*); \tag{20.37}$$

$$\overline{R(A)}^{\perp\perp} = \overline{R(A)} = N(A^*)^{\perp}; \tag{20.38}$$

$$\overline{R(A^*)}^{\perp} = N(A); \tag{20.39}$$

$$\overline{R(A^*)} = N(A)^{\perp}. \tag{20.40}$$

With A as in Theorem 20.7, we now establish a new theorem.

THEOREM 20.8. $\|A^*A\| = \|A\|^2$.

Proof. We note that

$$\|A^*A\| \le \|A^*\| \|A\| = \|A\| \|A\| = \|A\|^2,$$

while on the other hand,

$$\|A\|^2 = \left(\sup_{\|x\|=1} \|Ax\| \right)^2 = \sup_{\|x\|=1} \|Ax\|^2$$

$$= \sup_{\|x\|=1} (Ax, Ax)$$

$$= \sup_{\|x\|=1} (A^*Ax, x)$$

$$\le \sup_{\|x\|=1} \|A^*Ax\| \cdot \|x\| = \|A^*A\|.$$

The theorem now follows.

One recalls that, in the general setting of an arbitrary metric space, an isometric mapping is one that preserves distances. Viewing this in the more special setting of normed space, we saw that being an isometry was equivalent to preserving norms. Specifically if X and Y are inner product spaces, and

$$X \supset D_A \overset{A}{\to} Y,$$

$$x \to Ax,$$

we required that

$$\|Ax\| = \|x\| \qquad \text{(for all } x \in D_A\text{)}.$$

Using the following *polarization identity* for the case when X and Y are inner product spaces,

$$(x, y) = \tfrac{1}{4}[\|x + y\|^2 - \|x - y\|^2 + i\|x + iy\|^2 - i\|x - iy\|^2]. \qquad (20.41)$$

it is easily verified that the linear transformation A is an isometry if, and only if,

$$(Ax, Ay) = (x, y) \qquad (20.42)$$

for all $x, y \in D_A$ (see also Exercise 1.3). From now on when we are working with inner product spaces we shall use (20.42) as the defining property of an isometry.

If A is a linear transformation, X is a normed space, and

$$A : X \to X,$$

that is also an isometry onto X then A is called *unitary*. In simpler spaces (for example, euclidean spaces) the unitary transformations correspond to rotations and reflections about the origin. We now wish to link the notions of isometric transformations and unitary transformations. To this end let X and Y be complex Hilbert spaces and let U be a linear transformation such that

$$U : X \to Y. \qquad (20.43)$$

THEOREM 20.9. U is an isometry if and only if $U^*U = 1$.

Proof. If U is an isometry, then

$$(Ux, Uy) = (x, y) \qquad \text{(for all } x, y \in X\text{)}$$

$$\Leftrightarrow Uy \in D_{U^*}, \qquad U^*(Uy) = y \qquad \text{(for all } y \in X\text{)}$$

$$\Leftrightarrow U^*U = 1.$$

Suppose now that

$$U^*U = 1 \qquad (20.44)$$

and

$$UU^* = 1. \qquad (20.45)$$

By Theorem 20.9, Eq. (20.44) implies that U is an isometry, whereas (20.45) implies that every $y \in Y$ admits of the representation

$$y = U(U^*y),$$

which means that the mapping U is onto. Noting that isometries are always $1 : 1$ mappings, this argument can easily be reversed to yield a theorem.

THEOREM 20.10. U is an isometry onto Y if and only if

$$UU^* = 1 \quad \text{and} \quad U^*U = 1.$$

Combining the above results we now obtain the desired linkage between unitary transformations and isometries.

THEOREM 20.11. If $Y = X$, U is unitary if and only if $U^*U = UU^* = 1$. (Compare this to the discussion of unitary transformations on finite-dimensional spaces in Sec. 2.3: in particular, compare these results to Theorem 2.9.)

In an exactly analogous fashion to the way normal transformations were defined over finite-dimensional spaces in Sec. 1.3, we now let $A \in L(X, X)$, where X is a Hilbert space, and call A *normal* if $AA^* = A^*A$.

The Cayley Transform

The theorem below (which will be instrumental in proving the spectral theorem for unbounded, self-adjoint operators in Chapter 29) and many others to follow indicate a strong analogy between linear transformations and complex numbers. As one will see, in this analogy, self-adjoint transformations play the role of real numbers, positive transformations (cf. Sec. 22.3) correspond to the nonnegative real numbers, and unitary operators play the role of complex numbers on the unit circle. One recalls that the Moebius transformation

$$w = \frac{z - i}{z + i}$$

maps the real axis R into the unit circle (strictly into, because no real number maps into 1). Thus, when z is real, $|w| = 1$. Keeping the analogy mentioned above in mind, the following theorem seems rather to be expected.

THEOREM 20.12. Let X be a Hilbert space. If A is a self-adjoint transformation (not necessarily bounded) such that

$$X \supset D_A \xrightarrow{A} X,$$

where $\bar{D}_A = X$, then the transformation

$$U = (A - i)(A + i)^{-1}$$

is defined on all of X and is unitary. U is called the *Cayley transform of A*.

Proof. First it will be shown that the mappings $(A \pm i)$ are $1:1$; the inverse functions therefore exist on $R(A \pm i)$. We note that since A is self-adjoint,

$$
\begin{aligned}
\|Ax \pm ix\|^2 &= (Ax \pm ix, Ax \pm ix) \\
&= \|Ax\|^2 \pm (Ax, ix) \pm (ix, Ax) + \|x\|^2 \\
&= \|Ax\|^2 \pm (Ax, ix) \mp (Ax, ix) + \|x\|^2 \\
&= \|Ax\|^2 + \|x\|^2.
\end{aligned}
\tag{20.46}
$$

If $(A \pm i)x = 0$, $\|Ax\|^2 + \|x\|^2 = 0$ which implies $\|x\| = 0$ and the $1:1$-ness has been demonstrated.

Next it is shown that $\overline{R(A + i)} = X$. To this end, suppose

$$z \perp R(A + i) \qquad \text{and} \qquad x \in D_A.$$

In this case

$$(Ax + ix, z) = (Ax, z) + i(x, z) = 0,$$

which implies

$$(Ax, z) = (x, iz)$$

for all $x \in D_A$. Therefore,

$$z \in D_{A^*} \qquad \text{and} \qquad A^*z = iz.$$

Since A is self-adjoint, this is the same as

$$z \in D_A \qquad \text{and} \qquad Az = iz.$$

But, since $(A - i)$ is $1:1$, $(A - i)z = 0$ implies that $z = 0$. In summary, then,

$$z \perp \overline{R(A + i)} \Rightarrow z = 0,$$

and by Theorem 10.6 it follows that $\overline{R(A + i)} = X$.

We now show that $\overline{R(A + i)} = R(A + i)$. By the denseness proved above, for any $y \in X$ there exists a sequence of vectors $\{y_n\}$ from $R(A + i)$ such that $y_n \to y$. Clearly $\{y_n\}$ is a Cauchy sequence and each y_n must be of the form

$$y_n = Ax_n + ix_n$$

where $x_n \in D_A$; hence, since

$$\|y_n - y_m\|^2 = \|A(x_n - x_m) + i(x_n - x_m)\|^2$$

which by (20.46) is equal to

$$\|A(x_n - x_m)\|^2 + \|x_n - x_m\|^2,$$

the sequences $\{Ax_n\}$ and $\{x_n\}$ must be Cauchy sequences. Since X is a Hilbert space, there must exist vectors x and z belonging to X such that

$$x_n \to x \qquad \text{and} \qquad Ax_n \to z.$$

The fact that $A = A^*$ is closed implies that

$$x \in D_A \qquad \text{and} \qquad Ax = z.$$

Collecting these results, we have shown that

$$Ax_n + ix_n \to Ax + ix = y \in R(A + i)$$

and proved that $R(A + i)$ is closed. In view of this and the fact that $R(A + i)$ is dense in X, $(A + i)^{-1}$ exists on the whole space. By interchanging a few minus and plus signs, one can easily see that a similar result holds for $A - i$; that is, $(A - i)^{-1}$ is defined on all of X. We can now apply $(A + i)^{-1}$ to any $x \in X$, and this operation must yield a vector in

$$D_{A+i} = D_{A-i} = D_A.$$

Since $(A + i)^{-1}$ can be applied to any x and since $R(A - i) = X$, any $x \in X$ is the image of some vector under U—the mapping U is onto all of X.

So far we have shown U to be defined on the whole space, linear (this property is clear), and onto; hence, it only remains to show that it is an isometry to complete the proof. To show that it is an isometry, consider any $y \in X$ and write y as $(A + i)x$, where $x \in D_A$. In this case

$$Uy = (A - i)(A + i)^{-1}(A + i)x = (A - i)x.$$

Using (20.46) again,

$$\|Uy\|^2 = \|(A - i)x\|^2 = \|Ax\|^2 + \|x\|^2$$
$$= \|Ax + ix\|^2 = \|y\|^2.$$

This completes the proof.

To make the analogy of the Cayley transform with the Moebius transformation more complete, we now show that, knowing U, we can recover A analogous to the way z can be recovered if w is known; that is, it is analogous to

$$z = \frac{1}{i} \frac{w + 1}{w - 1} = i \frac{w + 1}{1 - w}.$$

For any $y \in X$, there exists $x \in D_A$ such that

$$y = (A + i)x.$$

Applying U to y, we have

$$Uy = (A - i)x.$$

Adding and subtracting these two equations yields

$$(1 + U)y = 2Ax \quad \text{and} \quad (1 - U)y = 2ix.$$

This implies that if $(1 - U)y = 0$, $x = 0$; thus $(1 - U)$ is $1:1$ and its inverse exists on $R(1 - U)$. For $x \in R(1 - U)$, then,

$$(1 + U)(1 - U)^{-1} 2ix = 2Ax,$$

which implies

$$A = i(1 + U)(1 - U)^{-1},$$

it being simple to verify that $D_A = R(1 - U)$; in particular, any $y \in X$ has the property that $(1 - U)y = 2ix \in D_A$ and, conversely, for any $x \in D_A$, consider $y = (A + i)x$. In this case $2ix = (1 - U)y$ so that $x \in R(1 - U)$.

20.4 Sesquilinear Functionals

In this section a new kind of functional will be introduced—a *sesquilinear* functional. Then, in a fashion similar to the case of linear functionals, the notion of a *bounded* sesquilinear functional is defined. The analogy continues as we note that the class of bounded sesquilinear functionals forms a normed linear space, where the norm introduced is defined in pretty much the same way as it was for the conjugate space.

As will be immediately seen from our first definition, the prototype sesquilinear functional is an inner product.

DEFINITION 20.1. Let X be a complex linear space. A mapping,

$$f : X \times X \to C,$$

$$\langle x, y \rangle \to f\langle x, y \rangle$$

with the following properties is called a *sesquilinear functional*:

(1) $f\langle x_1 + x_2, y \rangle = f\langle x_1, y \rangle + f\langle x_2, y \rangle$;

(2) $f\langle \alpha x, y \rangle = \alpha f\langle x, y \rangle$;

[Conditions (1) and (2) mean that f is linear in the first argument.]

(3) $f\langle x, y_1 + y_2 \rangle = f\langle x, y_1 \rangle + f\langle x, y_2 \rangle$ (additivity in the second argument);

(4) $f\langle x, \alpha y \rangle = \bar{\alpha} f\langle x, y \rangle$.

As noted above, an inner product is an example of a sesquilinear functional. Also, if X is an inner product space and

$$A : X \to X$$

is linear, then

$$\langle x, y \rangle \to (Ax, y)$$

is a sesquilinear functional.

If f is a sesquilinear functional, then

$$g\langle x, y \rangle = \overline{f\langle y, x \rangle}$$

is a sesquilinear functional. If it happens that $g = f$, that is, if

$$f\langle x, y \rangle = \overline{f\langle y, x \rangle}$$

for all $x, y \in X$, then f is called a *symmetric* sesquilinear functional.

If, for an arbitrary sesquilinear functional f,

$$f\langle x, x \rangle \geq 0, \tag{20.47}$$

then f is called *positive*; if the strict inequality prevails whenever $x \neq 0$, then f is called *strictly positive* or *positive definite*. An inner product is of course strictly positive and symmetric. If in (20.47) f was an inner product, the quantity designated there leads to a norm—something of strong interest. $f\langle x, x \rangle$ is still a quantity of interest in the case when f is an arbitrary sesquilinear functional, and is called the *quadratic form associated with the sesquilinear functional f*. When inner products are present, it is possible to recover the inner product, just knowing the norm, as is illustrated by the polarization identity, Eq. (20.41). A very similar result prevails for arbitrary sesquilinear functionals; specifically, letting $\hat{f}(x) = f\langle x, x \rangle$,

$$f\langle x, y \rangle = \hat{f}(\tfrac{1}{2}(x + y)) - \hat{f}(\tfrac{1}{2}(x - y)) + i\hat{f}(\tfrac{1}{2}(x + iy)) - i\hat{f}(\tfrac{1}{2}(x - iy)). \tag{20.48}$$

This is a straightforward computation and will not be verified here.

As one might suspect from the above formula, knowledge of the associated quadratic form uniquely determines the sesquilinear functional. In other words, if f_1 and f_2 are sesquilinear functionals with associated quadratic forms $\hat{f}_1$ and $\hat{f}_2$ and $\hat{f}_1 = \hat{f}_2$, then f_1 must equal f_2. To verify this, one need only apply Eq. (20.48).

Our next theorem shows that a sesquilinear functional is symmetric when, and only when, the associated quadratic form takes on only real values.

THEOREM 20.13. Let f be a sesquilinear functional and let $\hat{f}$ be the associated quadratic form. Then f is symmetric if, and only if, $\hat{f}$ is real-valued.

Proof. Suppose first that f is symmetric, that is,

$$f\langle x, y \rangle = \overline{f\langle y, x \rangle}.$$

Hence

$$\hat{f}(x) = f\langle x, x \rangle = \overline{f\langle x, x \rangle} = \overline{\hat{f}(x)},$$

which proves the necessity of the condition. On the other hand, suppose $\hat{f}(x)$ is real. Define

$$g\langle x, y \rangle = \overline{f\langle y, x \rangle}.$$

Then

$$\hat{g}(x) = g\langle x, x \rangle = \overline{f\langle x, x \rangle} = \overline{\hat{f}(x)}$$
$$= \hat{f}(x),$$

which implies, by the discussion preceding this theorem, that

$$g = f.$$

Suppose A is a Hermitian linear transformation on an inner product space with domain D_A and consider the sesquilinear functional

$$f\langle x, y \rangle = (Ax, y) = (x, Ay)$$
$$= \overline{(Ay, x)}$$
$$= \overline{f\langle y, x \rangle}.$$

Therefore f is a symmetric, sesquilinear functional on $D_A \times D_A$. By Theorem 20.13, this means that (Ax, x) is real. We now show that the converse of this statement is also true—namely, if (Ax, x) is real, then A is Hermitian. So suppose (Ax, x) is real for all $x \in D_A$ and consider the sesquilinear functional

$$f\langle x, y \rangle = (Ax, y).$$

Applying Theorem 20.13, we see that f must be symmetric, since its associated quadratic form is real-valued. The symmetry of f is equivalent to A being Hermitian and we have proved the next theorem.

THEOREM 20.14. Consider the linear transformation $X \supset D_A \xrightarrow{A} X$, where X is an inner product space. Then A is Hermitian if, and only if, (Ax, x) is real for all $x \in D_A$.

Bounded Sesquilinear Functionals

In the subsequent discussion we assume X to be a Hilbert space.

DEFINITION 20.2. The sesquilinear functional f is called *bounded* if there is some positive constant k such that

$$|f\langle x, y\rangle| \leq k\|x\| \|y\| \quad \text{(for all } x, y \in X\text{)}. \tag{20.49}$$

For sesquilinear functionals f_1 and f_2, and $\alpha \in C$, if we define $f_1 + f_2$ and αf, point-wise as

$$(f_1 + f_2)\langle x, y\rangle = f_1\langle x, y\rangle + f_2\langle x, y\rangle \quad \text{and} \quad (\alpha f)\langle x, y\rangle = \alpha f\langle x, y\rangle,$$

the class of bounded sesquilinear functionals over the space X forms a linear space. Letting K denote the set of all positive constants k satisfying (20.49), the following norm can be introduced on this space: For a bounded sesquilinear functional f over X,

$$\|f\| = \inf_{k \in K} k. \tag{20.50}$$

Consider the set of all quadratic forms associated with sesquilinear functionals over X. Letting $\hat{f}$ be a particular associated quadratic form, we define $\hat{f}$ to be a *bounded associated quadratic* form if there is some positive k such that

$$|\hat{f}(x)| \leq k\|x\|^2 \quad \text{(for all } x \in X\text{)}. \tag{20.51}$$

Letting K denote the set of all k satisfying (20.51), the following "norm" can be introduced on this set:

$$\|\hat{f}\| = \inf_{k \in K} k. \tag{20.52}$$

We can now establish the following analogs of familiar results about bounded linear functionals and bounded linear transformations.

For a bounded sesquilinear functional f,

$$\|f\| = \sup_{\|x\| = \|y\| = 1} |f\langle x, y\rangle|. \tag{20.53}$$

For a bounded associated quadratic form $\hat{f}$,

$$\|\hat{f}\| = \sup_{\|x\| = 1} |\hat{f}(x)|. \tag{20.54}$$

The proofs of these statements are quite similar and for this reason only (20.53) will be proved here.

Proof. It is easy to verify that $\|f\|$ itself is a bound for f in the sense of (20.49), just using (20.50). So, for $\|x\| = \|y\| = 1$,

$$|f\langle x, y\rangle| \leq \|f\| \|x\| \|y\| = \|f\|,$$

which implies

$$\sup_{\|x\| = \|y\| = 1} |f\langle x, y\rangle| \leq \|f\|. \tag{20.55}$$

We now show that (20.55) also goes the other way around. Consider, for any nonzero x and y,

$$|f\langle x, y\rangle| = \left|f\left\langle \frac{x}{\|x\|}\|x\|, \frac{y}{\|y\|}\|y\|\right\rangle\right|$$

$$= \left|f\left\langle \frac{x}{\|x\|}, \frac{y}{\|y\|}\right\rangle\right|\|x\|\cdot\|y\|$$

$$\leq \left(\sup_{\|x\|=\|y\|=1}|f\langle x, y\rangle|\right)\|x\|\cdot\|y\|;$$

therefore, with K as in (20.50),

$$\sup_{\|x\|=\|y\|=1}|f\langle x, y\rangle| \in K,$$

so certainly

$$\sup_{\|x\|=\|y\|=1}|f\langle x, y\rangle| \geq \inf_{k\in K} k = \|f\|. \tag{20.56}$$

Combining this with (20.55), Eq. (20.53) follows.

Our next result links the boundedness of the sesquilinear functional f with the boundedness of its associated quadratic form $\hat{f}$. One may have suspected that the two conditions were not independent and, indeed, they are equivalent.

THEOREM 20.15. f is bounded if and only if $\hat{f}$ is bounded; further, if f (and $\hat{f}$) are bounded, then

$$\|\hat{f}\| \leq \|f\| \leq 2\|\hat{f}\|. \tag{20.57}$$

Proof. If f is bounded, then

$$|\hat{f}(x)| = |f\langle x, x\rangle| \leq \|f\|\,\|x\|\,\|x\|,$$

which implies that not only is $\hat{f}$ bounded but also that

$$\|\hat{f}\| \leq \|f\|. \tag{20.58}$$

Suppose that $\hat{f}$ is bounded now. By Eq. (20.48),

$$f\langle x, y\rangle = \hat{f}\left(\tfrac{1}{2}(x+y)\right) - \hat{f}\left(\tfrac{1}{2}(x-y)\right) + i\hat{f}\left(\tfrac{1}{2}(x+iy)\right) - i\hat{f}\left(\tfrac{1}{2}(x-iy)\right).$$

Applying the triangle inequality for absolute values and using the fact that $\hat{f}$ is bounded, we find

$$|f\langle x, y\rangle| \leq \tfrac{1}{4}\|\hat{f}\|\,(\|x+y\|^2 + \|x-y\|^2 + \|x+iy\|^2 + \|x-iy\|^2).$$

Using the Parallelogram law (Theorem 1.4), the term on the right reduces to

$$\tfrac{1}{4}\|\hat{f}\|(2\|x\|^2 + 2\|y\|^2 + 2\|x\|^2 + 2\|iy\|^2) = \|\hat{f}\|\,(\|x\|^2 + \|y\|^2),$$

which enables us to conclude, for $\|x\| = \|y\| = 1$, that

$$\sup_{\|x\|=\|y\|=1}|f\langle x, y\rangle| \leq 2\|\hat{f}\|. \tag{20.59}$$

The fact that

$$\sup_{\|x\|=\|y\|=1}|f\langle x, y\rangle|$$

is finite implies that f is bounded, and (20.59) implies

$$\|f\| \le 2\|\hat{f}\|.$$

Combining this with (20.58), the theorem is proved.

Some strength can be added to (20.57) if f is symmetric; in particular, we now prove the next theorem.

THEOREM 20.16. If f is a bounded, symmetric, sesquilinear functional, then $\|f\| = \|\hat{f}\|$.

Proof. By Eq. (20.48) and Theorem 20.13, we see that

$$|\operatorname{Re} f\langle x, y\rangle| \le |\hat{f}(\tfrac{1}{2}(x + y))| + |\hat{f}(\tfrac{1}{2}(x - y))|,$$

which, since $\hat{f}$ must be bounded, is less than or equal to

$$\tfrac{1}{4}\|\hat{f}\|(\|x + y\|^2 + \|x - y\|^2) = \tfrac{1}{4}\|\hat{f}\|(2\|x\|^2 + 2\|y\|^2).$$

Hence, if $\|x\| = \|y\| = 1$,

$$|\operatorname{Re} f\langle x, y\rangle| \le \|\hat{f}\|. \tag{20.60}$$

Write $f\langle x, y\rangle$ in polar form now:

$$f\langle x, y\rangle = re^{i\theta}.$$

Letting $\alpha = e^{-i\theta}$, we have

$$\alpha f\langle x, y\rangle = r = |f\langle x, y\rangle|.$$

Now, invoking (20.60), it follows that

$$\|f\| \ge |\operatorname{Re} f\langle \alpha x, y\rangle| = |\operatorname{Re} \alpha f\langle x, y\rangle| = |f\langle x, y\rangle|$$

for all x, y such that $\|x\| = \|y\| = 1$; consequently,

$$\|f\| = \sup_{\|x\| = \|y\| = 1} |f\langle x, y\rangle| \le \|\hat{f}\|. \tag{20.61}$$

This result in conjunction with (20.57) yields the theorem.

We now present a result for sesquilinear functionals generalizing the Cauchy–Schwarz inequality for inner products.

THEOREM 20.17. Let f be a positive sesquilinear functional on the complex vector space X. Then

$$|f\langle x, y\rangle|^2 \le \hat{f}(x)\hat{f}(y) \qquad \text{(for all } x, y \in X\text{)}.$$

Proof. The inequality is of course satisfied if $f\langle x, y\rangle = 0$; therefore, assume that $f\langle x, y\rangle \ne 0$. Then for arbitrary complex numbers α, β, we have

$$0 \le \hat{f}(\alpha x + \beta y) = f\langle \alpha x + \beta y, \alpha x + \beta y\rangle$$

$$= \alpha\bar{\alpha}\hat{f}(x) + \alpha\bar{\beta}f\langle x, y\rangle + \bar{\alpha}\beta f\langle y, x\rangle + \beta\bar{\beta}\ (y)$$

$$= \alpha\bar{\alpha}\hat{f}(x) + \alpha\bar{\beta}f\langle x, y\rangle + \bar{\alpha}\beta\overline{f\langle x, y\rangle} + \beta\bar{\beta}\hat{f}(y).$$

since f is positive. Now let $\alpha = t$ be real and take $\beta = f\langle x, y\rangle/|f\langle x, y\rangle|$. Then

$$\bar{\beta}f\langle x, y\rangle = |f\langle x, y\rangle| \quad \text{and} \quad \beta\bar{\beta} = 1.$$

Hence,

$$0 \le t^2\hat{f}(x) + 2t|f\langle x, y\rangle| + \hat{f}(y)$$

for arbitrary real t. Thus the discriminant,

$$4|f\langle x, y\rangle|^2 - 4\hat{f}(x)\hat{f}(y) \le 0,$$

which completes the proof.

EXERCISES 20

1. Let $A, B \in L(X, X)$, where X is a Hilbert space. Show that, if A and B are Hermitian, then AB is Hermitian if and only if A and B commute.

2. Let $A \in L(X, X)$, where X is a Hilbert space. Show that there exist two uniquely determined Hermitian transformations B and C such that $A = B + iC$. Show, in addition, that A is normal if and only if B and C commute.

3. Let X and Y be Hilbert spaces and $A : X \to Y$, where $A \in L(X, Y)$. Show that A is completely continuous if and only if A^*A is completely continuous.

4. Let X and Y be Hilbert spaces. If $A \in L(X, Y)$ and if A is completely continuous, show that A^* is also completely continuous.

5. Let X be a Hilbert space. If A is a symmetric operator, show that $\bar{A}$ exists and is also symmetric.

6. A symmetric transformation is said to be *maximal* if it has no proper symmetric extensions. Show that a self adjoint transformation is a maximal symmetric transformation.

7. (a) If A is a maximal symmetric transformation, then $A = A^{**}$. (b) Every maximal symmetric extension of a symmetric transformation A is also an extension of A^{**}. (c) If A is a symmetric transformation such that $A^* = A^{**}$ (a symmetric transformation with this property is called *essentially self-adjoint*), then A^* is the only maximal symmetric extension of A.

8. Show that A^* cannot be defined uniquely if $\bar{D}_A \ne X$.

9. Let A be a linear transformation such that A^* and A^{**} exist. Show that $(A^{**})^* = A^*$.

10. Show that, if A is a symmetric transformation, then A^{**} exists.

11. Let f be a positive sesquilinear functional on X and let $N = \{x \in X | \hat{f}(x) = 0\}$. Show that $N = \{x \in X | f\langle x, y \rangle = 0$ all $y \in X\}$. [*Hint:* use the fact that for such an f, $|f\langle x, y \rangle|^2 \leq \hat{f}(x)\hat{f}(y)$ for all $x, y \in X$.]

12. Give an example of a sesquilinear functional that is not symmetric.

13. Let X be a Hilbert space. Suppose $A, B : X \to X$ and are such that $(Ax, y) = (x, By)$ for all $x, y \in X$. Then show that $A \in L(X, X)$ and $B = A^*$.

14. If U is a unitary transformation, prove that $\|U\| = 1$.

15. Let X be a Banach space and let $f\langle x, y \rangle$ be a complex-valued sesquilinear functional that is continuous in x for every fixed y and vice versa. Prove that $f\langle x, y \rangle$ is continuous on $X \times X$.

REFERENCES

1. S. Berberian, "Introduction to Hilbert Space."
2. M. Naimark, "Normed Rings."
3. M. Stone, "Linear Transformations in Hilbert Space."

CHAPTER 21

Some Spectral Results for Normal and Completely Continuous Operators

The concept of a sesquilinear functional, introduced in the last chapter, will prove to be an extremely useful tool here. Needless to say, there may be other ways of establishing some of the results we will obtain, but the treatment using sesquilinear functionals, aside from its expediency, has a certain crisp quality to it that makes it, esthetically, quite pleasant.

In this chapter, as in the last, the setting of Hilbert spaces will be the most usual one. In some instances it might be possible to weaken this requirement and work in only an inner product space, however. Even at this stage of the game it is difficult to start tracing back to see what results apply only in Hilbert spaces, and, for this reason, we shall rarely try to avail ourselves of the greater generality some of our results have. Our main interest as well as the truth and proof of most of our theorems from now on will be in Hilbert spaces.

Our long-range goal is to obtain decomposition theorems in infinite-dimensional Hilbert spaces, such as the spectral decomposition theorem for normal transformations, proved in Sec. 2.2. To further this end, certain vital results about the spectra of normal, self-adjoint, and completely continuous transformations will be proved. In particular, it will be shown that the spectrum of a normal transformation (mapping a Hilbert space into itself) coincides with the approximate spectrum —an extremely useful result, because the approximate spectrum can be described by (seemingly) less stringent requirements. Further, it will be shown that the spectrum of a self-adjoint transformation is purely real. We then devote our attention to completely continuous operators. The results obtained for these transformations are quite similar to results we obtained for linear transformations on finite-dimensional spaces.

21.1 A New Expression for the Norm of $A \in L(X, X)$

We have already obtained several convenient expressions for computing the norm of a bounded linear transformation. Since working in the structurally powerful setting of Hilbert spaces usually makes life a little easier, it is hardly surprising that we can obtain still another expression for the norm of a bounded linear transformation in such a situation.

Before doing this we shall establish a one-to-one correspondence (actually an isometry) between the space of bounded sesquilinear functionals on a complex Hilbert space X, and $L(X, X)$.

THEOREM 21.1. Let X be a complex Hilbert space, and let $A \in L(X, X)$. If

$$f\langle x, y \rangle = (Ax, y),$$

then f is a bounded sesquilinear functional and $\|f\| = \|A\|$. Conversely, if f is a bounded sesquilinear functional, there exists a unique $A \in L(X, X)$ such that $f\langle x, y \rangle = (Ax, y)$.

Proof. For arbitrary linear transformations A_1 and A_2, each mapping X into itself, if

$$(A_1 x, y) = (A_2 x, y) \tag{21.1}$$

for all $x, y \in X$ then, clearly,

$$A_1 = A_2.$$

Suppose that A_1 and A_2 satisfy the weaker condition that

$$(A_1 x, x) = (A_2 x, x) \qquad \text{(for all } x\text{).} \qquad (21.2)$$

Even this condition implies that $A_1 = A_2$, for consider the sesquilinear functionals,

$$f\langle x, y \rangle = (A_1 x, y) \qquad \text{and} \qquad g\langle x, y \rangle = (A_2 x, y).$$

Equation (21.2) implies that the associated quadratic forms of f and g are equal; therefore [see Sec. 20.4, especially Eq. (20.48)], f and g must be equal or, for all $x, y \in X$,

$$(A_1 x, y) = (A_2 x, y).$$

As we have already noted, this means that $A_1 = A_2$. We now prove the first part of the theorem. Let $A \in L(X, X)$ and consider $f\langle x, y \rangle = (Ax, y)$. The Cauchy–Schwarz inequality implies that, for all $x, y \in X$,

$$|f\langle x, y \rangle| = |(Ax, y)| \le \|A\| \, \|x\| \, \|y\|,$$

which means that f is a bounded sesquilinear functional and that

$$\|f\| \le \|A\|. \qquad (21.3)$$

Since, for every x,

$$\|Ax\|^2 = (Ax, Ax) = f\langle x, Ax \rangle = |f\langle x, Ax \rangle| \le \|f\| \, \|x\| \, \|Ax\|,$$

we see that

$$\|Ax\| \le \|f\| \, \|x\|,$$

which implies

$$\|A\| \le \|f\| \qquad (21.4)$$

and proves that $\|f\| = \|A\|$.

Conversely, suppose f is a bounded sesquilinear functional and consider the functional

$$g_x(y) = \overline{f\langle x, y \rangle}.$$

Clearly, g_x is a linear functional and, to see that it is also bounded, consider

$$|g_x(y)| = |f\langle x, y \rangle| \le \|f\| \, \|x\| \, \|y\|;$$

hence g_x is bounded and

$$\|g_x\| \le \|f\| \, \|x\|. \qquad (21.5)$$

Since g_x is a bounded linear functional then, by the Riesz representation theorem (Sec. 12.4), there must be some unique vector z such that

$$g_x(y) = \overline{f\langle x, y \rangle} = (y, z)$$

and

$$\|g_x\| = \|z\|. \qquad (21.6)$$

Denote the vector z by Ax now. In this notation we have

$$f\langle x, y \rangle = (Ax, y),$$

and to complete the proof it only remains to show that A is a bounded linear transformation. As for the linearity of A, consider

$$g_{x_1+x_2}(y) = (y, A(x_1 + x_2)).$$

Since

$$\begin{aligned}
g_{x_1+x_2}(y) &= \overline{f\langle x_1 + x_2, y\rangle} \\
&= \overline{f\langle x_1, y\rangle} + \overline{f\langle x_2, y\rangle} \\
&= g_{x_1}(y) + g_{x_2}(y) \\
&= (y, Ax_1) + (y, Ax_2) \\
&= (y, Ax_1 + Ax_2),
\end{aligned}$$

for all $y \in X$, we see that

$$A(x_1 + x_2) = Ax_1 + Ax_2.$$

The homogeneity of A follows similarly; thus A is linear. As for the boundedness of A we see that (21.5) and (21.6) imply

$$\|g_x\| = \|z\| = \|Ax\| \le \|f\| \, \|x\|.$$

Having established this isometry between the space of bounded sesquilinear functionals over a Hilbert space X and $L(X, X)$, then given any $A \in L(X, X)$, we can speak of its associated bounded sesquilinear functional, this functional being obtained by the procedure given in Theorem 21.1. Thus with A and f as bounded linear transformation and associated bounded sesquilinear functional, we see that we can now write a new expression for the norm of A [see Eq. (20.53)]:

$$\|A\| = \sup_{\|x\|=\|y\|=1} |f\langle x, y\rangle| = \sup_{\|x\|=\|y\|=1} |(Ax, y)|. \tag{21.7}$$

Suppose that A is a bounded, self-adjoint transformation now, and consider the bounded sesquilinear functional associated with it, f. Since A is self-adjoint, f must be symmetric (Theorems 20.13 and 20.14; see also the remarks preceding Theorem 20.4). Since f is symmetric, then, by Theorem 20.16, denoting its associated quadratic form by $\hat{f}$, we have

$$\|f\| = \|\hat{f}\|.$$

We can compute $\|\hat{f}\|$, however, as follows [Eq. (20.54)]:

$$\|\hat{f}\| = \sup_{\|x\|=1} |\hat{f}(x)|.$$

Hence, for a bounded, self-adjoint transformation A,

$$\|A\| = \sup_{\|x\|=1} |(Ax, x)|. \tag{21.8}$$

In the next section it will be shown that (21.8) holds even for normal transformations.

21.2 Normal Transformations

Throughout this section, we assume that X is a complex Hilbert space and that $A \in L(X, X)$. The first result is an extension of a familiar result that was obtained in Lemma 2.1 for the case when X was finite-dimensional. The proof here, as in Chapter 2, relies on the Riesz representation theorem, which allows us to guarantee the existence of A^*.

THEOREM 21.2. A is normal if and only if $\|A^*x\| = \|Ax\|$.

Proof. Consider

$$\|Ax\|^2 = (Ax, Ax) = (A^*Ax, x) \tag{21.9}$$

and

$$\|A^*x\|^2 = (A^*x, A^*x) = (A^{**}A^*x, x) = (AA^*x, x), \tag{21.10}$$

the fact that $A^{**} = A$ following from Theorem 20.3 (3). Necessity is obvious; the sufficiency of the conditions follows from the observations concerning Eq. (21.2).

As can be seen from Sec. 19.5 [especially Eq. (19.21)], the spectral radius of A, $r_\sigma(A)$, is always less than or equal to $\|A\|$. We now show that equality prevails if A is normal.

THEOREM 21.3. If A is normal, $r_\sigma(A) = \|A\|$; moreover, there is some $\lambda \in \sigma(A)$ such that $|\lambda| = \|A\|$.

Proof. First replace x by Ax in the preceding theorem. The result of that theorem implies

$$\|A^2x\| = \|A^*Ax\|;$$

so $\|A^2\| = \|A^*A\|$.
This in turn implies, using Theorem 20.8,

$$\|A^2\| = \|A^*A\| = \|A\|^2.$$

Hence,

$$\|A^2\| = \|A\|^2. \tag{21.11}$$

Since $(A^j)^* = (A^*)^j$ [see Eq. (20.36)], if A is normal then so is A^n. Hence it follows by induction that

$$\|A^m\| = \|A\|^m \tag{21.12}$$

for all integers m of the form 2^k. Now, using Eq. (19.22), we can write

$$r_\sigma(A) = \lim_n \|A^n\|^{1/n}$$

$$= \lim_n \|A^{2^n}\|^{1/2^n}$$

$$= \lim_n \|A\|$$

$$= \|A\|.$$

Since $|\lambda|$ is a continuous function of λ and $\sigma(A)$ is a compact set (Sec. 19.3), $|\lambda|$ must attain a maximum value on $\sigma(A)$; in other words,

$$\|A\| = r_\sigma(A) = \sup_{\lambda \in \sigma(A)} |\lambda| = \max_{\lambda \in \sigma(A)} |\lambda|,$$

and we see that there must be some $\lambda \in \sigma(A)$ such that $|\lambda| = \|A\|$. The proof is now complete.

In Sec. 18.3 the notion of the approximate spectrum of A, $\pi(A)$, was introduced. It was noted there that

$$\pi(A) \subset \sigma(A), \tag{21.13}$$

in general. For normal transformations, the inclusion can be reversed, which enables us to conclude that the spectrum and approximate spectrum coincide in this case. We now prove this result.

THEOREM 21.4. If A is normal, then $\pi(A) = \sigma(A)$.

Proof. By virtue of the preceding discussion, it only remains to show that $\sigma(A) \subset \pi(A)$. To prove this, it suffices to show that the complement of $\pi(A)$ is contained in the complement of $\sigma(A)$, $\rho(A)$. To this end, suppose that $\lambda \notin \pi(A)$. This means that, for some $\varepsilon > 0$,

$$\|Ay - \lambda y\| \geq \varepsilon \|y\| \tag{21.14}$$

for all $y \in X$. Equation (21.14) implies that $A - \lambda$ has a bounded inverse (Theorem 14.9). If we can now show that the range of $A - \lambda$ is dense in X, we can conclude that $\lambda \in \rho(A)$—the desired result. That is, we wish to show that

$$\overline{R(A - \lambda)} = X. \tag{21.15}$$

One can easily demonstrate that (21.15) is equivalent to the following condition:

$$R(A - \lambda)^\perp = \{0\}. \tag{21.16}$$

[To prove the equivalence of (21.15) and (21.16), one need only take orthogonal complements and use observation (7) of Sec. 10.4.]

Clearly if A is normal then so is $A - \lambda$. In view of the normality of $A - \lambda$, Eq. (21.14) coupled with Theorem 21.2 imply that

$$\|A^*y - \bar{\lambda} y\| \geq \varepsilon \|y\|. \tag{21.17}$$

Suppose now that

$$y \in R(A - \lambda)^\perp = N(A^* - \bar{\lambda})$$

[Eq. (20.37)]. This implies that

$$A^*y - \bar{\lambda} y = 0$$

which, together with (21.17), implies that $y = 0$. Thus (21.16) has been established. It now follows that $\lambda \in \rho(A)$, which in turn implies

$$C\pi(A) \subset \rho(A)$$

and establishes the theorem.

The importance of this result will be realized in the proof of the next theorem, for we now have a more convenient way of characterizing the spectral values of a normal transformation. The main thing that our next theorem does is allow us to use Eq. (21.8) for normal transformations.

THEOREM 21.5. For a bounded linear transformation A, $A \in L(X, X)$, the following statements are equivalent:

(1) There exists $\lambda \in \pi(A)$ such that $|\lambda| = \|A\|$;

(2) $\|A\| = \sup\limits_{\|x\| = 1} |(Ax, x)|$.

Proof (1) $\Rightarrow$ (2). Let λ be as in (1). If we can show that

$$|\lambda| \in \overline{\{|(Ax, x)| \mid \|x\| = 1\}},$$

it will follow that

$$\|A\| = |\lambda| \leq \sup_{\|x\| = 1} |(Ax, x)| \leq \sup_{\|x\| = \|y\| = 1} |(Ax, y)| = \|A\|,$$

and the result will be proved. Since $\lambda \in \pi(A)$, there must exist $x_n \in X$, $\|x_n\| = 1$, such that

$$\|Ax_n - \lambda x_n\| \to 0.$$

Since $\|x_n\| = 1$, we can write

$$|(Ax_n, x_n) - \lambda| = |(Ax_n, x_n) - \lambda(x_n, x_n)| = |(Ax_n - \lambda x_n, x_n)|$$

$$\leq \|Ax_n - \lambda x_n\| \to 0,$$

which implies

$$|(Ax_n, x_n)| \to |\lambda|$$

and, by the preceding remarks, the proof that (1) $\Rightarrow$ (2) is complete.

Proof (2) $\Rightarrow$ (1). If (2) holds, there must be vectors x_n, $\|x_n\| = 1$, such that

$$|(Ax_n, x_n)| \to \|A\|. \tag{21.18}$$

The sequence $\{(Ax_n, x_n)\}$ is therefore a bounded sequence of complex numbers and as such must have a convergent subsequence:

$$(Ax_{n_k}, x_{n_k}) \to \lambda. \tag{21.19}$$

Equation (21.18) implies that

$$|\lambda| = \|A\|;$$

hence to complete the proof, it suffices to show

$$\|Ax_{n_k} - \lambda x_{n_k}\| \to 0.$$

To prove this, consider (noting that $\|Ax\| \leq \|A\| \|x\| = |\lambda| \|x\|$)

$$\|Ax_{n_k} - \lambda x_{n_k}\|^2 = \|Ax_{n_k}\|^2 - (Ax_{n_k}, \lambda x_{n_k}) - (\lambda x_{n_k}, Ax_{n_k}) + |\lambda|^2$$

$$\leq |\lambda|^2 \|x_{n_k}\|^2 - \bar{\lambda}(Ax_{n_k}, x_{n_k}) - \lambda(x_{n_k}, Ax_{n_k}) + |\lambda|^2.$$

Equation (21.19) implies that the term on the right goes to

$$|\lambda|^2 - |\lambda|^2 - |\lambda|^2 + |\lambda|^2 = 0;$$

the proof is now complete.

Theorem 21.3 implies, for a normal transformation A, that there is always some $\lambda \in \sigma(A)$ such that $|\lambda| = \|A\|$. Theorem 21.4, however, states that the approximate spectrum and the spectrum coincide for normal transformations; therefore condition (1) of Theorem 21.5 is always satisfied for normal transformations. So, by Theorem 21.5, the norm of a normal, bounded linear transformation can be computed via

$$\|A\| = \sup_{\|x\|=1} |(Ax, x)|. \tag{21.20}$$

Certainly (2) holds for self-adjoint transformations, by Eq. (21.8). If one considers (21.19) in the case when A is self-adjoint, it is evident that $\lambda = \bar{\lambda}$ or that λ is real. Thus, if A is self-adjoint, then either $\|A\|$ or $-\|A\|$ belongs to $\pi(A)$.

In Theorem 1.10 we showed that if X was finite-dimensional and A was self-adjoint, all eigenvalues of A were real. We shall now extend this result and show, for any Hilbert space that, if A is self-adjoint, its entire spectrum is real.

THEOREM 21.6. If A is self-adjoint, $\sigma(A) = \pi(A) \subset R$.

REMARK. This means that the entire spectrum of a self-adjoint transformation A lies within the closed interval $[-\|A\|, \|A\|]$.

Proof. For any complex number λ with nonzero imaginary part and any $x \neq 0$,

$$\begin{aligned}
0 < |\lambda - \bar{\lambda}| \|x\|^2 &= |((A - \lambda)x, x) - ((A - \bar{\lambda})x, x)| \\
&= |((A - \lambda)x, x) - (x, (A^* - \lambda)x)| \\
&= |((A - \lambda)x, x) - (x, (A - \lambda)x)| \\
&\leq 2\|(A - \lambda)x\| \|x\|.
\end{aligned}$$

If $\lambda \in \sigma(A) = \pi(A)$, there must exist x_n, $\|x_n\| = 1$, such that

$$\|(A - \lambda)x_n\| \to 0.$$

Since $2\|(A - \lambda)x_n\|$ must be greater than or equal to $|\lambda - \bar{\lambda}| > 0$, if λ has a nonzero imaginary part, we conclude that if $\lambda \in \sigma(A)$, then $\lambda = \bar{\lambda}$. This proves the theorem.

As a rule it is difficult to get one's hands on the residual spectrum of a transformation. If the transformation is normal, such difficulties are done away with because the residual spectrum is empty, a property which makes normal transformations very desirable to work with. We now prove this result.

THEOREM 21.7. If A is normal, $R\sigma(A) = \emptyset$.

Proof. For any scalar λ, if $R(A - \lambda)$ is not dense in X, either $\lambda \in R\sigma(A)$ or $\lambda \in P\sigma(A)$. In this proof we show that anytime the range of $A - \lambda$ is not dense in X, then λ must belong to the point spectrum. Let λ be such that

$$\overline{R(A - \lambda)} \neq X.$$

Hence

$$\overline{R(\lambda - A)}^{\perp} \neq \{0\}.$$

But [Eq. (20.37)]

$$\overline{R(\lambda - A)}^{\perp} = N(A^* - \bar{\lambda}).$$

Since A is normal, Theorem 21.2 implies that (noting that $A - \lambda$ is normal too)

$$N(A^* - \bar{\lambda}) = N(A - \lambda) \neq \{0\},$$

so that there must be some nonzero vector x, such that

$$(A - \lambda)x = 0$$

which implies that $\lambda \in P\sigma(A)$. The theorem now follows.

Using this result and the fact that $P\sigma(A) \cup C\sigma(A) \subset \pi(A)$ (see Exercise 18.2), an alternate proof of Theorem 21.4, namely, $\sigma(A) = \pi(A)$ for normal transformations, is available.

21.3 Some Spectral Results for Completely Continuous Operators

Many of the results that held for linear transformations on finite-dimensional spaces get through unscathed in the infinite-dimensional case, provided the additional hypothesis of complete continuity is imposed. Some results in this spirit will be proved in this section. As for notation, X will denote a complex Hilbert space and A a member of $L(X, X)$, throughout this section, with additional assumptions about A applying where indicated.

THEOREM 21.8. If A is completely continuous and λ is a nonzero scalar, then $N(A - \lambda)$ is finite-dimensional.

Proof. We shall prove the contrapositive of this result. Suppose, for $\lambda \neq 0$, $N(A - \lambda)$ is infinite-dimensional. If so, we can select an infinite collection of linearly independent vectors,

$$x_1, x_2, \ldots, x_n, \ldots$$

from $N(A - \lambda)$; moreover, in view of the Gram–Schmidt process (Theorem 1.7), this set of vectors can be assumed to be an orthonormal set. For any $n \neq m$,

$$\|Ax_n - Ax_m\|^2 = \|\lambda x_n - \lambda x_m\|^2 = |\lambda|^2 \|x_n - x_m\|^2 = 2|\lambda|^2,$$

which means that no subsequence of $\{Ax_n\}$ can possibly be a Cauchy sequence. This precludes the possibility of A being completely continuous and proves the theorem.

In the preceding section, we showed that the spectrum and the approximate spectrum of a normal transformation coincide. For completely continuous transformations, a somewhat similar result prevails. In this case one can almost say that the approximate spectrum coincides with the *point* spectrum.

THEOREM 21.9. If A is completely continuous and $\lambda \neq 0$, then

$$\lambda \in \pi(A) \Rightarrow \lambda \in P\sigma(A).$$

REMARK. Since $P\sigma(A) \subset \pi(A)$, in general, this implies that $\pi(A) - \{0\} = P\sigma(A) - \{0\}$.

Proof. Let λ be a nonzero member of $\pi(A)$ and choose a sequence of vectors $\{x_n\}$, each with norm one such that

$$\|Ax_n - \lambda x_n\| \to 0.$$

The complete continuity of A implies that $\{Ax_n\}$ contains a convergent subsequence:

$$Ax_{n_k} \to y.$$

Since any subsequence of a convergent sequence must converge to the same thing that the sequence converges to,

$$\|Ax_{n_k} - \lambda x_{n_k}\| \to 0,$$

too. Thus, since

$$\|y - \lambda x_{n_k}\| \leq \|y - Ax_{n_k}\| + \|Ax_{n_k} - \lambda x_{n_k}\|,$$

we see that

$$\lim_k \lambda x_{n_k} = y.$$

Applying A to both sides of this expression and using the continuity of A, we have

$$Ay = A \lim_k \lambda x_{n_k} = \lambda \lim_k Ax_{n_k} = \lambda y.$$

In view of this, if we can show that y is not zero, it will follow that $\lambda \in P\sigma(A)$. Since, by the continuity of the norm,

$$\|y\| = \lim_k \|\lambda x_{n_k}\| = |\lambda| > 0,$$

we see that y is, indeed, not zero; hence $\lambda \in P\sigma(A)$ and the theorem is proved.

In general,

$$P\sigma(A) \cup C\sigma(A) \subset \pi(A). \tag{21.21}$$

Just taking part of this result, we have, in general, that

$$P\sigma(A) \subset \pi(A).$$

If A is completely continuous, however, we have just shown that

$$\pi(A) - \{0\} \subset P\sigma(A) - \{0\}.$$

Consequently, if A is completely continuous,

$$\pi(A) - \{0\} = P\sigma(A) - \{0\}. \tag{21.22}$$

Also, using (21.21), we see that for completely continuous transformations A, only 0 could belong to the continuous spectrum $C\sigma(A)$.

One says that the subspace M is *invariant under* A if $A(M) \subset M$. Also the pair of subspaces M and N is said to *reduce* A if

(1) M and N are complementary subspaces, that is, $X = M \oplus N$;
(2) M and N are each invariant under A.

Note that even though a subspace M has many complementary subspaces, in general, it may not be possible to find one that is also invariant under A. These two definitions are rather familiar ones and have been previously encountered in linear algebra. For our purposes it is advantageous to redefine the second slightly in the special case when X is a Hilbert space. In this connection we state a definition.

DEFINITION 21.1. The subspace M *reduces* A if M and $M^\perp$ are each invariant under A. We remark that in most of the results to follow, we shall assume that M is a *closed* subspace, the main reason for this being that for such subspaces, $M = M^{\perp\perp}$ (in a Hilbert space).

A convenient, equivalent description of invariance of closed subspaces is provided by the following result:

(A) The closed subspace M is invariant under A if and only if $M^\perp$ is invariant under A^*.

Proof. Suppose $M^\perp$ is invariant under A^*. Then for $x \in M$ and $y \in M^\perp$:

$$(x, A^*y) = 0 = (Ax, y).$$

This implies that

$$Ax \in M^{\perp\perp} = M.$$

Hence M is invariant under A. The converse follows similarly.

Our next result provides an equivalent description of when M reduces A.

(B) The closed subspace M reduces A if and only if M is invariant under A and A^*.

Proof. Suppose that M reduces A; that is, that M is invariant under A and that $M^\perp$ is invariant under A. The last statement means, using result (A) above, that $M^{\perp\perp} = M$ is invariant under A^*. On the other hand, suppose that M is invariant under A and A^*. Since M is invariant under A^*, $M^\perp$ is invariant under $A^{**} = A$, and M is seen to reduce A.

By the results of Theorem 19.7, we know that the spectrum of A is never empty. For completely continuous, normal transformations one can go a step further and say that the point spectrum is never empty—completely continuous, normal transformations always possess eigenvalues. Also for normal transformations A, we know by Theorem 21.3 that there is some $\lambda \in \sigma(A)$ such that $|\lambda| = \|A\|$. If A is completely continuous as well as normal, we can strengthen this result to assert the existence of $\lambda \in P\sigma(A)$ such that $|\lambda| = \|A\|$.

THEOREM 21.10. If A is completely continuous and normal, then $P\sigma(A) \neq \emptyset$, and there is some $\lambda \in P\sigma(A)$ such that $|\lambda| = \|A\|$.

Proof. Since A is normal we know that†

$$\pi(A) = \sigma(A),$$

so that certainly

$$\sigma(A) - \{0\} = \pi(A) - \{0\}.$$

By Theorem 21.9, however,

$$\pi(A) - \{0\} = P\sigma(a) - \{0\}. \tag{21.23}$$

Since one must always have $P\sigma(A) - \{0\} \subset \sigma(A) - \{0\}$, it follows that

$$P\sigma(A) - \{0\} = \sigma(A) - \{0\}.$$

Clearly if $A = 0$, we are done. On the other hand, if A is not zero, $\|A\| > 0$, and since A is normal‡ $r_\sigma(A) = \|A\|$, which implies $\sigma(A) - \{0\} \neq \emptyset$. Also‡ there must be some $\lambda \in \sigma(A)$ such that $|\lambda| = \|A\|$. Since $\lambda \neq 0$, $\lambda \in P\sigma(A)$, and the theorem is proved.

21.4 Numerical Range

Let $A \in L(X, X)$, where X is a Hilbert space; also let

$$W(A) = \{(Ax, x) \mid \|x\| = 1\}.$$

The set $W(A)$ of complex numbers is called the *numerical range* of A, and in this section we shall investigate a few properties of the numerical range of a bounded linear transformation.

THEOREM 21.11. Let $A \in L(X, X)$, where X is a Hilbert space. Then $\sigma(A) \subset \overline{W(A)}$ and, if $d = d(\lambda, \overline{W(A)}) > 0$ (d, as usual, denotes the distance function derived from the norm), then $\|(\lambda - A)^{-1}\| \leq 1/d$.

Proof. If $\lambda \notin \overline{W(A)}$, then of course $d = d(\lambda, \overline{W(A)}) > 0$. Thus, recalling the definition of $W(A)$, we must have

$$d \leq |(Ax, x) - \lambda|$$

for all x such that $\|x\| = 1$; that is,

$$d \leq |(Ax, x) - \lambda(x, x)| = |(Ax - \lambda x, x)|,$$

for all x such that $\|x\| = 1$. But this implies clearly that

$$d\|x\|^2 \leq |(Ax - \lambda x, x)| \qquad \text{(for all } x \in X).$$

Thus, applying the Cauchy–Schwarz inequality, we get

$$d\|x\|^2 \leq \|Ax - \lambda x\| \|x\|,$$

---

† Theorem 21.4. If A is normal, then $\pi(A) = \sigma(A)$.
‡ Theorem 21.3. If A is normal, $r_\sigma(A) = \|A\|$ and there exists $\lambda \in \sigma(A)$ such that $|\lambda| = \|A\|$.

so $d\|x\| \leq \|Ax - \lambda x\|$. This implies (see Theorem 14.9) that $(A - \lambda)^{-1}$ exists and is continuous; moreover,

$$\|(A - \lambda)^{-1}y\| \leq d^{-1}\|y\| \tag{21.24}$$

for all $y \in R(\lambda - A)$. There are, therefore, only two possibilities: Either $\lambda \in \rho(A)$ or $\lambda \in R\sigma(A)$. Suppose $\lambda \in R\sigma(A)$. We recall that $\overline{R(A - \lambda)}^{\perp} = N(A^* - \bar{\lambda})$. Thus if $\lambda \in R\sigma(A)$, then $\overline{R(A - \lambda)}^{\perp} \neq \{0\}$, so $N(A^* - \bar{\lambda}) \neq \{0\}$, and $\bar{\lambda}$ is an eigenvalue of A^*. Thus let $x \in X$, $\|x\| = 1$, such that $A^*x = \bar{\lambda}x$. Then

$$(Ax, x) = (x, A^*x) = (x, \bar{\lambda}x) = \lambda,$$

which implies that $\lambda \in W(A)$—a contradiction. Hence if $\lambda \notin \overline{W(A)}$, then $\lambda \in \rho(A)$, or, equivalently, $\lambda \in \sigma(A)$ implies that $\lambda \in \overline{W(A)}$. This together with (21.24) completes the proof of the theorem.

Now let us assume that X is just a complex vector space. We wish to prove a rather general theorem concerning convexity due to Stone (see Ref. 4 or Ref. 5), which will have, as a particular consequence, that the numerical range is a convex set.

THEOREM 21.12. Let f_1 be a sesquilinear functional on X and let $f_2 \neq 0$ be a positive, and consequently symmetric (Theorem (20.13)), sesquilinear functional on X. If $V = \{x|\hat{f}_2(x) = 1\}$, then $W = \hat{f}_1(V)$ is convex.

Proof. W is not empty, for let $\hat{f}_2(x) = \alpha \neq 0$. Then $\hat{f}_2(x/\alpha^{1/2}) = 1$. If W contains just a single point, we are then done. Thus let α_1 and α_2 be distinct complex numbers belonging to the set W. Then there exist vectors x_1 and x_2 in X such that $\hat{f}_1(x_1) = \alpha_1$, $\hat{f}_1(x_2) = \alpha_2$, and $\hat{f}_2(x_1) = 1$, $\hat{f}_2(x_2) = 1$.

We now consider the functions

$$g_1\langle\alpha, \beta\rangle = \hat{f}_1(\alpha x_1 + \beta x_2) = f_1\langle\alpha x_1 + \beta x_2, \alpha x_1 + \beta x_2\rangle$$
$$= \alpha\bar{\alpha}\alpha_1 + \alpha\bar{\beta}f_1\langle x_1, x_2\rangle + \bar{\alpha}\beta f_1\langle x_2, x_1\rangle + \beta\bar{\beta}\alpha_2$$

and

$$g_2\langle\alpha, \beta\rangle = \hat{f}_2(\alpha x_1 + \beta x_2) = f_2\langle\alpha x_1 + \beta x_2, \alpha x_1 + \beta x_2\rangle$$
$$= \alpha\bar{\alpha} + \alpha\bar{\beta}f_2\langle x_1, x_2\rangle + \bar{\alpha}\beta f_2\langle x_2, x_1\rangle + \beta\bar{\beta}.$$

We shall now show that, when we make the restriction $g_2\langle\alpha, \beta\rangle = 1$, g_1 assumes all values in the line segment connecting α_1 and α_2, that is, we restrict

$$g_2\langle\alpha, \beta\rangle = 1 = \hat{f}_2(\alpha x_1 + \beta x_2)$$

and consider

$$g_1\langle\alpha, \beta\rangle = \hat{f}_1(\alpha x_1 + \beta x_2).$$

If we can establish the above contention, then clearly the line segment connecting α_1 and α_2 belongs to W; hence W is convex. Thus we just prove the above contention. For g_1 to assume all values on the segment connecting α_1 and α_2, it suffices to show (assuming $g_2 = 1$) that

$$g\langle\alpha, \beta\rangle = \frac{g_1\langle\alpha, \beta\rangle - \alpha_2 g_2\langle\alpha, \beta\rangle}{\alpha_1 - \alpha_2} = \bar{\alpha}\alpha + a_{12}\bar{\alpha}\beta + a_{21}\alpha\bar{\beta}$$

assumes all real values between 0 and 1. Now let

$$\gamma = \begin{cases} \pm 1 & (\text{if } \bar{a}_{12} = a_{21}), \\ \pm \dfrac{\bar{a}_{12} - a_{21}}{|\bar{a}_{12} - a_{21}|} & (\text{if } \bar{a}_{12} \neq a_{21}), \end{cases}$$

and where the plus or minus sign is chosen in either case so that $c = \text{Re}(\gamma f_2 \langle x_2, x_1 \rangle)$ is not negative.

Substituting $\alpha = a$ and $\beta = \gamma b$ (where a and b are real) in $g\langle \alpha, \beta \rangle$ and $g_2 \langle \alpha, \beta \rangle$, it is easily checked that

$$g\langle a, \gamma b \rangle = a^2 + dab$$

where

$$d = \pm(a_{12} + a_{21}) \quad \text{or} \quad d = \pm \frac{(a_{12}\bar{a}_{12} - a_{21}\bar{a}_{21})}{|\bar{a}_{12} - a_{21}|},$$

depending on whether $\bar{a}_{12} = a_{21}$ or not. In either case, d is clearly real. Also,

$$g_2\langle a, b \rangle = a^2 + 2cab + b^2.$$

Now, by the choice of γ, $c \geq 0$; whereas

$$|c| = |\text{Re}(\gamma f_2 \langle x_2, x_1 \rangle)| \leq |\gamma| \, |f_2 \langle x_2, x_1 \rangle|,$$

but $|\gamma| = 1$, and by Theorem 20.17,

$$|f_2 \langle x_2, x_1 \rangle| \leq \hat{f}_2(x_1)^{1/2} \hat{f}_2(x_2)^{1/2} = 1.$$

Thus $0 \leq c \leq 1$. Hence to satisfy the equation $g_2 \langle a, b \rangle = 1$, we can take

$$b = -ca + (1 - (1 - c^2)a^2)^{1/2},$$

which we note is real for $0 \leq a \leq 1$. Substituting this value of b into $g\langle a, b \rangle$ yields

$$g\langle a, b \rangle = (1 - dc)a^2 + da(1 - (1 - c^2)a^2)^{1/2}.$$

When $a = 0$, $g\langle a, b \rangle = 0$, and when $a = 1$, $g\langle a, b \rangle = 1$. Moreover, for $0 \leq a \leq 1$, we see that $g\langle a, b \rangle$ is a real-valued continuous function; hence as a varies between 0 and 1, $g\langle a, b \rangle$ assumes all values between 0 and 1, which, as observed earlier, completes the proof.

As a special case, let X be a Hilbert space and let $A \in L(X, X)$. Consider

$$f_1\langle x, y \rangle = (Ax, y) \quad \text{and} \quad f_2\langle x, y \rangle = (x, y).$$

Then $W = \{(Ax, x) \,|\, \|x\| = 1\} = W(A)$, the numerical range of A. Hence by the theorem $W(A)$ is convex, as is stated in the following corollary.

COROLLARY. Let $A \in L(X, X)$, where X is a Hilbert space; then $W(A)$, the numerical range of A, is a convex set.

Since $W(A)$ is convex, it follows that $\overline{W(A)}$ is convex, too by Theorem 8.10. Moreover since $\overline{W(A)} \supset \sigma(A)$ by Theorem 21.11, it follows that $\overline{W(A)}$ contains the closed convex hull of $\sigma(A)$. In conclusion it is noted that if A is normal, it can be shown that $\overline{W(A)}$ coincides with the closed convex hull of $\sigma(A)$. We refer the reader to Ref. 4 for a proof or to Ref. 6 for a more recent proof.

EXERCISES 21

In all the exercises, X denotes a complex Hilbert space and $A \in L(X, X)$.

1. Show that if $A^{-1} \in L(X, X)$, then $\sigma(A^{-1}) = \sigma(A)^{-1}$.

2. Show that $\sigma(A^*) = \overline{\sigma(A)} = \{\bar{\lambda} | \lambda \in \sigma(A)\}$.

3. If A is normal, prove that $X = \overline{R(A)} \oplus N(A)$.

4. If A is unitary, show that $\lambda \in \sigma(A) \Rightarrow |\lambda| = 1$.

5. Let $A \in L(X, X)$. Prove that there exists a $\lambda \in P\sigma(A)$ such that $|\lambda| = \|A\|$, if and only if there exists a vector $x \in X$ such that $\|x\| = 1$ and $|(Ax, x)| = \|A\|$.

6. Show that, if $A \in L(X, X)$ is a completely continuous normal operator, there exists a vector x with $\|x\| = 1$ and such that $|(Ax, x)| = \|A\|$.

7. Let $A \in L(X, X)$ be normal and such that $\sigma(A) = \{0\}$; then show that $A = 0$.

8. Show that there exist Hilbert spaces X and normal transformations A on X such that $P\sigma(A) = \emptyset$. [*Hint:* Consider a separable Hilbert space X with the complete orthonormal set $x_1, x_2, ..., x_n, ...$ Let $z_n = x_{2n+1}$ ($n = 0, 1, 2, ...,$ and $z_{-n} = x_{2n}$ ($n = 1, 2, ...$). Then for any $x \in X$, $x = \sum_{j=-\infty}^{\infty} \alpha_j z_j$. Define now $Ax = \sum_{j=-\infty}^{\infty} \alpha_j z_{j+1}$.]

9. Let $A \in L(X, X)$ be a normal transformation. Show that $\overline{R(A)} = \overline{R(A^*)}$.

10. Let X be a separable Hilbert space with the complete orthonormal set $x_1, x_2, ..., x_n, ...$ For each $x = \sum_{i=1}^{\infty} \alpha_i x_i$ in X define $Ax = \sum_{i=1}^{\infty} \alpha_i x_{i+1}$. Is the transformation A normal?

11. Let $A \in L(X, X)$, where X is a Hilbert space and let U be a unitary transformation of X. Show that $W(A) = W(UAU^{-1})$.

12. Using the same notation as in 11, determine the relationship between $W(\alpha A + \beta)$ and $W(A)$, where $\alpha \neq 0$ and β are complex numbers.

13. Using the same notation as Exercise 11 and assuming that $W(A)$ contains more than a single point, show that the derived set $W(A)'$ of $W(A)$ is a perfect set (coincides with its derived set) which is convex and contains $W(A)$.

The Fredholm Alternative Theorem and the Spectrum of a Completely Continuous Transformation

In this appendix we wish to consider the case of an arbitrary completely continuous transformation defined on just a normed linear space. We wish to obtain facts concerning certain null spaces and ranges and facts concerning the spectrum of the transformation. Some of these facts will be generalizations of those already proved (see, for example, Sec. 21.3) in the special case of a normal completely continuous transformation on a Hilbert space. This section will not be used in the sequel, and the reader interested primarily in the spectral theory in Hilbert space may omit this appendix and go directly to the next chapter.

A.I Motivation

To motivate some of the ensuing discussion, we begin by considering the following system of inhomogeneous linear equations:

$$
\begin{aligned}
a_{11}\alpha_1 + a_{12}\alpha_2 + \cdots + a_{1n}\alpha_n &= \beta_1 \\
a_{21}\alpha_1 + a_{22}\alpha_2 + \cdots + a_{2n}\alpha_n &= \beta_2 \\
&\;\;\vdots \\
a_{n1}\alpha_1 + a_{n2}\alpha_2 + \cdots + a_{nn}\alpha_n &= \beta_n
\end{aligned} \tag{A.1}
$$

The corresponding system of homogeneous equations is, of course, just

$$
\begin{aligned}
a_{11}\alpha_1 + a_{12}\alpha_2 + \cdots + a_{1n}\alpha_n &= 0 \\
a_{21}\alpha_1 + a_{22}\alpha_2 + \cdots + a_{2n}\alpha_n &= 0 \\
&\;\;\vdots \\
a_{n1}\alpha_1 + a_{n2}\alpha_2 + \cdots + a_{nn}\alpha_n &= 0
\end{aligned} \tag{A.2}
$$

By working in the n-dimensional vector space X of n-tuples written as column vectors, we can write these equations in vector form, namely, if we let

$$
x_1 = \begin{pmatrix} 1 \\ 0 \\ 0 \\ \vdots \\ 0 \end{pmatrix}, \quad
x_2 = \begin{pmatrix} 0 \\ 1 \\ 0 \\ \vdots \\ 0 \end{pmatrix}, \quad \ldots, \quad
x_n = \begin{pmatrix} 0 \\ 0 \\ \vdots \\ 0 \\ 1 \end{pmatrix}, \quad
v = \begin{pmatrix} \beta_1 \\ \beta_2 \\ \vdots \\ \beta_n \end{pmatrix};
$$

then, defining the linear transformation A by its effect on the basis vectors x_j as: $Ax_j = \sum_{i=1}^{n} a_{ij}x_i$ $(j = 1, 2, \ldots, n)$ we have for the vector

$$
w = \alpha_1 x_1 + \alpha_2 x_2 + \cdots + \alpha_n x_n
$$

that

$$
Aw = v. \tag{A.3}
$$

This is the vector form of (A.1), and the vector form of (A.2) is just

$$
Aw = 0. \tag{A.4}
$$

If (A.2) or (A.4) has only the trivial solution $w = 0$, then A is $1:1$ and consequently onto, since the space is finite-dimensional and therefore (A.3) [or (A.1)] has a unique solution for any vector v. Conversely, if (A.3) has a solution for arbitrary v, then A is onto and therefore $1:1$, so (A.4) has only the trivial solution.

Now let us investigate the case where (A.4) does *not* have just the trivial solution; in other words, we assume [denoting the null space of A by $N(A)$] $N(A) \neq \{0\}$. Let dim $N(A) = v(A) = $ *nullity* of A. But [denoting the rank of A by $\rho(A)$],

$$\rho(A) = \dim R(A),$$

where $R(A)$ denotes the range of A so that†

$$v(A) + \rho(A) = n,$$

which implies that (A.4) has $n - \rho(A)$ linearly independent solutions. Next, by Sec. 17.4, Result (3), we see that

$$N(A') = \overline{R(A)}^a = R(A)^a \qquad (X \text{ finite-dimensional})$$

so that

$$\dim N(A') = n - \rho(A) = v(A).$$

Thus the equation

$$A'z = 0$$

has also $n - \rho(A)$ linearly independent solutions.

Let us choose $y_1', \ldots, y_n'$ to be the dual basis in the conjugate space of X to the basis $x_1, \ldots, x_n$ [that is, $y_i'(x_j) = \delta_{ij}$, where δ_{ij} is the Kronecker delta]. The matrix of A' with respect to this basis is (α_{ij}'), where $\alpha_{ij}' = \alpha_{ji}$ and letting $z = \sum_{i=1}^{n} \alpha_i y_i$, (A.5) becomes

$$a_{11}\alpha_1 + a_{21}\alpha_2 + \cdots + a_{n1}\alpha_n = 0$$

$$a_{12}\alpha_1 + a_{22}\alpha_2 + \cdots + a_{n2}\alpha_n = 0 \qquad (A.5)$$

$$\vdots \qquad\qquad\qquad \vdots$$

$$a_{1n}\alpha_1 + a_{2n}\alpha_2 + \cdots + a_{nn}\alpha_n = 0$$

Finally, we wish to show that the original equation [(A.1) or (A.3)] possesses solutions for just those vectors v that are annihilated by $z_1', \ldots, z_v'$, where $v = v(A)$ and the z_i' are linearly independent vectors of $N(A')$. *In other words we contend that* (A.3) *possesses solutions just for those* $v \in {}^aN(A')$. Suppose first that there exists some w such that $Aw = v$; since $A'z_i' = 0$ $(i = 1, 2, \ldots, v)$, then (using the bracket notation below as in Sec. 12.2)

$$[x, A'z_i'] = 0 \qquad (i = 1, 2, \ldots, v \text{ and all } x \in X).$$

In particular,

$$[w, A'z_i'] = 0 \qquad (i = 1, 2, \ldots, v)$$

that is,

$$[Aw, z_i'] = [v, z_i'] = 0 \qquad (i = 1, 2, \ldots, v);$$

therefore, $v \in {}^aN(A')$. Conversely, suppose $v \in {}^aN(A')$. Then, since

$${}^aN(A') = \overline{R(A)} = R(A),$$

it follows that there exists a vector $w \in X$ such that $Aw = v$.

In summary we have the following theorem.

---

† A standard fact from linear algebra. Cf. Halmos (Ref. 7), for example. One might also note that $\rho(A)$ is also the rank of the matrix (a_{ij}).

THEOREM A.1 (alternative theorem). Either the homogeneous system (A.2) (or (A.4)) has only the trivial solution in which case the inhomogenous system (A.1) [or (A.3)] has a unique solution for arbitrary $\beta_1, \ldots, \beta_n$ (arbitrary v), or the homogeneous system has $v = n - \rho(A) = n - \text{rank}\,(a_{ij})$ linearly independent solutions in which case the adjoint system of A, (A.5), has v linearly independent solutions $z_1', \ldots, z_v'$, and then the original inhomogeneous system has solutions for just those v such that $v \in {}^a[\{z_1', \ldots, z_v'\}] = {}^aN(A')$.

A.2   The Fredholm Alternative Theorem

Our aim is first to extend this theorem in a natural fashion to certain special operators in an infinite-dimensional normed linear space X. In particular, we consider linear operators of the form $\lambda - A$, where $A : X \to X$ is completely continuous. After having done this, we shall investigate certain further properties of the spectrum of A. *Thus we assume throughout the rest of the appendix that $A : X \to X$, where A is completely continuous and X is a normed linear space.* We noted earlier that (A.3) has a unique solution for arbitrary $v \in X$, if and only if (A.4) has only the trivial solution. We generalize part of this, at first, to the operators under consideration.

THEOREM A.2.   If $y = \lambda x - Ax$ has a solution for $\lambda \neq 0$ where A is completely continuous and arbitrary $y \in X$, the equation $\lambda x - Ax = 0$ has only the trivial solution $x = 0$ (in other words, if $\lambda - A$ is onto, it is also $1 : 1$).

Proof. We assume that the contention of the theorem is false and let $x_1 \neq 0$ be a vector of X such that

$$\lambda x_1 - A x_1 = 0$$

or, letting $B = \lambda - A$,

$$B x_1 = 0.$$

Clearly,

$$N(B) \subset N(B^2) \subset \cdots \subset N(B^n) \subset \cdots,$$

and all these null spaces are closed. We contend, moreover, that all these inclusions are proper. To prove this, we note that by hypothesis, the equation $y = Bx$ has a solution for arbitrary $y \in X$. Thus, there exists an $x_2 \in X$ such that $x_1 = Bx_2$; similarly, there exists an $x_3 \in X$ such that $Bx_3 = x_2$. Continuing in this fashion, we obtain a sequence $x_1, x_2, \ldots, x_n, \ldots$, no element of which is equal to zero, since $x_1 \neq 0$, such that

$$B x_2 = x_1, \qquad B x_3 = x_2, \ldots, \qquad B x_n = x_{n-1}, \ldots.$$

Now $x_n \in N(B^n)$, since

$$B^n x_n = B^{n-1} x_{n-1} = \cdots = B x_1 = 0,$$

but $x_n \notin N(B^{n-1})$, since

$$B^{n-1} x_n = B^{n-2} x_{n-1} = \cdots = B x_2 = x_1 \neq 0.$$

Since $N(B^n)$ properly contains $N(B^{n-1})$, it follows from Riesz's Theorem (Theorem 8.8) that there exists a vector $y_{n+1} \in N(B^{n+1})$ such that

$$\|y_{n+1}\| = 1 \qquad \text{and} \qquad \|x - y_{n+1}\| \geq \tfrac{1}{2}$$

for all $x \in N(B^n)$. Thus, we obtain another sequence of vectors $\{y_n\}$. Finally, we consider the sequence $\{Ay_n\}$. For $n > m$,

$$\|Ay_n - Ay_m\| = \|\lambda y_n - (\lambda y_m + By_n - By_m)\|$$

$$= |\lambda| \left\| y_n - \left(y_m + \frac{1}{\lambda} By_n - \frac{1}{\lambda} By_m \right) \right\| \geq \frac{|\lambda|}{2}, \qquad \text{(A.6)}$$

since

$$\lambda y_m + By_n - By_m \in N(B^{n-1})$$

for

$$B^{n-1}(\lambda y_m + By_n - By_m) = \lambda B^{n-1} y_m + B^n y_n - B^n y_m = 0.$$

Hence

$$\frac{1}{\lambda}(\lambda y_m + By_n - By_m) = y_m + \frac{1}{\lambda} By_n - \frac{1}{\lambda} By_m \in N(B^{n-1}).$$

It follows immediately from (A.6) that the sequence $\{Ay_n\}$ does not contain a convergent subsequence. This, of course, contradicts the hypothesis that A is completely continuous, and completes the proof of the theorem.

It follows immediately from the theorem that the solution of the equation $y = Ax - \lambda x \, (\lambda \neq 0)$ must be unique since, as we saw, $\lambda - A$ is $1:1$, assuming it is onto.

We know, if A is completely continuous, that A' is also completely continuous (Theorem 17.8). Thus, applying the theorem to $\lambda - A' \, (\lambda \neq 0)$, we find that, if $\lambda - A'$ is onto, it is also $1:1$; that is, if $y' = \lambda x' - Ax' \, (\lambda \neq 0)$ is solvable for all y', then $\lambda x' - Ax' = 0$ has only the trivial solution.

If X is finite-dimensional, all subspaces are closed as we know; in particular, $R(\lambda - A)$ is a closed subspace. We next show that this is true even if X is not finite-dimensional, provided $\lambda \neq 0$.

THEOREM A.3. If A is completely continuous, mapping the normed linear space X into itself, and λ is a nonzero scalar, then $R(\lambda - A)$ is a closed subspace of X.

Proof. We argue again by contradiction. If $R(\lambda - A)$ is not closed, there exists a sequence of vectors of the form $(\lambda - A)x_n \, (n = 1, 2, \ldots)$ such that $(\lambda - A)x_n \to y$, but $y \notin R(\lambda - A)$. Since $y \notin R(\lambda - A)$, it follows that y cannot be the zero vector and consequently, $x_n \notin N(\lambda - A)$ for all n sufficiently large. We may, therefore, without loss of generality, assume that $x_n \notin N(\lambda - A)$ for any n, but $N(\lambda - A)$ is a closed subspace so the distances

$$d_n = d(x_n, N(\lambda - A))$$

are all strictly positive. Let $z_n \in N(\lambda - A)$ be such that $\|x_n - z_n\| < 2d_n$. We contend that $a_n = \|x_n - z_n\| \to \infty$. If $a_n \nrightarrow \infty$, the sequence $\{x_n - z_n\}$ must contain a bounded subsequence, which in turn implies by the complete continuity of A that the sequence $\{A(x_n - z_n)\}$ has a convergent subsequence. However, we can write

$$x_n - z_n = \lambda^{-1}[(\lambda - A)(x_n - z_n) + A(x_n - z_n)];$$

moreover,

$$(\lambda - A)(x_n - z_n) = (\lambda - A)x_n \to y.$$

This, together with the fact that $\{A(x_n - z_n)\}$ has a convergent subsequence implies that the sequence $\{x_n - z_n\}$ itself has a convergent subsequence. Designating the limit of this subsequence by x, we see that the sequence $\{(\lambda - A)(x_n - z_n)\}$ would have a subsequence converging to both y and $(\lambda - A)x$. Thus, $y = (\lambda - A)x$, but this contradicts the assumption that $y \notin R(\lambda - A)$. Therefore we must have that $a_n \to \infty$. Then for

$$w_n = \frac{x_n - z_n}{a_n},$$

we see that $\|w_n\| = 1$ for every n and

$$(\lambda - A)w_n = \frac{1}{a_n}(\lambda - A)x_n \to 0, \tag{A.7}$$

since $a_n \to \infty$ and since the sequence $\{(\lambda - A)x_n\}$ converges.

We can write

$$w_n = \lambda^{-1}((\lambda - A)w_n + Aw_n), \tag{A.8}$$

but the sequence $\{w_n\}$ is bounded, as we noted, and A is completely continuous; using these facts in conjunction with (A.7), we see from (A.8) that $\{w_n\}$ has a convergent subsequence with limit w, say. Since

$$(\lambda - A)w_n \to 0$$

we must have $(\lambda - A)w = 0$. Finally, let $y_n = z_n + a_n w$ $(n = 1, 2, \ldots)$. Since both z_n and w belong to $N(\lambda - A)$, we have that $y_n \in N(\lambda - A)$ for all n; consequently, $d_n \leq \|x_n - y_n\|$. But

$$x_n - y_n = x_n - z_n - a_n w$$

$$= a_n w_n - a_n w$$

$$= a_n(w_n - w);$$

hence

$$d_n \leq \|x_n - y_n\| < 2d_n\|w_n - w\|,$$

so

$$1 < 2\|w_n - w\| \qquad \text{(for any } n\text{).} \tag{A.9}$$

However, we saw previously that $\{w_n\}$ has a subsequence converging to w. This clearly is at variance with (A.9), and so we have that the assumption that $R(\lambda - A)$ is not closed has given rise to a contradiction; therefore, $R(\lambda - A)$ must be closed.

REMARK. It is a simple matter to show, under the same hypothesis as Theorem A.3, that all $R((\lambda - A)^n)$ $(n = 1, 2, \ldots)$ are closed. Since $(\lambda - A)^n = \lambda^n - n\lambda^{n-1}A + \cdots + (-1)^n A^n = \lambda^n - AB$, where B is bounded [that is, $B = -n\lambda^{n-1} + \cdots + (-1)^n A^{n-1}]$, AB must be completely continuous (Theorem 17.6), and we can apply the result of the theorem to AB and conclude that $R(\lambda - AB) = R((\lambda - A)^n)$ is closed.

In Theorem 21.8, it was shown that if A is completely continuous and X is a Hilbert space, for $\lambda \neq 0$, $N(\lambda - A)$ is finite-dimensional. We now generalize this result to the case when X is only a normed space.

THEOREM A.4. Let X be a normed linear space and let $A : X \to X$ be a completely continuous transformation. Then, for $\lambda \neq 0$, $N(\lambda - A)$ is finite-dimensional.

Proof. Recalling Exercise 8.15, we see that it suffices to show that every closed and bounded set in $N(\lambda - A)$ is compact. Thus let $F \subset N(\lambda - A)$ be closed and bounded. Since compactness and sequential compactness are equivalent in a metric space (Theorem 5.5), all we need show is that from every sequence $\{x_n\}$ in F we can extract a convergent subsequence (the limit is of course in F because F is closed). Since F is bounded, there must be some $M > 0$ such that $\|x_n\| \leq M$ for all n. Since $x_n \in F$ for all n, it follows that all $x_n \in N(\lambda - A)$. Thus

$$\lambda x_n - A x_n = 0$$

or

$$x_n = \lambda^{-1} A x_n;$$

but A is completely continuous and, therefore, it follows from this relation that $\{x_n\}$ contains a convergent subsequence.

REMARK. As in the remark after Theorem A.3, it follows, under the same hypothesis as Theorem A.4, that $N\big((\lambda - A)^n\big)$ is finite-dimensional for $n = 1, 2, \ldots$, by just writing $(\lambda - A)^n = \lambda^n - AB$, where AB is completely continuous.

We now return to the problem of generalizing the alternative theorem of the finite-dimensional case to the case of operators of the form $\lambda - A$, where $\lambda \neq 0$ and where A is completely continuous, and defined on an arbitrary normed linear space X.

THEOREM A.5. Let $A : X \to X$ be a completely continuous transformation (X an arbitrary normed linear space), and let λ be a nonzero scalar. Then one can find an x to solve $y = Ax - \lambda x$ if and only if $z'(y) = [y, z'] = 0$ for all bounded linear functionals z' satisfying $\lambda z' - A' z' = 0$. In other words $y = \lambda x - Ax$ is solvable for those and only those y belonging to $^a N(\lambda - A')$.

Proof. Clearly, $y = \lambda x - Ax$ if and only if $y \in R(\lambda - A)$. But by Theorem A.3, $R(\lambda - A) = \overline{R(\lambda - A)}$, and by Sec. 17.4, Result (4),

$$\overline{R(\lambda - A)} = {}^a N(\lambda - A').$$

Consequently, $y \in R(\lambda - A)$ if and only if $y \in {}^a N(\lambda - A')$, that is, if and only if $z'(y) = 0$ for all z' such that $\lambda z' - A' z' = 0$, which completes the proof of the theorem.

REMARK. If $N(\lambda - A') = \{0\}$, clearly $R(\lambda - A) = X$ under the hypothesis of the theorem, and so the equation $y = \lambda x - Ax$ has a solution for all $y \in X$.

To proceed further, we must improve upon the relation $R(\lambda - A') \subset N(\lambda - A)^a$ for the operators at hand. [Note that this result is generally true by virtue of Sec. 1.74. Result (6).] Ultimately, for completely continuous operators, we shall show that $R(\lambda - A') = N(\lambda - A)^a$, and, to aid us in this endeavor, we first prove the following lemma.

LEMMA. Let $A : X \to X$ be a completely continuous transformation, let X be a normed linear space, and let λ be a nonzero scalar. If, for $x \in X$, we let $d(x) = d(x, N(\lambda - A))$, there exists a constant $M > 0$ such that $d(x) \leq M \|(\lambda - A)x\|$.

Proof. Suppose that no such constant M exists; then there must exist a sequence of points $\{x_n\}$ of X such that

$$\frac{d(x_n)}{\|(\lambda - A)x_n\|} \to \infty,$$

where we may assume that no $x_n \in N(\lambda - A)$. Now $N(\lambda - A)$ is finite-dimensional by Theorem A.4 and, consequently, a closed subspace of X; therefore there exists† $w_n \in N(\lambda - A)$ such that $\|x_n - w_n\| = d(x_n)$. Next, we let

$$v_n = \frac{x_n - w_n}{d(x_n)}.$$

Clearly, $\|v_n\| = 1$, and

$$(\lambda - A)v_n = \frac{(\lambda - A)x_n}{d(x_n)} \to 0.$$

But $v_n = \lambda^{-1}((\lambda - A)v_n + Av_n)$ and, therefore, by the complete continuity of A and the fact that $(\lambda - A)v_n \to 0$, it follows that $\{v_n\}$ has a convergent-subsequence with limit v, say. Again, using the fact that $(\lambda - A)v_n \to 0$, we must have that $(\lambda - A)v = 0$. Hence, $w_n + d(x_n)v \in N(\lambda - A)$; therefore, we must have, for any n,

$$\|v_n - v\| = \frac{\|x_n - (w_n + d(x_n)v)\|}{d(x_n)} \geq 1,$$

which means that the sequence $\{v_n\}$ cannot contain a subsequence which converges to v. Thus we have arrived at a contradiction, and this completes the proof of the lemma.

As our first application of the lemma, we have the next theorem.

THEOREM A.6. Let $A : X \to X$ be a completely continuous transformation on the normed linear space X and let λ be a nonzero scalar. Then if $y \in R(\lambda - A)$, there exists an x such that $(\lambda - A)x = y$ and $\|x\| \leq M\|y\|$, where M is the constant of the lemma.

REMARK. One might think that, from this theorem, an appeal could be made to Theorem 14.9 to assert the existence of $(\lambda - A)^{-1}$. This cannot be done, however, because the present theorem only yields the existence of *some* x such that $\|x\| \leq M\|(\lambda - A)x\|$; it is quite possible that there be a $z \in X$ such that $\|z\| > M\|(\lambda - A)z\|$.

Proof. Since $y \in R(\lambda - A)$, there exists an $x_0 \in X$ such that $y = (\lambda - A)x_0$. The subspace $N(\lambda - A)$ is finite-dimensional and therefore closed, so there exists an element $w \in N(\lambda - A)$ such that

$$d(x_0) = d(x_0, N(\lambda - A)) = \|x_0 - w\|.$$

† It is generally true that if M is a finite-dimensional subspace of the normed space X then, for any $x \in X$, there exists a $y \in M$ such that $d(x, M) = d(x, y)$. This follows from Ex.8.15 and Theorem 5.5.

Now let $x = x_0 - w$. We have, since $w \in N(\lambda - A)$, that $(\lambda - A)x = (\lambda - A)x_0 = y$ and, by the lemma,

$$\|x\| = \|x_0 - w\|$$

$$= d(x_0)$$

$$\leq M\|(\lambda - A)x_0\|$$

$$= M\|y\|.$$

We can now realize the goal announced prior to the lemma by proving the following theorem.

THEOREM A.7. Let X be a normed linear space, let $A : X \to X$ be a completely continuous linear transformation, and suppose that λ is a nonzero scalar. Then $R(\lambda - A') = N(\lambda - A)^a$.

Proof. We have already noted that $R(\lambda - A') \subset N(\lambda - A)^a$; so we must just show that $N(\lambda - A)^a \subset R(\lambda - A')$. Thus let g be any element of $N(\lambda - A)^a$. For $y \in R(\lambda - A)$, we define

$$f(y) = g(x),$$

where x is any element of X such that $(\lambda - A)x = y$. Clearly, there is some doubt as to whether the functional f is well-defined, and we shall deal with this issue first. Suppose that $x_1 \in X$ is such that $(\lambda - A)x_1 = y$. Then $x - x_1 \in N(\lambda - A)$; hence, $g(x - x_1) = 0$, since $g \in N(\lambda - A)^a$, which implies $g(x) = g(x_1)$. Thus, f is well-defined and is clearly a linear functional defined on $R(\lambda - A)$. By Theorem A.6, there exists an $x \in X$ such that $(\lambda - A)x = y$ and $\|x\| \leq M\|y\|$; therefore,

$$|f(y)| = |g(x)| \leq \|g\| \, \|x\| \leq M\|g\| \, \|y\|,$$

so f is a bounded linear functional on $R(\lambda - A)$. By the Hahn–Banach theorem (Theorem 11.2), f can be extended to a bounded linear functional F defined on all of X. Then

$$F((\lambda - A)x) = g(x)$$

for all $x \in X$, or, in bracket notation,

$$[x, g] = [(\lambda - A)x, F] = [x, (\lambda - A')F];$$

so $g = (\lambda - A')F$, that is, $g \in R(\lambda - A')$, and this completes the proof.

It is now a simple matter, using Theorem A.7, to establish the corresponding result of Theorem A.5. for the adjoint equation.

THEOREM A.8. Let X be a normed linear space, let $A : X \to X$ be a completely continuous linear transformation, and let λ be a nonzero scalar. There exists a bounded linear functional x' satisfying $y' = \lambda x' - A'x'$ if and only if $y' \in N(\lambda - A)^a$; that is, $y'(x) = 0$ for all x satisfying $\lambda x - Ax = 0$.

Proof. $y' = \lambda x' - A'x'$ is solvable if and only if $y' \in R(\lambda - A')$, but this is equivalent to, by Theorem A.7, $y' \in N(\lambda - A)^a$; that is

$$y'(x) = [x, y'] = 0$$

for all $x \in N(\lambda - A)$ or, in other words, for all x such that $\lambda x - Ax = 0$, which completes the proof.

REMARK. If $N(\lambda - A) = \{0\}$, then $R(\lambda - A') = \tilde{X}$, the conjugate space, and so the equation $y' = \lambda x' - A'x'$ is solvable for all $y' \in \tilde{X}$.

At long last we establish the converse of Theorem A.2.

THEOREM A.9. Let X be a normed linear space, let $A : X \to X$ be completely continuous, and let λ be a nonzero scalar. If the equation $\lambda x - Ax = 0$ has only the trivial solution, the equation $\lambda x - Ax = y$ is solvable for arbitrary $y \in X$ (or, in other words, if $\lambda - A$ is $1:1$, it is onto).

Proof. By hypothesis, $N(\lambda - A) = \{0\}$; therefore by the Remark to Theorem A.8, $R(\lambda - A') = \tilde{X}$, the conjugate space of X. It follows immediately now by the comment following Theorem A.2 that $N(\lambda - A') = \{0\}$ and finally, by the Remark following Theorem A.5, we have that $R(\lambda - A) = X$, which is what we wanted to prove.

The final fact, concerning the finite-dimensional situation, we want to generalize to get the complete analog of the alternative theorem for the operators we have been considering is the equality of the dimension of the null space of the operator and of its adjoint. That is, suppose as usual that $A : X \to X$ is completely continuous (linear of course) on the normed linear space X and that $\lambda \neq 0$. We know, by Theorem A.4, that both $N(\lambda - A)$ and $N(\lambda - A')$ are finite-dimensional. We contend that the dimensions are the same. Before formally stating and proving this, we shall establish two preparatory results.

Recall (cf. Theorem 12.3) that, if X is a normed linear space and M is a closed subspace of X, for any $x \in X$ such that $d = d(x, M) > 0$, there exists an $f \in \tilde{X}$ such that $f(x) = d$, $\|f\| = 1$, and $f(y) = 0$ for all $y \in M$. Suppose now that $x_1, x_2, \ldots, x_m$ are linearly independent vectors of X. Let $M = [\{x_1, \ldots, x_{j-1}, x_{j+1}, \ldots, x_m\}]$. Then M is a closed subspace of X and $x_j \notin M$. Hence there exists an $f_j \in \tilde{X}$ such that $f_j(x_j) = d = d(x_j, M)$ and $f_j(x_i) = 0$ for $i \neq j$. Letting $g_j = (1/d)f_j$, we have established the lemma.

LEMMA I. If $x_1, \ldots, x_m$ are linearly independent vectors of the normed linear space X, there exist elements $g_1, \ldots, g_m$ in $\tilde{X}$ such that $g_i(x_j) = \delta_{ij}$ $(i, j = 1, 2, \ldots, m)$.

The second preparatory result we need is just a dual type statement of Lemma 1. First, suppose that $f_1, \ldots, f_n$ are linearly independent elements of $\tilde{X}$ and that $g \in \tilde{X}$. Assume in addition that

$$^a[\{f_1, \ldots, f_n\}] \subset {}^a[g].$$

Then we claim that g is a linear combination of $f_1, \ldots, f_n$. This is most readily established by induction. Let $n = 1$ and write

$$x = \lambda y + z \qquad \text{(for } x \in X\text{)},$$

where z belongs to the null space of f_1 (see Exercise 12.4), the fixed vector y does not belong to this null space and $f_1(y) = \beta \neq 0$. Then, since

$$^a[f_1] \subset {}^a[g],$$

we have, letting $g(y) = \alpha$,

$$g(x) = \lambda g(y) = \lambda \alpha \beta^{-1} f_1(x),$$

which implies that $g = \alpha \beta^{-1} f_1$.

Now suppose the statement is true for $n - 1$ linearly independent elements of $\tilde{X}$. Then there exists an $x_j \in X$ such that $f_j(x_j) = \alpha_j \neq 0$, but $f_i(x_i) = 0$ for $i \neq j$, for otherwise $^a[f_1, ..., f_{j-1}, f_{j+1}, ..., f_n] \subset {}^a[f_j]$, which implies by the induction hypothesis that f_j is a linear combination of $f_1, ..., f_{j-1}, f_{j+1}, ..., f_n$, and this contradicts the linear independence of these functionals; therefore, such x_j exist. Replacing the x_j by $(1/\alpha_j) x_j$, we see that there exist vectors $x_j (j = 1, ..., n)$ in X such that $f_i(x_j) = \delta_{ij} (i, j = 1, 2, ..., n)$.

Let

$$y = x - \sum_{j=1}^{n} f_j(x) x_j,$$

where x is any vector in X. Then

$$f_i(y) = f_i(x) - \sum_{j=1}^{n} f_j(x) f_i(x_j) = f_i(x) - f_i(x) = 0,$$

which implies that $g(y) = 0$; hence,

$$g(x) = \sum_{j=1}^{n} f_j(x) g(x_j),$$

that is,

$$g = \sum_{j=1}^{n} g(x_j) f_j,$$

which finishes the proof. We see immediately from this discussion that the following lemma is valid.

LEMMA 2. Let $f_1, ..., f_n$ be linearly independent elements of $\tilde{X}$, the conjugate space of the normed linear space X. Then there exist vectors $x_1, ..., x_n$ of X such that $f_i(x_j) = \delta_{ij} (i, j = 1, ..., n)$.

We are now ready to prove the theorem referred to prior to proving Lemma 1.

THEOREM A.10. Let X be a normed linear space, let $A : X \to X$ be completely continuous, and let λ be a nonzero scalar. Then

$$\dim N(\lambda - A) = \dim N(\lambda - A').$$

Proof. Let $\dim N(\lambda - A) = m$ and $\dim N(\lambda - A') = n$. If $m = 0$, then $N(\lambda - A) = \{0\}$ and, therefore by Theorem A.9, $R(\lambda - A) = X$, so $N(\lambda - A') = \overline{R(\lambda - A)}^a = X^a = \{0\}$ or $n = 0$. Similarly, $n = 0$ implies $m = 0$. Thus we assume now that m and n are each greater than or equal to 1.

Next choose a basis $x_1, ..., x_m$ of $N(\lambda - A)$ and a basis $f_1, ..., f_n$ of $N(\lambda - A')$. Using Lemmas 1 and 2, we know that there exist elements $y_1, y_2, ..., y_n$ in X and $g_1, g_2, ..., g_m$ in $\tilde{X}$ such that

$$f_i(y_j) = \delta_{ij} \tag{A.10}$$

and

$$g_i(x_j) = \delta_{ij}. \tag{A.11}$$

Suppose now that $m < n$. Let

$$Cx = Ax + \sum_{i=1}^{m} g_i(x)y_i.$$

C is the sum of the two completely continuous operators A and B, where $Bx = \sum_{i+1}^{m} g_i(x)y_i$. (Note that B is bounded and finite-dimensional and therefore completely continuous.) Hence C is completely continuous (cf. Theorem 17.4). Also since $\lambda \neq 0$, Theorem A.3 and Result (3) of Sec. 17.4 imply that

$$R(\lambda - A)^a = \overline{R(\lambda - A)}^a = N(\lambda - A').$$

But $(\lambda - A)x \in R(\lambda - A)$. Thus,

$$f_j((\lambda - A)x) = 0 \qquad \text{(for all } x \in X) \tag{A.12}$$

since each $f_j \in N(\lambda - A')$. Recalling the definition of C, we have

$$(\lambda - C)x = (\lambda - A)x - \sum_{i=1}^{m} g_i(x)y_i. \tag{A.13}$$

Hence $(\lambda - C)x = 0$ implies that

$$(\lambda - A)x = \sum_{i=1}^{m} g_i(x)y_i;$$

so, by (A.12),

$$0 = f_j((\lambda - A)x) = \sum_{i=1}^{m} g_i(x)f_j(y_i) = g_j(x) \qquad (j = 1, 2, ..., n),$$

where the last equality results from (A.10). However, substituting the result that each $g_j(x) = 0$ $(j = 1, ..., n)$ into (A.13), yields

$$(\lambda - A)x = 0$$

so $x \in N(\lambda - A)$ and, therefore, $x = \sum_{i=1}^{m} \alpha_i x_i$. Consequently,

$$0 = g_j(x) = \sum_{i=1}^{m} \alpha_i g_j(x_i),$$

and it follows from (A.11) that $\alpha_j = 0$ $(j = 1, ..., m)$ so $x = 0$. Thus $(\lambda - C)$ is $1:1$ and, since C is completely continuous, Theorem A.9 implies that $R(\lambda - C) = X$. Hence, there exists an element $w \in X$ such that

$$(\lambda - C)w = y_{m+1}$$

(recall that we are assuming $n > m$). Then

$$1 = f_{m+1}(y_{m+1})$$

$$= f_{m+1}((\lambda - C)w)$$

$$= f_{m+1}((\lambda - A)w) - \sum_{i=1}^{m} g_i(w)f_{m+1}(y_i)$$

$$= 0 \qquad \text{[by (A.12) and (A.10)]}.$$

Hence we conclude that $n \leq m$.

To show that $m > n$ is impossible, we argue quite similarly (or perhaps dually is the better word). Namely, suppose $m > n$. Let

$$Tf = A'f + \sum_{i=1}^{n} f(y_i)g_i.$$

Since A' is completely continuous (cf. Theorem 17.8) and $T - A'$ is continuous and finite-dimensional, hence completely continuous by Theorem 17.3, it follows that T is completely continuous. Using the complete continuity of A' and the fact that λ is not 0, we see that

$$^aR(\lambda - A') = \overline{^aR(\lambda - A')} = N(\lambda - A).$$

But this implies that $x_{n+1} \in {^aR(\lambda - A')}$, since $x_{n+1} \in N(\lambda - A)$. Suppose $(\lambda - T)f = 0$. Then

$$(\lambda - A')f = \sum_{i=1}^{m} f(y_i)g_i,$$

so

$$0 = ((\lambda - A')f)(x_j) = f(y_j) \qquad (j = 1, 2, ..., n)$$

using (A.11). Thus $(\lambda - A')f = 0$, so $f \in N(\lambda - A')$. Therefore, $f = \sum_{i=1}^{n} \beta_i f_i$. From this it follows that $0 = f(y_j) = \beta_j (j = 1, ..., n)$ by (A.10), but this means that $f = 0$. Thus $(\lambda - T)$ is $1:1$, and so $R(\lambda - T) = \tilde{X}$ by Theorem A.9. Choose h such that $(\lambda - T)h = g_{n+1}$ (recall that we are assuming $m > n$). Then,

$$
\begin{aligned}
1 &= g_{n+1}(x_{n+1}) \\
&= ((\lambda - T)h)(x_{n+1}) \\
&= (\lambda - A')h(x_{n+1}) - \sum_{i=1}^{n} h(y_i)g_i(x_{n+1}) \\
&= 0,
\end{aligned}
$$

since $x_{n+1} \in {^aR(\lambda - A')}$ and since $g_i(x_{n+1}) = 0$ $(i = 1, ..., n)$. Since $m > n$ has led to a contradiction, we must finally conclude that $m = n$, which completes the proof of the theorem.

In a form of partial summary, we have from Theorems A.9, A.4, A.10, and A.5 the following Fredholm Alternative Theorem analogous to the finite-dimensional situation which motivated this entire discussion.

THEOREM A.11 (Fredholm alternative theorem). Let λ be a nonzero scalar, let X be a normed linear space, and let A be a completely continuous transformation mapping X into X. Then either the homogeneous equation

$$\lambda x - Ax = 0$$

has only the trivial solution, in which case the inhomogeneous equation

$$\lambda x - Ax = y$$

has a unique solution for arbitrary $y \in X$, *or* the homogeneous system has a finite number $v = \dim N(\lambda - A)$ of linearly independent solutions, in which case the adjoint homogeneous equation

$$\lambda x - A'x = 0$$

has v linearly independent solutions, too; then the inhomogeneous equation has solutions for just those $y \in {}^aN(\lambda - A')$.

As an application, we consider the Fredholm integral equation of the second kind; namely,

$$g(t) = f(t) - \lambda \int_a^b k(s, t)f(s)\, ds, \qquad (A.14)$$

where in this example $X = L_2(a, b)$, $k(s, t)$, the *kernel*, is given such that

$$\int_a^b \int_a^b k^2(s, t)\, ds\, dt < \infty \qquad (A.15)$$

and $g \in L_2(a, b)$ is also given; a solution $f \in L_2(a, b)$ is sought.

Equation (A.14) can be written in the form

$$g = f - \lambda A f,$$

where

$$(Af)(t) = \int_a^b k(s, t)f(s)\, ds$$

is a completely continuous transformation in view of the conditions imposed on $k(s, t)$. [The reader should attempt to prove this by first proving it for continuous kernels and then for kernels of the type stated, by using the denseness (in the mean) of the continuous functions in the square $\{(x, y)|a \le x \le b, a \le y \le b$ in the space of square integrable functions on that square, and finally recalling (cf. Theorem 17.5) that the uniform limit of a sequence of completely continuous transformations is completely continuous).]

We can write our equation in the standard form

$$\lambda^{-1}g = \lambda^{-1}f - Af$$

(clearly we may assume $\lambda \ne 0$).

Applying the alternative theorem, we know that either (A.14) has a unique solution for arbitrary $g \in L_2(a, b)$ or the homogeneous equation

$$0 = f(t) - \lambda \int_a^b k(s, t)f(s)\, ds$$

has v linearly independent solutions. In this case, the adjoint equation

$$0 = f(t) - \lambda \int_a^b k(t, s)f(s)\, ds \qquad (A.16)$$

[Note: The reader should prove that the adjoint A' is indeed given by $(A'f)(t) = \int_a^b k(t, s)f(s)\, ds$] has v linearly independent solutions, and then the original inhomogeneous equation (A.14) has solutions for just those g such that $g \in {}^aN(1/\lambda - A')$.

Before concluding the appendix, we want to give a more heavily spectral interpretation to some of our results. We see directly from Theorem A.6 that, if A is completely continuous and $\lambda \neq 0$, then if $(\lambda - A)^{-1}$ exists, it is bounded. We also know by Theorem A.9 that, if $(\lambda - A)^{-1}$ exists, then $R(\lambda - A) = X$. Thus, for $\lambda \neq 0$, there are just two possibilities: Either $\lambda \in P\sigma(A)$ or $(\lambda - A)^{-1}$ exists, in which case, as we have just seen, $(\lambda - A)^{-1}$ is bounded and $R(\lambda - A) = X$, which implies that $\lambda \in \rho(A)$.

Now let us investigate the value $\lambda = 0$, where X is infinite-dimensional. If $0 \in \rho(A)$, then $\overline{R(A)} = X$ and A has a bounded inverse defined on $R(A)$. We already noted in Sec. 17.5 that the identity transformation on an infinite-dimensional normed space is not completely continuous; in Theorem 17.6, however, it is shown that if A is completely continuous and $B \in L(X, X)$, then AB and BA are also completely continuous. In view of these two results, if A^{-1} were defined on all of X instead of just $R(A)$, it would follow that the identity is completely continuous— a contradiction.

One's first thought now is to extend A^{-1} to all of X using, somehow, the fact that $R(A)$ is dense in X. To carry out this program, first consider the completion of $X, \hat{X}$. By a simple application of the triangle inequality one sees that $R(A)$ is dense in $\hat{X}$, too. Using the results of Sec. 17.2, we can now extend the bounded transformation A^{-1} to the bounded transformation $\widehat{A^{-1}}$ defined on all of $\hat{X}$ (extension by continuity). Similarly, one can extend the completely continuous transformation A to the bounded transformation $\hat{A}$, defined on $\hat{X}$. There is some doubt as to whether $\hat{A}$ is completely continuous, though, and we shall verify that it is now.

Let $\{x_n\}$ be a bounded sequence of vectors from $\hat{X}$. Since X is dense in $\hat{X}$, there exists a sequence of vectors $\{y_n\}$ from X such that

$$\|x_n - y_n\| < 1/n.$$

The boundedness of the sequence $\{x_n\}$ clearly implies the boundedness of the sequence $\{y_n\}$ by the triangle inequality. Since A is completely continuous on X, there must be some subsequence $\{y_{n_k}\}$ such that $\{\hat{A}y_{n_k}\}$ converges to some $y \in X$. But since

$$\|\hat{A}x_{n_k} - y\| \leq \|\hat{A}x_{n_k} - \hat{A}y_{n_k}\| + \|\hat{A}y_{n_k} - y\|,$$

it follows that

$$\hat{A}x_{n_k} \to y$$

and proves the complete continuity of $\hat{A}$.

By the previously quoted theorems, it now follows that $\widehat{A^{-1}}\hat{A}$ is completely continuous. Showing that $\widehat{A^{-1}}\hat{A} = 1$ will therefore complete the proof. To show this, let $x \in \hat{X}$ and consider a sequence of vectors $\{x_n\}$ from X such that $x_n \to x$. We now have

$$\widehat{A^{-1}}\hat{A}x_n = x_n \quad \text{(for every } n\text{)},$$

and, using the continuity of $\widehat{A^{-1}}\hat{A}$,

$$x = \lim x_n = \lim \widehat{A^{-1}}\hat{A}x_n = \widehat{A^{-1}}\hat{A}(\lim x_n) = \widehat{A^{-1}}\hat{A}x.$$

This completes the argument and we now summarize our results in the following theorem.

THEOREM A.12. Let X be a normed linear space and let A be a completely continuous transformation mapping X into X. If $\lambda \neq 0$, either $\lambda \in P\sigma(A)$ or $\lambda \in \rho(A)$. If X is infinite-dimensional, then $\lambda = 0 \in \sigma(A)$.

We note in conclusion that it is possible for a completely continuous transformation A that $\sigma(A) = \{0\}$ and that $P\sigma(A) = \emptyset$ (see Exercise 18.5). Also any of the three possibilities $0 \in P\sigma(A)$, $0 \in C\sigma(A)$, or $0 \in R\sigma(A)$ can occur (see, for example, Exercises 18.12, 18.11, and 18.5, respectively).

REFERENCES

1. S. Berberian, "Introduction to Hilbert Space."
2. A. Taylor, "Introduction to Functional Analysis."
3. P. Halmos, "Introduction to Hilbert Space."
4. M. Stone, "Linear Transformations in Hilbert Space."
5. M. Stone, "Hausdorff's Theorem Concerning Hermitian Forms," *Bull. A.M.S.*, **36** (4), April, 1930.
6. S. Berberian, "The Numerical Range of a Normal Operator," *Duke Math. J.*, **31** No.3, September, 1964.
7. P. Halmos, "Finite-Dimensional Vector Spaces."

CHAPTER 22

**Orthogonal Projections
and Positive Operators**

In the setting of Hilbert space, we now return to the study of orthogonal projections, introduced in Sec. 2.2, redefining the notion slightly for the infinite-dimensional case. Since these mappings are so simple, it is desirable to be able to express other mappings in terms of them, and this is our ultimate goal for certain transformations.

In this chapter we investigate the properties of orthogonal projections—properties they possess and equivalent ways of describing them—and also what happens when orthogonal projections are combined; that is, we investigate products and linear combinations of orthogonal projections. By requiring, for example, that two orthogonal projections commute, we show that their product is also an orthogonal projection. Differences and sums of orthogonal projections are a little more difficult to treat and are best handled after the notion of a *positive* operator has been introduced. Using this notion, it becomes easier to describe the conditions under which differences and sums of projections will also be projections.

22.1 Properties of Orthogonal Projections

In this section the notion of an orthogonal projection on a Hilbert space is introduced, and some properties of orthogonal projections that follow quickly from the definition are obtained. In particular, in Theorem 22.1 we get an equivalent way of describing an orthogonal projection. We assume throughout that X is a complex Hilbert space and that M is a closed subspace of X.

Consider the direct sum decomposition of X, $X = M \oplus M^\perp$. It follows that any vector $z \in X$ can be written uniquely as

$$z = x + y, \tag{22.1}$$

where $x \in M$ and $y \in M^\perp$.

DEFINITION 22.1. The mapping $E : X \to X$, defined by taking $Ez = x$ is called the *orthogonal projection on M*.

It immediately follows that

(1) E is linear;
(2) E is idempotent ($E^2 = E$);
(3) the range of E, $R(E)$, is given by

$$M = \{z \mid Ez = z\}$$

whereas the null space of E, $N(E)$, is just

$$M^\perp = \{z \mid Ez = 0\}.$$

Throughout the rest of the section we exclude from consideration the *trivial* projection defined on the subspace consisting only of the zero vector. For any non-trivial projection E, we now show that E is bounded and that the norm of E is 1. To show E is bounded consider, with z as in (22.1),

$$Ez = x;$$

hence,

$$\|Ez\|^2 = \|x\|^2 \le \|x\|^2 + \|y\|^2 = \|z\|^2,$$

which implies E is bounded and that

$$\|E\| \le 1. \tag{22.2}$$

Clearly the strict inequality cannot prevail in (22.2), because for any nonzero $x \in M$, we must have $Ex = x$; thus we have

(4) $\|E\| = 1$.

Now consider any two vectors z_1 and z_2 in X. From the decomposition, we can write

$$z_1 = x_1 + y_1, \qquad z_2 = x_2 + y_2,$$

where $x_1, x_2 \in M$ and $y_1, y_2 \in M^\perp$. Since

$$(Ez_1, z_2) = (x_1, z_2) = (x_1, x_2)$$

and

$$(z_1, Ez_2) = (z_1, x_2) = (x_1, x_2),$$

we see that

(5) E is self-adjoint.

We now wish to reverse the above trend and prove that if a linear transformation E is idempotent and self-adjoint, it must be an orthogonal projection on

$$M = \{z \mid Ez = z\}.$$

First we show that M is closed, it being clear that M is a subspace. For $z \in \overline{M}$, we consider the sequence of points in M, $\{z_n\}$, such that $z_n \to z$. Since $E = E^*$, it follows that E must be a *bounded* linear transformation. (Proof: Any adjoint is a closed transformation; then apply Theorem 16.7). Since E is continuous,

$$Ez_n = z_n \to Ez,$$

which proves that $Ez = z$ and therefore $z \in M$.

Any $z \in X$ can be written

$$z = Ez + (1 - E)z.$$

Since E is idempotent,

$$E(Ez) = Ez;$$

therefore, $Ez \in M$. On the other hand, for any $w \in M$,

$$\big(w, (1 - E)z\big) = \big((1 - E)w, z\big) = (0, z) = 0,$$

which implies

$$(1 - E)z \in M^\perp,$$

and we have proved that E is the orthogonal projection on M. In summary, we have established a theorem.

THEOREM 22.1. A linear transformation E is an orthogonal projection if and only if it is idempotent and self-adjoint.

Again letting E be a projection on M, we show

(6) If $\|Ez\| = \|z\|$, then $Ez = z$.

Proof. Noting that

$$z = Ez + (1 - E)z,$$

where $Ez \in M$ and $(1 - E)z \in N(E) = M^{\perp}$, the Pythagorean theorem implies that

$$\|z\|^2 = \|Ez\|^2 + \|z - Ez\|^2.$$

Since $\|z\|^2 = \|Ez\|^2$, we have $\|z - Ez\| = 0$ and the result follows.
    Since

$$(Ez, z) = (E^2z, z) = (Ez, Ez) = \|Ez\|^2,$$

we see that, for orthogonal projections,

(7) $\|Ez\|^2 = (Ez, z)$.

In our next two results we assume that $A \in L(X, X)$ and investigate the questions: Is the closed subspace M invariant under A? Does M reduce A? As we shall see, this investigation will be greatly facilitated by introducing the orthogonal projection E on M.

(8) M is invariant under A if and only if $AE = EAE$.

We shall first prove sufficiency; hence suppose $AE = EAE$ and that x is any vector in M. In this case $x = Ex$ and

$$Ax = AEx = E(AEx) = E(Ax),$$

which implies that $Ax \in M$. Conversely, suppose that M is invariant under A. For any $x \in X$, $Ex \in M$. By the invariance of M under A, $AEx \in M$ as well. Hence

$$E(AEx) = AEx.$$

Since x was any vector in the space, the fact that $EAE = AE$ now follows. Although we shall not have occasion to use it, we note that this result is valid for any linear transformation A mapping X into X. In our next result the continuity of A is needed, because we wish to be guaranteed of the existence of A^*.

(9) M reduces A if and only if A commutes with E.

If $AE = EA$, then $EAE = E^2A = EA = AE$, which, by the preceding result, implies that M is invariant under A. From the hypothesis it also follows that $EA^* = A^*E$; hence $A^*E = EA^*E$, and we see that M is also invariant under A^*. By Result (B) of Sec. 21.3 we see that M reduces A. The proof of the converse follows in a similar fashion and will be omitted.

22.2 Products of Projections

In this section our main result shows that the product of projections is also a projection, provided the projections commute. As in the preceding section, we assume the underlying space to be a complex Hilbert space.

THEOREM 22.2. Let E_1 and E_2 be the orthogonal projections on the closed subspaces M_1 and M_2, respectively. Then E_1E_2 is an orthogonal projection if and only if $E_1E_2 = E_2E_1$. In this case $E = E_1E_2$ is the orthogonal projection on $M_1 \cap M_2$.

Proof. If E_1E_2 is an orthogonal projection, it must be self-adjoint; therefore,

$$(E_1E_2) = (E_1E_2)^* = E_2^*E_1^* = E_2E_1,$$

and the necessity of the condition is established. On the other hand, suppose that E_1 and E_2 commute. To show that $E = E_1E_2$ is a projection, we must show that it is idempotent and self-adjoint; Eqs. (22.3) and (22.4), show both of these requirements to be satisfied.

$$(E_1E_2)^* = E_2^*E_1^* = E_2E_1 = E_1E_2. \tag{22.3}$$

$$(E_1E_2)^2 = E_1E_2E_1E_2 = E_1^2E_2^2 = E_1E_2. \tag{22.4}$$

We now show that, if E is a projection, its range is $M_1 \cap M_2$. If $x \in M = M_1 \cap M_2$.

$$E_1x = x \qquad \text{and} \qquad E_2x = x;$$

therefore,

$$E_1E_2x = x;$$

consequently, $M \subset R(E)$. If $x \in R(E)$,

$$Ex = x = E_1(E_2x), \tag{22.5}$$

and

$$x = E_2(E_1x). \tag{22.6}$$

Equation (22.5) implies that $x \in R(E_1) = M_1$, whereas (22.6) implies $x \in R(E_2) = M_2$; thus $x \in M$, and the proof is complete.

DEFINITION 22.2. Two orthogonal projections, E_1 and E_2, will be called *orthogonal to each other*, written $E_1 \perp E_2$, if $E_1E_2 = 0$. (Taking adjoints we see that this equivalent to $E_2E_1 = 0$.)

The following theorem provides us with some ways of restating this definition. In particular, it implies that E_1 and E_2 will be orthogonal if and only if the range of E_1 is orthogonal to the range of E_2.

THEOREM 22.3. Let E_1 and E_2 be the orthogonal projections on the closed subspaces M_1 and M_2. Then the following statements are equivalent:

(1) $M_1 \perp M_2$;

(2) $E_1E_2 = 0$;

(3) $E_2E_1 = 0$;

(4) $E_1(M_2) = \{0\}$;

(5) $E_2(M_1) = \{0\}$.

Proof (1) $\Rightarrow$ (2). If $M_1 \perp M_2$, then $M_2 \subset M_1^\perp = N(E_1)$. Hence, since for any $x \in X$, $E_2 x \in M_2$, we see that

$$E_1(E_2 x) = 0.$$

Consequently, $E_1 E_2 = 0$.

The equivalence of (2) and (3) is easily demonstrated by taking adjoints. We now prove that (2) implies (4). Suppose (2) holds and let $x \in M_2$. In this case we can write x as $E_2 x$, and it follows that $E_1 x = 0$. Result (4) now follows.

Proof (4) $\Rightarrow$ (5). If (4) holds, $M_2 \subset M_1^\perp$, which implies $M_1 \subset M_2^\perp = N(E_2)$; therefore, $E_2(M_1) = 0$. To prove (5) $\Rightarrow$ (4), we merely interchange subscripts in the above argument. Last, we prove that (4) $\Rightarrow$ (1).

Proof (4) $\Rightarrow$ (1). If $E_1(M_2) = \{0\}$, $M_2 \subset M_1^\perp = N(E_1)$. This implies that $M_2 \perp M_1$. The proof of the theorem is now complete.

We now have, in particular, that

$$E_1 \perp E_2 \Leftrightarrow M_1 \perp M_2 \Leftrightarrow E_1(M_2) = \{0\}. \tag{22.7}$$

22.3 Positive Operators

In Sec. 22.4 we shall wish to consider sums and differences of orthogonal projections and see under what conditions these quantities are also orthogonal projections. It will be most convenient to state the requirements that are needed using the terminology we introduce in this section—that of *positive operators*.

Let A be a linear transformation mapping the complex Hilbert space X into itself. A is called *positive* if

$$(Ax, x) \geq 0 \tag{22.8}$$

for all $x \in X$. If $(Ax, x) > 0$ for all $x \neq 0$, then A is called *positive definite*. One immediately sees that the only way a linear transformation A has a chance of satisfying (22.8) is if the quantity (Ax, x) is always real; that is, a necessary condition that A be positive is that (Ax, x) is always real. Hence, by Theorem 20.14, A must necessarily be self-adjoint. In view of this, nothing is lost by restating the definition as follows:

DEFINITION 22.3. If A is self-adjoint and (22.8) is satisfied for every $x \in X$, then A is called positive, denoted by $A \geq 0$.

For self-adjoint transformations A and B, we now define $A \geq B$, $B \leq A$, if and only if $A - B \geq 0$.

THEOREM 22.4. Let E_1 and E_2 be the orthogonal projections on the closed subspaces M_1 and M_2, respectively. The following conditions are then equivalent.

(1) $E_1 \leq E_2$;
(2) $\|E_1 x\| \leq \|E_2 x\|$ (for all $x \in X$);
(3) $M_1 \subset M_2$;
(4) $E_2 E_1 = E_1$;
(5) $E_1 E_2 = E_1$.

Proof (1) $\Rightarrow$ (2). By (7) of Sec. 22.1, assuming (1) to hold, we have

$$\|E_1 x\|^2 = (E_1 x, x) \le (E_2 x, x) = \|E_2 x\|^2.$$

Proof (2) $\Rightarrow$ (3). Let $x \in M_1$ and suppose (2) holds. In this case, using (4) of Sec. 22.1,

$$\|x\| = \|E_1 x\| \le \|E_2 x\| \le \|x\|;$$

therefore,

$$\|x\| = \|E_2 x\|.$$

By (6) of Sec. 22.1, this implies $x = E_2 x$ and (3) is established.

Proof (3) $\Rightarrow$ (4). Suppose (3) holds and consider

$$z = x + y, \qquad x \in M_1, \qquad y \in M_1^{\perp},$$

where z is arbitrary.

$$E_1 z = x \in M_1 \subset M_2.$$

Therefore,

$$E_2 E_1 z = E_2 x = x = E_1 z,$$

and it follows that

$$E_2 E_1 = E_1.$$

Taking adjoints in this expression we see that (4) $\Leftrightarrow$ (5).

Proof (5) $\Rightarrow$ (1). If $E_1 E_2 = E_2 E_1 = E_1$, then, for any x,

$$(E_2 x, x) - (E_1 x, x) = (E_2 x, x) - (E_2 E_1 x, x)$$
$$= (E_2 (1 - E_1) x, x).$$

Since E_1 and E_2 commute, so does E_2 and $(1 - E_1)$; therefore, by Theorem 22.2 $E_2(1 - E_1)$ is an orthogonal projection. Letting $E_2(1 - E_1) = E$, we have

$$(E_2 (1 - E_1) x, x) = (E x, x) = \|E x\|^2 \ge 0.$$

This completes the proof.

22.4 Sums and Differences of Orthogonal Projections

Having treated products of orthogonal projections and introduced positive operators, we are now in the best position to handle sums and differences of orthogonal projections. We shall show in this section that the sum or difference of orthogonal projections E_1 and E_2 is also an orthogonal projection if and only if E_1 is orthogonal to E_2; also it is demonstrated that $E_1 - E_2$ is an orthogonal projection if and only if $E_1 \ge E_2$.

We assume throughout the section that E_1 and E_2 are the orthogonal projections on M_1 and M_2, respectively, and that the underlying space is a complex Hilbert space.

The difference of orthogonal projections is treated first.

THEOREM 22.5. $E = E_1 - E_2$ is an orthogonal projection if and only if $E_1 \geq E_2$. In this case the range of E is given by $M_1 \cap M_2^{\perp}$.

Proof. Suppose $E_1 - E_2$ is an orthogonal projection. Then, by (7) of Sec. 22.1, we must have

$$((E_1 - E_2)x, x) = \|(E_1 - E_2)x\|^2 \geq 0,$$

which proves that the condition is necessary. To prove its sufficiency we must show $E_1 - E_2$ to be idempotent and self-adjoint. Noting that, by the theorem of the preceding section, $E_1 E_2 = E_2 = E_2 E_1$, we see that $E_1 - E_2$ is idempotent; namely,

$$(E_1 - E_2)^2 = E_1{}^2 - E_1 E_2 - E_2 E_1 + E_2{}^2 = E_1 - E_2. \tag{22.9}$$

That it is also self-adjoint is demonstrated by Eq. (22.10):

$$(E_1 - E_2)^* = E_1{}^* - E_2{}^* = E_1 - E_2. \tag{22.10}$$

We now show that in the event that E is a projection, the range of E is given by $M_1 \cap M_2^{\perp}$. Since E_1 and E_2 commute, so do E_1 and $(1 - E_2)$; hence, $E_1(1 - E_2)$ is an orthogonal projection with range given by

$$R(E_1) \cap R(1 - E_2) = R(E_1) \cap R(E_2)^{\perp}.$$

Noting that, by Theorem 22.4,

$$E_1 - E_2 = E_1(1 - E_2),$$

the proof is seen to be complete.

THEOREM 22.6. $E = E_1 + E_2$ is an orthogonal projection if and only if $E_1 \perp E_2$. In this case the range of E is given by $\overline{[M_1 \cup M_2]} = M_1 + M_2$.

Proof. We suppose $E_1 \perp E_2$ and show $E_1 + E_2$ to be idempotent and self-adjoint.

idempotent: $\qquad (E_1 + E_2)^2 = E_1{}^2 + E_2{}^2 = E_1 + E_2.$

self-adjoint: $\qquad (E_1 + E_2)^* = E_1{}^* + E_2{}^* = E_1 + E_2.$

Conversely, suppose $E = E_1 + E_2$ is an orthogonal projection. Since this implies $\|E\| = 1$, we can write, for any $x \in M_1$,

$$\|x\|^2 \geq \|Ex\|^2 = (Ex, x) = (E_1 x, x) + (E_2 x, x)$$
$$= \|E_1 x\|^2 + \|E_2 x\|^2 \tag{22.11}$$
$$= \|x\|^2 + \|E_2 x\|^2 \geq \|x\|^2.$$

Therefore, $E_2 x = 0$ which, since x was any vector in M_1, implies

$$E_2(M_1) = \{0\}.$$

Applying Theorem 22.3, we see that this implies

$$E_1 \perp E_2.$$

Equation (22.11) also implies that, for $x \in M_1$,

$$\|Ex\| = \|x\|,$$

which in turn implies that $x = Ex$ or

$$x \in R(E).$$

Hence,

$$M_1 \subset R(E).$$

Since the argument is completely symmetric, we also must have

$$M_2 \subset R(E),$$

which implies

$$M_1 \cup M_2 \subset R(E).$$

It follows that

$$\overline{[M_1 \cup M_2]} \subset R(E).$$

For any $x \in R(E)$, we can write

$$x = Ex = E_1 x + E_2 x.$$

Since $E_1 x \in R(E_1)$ and $E_2 x \in R(E_2)$, we see that

$$x \in M_1 + M_2 = \{y + z \mid y \in M_1, z \in M_2\}.$$

Therefore,

$$R(E) \subset M_1 + M_2 \subset \overline{[M_1 \cup M_2]},$$

and the proof is seen to be complete.

Having treated sums, differences, and products of orthogonal projections, we now form a slightly more complicated combination.

THEOREM 22.7. If E_1 and E_2 commute, then $E = E_1 + E_2 - E_1 E_2$ is an orthogonal projection on $\overline{[M_1 \cup M_2]}$.

Proof. Write E as

$$E = E_1 + (1 - E_1)E_2.$$

Since $1 - E_1$ and E_2 must commute, $(1 - E_1)E_2$ is the orthogonal projection on $M_1^{\perp} \cap M_2$. Since

$$E_1(1 - E_1)E_2 = (E_1 - E_1^2)E_2 = 0,$$

we see that

$$E_1 \perp (1 - E_1)E_2,$$

and the preceding theorem can be applied, which enables us to assert that E is a projection on

$$R(E) = \overline{[M_1 \cup (M_1^{\perp} \cap M_2)]} \subset \overline{[M_1 \cup M_2]}. \tag{22.12}$$

Since we can write

$$E = E_2 + E_1(1 - E_2),$$

we must also have

$$R(E) = \overline{[M_2 \cup (M_2^{\perp} \cap M_1)]}. \tag{22.13}$$

Equations (22.12) and (22.13) imply that $R(E)$ contains M_1 and M_2 and, therefore, $M_1 \cup M_2$. Since $R(E)$ is a closed subspace, we have

$$R(E) \supset \overline{[M_1 \cup M_2]},$$

and the proof is complete.

22.5   The Product of Positive Operators

In this section we show the product of positive operators to be positive, provided the operators commute. The underlying space is assumed to be a complex Hilbert space.

THEOREM 22.8.   Suppose $A, B \in L(X, X)$ and that A and B commute. Then $A \geq 0$ and $B \geq 0$ implies that $AB \geq 0$.

Proof. We first note that, since A is self-adjoint, $A^2 \geq 0$ for

$$(A^2 x, x) = (Ax, Ax) \geq 0.$$

Assuming that A is not the zero linear transformation, we form the operators

$$A_1 = \frac{A}{\|A\|}$$

$$A_2 = A_1 - A_1{}^2$$

$$A_3 = A_2 - A_2{}^2$$

$$\vdots \qquad \vdots$$

$$A_{n+1} = A_n - A_n{}^2$$

and note that each is self-adjoint. In addition we now make the following statement.

CONTENTION. $0 \leq A_n \leq 1$ for every n.
The contention will now be proved by induction. For $n = 1$, we clearly have

$$A_1 \geq 0.$$

To show $1 - A_1 \geq 0$, consider

$$((1 - A_1)x, x) = (x, x) - (A_1 x, x) = (x, x) - \frac{1}{\|A\|}(Ax, x)$$

$$\geq (x, x) - \frac{1}{\|A\|}\|A\|\|x\|^2 = 0.$$

Assume the contention to be true for n now. We must show that

$$A_{n+1} = A_n - A_n{}^2$$

satisfies the contention. To this end, consider

$$(A_n{}^2(1 - A_n)x, x) = ((1 - A_n)A_n x, A_n x).$$

This quantity must always be nonnegative by the induction assumption; therefore,

$$A_n^2(1 - A_n) \geq 0.$$

In similar fashion we can also show

$$A_n(1 - A_n)^2 \geq 0.$$

Since the sum of positive operators must also be positive, we have

$$A_n^2(1 - A_n) + A_n(1 - A_n)^2 \geq 0.$$

Multiplying out, we see that this is just

$$A_{n+1} = A_n - A_n^2 \geq 0.$$

We now show $A_{n+1} \leq 1$ by noting that, since $1 - A_n$ and A_n^2 are positive

$$1 - A_{n+1} = 1 - A_n + A_n^2 \geq 0.$$

The contention is now established.

From the definition of A_2 it follows that

$$A_1 = A_1^2 + A_2.$$

Similarly,

$$A_1 = A_1^2 + A_2^2 + A_3$$

and

$$A_1 = A_1^2 + A_2^2 + \cdots + A_n^2 + A_{n+1}$$

or

$$\sum_{i=1}^{n} A_i^2 = A_1 - A_{n+1} \leq A_1. \tag{22.14}$$

Thus, since each A_i is self-adjoint,

$$\left(\sum_{i=1}^{n} A_i^2 x, x \right) = \sum_{i=1}^{n} (A_i^2 x, x) = \sum_{i=1}^{n} (A_i x, A_i x) \leq (A_1 x, x),$$

for every n. This implies that the monotone sequence $\{\sum_{i=1}^{n} \|A_i x\|^2\}$ is bounded above; therefore its limit exists. The limit existing, however, implies that

$$\|A_n x\| \to 0.$$

Consequently, using (22.14),

$$\sum_{i=1}^{n} A_i^2 x = A_1 x - A_{n+1} x \to A_1 x. \tag{22.15}$$

Note that so far we have made no use of the fact that B commutes with A. We do so now and complete the proof. Since B commutes with A, it must commute with every A_n; this enables us to write

$$(ABx, x) = \|A\|(BA_1 x, x)$$

$$= \|A\| \left(B \lim_n \sum_{i=1}^{n} A_i^2 x, x \right)$$

$$= \|A\| \lim_n \sum_{i=1}^{n} (BA_i^2 x, x)$$

$$= \|A\| \lim_n \sum_{i=1}^{n} (BA_i x, A_i x).$$

Since $B \geq 0$ implies that

$$(BA_i x, A_i x) \geq 0 \qquad \text{(for every } i\text{)},$$

the last equation implies that

$$(ABx, x) \geq 0$$

and the proof is seen to be complete.

EXERCISES 22

In the following exercises, X will designate a Hilbert space.

1. Let $A, B \in L(X, X)$. Show that, if $A \leq B$ and $B \leq A$, then $A = B$. If $C \in L(X, X)$, show that $A \leq B$ and $B \leq C$ implies $A \leq C$.

2. Let $A, B \in L(X, X)$. If $0 \leq A$ and $0 \leq B$ and $A + B = 0$, show that $A = B = 0$.

3. Let $A, B \in L(X, X)$ be such that $A \geq 0$, $B \geq 0$, and $AB \geq 0$; show that A and B commute.

4. Let $A, B \in L(X, X)$ and suppose $B \geq 0$; show that $A^*BA \geq 0$.

5. Let $A \in L(X, X)$. Show that A is a projection (orthogonal, of course) if and only if $A = A^*A$.

6. If E_1 and E_2 are nonzero orthogonal projections and $E_1 E_2 = 0$, show that $\|E_1 + E_2\| < \|E_1\| + \|E_2\|$.

7. Let $A \in L(X, X)$. A is called *hyponormal* if $\|A^*x\| \leq \|Ax\|$ for all $x \in X$. Prove (a) A is hyponormal if and only if $AA^* \leq A^*A$. (b) if $\lambda \in \pi(A)$, then $\bar{\lambda} \in \pi(A^*)$, when A is hyponormal.

8. Give an example of a hyponormal operator that is not normal.

9. Let E be an orthogonal projection. Determine $\sigma(E)$.

10. Given two orthogonal projections, E_1 and E_2, on X, under what conditions does the relation $\sigma(E_1) \cup \sigma(E_2) = \sigma(E_1 + E_2)$ hold?

11. Suppose that $A \leq B$ and C is positive such that C commutes with both A and B. Then show that $AC \leq BC$.

12. Let A be a closed operator with domain D_A, where D_A is assumed to be dense in the Hilbert space X. Show that A^*A is a positive operator.

REFERENCES

1. F. Riesz and B. Sz.-Nagy, "Functional Analysis."
2. P. Halmos, "Introduction to Hilbert Space."
3. L. Liusternik and W. Sobolev, "Elements of Functional Analysis."

CHAPTER 23

Square Roots and a Spectral Decomposition Theorem

In certain respects self-adjoint transformations behave very much like real numbers. The reader has probably observed this in some of the earlier theorems. Further evidence of this fact will be displayed in the first theorem of this chapter. We shall also show that, if the squares of two positive, self-adjoint transformations are equal, the transformations themselves are equal.

To show the last property, we shall make use of the notion of the *square root* of a positive operator. This is defined exactly as one might suspect: For $A \geq 0$ and B such that $B^2 = A$, B is called a square root of A. When square roots of real numbers are spoken of, there is usually some ambiguity about the square root. This is eliminated if one agrees to always take the *positive* square root, however. Similarly one can show (Theorem 23.2) that for each positive operator, there is a *unique positive square root*.

The existence and uniqueness of a positive square root for each positive operator will enable us to prove a very important theorem later on (a spectral decomposition theorem for bounded, self-adjoint operators) and, indeed, this is the main reason for introducing the notion.

In this chapter we shall obtain a spectral decomposition theorem for bounded, normal, finite-dimensional operators, which will be generalized in the next chapter to the case of normal, completely continuous operators.

23.1 Square Root of Positive Operators

We assume throughout this section that X is a complex Hilbert space. In this setting, then, our first result provides an analog for self-adjoint transformations from $L(X, X)$ to the following well-known result: A bounded, monotone sequence of real numbers has a limit.

THEOREM 23.1. Let $\{A_n\}$ be a sequence of self-adjoint, commuting transformations from $L(X, X)$. Let $B \in L(X, X)$ be a self-adjoint transformation such that $A_j B = B A_j$, for all j, and further suppose that

$$A_1 \leq A_2 \leq \cdots \leq A_n \leq \cdots \leq B,$$

Then there exists a self-adjoint, bounded linear transformation A such that $A_n \xrightarrow{s} A$, (see Sec. 15.1, Table II) and $A \leq B$.

REMARK. It is easy to formulate a corresponding statement about monotone decreasing sequences.

Proof. Consider the sequence of linear transformations $\{C_n\}$ formed by taking

$$C_n = B - A_n \geq 0.$$

We now show that, for any x, the sequence $\{C_n x\}$ converges. Then, since

$$C_n x = (B - A_n)x \tag{23.1}$$

we shall be able to conclude that $\lim_n A_n x$ exists for every x. Taking $Ax = \lim_n A_n x$ will then yield the theorem.

Clearly, the C_n commute with each other. Further, they are seen to be a monotone decreasing sequence, viz., for $n > m$, $A_n - A_m \geq 0$, which implies

$$C_m = B - A_m \geq B - A_n = C_n.$$

Hence, since C_m and $C_m - C_n$ are each positive and commute with each other, their product is positive:

$$(C_m - C_n)C_m \geq 0.$$

Similarly,

$$C_n(C_m - C_n) \geq 0.$$

Combining these two statements, we have, for any x,

$$(C_m^2 x, x) \geq (C_n C_m x, x) \geq (C_n^2 x, x) \geq 0, \tag{23.2}$$

which implies that $\{(C_n^2, x\ x)\}$ is a monotone decreasing sequence of real numbers bounded below by zero; therefore the limit

$$\lim_n (C_n^2 x, x) = \alpha \tag{23.3}$$

exists. Since the C_n are self-adjoint and commute, we can now write, using (23.2) and (23.3),

$$\begin{aligned}
\|(C_n - C_m)x\|^2 &= \big((C_n - C_m)x, (C_n - C_m)x\big) \\
&= \big((C_n - C_m)(C_n - C_m)x, x)\big) \\
&= (C_n^2 x, x) - 2(C_n C_m x, x) + (C_m^2 x, x) \\
&\to \alpha - 2\alpha + \alpha = 0.
\end{aligned}$$

We have now proved that $\{C_n x\}$ is a Cauchy sequence. Since X is a Hilbert space, $\lim_n C_n x$ exists, which implies by (23.1) that $\lim_n A_n x$ exists for all x. Clearly, the transformation A defined by

$$Ax = \lim_n A_n x$$

has the property that $A_n \xrightarrow{s} A$, while the linearity of A is obvious. To show that A is bounded consider

$$\|Ax\| = \lim_n \|A_n x\|. \tag{23.4}$$

Since $\lim_n A_n x$ exists for every x, $\sup_n \|A_n x\|$ must be finite for every x. By the principle of uniform boundedness (Sec. 15.2) then, $\sup_n \|A_n\| < \infty$, and there must be some M such that $\|A_n\| \leq M$ for all n. Hence for any n,

$$\|A_n x\| \leq M\|x\|,$$

which implies

$$\lim_n \|A_n x\| \leq M\|x\|$$

and proves that A is bounded. (Note that we actually proved this in Theorem 15.3.) It remains to prove that A is self-adjoint and that $A \leq B$. As for the former, consider, using the continuity of the inner product,

$$(Ax, y) = \lim_n (A_n x, y) = \lim_n (x, A_n y) = (x, Ay).$$

Last, since

$$(A_n x, x) \le (Bx, x)$$

for all n, it follows that $A \le B$, and the proof is complete.

DEFINITION 23.1. The self-adjoint operator B is called a *square root* of the positive operator A if $B^2 = A$. When, in addition, $B \ge 0$, we often denote this relationship by writing $B = \sqrt{A}$ or $B = A^{1/2}$.

Using Theorem 23.1, we can now prove a theorem about the existence and uniqueness of square roots.

THEOREM 23.2 Suppose $A \in L(X, X)$ is positive. If so, there exists a unique $B \ge 0$, $B \in L(X, X)$, such that $B^2 = A$; B commutes with any $C \in L(X, X)$ that commutes with A.

Proof. Our approach in this theorem is similar to the outlook one takes when trying to compute the square root of a real number by using the binomial expansion —similar in that we too shall obtain the square root as the limit of a sequence. We shall be sure that the limit of the sequence defined exists by virtue of the preceding theorem; that is, we shall construct a monotone sequence of self-adjoint, commuting transformations, "bounded" above by the identity transformation.

If $A = 0$, then clearly $B = 0$ works. With this trivial case out of the way, we now claim that there is no loss of generality in assuming $A \le 1$. To show that we lose nothing, consider, for any positive A,

$$(Ax, x) \le \|A\| \|x\| \|x\| = \|A\|(x, x),$$

which implies

$$\left(\frac{A}{\|A\|} x, x \right) \le (1x, x),$$

and we see that

$$\frac{A}{\|A\|} \le 1.$$

Assuming we had already proved the theorem for this case, we could assert the existence of a positive operator C such that

$$C^2 = \frac{A}{\|A\|},$$

and we see that taking $B = \|A\|^{1/2} C$ would provide a positive square root of A.

Consider now the sequence of transformations $B_0, B_1, ..., B_n, ...$, where $B_0 = 0$ and

$$B_{n+1} = B_n + \tfrac{1}{2}(A - B_n^2) \qquad (n = 1, 2, ...).$$

Since each B_n is a polynomial in A we see that

$$\text{each } B_n \text{ is self-adjoint;} \tag{23.5}$$

$$\begin{array}{c} B_n \text{ commutes with anything that } A \text{ commutes with; in particular,} \\ \text{the } B_n \text{ all commute with each other.} \end{array} \tag{23.6}$$

We next show that $\{B_n\}$ is a monotone increasing sequence, bounded above by 1. To show that $B_n \leq 1$ for all n, first note it to be true for $n = 0$. Assume it is true for B_n now. Since $A \leq 1$ and $1 - B_n$ is positive, hence, self-adjoint, it follows that

$$1 - B_{n+1} = \tfrac{1}{2}(1 - B_n)^2 + \tfrac{1}{2}(1 - A)$$

is positive therefore,

$$B_n \leq 1 \qquad \text{(for all } n\text{).} \tag{23.7}$$

As for the monotonicity, we first observe that $B_0 \leq B_1$. Assuming $B_{n-1} \leq B_n$, this implies that

$$B_{n+1} - B_n = \tfrac{1}{2}((1 - B_{n-1}) + (1 - B_n))(B_n - B_{n-1})$$

is also positive; hence,

$$0 = B_0 \leq B_1 \leq B_2 \leq \cdots \leq 1. \tag{23.8}$$

We note that (23.8) also implies that each B_n is positive. By the preceding theorem, there exists a self-adjoint transformation B such that $B_n \xrightarrow{s} B$. Since, for every n, and every x, $(B_n x, x) \geq 0$, it follows that for every x,

$$\lim_n (B_n x, x) \geq 0$$

and proves that $B \geq 0$. To show that $B^2 = A$, consider

$$B_{n+1} x = B_n x + \tfrac{1}{2}(Ax - B_n^2 x).$$

Noting that $\lim_n B_n x = Bx$ and letting $n \to \infty$ in the above expression, we have $A = B^2$. So much for the existence of the square root. We now prove that the positive square root is unique. First, we show that B commutes with anything that commutes with A. Let $C \in L(X, X)$ be any transformation that commutes with A. By (23.6),

$$CB_n = B_n C \qquad \text{(for every } n\text{).}$$

Since $B_n \xrightarrow{s} B$,

$$B_n Cx \to BCx.$$

Using the continuity of C, we now have

$$\lim_n CB_n x = C \lim_n B_n x = CBx = \lim_n B_n Cx = BCx.$$

To prove the uniqueness of B, consider any positive C now such that $C^2 = A$. In this case

$$CA = CC^2 = C^2 C = AC,$$

and we see that C commutes with A—hence with B. Let x be any vector in X and call $(B - C)x = y$. In this case

$$(By, y) + (Cy, y) = ((B + C)y, y)$$

$$= ((B + C)(B - C)x, y).$$

Since B and C commute, $(B + C)(B - C) = B^2 - C^2 = 0$, and the last term on the right is zero. Since (By, y) and (Cy, y) are each nonnegative, each of these terms must be zero:

$$(By, y) = (Cy, y) = 0. \tag{23.9}$$

By the first part of the proof, since $B \geq 0$, there must be some $D \geq 0$ such that $D^2 = B$. With y as above then, using (23.9), we have

$$\|Dy\|^2 = (Dy, Dy) = (D^2y, y) = (By, y) = 0,$$

which implies $Dy = 0$; hence,

$$D^2y = By = 0. \tag{23.10}$$

Similarly,

$$Cy = 0. \tag{23.11}$$

In view of (23.10) and (23.11) we have, for any vector x,

$$\|Bx - Cx\|^2 = ((B - C)(B - C)x, x)$$
$$= ((B - C)y, x)$$
$$= 0;$$

therefore $B = C$, and we have proved that B is unique.

The squares of two real numbers may easily be equal without the numbers themselves being equal; the closest statement to equality we can make about the numbers themselves is to within a minus sign. A similar situation prevails here. For example, suppose $A \in L(X, X)$ is self-adjoint but *not* positive. Since it is self-adjoint, its square A^2 will be positive and, as such, will have a *positive* square root, B. Thus, in this situation, we shall have $A^2 = B^2$ but shall definitely not have $A = B$, because B is positive and A is not. The following corollary is now apparent.

COROLLARY. Let $A, B \in L(X, X)$. Then if $A \geq 0$, $B \geq 0$, and $A^2 = B^2$, it follows that $A = B$.

We treat the issue of, "what can we say about A and B if $A^2 = B^2$?" further in our next theorem.

THEOREM 23.3. Suppose $A, B \in L(X, X)$ are self-adjoint, commute with each other, and $A^2 = B^2$. Then there exists an orthogonal projection E such that:

(1) E commutes with any transformation that commutes with $A - B$;
(2) $Ax = 0$ implies that $Ex = x$; equivalently, $N(A) \subset R(E)$;
(3) $A = (2E - 1)B$.

We can also prove the following result with the same hypothesis: There exists an orthogonal projection F such that

(1') F commutes with any transformation that commutes with $A + B$;
(2') $Ax = 0$ implies that $Fx = x$; equivalently, $N(A) \subset R(F)$;
(3') $A = (1 - 2F)B$.

Proof. Consider the closed subspace $M = N(A - B)$ and let E be the orthogonal projection on M. [To prove the alternate statement we would let F be the orthogonal projection on $N(A + B)$.] Suppose that C commutes with $A - B$. This implies that M is invariant under C:

$$C(M) \subset M.$$

Similarly, since

$$C(A - B) = (A - B)C \Rightarrow (A - B)C^* = C^*(A - B),$$

we see that C^* commutes with $A - B$, which implies

$$C^*(M) \subset M;$$

therefore M reduces C. By Result (9) of Sec. 22.1, it follows that

$$CE = EC,$$

and proves part (1).

Proof (2). If $Ax = 0$, then

$$\|Bx\|^2 = (Bx, Bx) = (B^2x, x) = (A^2x, x) = \|Ax\|^2 = 0,$$

that is, $Bx = 0$, so certainly $(A - B)x = 0$. This, however, is the statement that

$$x \in N(A - B) = M = R(E).$$

Proof (3). Since, for any x,

$$(A - B)(A + B)x = (A^2 - B^2)x = 0,$$

it follows that

$$(A + B)x \in N(A - B) = R(E);$$

consequently,

$$E(A + B)x = (A + B)x.$$

Since this is so for every x, we must have

$$E(A + B) = A + B. \tag{23.12}$$

Let z be any vector in X and write it as $z = x + y$, where $x \in M$ any $y \in M^{\perp}$. In this case

$$E(A - B)z = E(A - B)x + E(A - B)y.$$

The first term on the right is zero, because $x \in M = N(A - B)$, and the second is zero because E commutes with $A - B$ and y is in the null space of E, $M^{\perp}$, so $E(A - B)z = 0$ for all $z \in X$, which implies

$$E(A - B) = 0. \tag{23.13}$$

Using (23.12) and (23.13), we have

$$E(A + B) - E(A - B) = A + B,$$

$$EA + EB - EA + EB = A + B$$

or

$$A = 2EB - B = (2E - 1)B.$$

23.2 Spectral Theorem for Bounded, Normal, Finite-Dimensional Operators

Since we shall be dealing mainly with completely continuous, normal transformations in this section, we ask the reader to recall the following facts about them:

if A is completely continuous and λ is a nonzero scalar, then $N(A - \lambda)$ is finite-dimensional (Theorem 21.8);   (23.14)

if A is completely continuous, then $P\sigma(A)$ is at most countable and zero is its only possible limit point (Theorem 18.2).   (23.15)

Establishing our first theorem really allows the second—the decomposition theorem—to be brought to a swift ending.

THEOREM 23.4. Let $A \in L(X, X)$, X a complex Hilbert space, be a completely continuous normal transformation. Then if $x \perp N(A - \lambda)$ for every λ, $x = 0$; that is, $\bigcap_{\lambda \in C} N(A - \lambda)^{\perp} = \{0\}$.

REMARK. By virtue of (23.15), we note that $N(A - \lambda) \neq \{0\}$ for at most a countable number of λ.

Proof. Let

$$L = \bigcup_{\lambda \in C} N(A - \lambda).$$

Since

$$L^{\perp} = \left(\bigcup_{\lambda \in C} N(A - \lambda) \right)^{\perp}$$

$$= \overline{\left[\bigcup_{\lambda \in C} N(A - \lambda) \right]}^{\perp}$$

$$= \bigcap_{\lambda \in C} N(A - \lambda)^{\perp},$$

if we can show that $L^{\perp} = \{0\}$, we shall be done. We now proceed to show this. For any λ, A and A^* commute with $A - \lambda$, so

$$A\big(N(A - \lambda)\big) \subset N(A - \lambda)$$

and

$$A^*\big(N(A - \lambda)\big) \subset N(A - \lambda).$$

This implies that

$$A(L) \subset L \quad \text{and} \quad A^*(L) \subset L. \tag{23.16}$$

Let z be any vector in L and let y be any vector in $L^{\perp}$. By virtue of (23.16),

$$(z, Ay) = (A^*z, y) = 0;$$

hence,

$$A(L^{\perp}) \subset L^{\perp}. \tag{23.17}$$

Similarly $(z, A^*y) = (Az, y) = 0$, which proves

$$A^*(L^\perp) \subset L^\perp. \tag{23.18}$$

Since $L^\perp$ is a closed subspace, (23.17) and (23.18) imply that $L^\perp$ reduces A. We now make one assumption [Eq. (23.19)] that will lead to a contradiction: Suppose

$$L^\perp \neq \{0\}. \tag{23.19}$$

Consider the restriction of A to $L^\perp$; that is, let $B = A|_{L^\perp}$. Since $L^\perp$ reduces A,

$$A : L^\perp \to L^\perp \quad \text{and} \quad A^* : L^\perp \to L^\perp,$$

and it makes sense to talk about the normality of $A|_{L^\perp}$. In view of this and the fact that A is normal, we conclude that B is normal. Consider a sequence of vectors $\{x_n\}$ from $L^\perp$ such that $\|x_n\| \leq M$ for all n. Since $Bx_n = Ax_n$ for all n and A is completely continuous, the sequence $\{Bx_n\} = \{Ax_n\}$ must have a convergent subsequence

$$Bx_{n_k} = Ax_{n_k} \to y.$$

Since $L^\perp$ is closed, $y \in L^\perp$. Therefore B is completely continuous. Collecting these last few facts, we have shown B to be a completely continuous, normal transformation on the nontrivial space $L^\perp$. Since $L^\perp \neq \{0\}$, it is possible for B to possess eigenvectors and, indeed, by Theorem 21.10, B must possess an eigenvector. Hence there must be a nonzero vector $x \in L^\perp$ and associated eigenvalue λ such that

$$Bx = Ax = \lambda x \Rightarrow x \in N(A - \lambda) \subset L. \tag{23.20}$$

Consequently, $x \in L \cap L^\perp = \{0\}$, contrary to the way in which x was chosen. The theorem is now proved.

Before proceeding to the spectral decomposition theorem for bounded, normal, finite-dimensional operators, we wish to note how very similar this consideration is to the finite-dimensional case. When dim $X < \infty$ and A is a linear transformation mapping X into X, A has only a point spectrum consisting of a finite number of eigenvalues. The same thing happens when X is an infinite-dimensional Hilbert space and A is bounded, normal and finite-dimensional. Since A is normal, its residual spectrum is empty (Theorem 21.7); since it is bounded and finite-dimensional, therefore completely continuous, only zero can belong to its continuous spectrum.† For zero to belong to $C\sigma(A)$, however, $A(X)$ must be dense in X. Since $A(X)$ is finite-dimensional, therefore closed, and X is infinite-dimensional, we see that this cannot happen. Therefore the continuous spectrum of A is empty. Knowing that the spectrum of A reduces to just the point spectrum, we now prove that the point spectrum can have only a finite number of elements. The second part of the theorem below yields the desired decomposition theorem.

THEOREM 23.5. Let $A \in L(X, X)$, X a complex Hilbert space, be a normal transformation.

(1) If A is finite-dimensional, its point spectrum is a finite set.
(2) If A is completely continuous and has a finite point spectrum, then A is finite-dimensional. In this case, letting $P\sigma(A) = \{\lambda_1, ..., \lambda_k\}$:

† Section 21.3, especially Eq. (21.22).

(a) $A = \sum_{i=1}^{k} \lambda_i E_i$, where E_i is the orthogonal projection on $N(A - \lambda_i)$;
(b) $E_i \perp E_j$ for $i \neq j$;
(c) $\sum_{i=1}^{k} E_i = 1$.

Proof. We first assume A to be bounded, normal, and finite-dimensional and prove that it has a finite point spectrum. The proof will be by contradiction. Suppose

$$\lambda_1, \lambda_2, \ldots, \lambda_n, \ldots$$

are all distinct eigenvalues of A with associated eigenvectors

$$x_1, x_2, \ldots, x_n, \ldots$$

Since eigenvectors associated with distinct eigenvalues of a normal transformation are orthogonal—therefore linearly independent—and we can write,

$$A(\lambda_n^{-1} x_n) = x_n$$

for all but at most one λ_n (the λ_n were assumed to be distinct), it follows that $A(X)$ is infinite-dimensional, thus contradicting the finite-dimensionality of A.

Now, suppose $P\sigma(A) = \{\lambda_1, \ldots, \lambda_k\}$, where A is normal and completely continuous. We now show that this implies A is finite-dimensional. Since A is normal, eigenvectors associated with distinct eigenvalues are orthogonal, which implies that the subspaces $N(A - \lambda_i)(i = 1, 2, \ldots, \kappa)$ are mutually orthogonal. The space spanned by the union of these subspaces

$$M = \left[\bigcup_{i=1}^{k} N(A - \lambda_i) \right]$$

is just

$$N(A - \lambda_1) + N(A - \lambda_2) + \cdots + N(A - \lambda_k) = \left\{ \sum_{i=1}^{k} x_i \mid x_i \in N(A - \lambda_i) \right\},$$

and since the subspaces are mutually orthogonal, we can write

$$M = N(A - \lambda_1) \oplus N(A - \lambda_2) \oplus \cdots \oplus N(A - \lambda_k).$$

We now wish to show that $X = M$. Since $M_1 \perp M_2$ (M_1 and M_2 closed subspaces) implies that $M_1 + M_2$ is a closed subspace (Theorem 10.7) it follows by induction that M is a closed subspace. In view of this, if we can show that $M^\perp = \{0\}$, we shall have $X = M$. Let

$$L = \bigcup_{i=1}^{k} N(A - \lambda_i)$$

and note that $L^\perp = M^\perp$; therefore we have further reduced the problem to showing that $L^\perp = \{0\}$. For any $\lambda \notin P\sigma(A)$, we have $N(A - \lambda) = \{0\}$; therefore,

$$L = \bigcup_{i=1}^{k} N(A - \lambda_i) = \bigcup_{\lambda \in C} N(A - \lambda)$$

and Theorem 23.4 implies that the orthogonal complement of this set is the set consisting of just the zero vector. Hence $X = M$ and any $x \in X$ can be written uniquely as

$$x = \sum_{i=1}^{k} x_i, \qquad [x_i \in N(A - \lambda_i)]. \tag{23.21}$$

For $i \neq j$,

$$x_i \in N(A - \lambda_i) \subset N(A - \lambda_j)^{\perp},$$

which implies

$$E_j x_i = 0,$$

for $i \neq j$. Therefore, $E_i \perp E_j$ and we can write

$$E_j x = \sum_{i=1}^{k} E_j x_i = E_j x_j = x_j.$$

Substituting $E_i x$ for x_i in (23.21), we have

$$x = \sum_{i=1}^{k} E_i x = \left(\sum_{i=1}^{k} E_i \right) x, \qquad (23.22)$$

which implies

$$\sum_{i=1}^{k} E_i = 1$$

and proves part (c).

Applying A to both sides of (23.21), and using (23.22), we have

$$Ax = \sum_{i=1}^{k} Ax_i = \sum_{i=1}^{k} \lambda_i x_i$$

$$= \sum_{i=1}^{k} \lambda_i E_i x$$

$$= \left(\sum_{i=1}^{k} \lambda_i E_i \right) x;$$

hence,

$$A = \sum_{i=1}^{k} \lambda_i E_i$$

and part (a) is proved.

Suppose $\lambda_1, \lambda_2, \ldots, \lambda_k$ to be distinct and delete zero, if it occurs. Clearly nothing is lost by writing

$$Ax = \sum_{i=1}^{k} \lambda_i E_i x, \qquad (\lambda_i \neq 0).$$

This means that

$$Ax \in N(A - \lambda_1) \oplus N(A - \lambda_2) \oplus \cdots \oplus N(A - \lambda_k), \qquad (\lambda_i \neq 0).$$

By (23.14), each of the spaces $N(A - \lambda_i)$ is finite-dimensional for $\lambda_i \neq 0$ (therefore their sum is too) and we have proved $A(X)$ to be finite-dimensional or, equivalently, have proved A to be a finite-dimensional operator. The proof of the theorem is now complete.

Let us summarize what we have done. If we start with $A \in L(X, X)$, X as in the theorem, A normal and finite-dimensional, we observe the following:

(i) A is completely continuous because it is bounded and finite-dimensional;

(ii) A has a finite-point spectrum because it is finite-dimensional.

Parts (2,a,b,c) are now applicable to A, and we have indeed achieved a spectral decomposition for A just like the one in Sec. 2.2 for normal transformations on finite-dimensional spaces. After proving that theorem, we noted that there was a uniqueness statement going along with it which we proved as Theorem 2.7. In proving the uniqueness statement, however, nothing about the finite-dimensionality was needed—only the finiteness of the spectrum was used; therefore the same uniqueness statement applies in the general, infinite-dimensional case. For future reference we now state a theorem.

THEOREM 23.6.

(1) Let X be a complex Hilbert space and let $A \in L(X, X)$. If A is normal and finite-dimensional, then

(a) A has a finite point spectrum: $\lambda_1, \lambda_2, ..., \lambda_k$.
(b) $A = \sum_{i=1}^{k} \lambda_i E_i$, where E_i is the orthogonal projection on $N(A - \lambda_i)$;
(c) $E_i \perp E_j$ (for $i \neq j$);
(d) $\sum_{i=1}^{k} E_i = 1$.

(2) If A is a normal linear transformation mapping X into X, $\lambda_1, \lambda_2, ..., \lambda_k$ are distinct complex numbers, and $E_1, E_2, ..., E_k$ are nonzero linear transformations such that

(a) $A = \sum_{i=1}^{k} \lambda_i E_i$;
(b) $E_i E_j = 0$ for $i \neq j$;
(c) $\sum_{i=1}^{k} \lambda_i E_i = 1$;

then $\lambda_1, ..., \lambda_k$ are the distinct eigenvalues of A, and E_i is the orthogonal projection on $N(A - \lambda_i)$, for $i = 1, 2, ..., k$.

EXERCISES 23

1. If $\{S_\alpha\}_{\alpha \in \wedge}$ is a collection of subsets of the inner product space X, prove that $\overline{[\bigcup_\alpha S_\alpha]}^\perp = \bigcap_\alpha S_\alpha^\perp$.

2. Let X be a Hilbert space and $A \in L(X, X)$ be a normal completely continuous transformation; also, let $B \in L(X, X)$. Show that all $N(A - \lambda)$ are invariant under B if and only if $BA^* = A^*B$. Show that this is also equivalent to the fact that all $N(A - \lambda)$ reduce B.

3. Let X be a Hilbert space, and $A \in L(X, X)$ be such that $0 \leq A \leq 1$. Prove that $0 \leq A^2 \leq A$.

4. Let X be a Hilbert space and let $A, B \in L(X, X)$ be such that $A \geq 0$ and $B \geq 0$; also, suppose that $AB = BA$. Then prove that $\sqrt{AB} = \sqrt{A}\sqrt{B}$.

5. Let X be a Hilbert space and let $A \in L(X, X)$ be self-adjoint. Prove that there exist positive operators A^+ and A^- such that $A = A^+ - A^-$ and $A^+A^- = 0$. [*Hint:* let $A^+ = \frac{1}{2}(\sqrt{A^2} + A)$ and $A^- = \frac{1}{2}(\sqrt{A^2} - A)$.]

6. Suppose A is a normal completely continuous transformation defined on the Hilbert space X. If A has only finitely many eigenvalues and if A is $1:1$, prove that X is finite-dimensional.

7. Give an example of an operator that does not possess a square root.

8. Let X be a Hilbert space and let $A \in L(X, X)$ be a positive operator. What can be said about $\sigma(\sqrt{A})$?

9. Using the functional calculus as in the case of finite-dimensional spaces (Sec. 2.3), show that if X is a Hilbert space and U is a unitary operator belonging to $L(X, X)$ such that U is also a finite-dimensional operator, there exists a self-adjoint transformation A such that $U = e^{iA}$.

10. Let A be a finite-dimensional operator belonging to $L(X, X)$, where X is a Hilbert space. Show that there exists a sequence $\{A_n\}$ of invertible linear transformations such that $A_n \to A$.

REFERENCES

1. F. Riesz and B. Sz.-Nagy, "Functional Analysis."
2. L. Liusternik and W. Sobolev, "Elements of Functional Analysis."
3. S. Berberian, "Introduction to Hilbert Space."

CHAPTER **24**

Spectral Theorem for Completely
Continuous Normal Operators

Having obtained a spectral decomposition theorem for bounded, normal, finite-dimensional operators in the preceding chapter, we now wish to generalize this result to the case of completely continuous, normal operators. The main weakening of the previous hypothesis lies in dropping the assumption of a finite-point spectrum. The omission of this assumption shows itself in the result in that, where finite sums appeared, "infinite series" replace them. This is a reflection of the fact that completely continuous operators possess point spectra consisting of, at most, a denumerable set of eigenvalues. As the chain of generalizations increases to the consideration of operators whose point spectra do not obey this cardinality requirement, one might wonder what will replace the "infinite series" in the spectral theorem. Indeed, a reasonable guess would be some sort of integral, and this is precisely what happens.

Our first results provide the necessary framework to treat the spectral theorem. For example, if $\{A_\alpha | \alpha \in \Lambda\}$ is a collection of linear transformations, the notion of $\{A_\alpha\}$ being *summable to the transformation* A is formulated precisely. After introducing this and related concepts and proving certain results about them, proving the spectral decomposition theorem for completely continuous, normal operators is a relatively simple task.

24.1 Infinite Orthogonal Direct Sums; Infinite Series of Transformations

Let X be a complex Hilbert space and let $\{M_\alpha | \alpha \in \Lambda\}$ be a collection of closed subspaces of X. We define $\sum_\alpha M_\alpha$ to be all commutatively convergent (Sec. 9.1) sums $\sum_\alpha x_\alpha$, where $x_\alpha \in M_\alpha$. We remark that $\sum_\alpha M_\alpha$ is a subspace and further note that it *need not be closed*.† Our first two theorems deal with alternate ways of expressing $\overline{[\bigcup_\alpha M_\alpha]}$, the smallest closed subspace containing $\bigcup_\alpha M_\alpha$. Note that in our first theorem (and only in the first theorem) the assumptions that X is a *Hilbert* space, and each M_α is a *closed* subspace can be done without—the result is true in any inner product space for any collection of subspaces, closed or not.

THEOREM 24.1. Letting $M = \sum_\alpha M_\alpha$,

$$\overline{[\cup M_\alpha]} = \overline{\sum_\alpha M_\alpha} = \overline{M}.$$

REMARK. For the case of two subspaces M_1 and M_2, we have already noted that

$$\overline{[M_1 \cup M_2]} = \overline{M_1 + M_2}.$$

The theorem is just a generalization of this fact.

Proof. Certainly, since $\overline{M}$ is a closed subspace containing $\bigcup_\alpha M_\alpha$,

$$\overline{M} \supset \left[\bigcup_\alpha M_\alpha\right].$$

Conversely, let

$$S = \left\{\sum_\alpha x_\alpha \mid x_\alpha \in M_\alpha, \text{ almost all } x_\alpha = 0\right\}.$$

† For a counterexample see Halmos, *Introduction to Hilbert Space*, pp. 28–29.

We have

$$S \subset \left[\bigcup_\alpha M_\alpha \right],$$

which implies

$$\bar{S} \subset \left[\bigcup_\alpha M_\alpha \right].$$

Since the terms in M are all limits of members of S,

$$M \subset \bar{S},$$

which implies

$$\bar{M} \subset \bar{S}$$

and the theorem is proved.

The added hypothesis of the next theorem allows the bar over $\sum_\alpha M_\alpha$ in the preceding result to be removed.

THEOREM 24.2. If $\{M_\alpha\}$ is a pair-wise orthogonal, family of closed subspaces, then

$$\overline{\left[\bigcup_\alpha M_\alpha \right]} = \sum_\alpha M_\alpha.$$

Since the family $\{M_\alpha\}$ is an orthogonal family, we shall denote $\sum_\alpha M_\alpha$ by $\sum_\alpha \oplus M_\alpha$. In this notation the conclusion of the theorem can be written

$$\overline{\left[\bigcup_\alpha M_\alpha \right]} = \sum_\alpha \oplus M_\alpha.$$

Proof. By the preceding theorem,

$$\overline{\left[\bigcup_\alpha M_\alpha \right]} = \overline{\sum_\alpha \oplus M_\alpha} \supset \sum_\alpha \oplus M_\alpha.$$

To get the inclusion going the other way, we must make full use of the closedness and pair-wise orthogonality of the M_α. Let

$$x \in \overline{\left[\bigcup_\alpha M_\alpha \right]}.$$

Since each M_α is closed,

$$X = M_\alpha \oplus M_\alpha^\perp,$$

and we can write x as

$$x = x_\alpha + y_\alpha,$$

where $x_\alpha \in M_\alpha$ and $y_\alpha \in M_\alpha^\perp$; thus we have a myriad of representations for x. Consider those indices Λ' such that $x_\alpha \neq 0$. For any of those

$$\left(x, \frac{x_\alpha}{\|x_\alpha\|} \right) = \left(x_\alpha + y_\alpha, \frac{x_\alpha}{\|x_\alpha\|} \right) = \frac{(x_\alpha, x_\alpha)}{\|x_\alpha\|} = \|x_\alpha\|.$$

Therefore, by the Bessel inequality (since the collection $\{x_\alpha/\|x_\alpha\| \mid x_\alpha \neq 0\}$ is an orthonormal set),

$$\sum_\alpha \|x_\alpha\|^2 = \sum_\alpha |(x, x_\alpha/\|x_\alpha\|)|^2 \leq \|x\|^2,$$

and we see that

$$\sum_\alpha \|x_\alpha\|^2 < \infty.$$

By the proof of Theorem 9.1, for any $\varepsilon > 0$, there must exist a finite set $K \subset \Lambda$ such that, for any finite set $J \subset \Lambda$, $J \cap K = \emptyset$,

$$\sum_{\alpha \in J} \|x_\alpha\|^2 < \varepsilon.$$

By the Pythagorean theorem,

$$\left\| \sum_{\alpha \in J} x_\alpha \right\|^2 = \sum_{\alpha \in J} \|x_\alpha\|^2 < \varepsilon,$$

which implies that

$$\sum_{\alpha \in \Lambda'} x_\alpha = \sum_{\alpha \in \Lambda} x_\alpha$$

converges (Exercise 9.1) to some $w \in \sum_\alpha \oplus M_\alpha \subset \overline{[\bigcup_\alpha M_\alpha]}$; consequently,

$$x - w \in \overline{\left[\bigcup_\alpha M_\alpha\right]}. \tag{24.1}$$

Let $y \in M_\beta$, where β is an arbitrary element of Λ. In this case

$$(x - w, y) = \left(x_\beta + y_\beta - \sum_{\alpha \in \Lambda} x_\alpha, y \right)$$

$$= (x_\beta, y) - \left(\sum_{\alpha \in \Lambda} x_\alpha, y \right)$$

$$= (x_\beta, y) - \sum_{\alpha \in \Lambda} (x_\alpha, y)$$

$$= (x_\beta, y) - (x_\beta, y)$$

$$= 0;$$

hence,

$$x - w \perp M_\alpha \qquad \text{(for every } \alpha \in \Lambda).$$

This implies

$$x - w \perp \overline{\left[\bigcup_\alpha M_\alpha\right]}. \tag{24.2}$$

Combining (24.1) and (24.2), we have

$$x - w = 0$$

or

$$x = w = \sum_{\alpha \in \Lambda'} x_\alpha \in \sum_\alpha \oplus M_\alpha. \qquad \text{Q.E.D.}$$

In the event that $X = \sum_\alpha \oplus M_\alpha$, this means that any $x \in X$ can be written *uniquely* as

$$x = \sum_\alpha x_\alpha,$$

where $x_\alpha \in M_\alpha$. Demonstrating that this representation is unique is equivalent to showing that

$$\sum_\alpha x_\alpha = 0 \Rightarrow x_\alpha = 0 \qquad \text{(for all } \alpha\text{).}$$

But if $\sum_\alpha x_\alpha = 0$, then

$$0 = \left\| \sum_\alpha x_\alpha \right\|^2 = \sum_\alpha \|x_\alpha\|^2,$$

which does indeed imply that each x_α must be zero.

In our next theorem we continue to be concerned with orthogonal families of closed subspaces. Before proceeding to it, we wish to recall some facts from Chapter 9, where the notion of a complete orthonormal set, $\{x_\alpha\} = A$, was defined by requiring that

$$x \perp A \Rightarrow x = 0.$$

We saw that, in a Hilbert space X, this was equivalent to requiring that $\overline{[A]} = X$. These results have an obvious analog in the following theorem.

THEOREM 24.3. Suppose $\{M_\alpha\}$ is an orthogonal family of closed subspaces. If

$$x \perp \bigcup_\alpha M_\alpha \Rightarrow x = 0,$$

then $X = \overline{[\bigcup_\alpha M_\alpha]} = \sum_\alpha \oplus M_\alpha$ and conversely.

Proof. Suppose the condition holds. Clearly we can always write

$$X = \overline{\left[\bigcup_\alpha M_\alpha \right]} \oplus \overline{\left[\bigcup_\alpha M_\alpha \right]}^\perp$$

If $x \in \overline{[\bigcup_\alpha M_\alpha]}^\perp$, then certainly

$$x \perp \bigcup_\alpha M_\alpha,$$

which, by the condition, implies $x = 0$; hence

$$\overline{\left[\bigcup_\alpha M_\alpha \right]}^\perp = \{0\}$$

and sufficiency is established.

Conversely if $x = \sum_\alpha \oplus M_\alpha$, any $x \in X$ can be written

$$x = \sum_\alpha x_\alpha,$$

where $x_\alpha \in M_\alpha$. If $x \perp M_\alpha$ for every α, then

$$(x, x) = \left(\sum_\alpha x_\alpha, x \right) = \sum_\alpha (x_\alpha, x) = 0;$$

therefore $x = 0$ and the theorem is proved.

We now wish to demonstrate the validity of the interchange of arbitrary convergent summation and continuous linear operation. To this end let $A \in L(X, X)$ and suppose

$$\sum_\alpha x_\alpha = x.$$

We claim that

$$Ax = \sum_\alpha Ax_\alpha .$$

Since $\{x_\alpha\}$ is assumed to be summable to x, for any $\varepsilon > 0$, there must exist a finite set of indices H such that if J is any finite set containing H,

$$\left\| x - \sum_{\alpha \in J} x_\alpha \right\| < \varepsilon.$$

To finish the proof of the above contention, we merely observe that

$$\left\| Ax - \sum_{\alpha \in J} Ax_\alpha \right\| = \left\| A\left(x - \sum_{\alpha \in J} x_\alpha \right) \right\| \le \| A \| \varepsilon.$$

DEFINITION 24.1. The collection $\{A_\alpha\}$ from $L(X, X)$ is said to be *summable to A* if

$$\sum_\alpha A_\alpha x = Ax$$

for all $x \in X$. We denote this relation by writing

$$A = \sum_\alpha A_\alpha .$$

Note how similar this is to strong convergence of a sequence of bounded linear transformations as defined in Chapter 15. We are now in a position to generalize the following result: If E_1 and E_2 are orthogonal projections, then $E_1 + E_2$ is an orthogonal projection if, and only if, $E_1 \perp E_2$; moreover, in this case the range of $E_1 + E_2$ is given by $\overline{[R(E_1) \cup R(E_2)]} = R(E_1) \oplus R(E_2)$.

THEOREM 24.4. If $\{E_\alpha | \alpha \in \Lambda\}$ is a pair-wise orthogonal family of orthogonal projections, then

$$\sum_\alpha E_\alpha = E$$

is the orthogonal projection on

$$\overline{\left[\bigcup_\alpha M_\alpha \right]} = \sum_\alpha \oplus M_\alpha ,$$

where $M_\alpha = R(E_\alpha)$.

REMARK. The converse of this theorem is also true in this, the infinite case. The reader is asked to prove it in Exercise 24.1.

Proof. Let E be the orthogonal projection on $M = \sum_\alpha \oplus M_\alpha$. For any $z \in X$ we can write

$$z = x + y,$$

where $x \in M$ and $y \in M^\perp$. In this case $x = \sum_\alpha x_\alpha$ and, for any β,

$$E_\beta x = x_\beta$$

and

$$E_\beta y = 0.$$

(Note that $y \in M^{\perp} \subset M_{\beta}^{\perp}$.) Thus $E_{\beta}z = x_{\beta}$ and

$$Ez = x = \sum_{\alpha} x_{\alpha} = \sum_{\alpha} E_{\alpha}z,$$

which implies

$$E = \sum_{\alpha} E_{\alpha}.$$

24.2 Spectral Decomposition Theorem for Completely Continuous Normal Operators

Having introduced the necessary framework in the preceding section, it is a relatively simple matter to get the spectral theorem. Note how very similar it is to the spectral theorem obtained in the last chapter. After proving the spectral theorem some alternate ways of expressing Ax, where A is completely continuous and normal, and x is any vector, are obtained. These alternate forms are often quite useful.

THEOREM 24.5. (Spectral theorem for completely continuous, normal operators). Let $A \in L(X, X)$, X a complex Hilbert space, be completely continuous and normal. Then

(1)
$$X = N(A) \oplus \sum_{\substack{\lambda \in P\sigma(A) \\ \lambda \neq 0}} \oplus N(A - \lambda) = N(A) \oplus \sum_{\substack{\lambda \in C \\ \lambda \neq 0}} \oplus N(A - \lambda)$$

$$= \sum_{\lambda \in C} \oplus N(A - \lambda),$$

or
$$X = \sum_{\lambda \in C} \oplus N(A^* - \bar{\lambda});$$

(2)
$$\overline{R(A)} = \overline{R(A^*)} = \sum_{\lambda \neq 0} \oplus N(A - \lambda) = \sum_{\lambda \neq 0} \oplus N(A^* - \bar{\lambda});$$

(3)
$$A = \sum_{\lambda \in C} \lambda E_{\lambda} = \sum_{\lambda \in P\sigma(A)} \lambda E_{\lambda} \quad \text{and} \quad 1 = \sum_{\lambda \in C} E_{\lambda},$$

where E_{λ} is the orthogonal projection on $N(A - \lambda)$.

REMARK. Since A is completely continuous, all but a countable number of $N(A - \lambda)$ will consist solely of the zero vector.

Proof. Since A is normal the family $\{N(A - \lambda)|\lambda \in C\}$ is an *orthogonal* family of closed subspaces. By Theorem 23.4, since A is completely continuous and normal,

$$x \perp N(A - \lambda) \quad \text{for all } \lambda \Rightarrow x = 0.$$

Applying Theorem 24.3, this implies that

$$X = \left[\bigcup_{\lambda \in C} N(A - \lambda) \right] = \sum_{\lambda \in C} \oplus N(A - \lambda), \tag{24.3}$$

and proves (1).

Using the fact that, for normal transformations, $N(A) = N(A^*)$, we see that [Eq. (20.38)]

$$\overline{R(A)} = N(A^*)^\perp = N(A)^\perp.$$

Thus we have the orthogonal direct sum decomposition

$$X = N(A) \oplus \overline{R(A)},$$

whereas rewriting (24.3) yields the orthogonal direct sum

$$X = N(A) \oplus \sum_{\lambda \neq 0} \oplus N(A - \lambda).$$

Since the orthogonal complement of a subspace is unique, it follows that

$$\overline{R(A)} = \sum_{\lambda \neq 0} \oplus N(A - \lambda),$$

which implies (since A^* is also completely continuous and normal)

$$\overline{R(A^*)} = \sum_{\lambda \neq 0} \oplus N(A^* - \bar\lambda)$$

and proves (2), because, since $A - \lambda$ is normal, $N(A - \lambda) = N((A - \lambda)^*) = N(A^* - \bar\lambda)$.

Since the family $\{N(A - \lambda) | \lambda \in C\}$ is an orthogonal family, the collection $\{E_\lambda | \lambda \in C\}$ must be an orthogonal family of orthogonal projections. Since they are an orthogonal family, by Theorem 24.4, they must be summable to the orthogonal projection on

$$\overline{\left[\bigcup_{\lambda \in C} N(A - \lambda) \right]} = \sum_{\lambda \in C} \oplus N(A - \lambda) = X.$$

The orthogonal projection on the whole space, however, is just the identity transformation; hence,

$$\sum_{\lambda \in C} E_\lambda = 1.$$

Using (24.3), we can write any $x \in X$

$$x = \sum_{\lambda \in C} x_\lambda,$$

where $x_\lambda \in N(A - \lambda)$, and, since A is continuous, it follows that

$$Ax = \sum_{\lambda \in C} Ax_\lambda = \sum_{\lambda \in C} \lambda x_\lambda.$$

For any $v \in C$, we can write

$$E_v x = \sum_{\lambda \in C} E_v x_\lambda = x_v.$$

Therefore,

$$Ax = \sum_{\lambda \in C} \lambda E_\lambda x,$$

which implies

$$A = \sum_{\lambda \in C} \lambda E_\lambda.$$

This completes the proof.

By the results of the preceding chapter, if A is completely continuous and normal but not finite-dimensional, it must have an infinite point spectrum. Since $P\sigma(A) \subset \sigma(A)$ and $\sigma(A)$ is a bounded set (it lies within the circle bounded by $\|A\|$) it follows by the Bolzano–Weierstrass theorem that $P\sigma(A)$ must have a limit point. We noted in Theorem 18.2 that zero is the only possible limit point of the point spectrum of a completely continuous transformation; in this case, therefore, zero *must* be a limit point of $P\sigma(A)$. Suppose we ordered the points of $P\sigma(A)$ according to diminishing absolute value; that is, we write

$$\lambda_1, \lambda_2, ..., \lambda_n, ...,$$

where

$$|\lambda_1| \geq |\lambda_2| \geq \cdots \geq |\lambda_n| \geq \cdots.$$

The sequence $\{\lambda_n\}$ will converge to zero. In any event, since we can write

$$A = \sum_{i=1}^{\infty} \lambda_i E_{\lambda_i} \Leftrightarrow Ax = \sum_{i=1}^{\infty} \lambda_i E_{\lambda_i} x = \lim_n \sum_{i=1}^{n} \lambda_i E_{\lambda_i} x \qquad \text{(for all } x\text{),}$$

then, letting

$$A_n = \sum_{i=1}^{n} \lambda_i E_{\lambda_i},$$

we see that

$$A_n \overset{s}{\to} A.$$

More than strong convergence is true, though: In fact we have uniform convergence,

$$A_n \to A,$$

and we shall prove this now.

Using (24.3), we can write any $x \in X$

$$x = x_0 + \sum_{i=1}^{\infty} y_i, \tag{24.4}$$

where $x_0 \in N(A)$ and $y_i \in N(A - \lambda_i)$. Since, for $i = 1, 2, ...$, each $N(A - \lambda_i)$ is finite-dimensional, let us choose an orthonormal basis for each:

$$x_{i1}, x_{12}, ..., x_{in_i}.$$

We can now choose scalars α_{ij} such that

$$y_i = \sum_{j=1}^{n_i} \alpha_{ij} x_{ij}; \tag{24.5}$$

thus

$$x = x_0 + \sum_{i=1}^{\infty} \sum_{j=1}^{n_i} \alpha_{ij} x_{ij},$$

and by the orthogonality,

$$\|x\|^2 = \|x_0\|^2 + \sum_{i=1}^{\infty} \sum_{j=1}^{n_i} |\alpha_{ij}|^2. \tag{24.6}$$

We now compute Ax and $A_n x$:

$$Ax = 0 + \sum_{i=1}^{\infty} \sum_{j=1}^{n_i} \alpha_{ij} A x_{ij}$$

$$= \sum_{i=1}^{\infty} \sum_{j=1}^{n_i} \alpha_{ij} \lambda_i x_{ij}. \tag{24.7}$$

$$A_n x = 0 + A_n \sum_{i=1}^{\infty} y_i = \sum_{i=1}^{\infty} A_n y_i,$$

whereas

$$A_n y_i = \sum_{k=1}^{n} \lambda_k E_{\lambda_k} y_i = \begin{cases} 0 & (i > n), \\ \lambda_i y_i, & (i \leq n). \end{cases}$$

Therefore,

$$A_n x = \sum_{i=1}^{n} A_n y_i = \sum_{i=1}^{n} \lambda_i y_i$$

$$= \sum_{i=1}^{n} \sum_{j=1}^{n_i} \lambda_i \alpha_{ij} x_{ij}.$$

It follows that

$$\|Ax - A_n x\|^2 = \sum_{i=n+1}^{\infty} \sum_{j=1}^{n_i} |\lambda_i|^2 |\alpha_{ij}|^2. \tag{24.8}$$

Since $\{|\lambda_i|\}$ converges monotonically to zero we can choose an N such that $i > N$ implies $|\lambda_i| < \varepsilon$ for a prescribed $\varepsilon > 0$. By taking n bigger than N in (24.8), we can thus guarantee

$$\|Ax - A_n x\|^2 \leq \varepsilon^2 \sum_{i=n+1}^{\infty} \sum_{j=1}^{n_i} |\alpha_{ij}|^2,$$

which by (24.6), is less than or equal to

$$\varepsilon^2 \|x\|^2.$$

Since, for $n > N$, we have made

$$\|A - A_n\| \leq \varepsilon,$$

it follows that

$$A_n \to A.$$

Having noted this, we now wish to note the following alternate ways of expressing Ax when $A \in L(X, X)$ is completely continuous and normal. In the same notation as before, using (24.7), we can write

$$(Ax, x_{st}) = \sum_{i=1}^{\infty} \sum_{j=1}^{n_i} \alpha_{ij} \lambda_i (x_{ij}, x_{st}).$$

This expression evaluates only when $i = s$ and $j = t$; hence,

$$(Ax, x_{st}) = \alpha_{st} \lambda_s.$$

Using this fact in (24.7), we get

$$Ax = \sum_{i=1}^{\infty} \sum_{j=1}^{n_i} (Ax, x_{ij}) x_{ij}. \tag{24.9}$$

Since $(Ax, x_{ij}) = (x, \bar{\lambda}_i x_{ij}) = \lambda_i(x, x_{ij})$, Eq. (24.9) can be rewritten as

$$Ax = \sum_{i=1}^{\infty} \sum_{j=1}^{n_i} \lambda_i(x, x_{ij})x_{ij}. \qquad (24.10)$$

EXERCISES 24

1. Let $A \in L(X, X)$, where X is a Hilbert space. Suppose that $\{E_\alpha\}_{\alpha \in \Lambda}$ is a family of projections such that $A = \sum_\alpha E_\alpha$. Prove that A is a projection if and only if the family $\{E_\alpha\}$ is orthogonal. Show in this case that, if $R(E_\alpha) = M_\alpha$, then $R(A) = \overline{[\bigcup_\alpha M_\alpha]}$.

2. Let $A \in L(X, X)$, where X is a Hilbert space. Show that A is completely continuous if, and only if, there exists a sequence $\{A_n\}$ in $L(X, X)$ of finite-dimensional transformations such that $A_n \to A$.

3. Let $\{M_\alpha\}_{\alpha \in \Lambda}$ be a family of closed subspaces of the Hilbert space X. If each M_α is invariant under A (or reduces A), then show that $\overline{[\bigcup_\alpha M_\alpha]}$ and $\bigcap_\alpha M_\alpha$ are both invariant under A (or reduce A), where $A \in L(X, X)$.

4. Suppose A is a completely continuous normal transformation defined on the Hilbert space X. Show that A is self-adjoint if all eigenvalues of A are real. Show that A is positive if all eigenvalues are nonnegative.

5. Let A be a completely continuous normal transformation defined on the Hilbert space X. Then $Ax = \sum_{i=1}^{\infty} \sum_{j=1}^{n_i} \lambda_i(x, x_{ij})x_{ij}$, using the notation of the spectral theorem. Show that $R(A)$ consists of those elements $y \in \sum_{\lambda \neq 0} \oplus N(A - \lambda)$ such that $\sum_{i=1}^{\infty} \sum_{j=1}^{n_i} [(y, x_{ij})/\lambda_i] x_{ij}$ converges.

6. Let $A \in L(X, X)$, where X is a Hilbert space be self-adjoint and completely continuous with eigenvalues $\lambda_1, \ldots, \lambda_n, \ldots$ such that $|\lambda_1| > |\lambda_2| \ldots$. Prove that

$$|\lambda_k| = \pm \sup |(Ax, x)|,$$

where the sup is taken over all x with $\|x\| = 1$ and $x \perp N(A - \lambda_i)$, $(i = 1, \ldots, k)$.

7. Suppose A is a bounded normal operator such that $A = \sum_{i=1}^{\infty} \lambda_i E_{\lambda_i}$, where the E_{λ_i} are mutually orthogonal projections on finite-dimensional subspaces M_i and where the λ_i are distinct, no $\lambda_i = 0$, and $\lambda_i \to 0$. Then prove that A is completely continuous, λ_i are the distinct nonzero eigenvalues of A, and $M_i = N(A - \lambda_i)$.

8. Using the notation of Eq. (24.10), give a necessary and sufficient condition for the set $\{x_{ij}\}$ to be a complete orthonormal set.

9. Given a family $\{X_\alpha\}_{\alpha \in \Lambda}$ of Hilbert spaces, try to define an "external" direct sum X, of the X_α, in such a way that $X = \sum_\alpha \oplus M_\alpha$, where each M_α is congruent to X_α.

10. Using the notation of the spectral theorem, show that, if $A = \sum_{\lambda \in C} \lambda E_\lambda$, then $p(A) = \sum_{\lambda \in C} p(\lambda) E_\lambda$, where $p(\lambda)$ is any polynomial.

REFERENCES

1. P. Halmos, "Introduction to Hilbert Space."
2. S. Berberian, "Introduction to Hilbert Space."

CHAPTER **25**

Spectral Theorem for Bounded, Self-Adjoint Operators

In this chapter we shall obtain a spectral decomposition theorem for bounded, self-adjoint operators on a Hilbert space. Although we have dropped the weaker requirement of normality for the stronger self-adjoint, we compensate for this by doing away with the need for a countable point spectrum, as the assumption of complete continuity dictated in Chapter 24. Doing away with this is bound to make some changes in the results, however. We have already noticed how finite sums have given way to infinite series. We now make the final transition from infinite series to integrals. Something like this must be expected, though, because it is possible that there be a whole continuum of eigenvalues. For motivation, we first rephrase the results of the last chapter—the infinite series—in terms of a Riemann–Stieltjes integral. Specifically, where we had (for a completely continuous, *normal* transformation A)

$$Ax = \sum_{\lambda_i \in P\sigma(A)} \lambda_i E_{\lambda_i} x,$$

we shall get for completely continuous, *self-adjoint* transformations, an $F(\lambda)$ such that

$$Ax = \int_a^b \lambda \, dF(\lambda)x,$$

where a and b are two real numbers such that $\sigma(A) \subset (a, b)$. If we wished, we could also have dealt with A instead of Ax and gotten an $F(\lambda)$ such that

$$A = \int_a^b \lambda \, dF(\lambda).$$

This will not be done, however.

We have noted before that the entire spectrum of a self-adjoint transformation A must be wholly contained in $[-\|A\|, \|A\|]$. We wish to do away with the symmetry imposed by this observation and obtain a tighter bound on the spectrum. We shall show that it is contained in $[m, M]$, where

$$m = \inf_{\|x\|=1} (Ax, x) \quad \text{and} \quad M = \sup_{\|x\|=1} (Ax, x).$$

Then, relying heavily on the results of Chapter 13, we establish the spectral theorem for self-adjoint, bounded operators. To provide further insight into this theorem, two alternate proofs will be given.

25.1 A Special Case—the Self-Adjoint, Completely Continuous Operator

In this section we adhere to the notation of Theorem 24.5, and obtain a restatement of one of the results of that theorem for the case when A is self-adjoint. In that section we noted that, if A is not finite-dimensional, it must have an infinite point spectrum and zero must be a limit point of $P\sigma(A)$. We then arranged the eigenvalues of A according to decreasing absolute value to get a sequence of scalars converging to zero. With the added assumption that A is self-adjoint thrown in, we see that all these scalars must be real (Theorem 21.6), which enables us to define

$$F(\lambda) = \begin{cases} \displaystyle\sum_{\lambda_i \le \lambda} E_{\lambda_i} & (\lambda < 0), \\[2ex] 1 - \displaystyle\sum_{\lambda_i > \lambda} E_{\lambda_i} & (\lambda \ge 0), \end{cases}$$

where we take $\sum_{\lambda_i \le \lambda} E_{\lambda_i}$ and $\sum_{\lambda_i > \lambda} E_{\lambda_i}$ to be 0 if there are no $\lambda_i \le \lambda$ or $\lambda_i > \lambda$, respectively. For any λ, $F(\lambda)$ will be an orthogonal projection [Theorems 24.4, 22.5, and 24.5(3)]. Loosely speaking, $F(\lambda)$ is like a step function. In addition it also has the following properties:

(1) $\lambda_a \le \lambda_b \Rightarrow F(\lambda_a) \le F(\lambda_b)$;
(2) strong continuity from the right;
(3) for any λ_0, the limit from the left

$$\lim_{\lambda \to \lambda_0^-} F(\lambda)x = F(\lambda_0^-)x$$

exists;
(4) for $\lambda_i \in P\sigma(A)$,

$$F(\lambda_i) - F(\lambda_i^-) = E_{\lambda_i};$$

and, as we shall clarify later,

(5) $$Ax = \int_a^b \lambda \, dF(\lambda)x$$

for $x \in X$ and $\sigma(A) \subset (a, b)$.

We now set about proving these facts. To do a complete job, we should actually verify each assertion in the separate cases when $\lambda < 0$ or $\lambda \ge 0$. This, however, constitutes more of a bothersome prospect than an enlightening one, and in most instances we shall verify it in only one case. Suppose $\lambda_a < \lambda_b < 0$. This implies

$$F(\lambda_b) - F(\lambda_a) = \sum_{\lambda_a < \lambda_i \le \lambda_b} E_{\lambda_i} \ge 0,$$

and verifies (1). It says more though; indeed, by Theorem 22.4, it also implies that

$$F(\lambda_a)F(\lambda_b) = F(\lambda_b)F(\lambda_a) = F(\lambda_a). \tag{25.1}$$

We now wish to show that $F(\lambda)$ is strongly continuous from the right; that is, for any λ_0 and any $x \in X$,

$$\lim_{\lambda \to \lambda_0^+} F(\lambda)x = F(\lambda_0^+)x = F(\lambda_0)x.$$

First, using the completeness of X, it will be shown that $F(\lambda_0^+)x$ exists for every x. Let $\{\mu_n\}$ be a monotone sequence converging to λ_0 from the right. From (1) it follows that, as μ_n decreases toward λ_0, $(F(\mu_n)x, x)$ decreases monotonically and is bounded below by $(F(\lambda_0)x, x)$; therefore $\lim_n (F(\mu_n)x, x)$ exists. In view of this, since (assuming $n > m$)

$$\|F(\mu_n)x - F(\mu_m)x\|^2 = ((F(\mu_n) - F(\mu_m))x, x)$$
$$= (F(\mu_n)x, x) - (F(\mu_m)x, x),$$

we see that

$$\lim_{n,m} \|F(\mu_n)x - F(\mu_m)x\| = 0;$$

that is, $\{F(\mu_n)x\}$ is a Cauchy sequence. The completeness of X now yields the existence of the limit $F(\lambda_0{}^+)x$. We now wish to show $F(\lambda_0{}^+) = F(\lambda_0)$. Suppose that $\lambda_0 < 0$, choose ε so that $\lambda_0 + \varepsilon < 0$, and consider

$$F(\lambda_0 + \varepsilon) = \sum_{\lambda_i \leq \lambda_0 + \varepsilon < 0} E_{\lambda_i}.$$

Since zero is the only accumulation point of $P\sigma(A)$, there must be some $\varepsilon > 0$ such that

$$P\sigma(A) \cap (\lambda_0, \lambda_0 + \varepsilon) = \emptyset;$$

for such ε,

$$F(\lambda_0 + \varepsilon) = \sum_{\lambda_i \leq \lambda_0} E_{\lambda_i} = F(\lambda_0).$$

Assertion (2) now follows for $\lambda_0 < 0$ and similarly for $\lambda_0 > 0$. The case $\lambda_0 = 0$ requires special consideration, because the above argument cannot be applied. If $\lambda_0 = 0$,

$$F(0)x = x - \sum_{\lambda_i > 0} E_{\lambda_i} x$$

and, for $\lambda > 0$,

$$\|F(\lambda)x - F(0)x\| = \left\| \sum_{0 < \lambda_i \leq \lambda} E_{\lambda_i} x \right\|$$

$$\leq \left\| \sum_{0 < \lambda_i \leq \lambda} E_{\lambda_i} x \right\| + \left\| \sum_{-\lambda \leq \lambda_i < 0} E_{\lambda_i} x \right\|.$$

Since, for any x,

$$\sum_{i=1}^{\infty} E_{\lambda_i} x$$

converges to x, and $|\lambda_i|$ converges monotonically to zero, there exists for arbitrary $\lambda > 0$ an N such that for $i > N$, $\lambda_i \in [-\lambda, \lambda]$, and

$$\varepsilon > \left\| \sum_{i > N} E_{\lambda_i} x \right\|^2 = \sum_{\substack{\lambda_i < 0 \\ i > N}} \|E_{\lambda_i} x\|^2 + \sum_{\substack{\lambda_i > 0 \\ i > N}} \|E_{\lambda_i} x\|^2,$$

where $\varepsilon > 0$ is arbitrary. Thus, as $\lambda \to 0^+$,

$$\|F(\lambda)x - F(0)x\| \to 0.$$

Assertion (3) is easily established using an argument analogous to the proof of the existence of the strong limit from the right. We proceed to establish (4) now. For a nonzero $\lambda_0 \notin P\sigma(A)$, there must exist ε such that

$$(\lambda_0 - \varepsilon, \lambda_0 + \varepsilon) \cap P\sigma(A) = \emptyset$$

which implies

$$F(\lambda_0) - F(\lambda_0{}^-) = 0.$$

If $\lambda_0 \in P\sigma(A)$, $\lambda_0 \neq 0$, we choose ε such that

$$(\lambda_0 - \varepsilon, \lambda_0] \cap P\sigma(A) = \{\lambda_0\};$$

hence,

$$F(\lambda_0) - F(\lambda_0{}^-) = E_{\lambda_0}$$

in this case. Last, let $\lambda_0 = 0$. If we can show that

$$F(0^-) = \lim_{\lambda \to 0^-} \sum_{\lambda_i \leq \lambda} E_{\lambda_i} \tag{25.2}$$

is just $\sum_{\lambda_i < 0} E_{\lambda_i}$, it will follow that

$$F(0) - F(0^-) = E_0,$$

where we note that $E_0 = 0$ if $0 \notin P\sigma(A)$. To show (25.2), consider

$$\left\| \sum_{\lambda_i < 0} E_{\lambda_i} x - \sum_{\lambda_i \le \lambda} E_{\lambda_i} x \right\| = \left\| \sum_{\lambda < \lambda_i < 0} E_{\lambda_i} x \right\|,$$

which must go to zero because it is part of the "tail" of a convergent series.

We can now write

$$Ax = \sum_{\lambda_i \in P\sigma(A)} \lambda_i E_{\lambda_i} x$$

as

$$Ax = \sum_{\lambda_i \in P\sigma(A)} \lambda_i (F(\lambda_i) - F(\lambda_i^-)) x$$

$$= \sum_{\lambda \in C} \lambda (F(\lambda) - F(\lambda^-)) x$$

or as the following Riemann–Stieltjes integral:

$$Ax = \int_a^b \lambda \, dF(\lambda) x, \tag{25.3}$$

where a and b are chosen so that $\sigma(A) \subset [a, b]$ and the limit in (25.3) is defined as the natural extension of Riemann–Stieltjes integrals. Bear in mind that the implied limit in (25.3) is taken with respect to the norm on X. In addition, we remark that, using the uniform convergence of $A_n = \sum_{i=1}^n \lambda_i E_{\lambda_i}$, to A, we could also have shown that

$$A = \int_a^b \lambda \, dF(\lambda);$$

that is, a uniform-type convergence instead of the strong-type convergence asserted by (25.3).

25.2 Further Properties of the Spectrum of Bounded, Self-Adjoint Transformations

We assume that X is a complex Hilbert space and that A is a self-adjoint transformation belonging to $L(X, X)$. In this case [Eq. (21.8)]

$$\|A\| = \sup_{\|x\| = 1} |(Ax, x)|.$$

We define

$$m = \inf_{\|x\| = 1} (Ax, x) \quad \text{and} \quad M = \sup_{\|x\| = 1} (Ax, x).$$

Clearly,

$$\|A\| = \max(|m|, |M|). \tag{25.4}$$

We know from previous investigations that the purely real spectrum of A will be contained in the closed interval $[-\|A\|, \|A\|]$. Our next result strengthens this.

THEOREM 25.1. $\sigma(A) \subset [m, M]$.

Proof. For any $x \neq 0$,

$$\left(A \frac{x}{\|x\|}, \frac{x}{\|x\|}\right) \leq M,$$

which implies, for every x,

$$(Ax, x) \leq M(x, x).$$

Suppose $\lambda \notin [m, M]$. This can only mean $\lambda < m$ or $\lambda > M$. We assume the latter to hold and show $\lambda \notin \sigma(A)$. To show that the same conclusion follows from the former is a simple matter and will be omitted. If $\lambda > M$, then

$$\lambda = M + \varepsilon \qquad (\varepsilon > 0).$$

Thus,

$$((A - \lambda)x, x) = (Ax, x) - \lambda(x, x)$$

$$\leq M(x, x) - \lambda(x, x) = -\varepsilon(x, x) \leq 0,$$

or, taking absolute values,

$$|((A - \lambda)x, x)| \geq \varepsilon\|x\|^2.$$

Since

$$|((A - \lambda)x, x)| \leq \|(A - \lambda)x\|\, \|x\|,$$

it follows that

$$\|(A - \lambda)x\| \geq \varepsilon\|x\|$$

for every x, so, certainly,

$$\lambda \notin \pi(A),$$

where $\pi(A)$ is the approximate spectrum of A. For self-adjoint transformations, $\pi(A) = \sigma(A)$; a similar argument is applicable when $\lambda < m$, and the theorem follows.

The following result shows that the bound on $\sigma(A)$ cannot be tightened.

THEOREM 25.2. Both m and M belong to $\sigma(A)$.

Proof. There must be some real number v such that

$$M - v \geq m - v \geq 0.$$

In this case, using (25.4),

$$\sup_{\|x\|=1} ((A - v)x, x) = M - v = \|A - v\|.$$

If we could show that, under these conditions,†

$$M - v \in \sigma(A - v) = \{\lambda - v \mid \lambda \in \sigma(A)\}, \tag{25.5}$$

it will follow that

$$M \in \sigma(A).$$

---

† The full justification for the equality in (25.5) follows from the spectral mapping theorem for polynomials, Theorem 19.9.

Consequently there is no loss in generality in assuming $M \geq m \geq 0$ and showing $M \in \sigma(A)$. There must exist a sequence of vectors x_n, $\|x_n\| = 1$ for every n, such that

$$(Ax_n, x_n) \to M.$$

By the definition of M, $\{(Ax_n, x_n)\}$ must approach M from below, or

$$(Ax_n, x_n) = M - \varepsilon_n,$$

where $\varepsilon_n \to 0$ and $\varepsilon_n > 0$ for all n. Consider

$$\begin{aligned}
\|Ax_n - Mx_n\|^2 &= \|Ax_n\|^2 - 2M(Ax_n, x_n) + M^2\|x_n\|^2 \\
&\leq \|A\|^2 - 2M(Ax_n, x_n) + M^2 \\
&= 2M^2 - 2M(M - \varepsilon_n) \\
&= 2M\varepsilon_n \to 0;
\end{aligned}$$

therefore $M \in \pi(A) = \sigma(A)$. An analogous argument shows $m \in \sigma(A)$ and proves the theorem.

25.3 Spectral Theorem for Bounded, Self-Adjoint Operators

Now, for the first time, we obtain a spectral theorem for an operator which need not have a countable point spectrum. We assume that A is a self-adjoint transformation belonging to $L(X, X)$, where X is a complex Hilbert space, and let m and M carry the same meaning as in the preceding section. We shall be very concerned with the real Banach space, the real-valued, continuous functions with sup norm on $[m, M]$, $C[m, M]$, and the representation of bounded linear functionals on this space gotten in Chapter 13. Owing to the very different nature of the operator discussed here, the methods used in treating it will also be quite different from those used in obtaining the previous spectral theorems.

THEOREM 25.3. There exists a family of orthogonal projections $\{E(\lambda)\}$ defined for each real λ such that:

(1) for $\lambda_1 \leq \lambda_2$, $E(\lambda_1)E(\lambda_2) = E(\lambda_2)E(\lambda_1) = E(\lambda_1)$;
[Note that this is equivalent to $E(\lambda_1) \leq E(\lambda_2)$.]
(2) $\lim\limits_{\lambda \to \lambda_0^+} E(\lambda)x = E(\lambda_0^+)x = E(\lambda_0)x$;
(3) $E(\lambda) = 0$ for $\lambda < m$ and $E(\lambda) = 1$ for $\lambda \geq M$;
(4) $AE(\lambda) = E(\lambda)A$ [$R(E(\lambda))$ reduces A];
(5) for any $x, y \in X$ and any polynomial in the real variable λ with real coefficients, $p(\lambda)$, $(E(\lambda)x, y)$ is of bounded variation (as a function of λ on $[a, b]$) and

$$(p(A)x, y) = \int_a^b p(\lambda) \, d(E(\lambda)x, y)$$

for $a < m$ and $b \geq M$; in particular,

$$(Ax, y) = \int_{m^-}^M \lambda \, d(E(\lambda)x, y).$$

DEFINITION 25.1. The family $\{E(\lambda)\}$ is called the *resolution of the identity associated with A*.

Proof. With $p(\lambda)$ as in (5), since A is self-adjoint, $p(A)$ must be self-adjoint. Since it is self-adjoint its spectral radius must be equal to its norm (Theorem 21.3); symbolically,

$$\|p(A)\| = r_\sigma(p(A))$$

$$= \sup_{\lambda \in \sigma(p(A))} |\lambda|$$

$$= \sup_{\lambda \in \sigma(A)} |p(\lambda)|.$$

Viewing $p(\lambda)$ as an element of $C[m, M]$, we see that its norm is given by

$$\|p\| = \sup_{\lambda \in [m,M]} |p(\lambda)|.$$

Since, by the results of the preceding section, $\sigma(A) \subset [m, M]$,

$$\|p(A)\| = \sup_{\lambda \in \sigma(A)} |p(\lambda)| \le \sup_{\lambda \in [m,M]} |p(\lambda)| = \|p\|. \tag{25.6}$$

The space P of all such polynomials p is clearly a subspace of $C[m, M]$; moreover, by virtue of the Weierstrass theorem† it is a *dense* subspace. Consider the linear functional F, now. With x and y fixed vectors,

$$F : P \to C,$$

$$p \to (p(A)x, y).$$

In this case, using (25.6),

$$|F(p)| = |(P(A)x, y)| \le \|p(A)\| \, \|x\| \, \|y\|$$

$$\le \|p\| \, \|x\| \, \|y\|; \tag{25.7}$$

therefore, $F \in L(P, C)$ and $\|F\| \le \|x\| \|y\|$. By the results of Sec. 17.2, we can extend F uniquely to a new bounded linear functional with the same norm defined on $\bar{P} = C[m, M]$. We shall denote the extension of F by F, too. We now write F as a sum of its real and imaginary parts:

$$F = F_1 + iF_2;$$

clearly, the real-valued F_1 and F_2 are bounded linear functionals on $C[m, M]$. As such, by Theorem 13.2, there must exist real-valued normalized functions of bounded variation G_1 and G_2, such that, letting $G = G_1 + iG_2$ we have

$$F(p) = \int_m^M p(\lambda) \, dG(\lambda; x, y) \quad \text{and} \quad \|F_1\| = \|G_1\| = V(G_1), \ \|F_2\| = \|G_2\| = V(G_2).$$

[Since G depends on λ, x and y, albeit in different ways, we indicate its functional dependence by writing $G(\lambda; x, y)$.]

† See footnote, p. 81.

For an arbitrary interval $[a, b]$, if $g \in NBV[a, b]$ and

$$\int_a^b t^n \, dg(t) = 0 \qquad (\text{for } n = 0, 1, 2, \ldots,),$$

then $g(t)$ must be identically zero. For if the above integral vanishes for all t^n, it must also vanish for all polynomials. If it vanishes for all polynomials then, using the Weierstrass theorem again, it must vanish for any continuous function $x(t)$. Using the equivalence relation introduced in Chapter 13, it follows that g is equivalent to the zero function. Since $g(t)$ is normalized and the zero function is manifestly normalized and there is only one normalized function of bounded variation in each equivalence class, it follows that $g = 0$.

CLAIM. $G(\lambda; x, y)$ is a symmetric bounded sesquilinear functional on X for any λ.

We now prove this. It will be shown that $G(\lambda; x, y)$, viewed on the pair $\langle x, y \rangle$, is additive in the first argument. To this end consider the polynomial $p(\lambda) = \lambda^n$. In this case

$$(A^n(x_1 + x_2), y) = \int_m^M \lambda^n \, dG(\lambda; x_1 + x_2, y), \qquad (25.8)$$

which is equal to

$$(A^n x_1, y) + (A^n x_2, y) = \int_m^M \lambda^n \, dG(\lambda; x_1, y) + \int_m^M \lambda^n \, dG(\lambda; x_2, y)$$

$$= \int_m^M \lambda^n \, d\,[G(\lambda; x_1, y) + G(\lambda; x_2, y)]. \qquad (25.9)$$

Subtracting (25.9) from (25.8) and noting that the resulting expression vanishes for all λ^n, then by the remarks in the preceding paragraph, it follows that

$$G(\lambda; x_1 + x_2, y) = G(\lambda; x_1, y) + G(\lambda; x_2, y).$$

In identical fashion the homogeneity in the first argument, additivity, and conjugate homogeneity in the second argument follow. The symmetry of $G(\lambda; x, y)$ will now be demonstrated. Since λ is real and A is self-adjoint.

$$\overline{\int_m^M \lambda^n \, dG(\lambda; x, y)} = \int_m^M \lambda^n \, \overline{dG(\lambda; x, y)}$$

$$= \overline{(A^n x, y)}$$

$$= \overline{(x, A^n y)}$$

$$= (A^n y, x)$$

$$= \int_m^M \lambda^n \, dG(\lambda; y, x)$$

for every n. Subtracting the last expression from $\int_m^M \lambda^n \, dG(\lambda; y, x)$ and noting that the resulting quantity is zero for all λ^n, it follows that

$$\overline{G(\lambda; x, y)} = G(\lambda; y, x),$$

which proves that $G(\lambda; x, y)$ is a symmetric sesquilinear functional on X. From the definition of total variation, it follows that

$$V(G) \leq V(G_1) + V(G_2) = \|F_1\| + \|F_2\|,$$

which, by (25.7), is less than or equal to $2\|x\|\|y\|$. Since, recalling that G is normalized,

$$V(G) \geq |G(\lambda; x, y) - G(m; x, y)| = |G(\lambda; x, y)|,$$

the boundedness of the sesquilinear functional $G(\lambda; x, y)$ is proved.

To every bounded sesquilinear functional on a Hilbert space there is a unique, corresponding bounded linear transformation (Theorem 21.1.) so that in our case we can assert the existence of $B(\lambda) \in L(X, X)$ such that

$$G(\lambda; x, y) = (B(\lambda)x, y).$$

Moreover, since $G(\lambda; x, y)$ is symmetric, $B(\lambda)$ must be self-adjoint. Since $G(\lambda; x, y)$ is a normalized function of λ,

$$0 = G(m; x, y) = (B(m)x, y)$$

for all $x, y \in X$. This implies that

$$B(m) = 0. \tag{25.10}$$

Similarly, letting $p(\lambda) = 1$, for any $x, y \in X$,

$$(x, y) = \int_m^M dG(\lambda; x, y) = G(M; x, y) - G(m; x, y) = (B(M)x, y),$$

which implies that

$$B(M) = 1. \tag{25.11}$$

Thus $B(m) \leq B(M)$. A much stronger condition prevails, however: if $\lambda_a \leq \lambda_b$, then $B(\lambda_a) \leq B(\lambda_b)$. It is this fact—the monotonicity of the family $\{B(\lambda)\}$—that will be proved next. A result that will aid us in this endeavor is needed first.†

Let $f, g \in C[a, b]$, let $h \in BV[a, b]$, and form the function

$$w(t) = \int_a^t g(x) \, dh(x).$$

Then $w(t)$ is of bounded variation and

$$\int_a^b f(t) \, dw(t) = \int_a^b f(t)g(t) \, dh(t).$$

There is no doubt that $w(a) = 0$. Hence $w(t)$ has a chance of being normalized. Indeed, if $h(t)$ is normalized then so is $w(t)$, for consider

$$w(t^+) - w(t) = g(t)(h(t^+) - h(t)).$$

† For a proof see Widder, *The Laplace Transform.*

If $h(t)$ is normalized, then $h(t^+) - h(t) = 0$ for any $t \in (a, b)$ and the normality of $w(t)$ is evident.

Consider the function

$$C(\lambda; x, y) = \int_m^\lambda \mu^k \, dG(\mu; x, y). \tag{25.12}$$

Identifying $C(\lambda; x, y)$ with $w(t)$ in the previous result, we see that, since $G(\lambda; x, y)$ is normalized, $C(\lambda; x, y)$ is normalized; in addition the previous result also implies

$$\int_m^M \lambda^n \, dC(\lambda; x, y) = \int_m^M \lambda^{n+k} \, dG(\lambda; x, y)$$

$$= (A^{n+k}x, y) = (A^n x, A^k y)$$

$$= \int_m^M \lambda^n \, dG(\lambda; x, A^k y).$$

Since this is valid for every n, it follows by a now standard argument that (since C and G are each normalized)

$$C(\lambda; x, y) = G(\lambda; x, A^k y); \tag{25.13}$$

in effect we have carried out the integration in (25.12). We are now in a position to show, for $\lambda_a \le \lambda_b$, that

$$B(\lambda_a)B(\lambda_b) = B(\lambda_b)B(\lambda_a) = B(\lambda_a). \tag{25.14}$$

Since

$$G(\lambda; x, A^k y) = (B(\lambda)x, A^k y) = (A^k B(\lambda)x, y)$$

$$= \int_m^M \mu^k \, dG(\mu; B(\lambda)x, y)$$

$$= \int_m^M \mu^k \, d(B(\mu)B(\lambda)x, y),$$

we have, using (25.13) and (25.12),

$$C(\lambda; x, y) = \int_m^\lambda \mu^k \, d(B(\mu)x, y) = \int_m^M \mu^k \, d(B(\mu)B(\lambda)x, y). \tag{25.15}$$

We now introduce a normalized function $W(\mu; x, y)$ with the property that

$$\int_m^\lambda \mu^k \, d(B(\mu)x, y) = \int_m^M \mu^k \, dW(\mu; x, y).$$

In this case, (25.15) implies that $W(\mu; x, y) = (B(\mu)B(\lambda)x, y)$. Clearly, a suitable $W(\mu; x, y)$ is obtained by taking

$$W(\mu; x, y) = \begin{cases} (B(\mu)x, y) & (m \le \mu \le \lambda), \\ (B(\lambda)x, y) & (\lambda \le \mu \le M); \end{cases} \tag{25.16}$$

that is, $W(\mu; x, y)$ is constant from λ on. If $(B(\mu)x, y)$ happened to be real-valued, things might be as in Fig. 25.1. Note that, for $W(\mu; x, y)$ as in (25.16), $\int_\lambda^M dW(\mu; x, y) = 0$ and also that

$$\int_m^M \mu^k \, dW(\mu; x, y) = \int_m^M \mu^k \, d(B(\mu)B(\lambda)x, y).$$

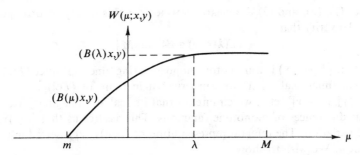

FIG. 25.1. A typical function $W(\mu; x, y)$ for the case where $(B(\mu) x, y)$ is real-valued.

As noted before, since $W(\mu; x, y)$ is clearly normalized, we conclude that

$$(B(\mu)B(\lambda)x, y) = W(\mu; x, y) = (B(\mu)x, y) \qquad \text{or} \qquad (B(\lambda)x, y)$$

as

$$\mu \leq \lambda \qquad \text{or} \qquad \lambda \leq \mu.$$

Consequently, letting $v = \min(\lambda, \mu)$, we have

$$B(\mu)B(\lambda) = B(v).$$

Taking adjoints, this is the same as

$$B(\lambda)B(\mu) = B(v).$$

If we let $\lambda = \mu$, we have

$$B(\lambda)^2 = B(\lambda);$$

or each $B(\lambda)$ is idempotent. Hence, since we have already noted that each $B(\lambda)$ is self-adjoint, $\{B(\lambda)\}$ is a family of orthogonal projections. Moreover, for $\lambda \neq m$,

$$\lim_{\mu \to \lambda^+} B(\mu)x = B(\lambda^+)x = B(\lambda)x;$$

that is, continuity from the right prevails everywhere except possibly at $\lambda = m$. To prove this, consider $\mu \geq \lambda > m$; $B(\mu) - B(\lambda)$ is an orthogonal projection because $B(\mu) \geq B(\lambda)$ and, therefore,

$$\|B(\mu)x - B(\lambda)x\|^2 = ([B(\mu) - B(\lambda)]x, x)$$

$$= (B(\mu)x, x) - (B(\lambda)x, x)$$

$$= G(\mu; x, x) - G(\lambda; x, x).$$

As $\mu \to \lambda^+$,

$$G(\mu; x, x) \to G(\lambda^+; x, x),$$

which must equal $G(\lambda; x, x)$ at each interior point of $[m, M]$, because $G(\lambda; x, y)$ is normalized. This proves the asserted continuity from the right. We now proceed to obtain the family $\{E(\lambda)\}$ mentioned in the statement of the theorem, by defining

$$E(\lambda) = \begin{cases} 0 & (\lambda < m), \\ B(\lambda^+) & (m \leq \lambda < M), \\ 1 & (\lambda \geq M). \end{cases}$$

Properties (1), (2), and (3) are clearly satisfied for this family of orthogonal projections. To verify that

$$(E(\lambda)x, y) \in BV[m, M],$$

we note that $(E(\lambda)x, x)$ is a monotone nondecreasing function; since $(E(\lambda)x, y)$ is a sesquilinear functional on X, we can write† it in terms of $(E(\lambda)[x + y], [x + y])$, $(E(\lambda)[x - y], [x - y])$, etc., which implies that its real and imaginary parts can be written as differences of monotone functions. This assures us that $(E(\lambda)x, y)$ is of bounded variation. The integral representation of $(p(A)x, y)$ asserted in (5) will be verified now. We wish to show

$$(p(A)x, y) = \int_a^b p(\lambda) \, d(E(\lambda)x, y). \tag{25.17}$$

Since $E(\lambda)$ is constant from M on,

$$\int_M^b p(\lambda) \, d(E(\lambda)x, y) = 0. \tag{25.18}$$

Allowing a possible jump in $E(\lambda)$ at $\lambda = m$, we have

$$\int_a^m p(\lambda) \, d(E(\lambda)x, y) + \int_m^M p(\lambda) \, d[(E(\lambda)x, y) - B(\lambda)x, y)]$$

$$= p(m)[(E(m)x, y) - (E(m^-)x, y)]$$
$$+ p(m)[(E(m^+)x, y) - (B(m^+)x, y)$$
$$- (E(m)x, y) + (B(m)x, y)].$$

Noting that $E(m) = E(m^+) = B(m^+)$ and $E(m^-) = B(m) = 0$, the above quantity is seen to be zero. Therefore,

$$\int_a^b p(\lambda) \, d(E(\lambda)x, y) = \int_m^M p(\lambda) \, d(B(\lambda)x, y)$$

$$= (p(A)x, y),$$

and (5) is proved. Last, (4) is proved by a direct computation; consider

$$(E(\mu)Ax, y) = (Ax, E(\mu)y)$$

$$= \int_a^b \lambda \, d(E(\lambda)x, E(\mu)y),$$

and compare this to

$$(AE(\mu)x, y) = \int_a^b \lambda \, d(E(\lambda)E(\mu)x, y)$$

$$= \int_a^b \lambda \, d(E(\mu)E(\lambda)x, y)$$

$$= \int_a^b \lambda \, d(E(\lambda)x, E(\mu)y).$$

† Equation (20.48), the polarization identity.

Since x and y are arbitrary,

$$AE(\mu) = E(\mu)A,$$

and the theorem is proved.

The family of projections $\{E(\lambda)\}$ is uniquely determined by A. If $\{F(\lambda)\}$ is a family of orthogonal projections also satisfying parts (1), (2), and (3) of the theorem such that, for all $p \in P$,

$$(p(A)x, x) = \int_a^b p(\lambda) \, d(F(\lambda)x, x), \tag{25.19}$$

then $F(\lambda)$ must equal $E(\lambda)$. Clearly, the function $(F(\lambda)x, y) \in NBV[a, b]$ by virtue of (1), (2), and (3). In view of (25.19),

$$(p(A)x, x) = \int_a^b p(\lambda) \, d(E(\lambda)x, x) = \int_a^b p(\lambda) \, d(F(\lambda)x, x) \qquad \text{(for all } p \in P).$$

Since $(E(\lambda)x, x)$ and $(F(\lambda)x, x)$ are each normalized, it follows that

$$(E(\lambda)x, x) = (F(\lambda)x, x) \qquad \text{(for every } x),$$

which implies

$$E(\lambda) = F(\lambda).$$

A stronger result than (5) of Theorem 25.3 prevails, however, and this constitutes our next theorem. Some notation pertaining to it is introduced first.

Let

$$a = \lambda_0 < \lambda_1 < \cdots < \lambda_n = b$$

be a partition of $[a, b]$, let

$$\xi_k \in [\lambda_{k-1}, \lambda_k],$$

let

$$\delta_k = \max_{\xi \in [\lambda_{k-1}, \lambda_k]} |p(\lambda) - p(\xi)|,$$

let $\varepsilon = \max_k \delta_k$, and let

$$B = \sum_{k=1}^n p(\xi_k)[E(\lambda_k) - E(\lambda_{k-1})].$$

In our next theorem we shall show that, as $\max(\lambda_k - \lambda_{k-1}) \to 0$, independently of the method of subdivision and the choice of ξ_k,

$$\|p(A) - B\| \to 0,$$

denoting the limit of B so obtained by

$$\int_a^b p(\lambda) \, dE(\lambda).$$

THEOREM 25.4. In the same notation as the preceding theorem and the above,

$$p(A) = \int_a^b p(\lambda) \, dE(\lambda).$$

Proof. With B as above and any x,

$$(Bx, x) = \sum_{k=1}^{n} \int_{\lambda_{k-1}}^{\lambda_k} p(\xi_k) \, d(E(\lambda)x, x).$$

[Note that

$$p(\xi_k) \int_{\lambda_{k-1}}^{\lambda_k} d(E(\lambda)x, x) = p(\xi_k)([E(\lambda_k) - E(\lambda_{k-1})]x, x)].$$

Since

$$(p(A)x, x) = \int_a^b p(\lambda) \, d(E(\lambda)x, x)$$

$$= \sum_{k=1}^{n} \int_{\lambda_{k-1}}^{\lambda_k} p(\lambda) \, d(E(\lambda)x, x),$$

we can write

$$((p(A) - B)x, x) = \sum_{k=1}^{n} \int_{\lambda_{k-1}}^{\lambda_k} (p(\lambda) - p(\xi_k)) \, d(E(\lambda)x, x),$$

which, since $(E(\lambda)x, x)$ is monotonic, is less than or equal to†

$$\varepsilon \sum_{k=1}^{n} ([E(\lambda_k) - E(\lambda_{k-1})]x, x) = \varepsilon(E(\lambda_n)x, x) - \varepsilon(E(\lambda_0)x, x)$$

$$= \varepsilon(x, x),$$

the last equality following because $E(\lambda_n) = 1$ and $E(\lambda_0) = 0$. We have established, for any x, that

$$([p(A) - B]x, x) \le \varepsilon(x, x),$$

which implies that

$$([p(A) - B]x, x) \le \varepsilon,$$

for all x with norm one; this implies

$$\inf_{\|x\|=1} ([p(A) - B]x, x) \le \sup_{\|x\|=1} ([p(A) - B]x, x) \le \varepsilon.$$

In similar fashion, since $(E(\lambda)x, x)$ is nondecreasing, we can show

$$-\varepsilon \le \inf_{\|x\|=1} ([p(A) - B]x, x).$$

In light of the self-adjointness of $p(A) - B$, these last results imply, by (25.4), that

$$\|p(A) - B\| \le \varepsilon.$$

By the continuity of $p(\lambda)$, ε can be made arbitrarily small and the theorem follows.

EXERCISES 25

X denotes a Hilbert space in all exercises.

1. Let $A \in L(X, X)$ be self-adjoint and let $m = \inf_{\|x\|=1}(Ax, x)$. Show that $m \in \sigma(A)$.

† See (II), Chapter 13. It is easy to verify that the analogous result holds here.

2. Let $\{F(\lambda)\}_{\lambda \in [a,b]}$ be a family of projections such that $F(\lambda) \le F(v)$ for $\lambda \le v$. Let $F(\lambda^+)x = \lim_{v \to \lambda^+} F(v)x$, where $a \le \lambda < b$. Prove that $F(\lambda^+)$ is a projection and that $F(\lambda^+) \le F(v)$ for $v > \lambda$. Show also that $\lim_{v \to \lambda^-} F(v)x$ exists for $a < \lambda \le b$ and defines a projection, denoted by $F(\lambda^-)$; also show $F(\lambda^-) \ge F(v)$ for $v < \lambda$.

3. Let $\mathscr{S}$ be a collection of subsets of R, the real axis, such that (a) $R \in \mathscr{S}$; (b) if $A \in \mathscr{S}$, then $CA \in \mathscr{S}$; (c) if $\{A_n\}$ is a countable sequence of sets of $\mathscr{S}$, then $\bigcup_n A_n \in \mathscr{S}$. Let E be a set function defined on $\mathscr{S}$ whose values are projections on X such that $E(R) = 1$ and $E(\bigcup_n A_n) = \sum_n E(A_n)$, where $\{A_n\}$ is a countable disjoint collection of sets of $\mathscr{S}$. Prove the following:
   (a) $A, B \in \mathscr{S}$ and $A \subset B$, then $E(A) \le E(B)$ and $E(B - A) = E(B) - E(A)$.
   (b) $A, B \in \mathscr{S}$, then $E(A \cup B) + E(A \cap B) = E(A) + E(B)$ and $E(A \cap B) = E(A)E(B)$.

4. If $B \in L(X, X)$ and $BE(\lambda) = E(\lambda)B$ for all λ, where $\{E(\lambda)\}$ is the resolution of identity of the self-adjoint operator A, show that $AB = BA$.

5. Consider $\int_a^b \lambda \, dE(\lambda)$ and assume that it exists in the uniform sense on $L(X, X)$. Show that it represents a self-adjoint transformation.

6. Show that, if $A \in L(X, X)$ is self-adjoint, then $A = B - C$, where B and C are bounded and positive.

7. Consider the operator A defined on $L_2[-1, 1]$ by $(Ax)(t) = tx(t)$ for $x \in L_2[-1, 1]$.
   (a) Show that A is bounded and self-adjoint.
   (b) Find m and M.
   (c) Show that $(E(\lambda)x)(t) = 0$ (for $t > \lambda$) or $x(t)$ (for $t < \lambda$) is a resolution of identity associated with A.

8. Let X be finite-dimensional and A self-adjoint with $\lambda_1, \dots, \lambda_k$ as distinct eigenvalues. Starting with Theorem 25.3, show that, if $E(\lambda)$, for $\lambda_i < \lambda < \lambda_{i+1}$ is the projection on $\sum_{j=1}^i N(A - \lambda_j)$ then $E(\lambda)$ is a resolution of identity for A and obtain the spectral theorem of Chapter 2 for self-adjoint transformations as a special case of Theorem 25.3.

9. Let $A \in L(X, X)$ be self-adjoint. Show that, if $\sigma(A)$ is nonnegative, then A is positive.

10. Using the notation of Theorem 25.3, show that if $f(\lambda)$ is continuous then $\int_a^b f(\lambda) \, d(E(\lambda)x, y)$ exists. Denote its value by $g(x, y)$. Prove that $g(x, y)$ is a bounded sesquilinear functional.

REFERENCES

1. A. Taylor, "Introduction to Functional Analysis."
2. P. Halmos, "Introduction to Hilbert Space."
3. D. V. Widder, "The Laplace Transform."

CHAPTER **26**

A Second Approach to the
Spectral Theorem for Bounded,
Self-Adjoint Operators

In this chapter a completely different proof of the spectral theorem for bounded, self-adjoint operators on a Hilbert space X will be given. In developing the proof a mapping from the class C_2 of differences of nonnegative, upper semicontinuous functions into $L(X, X)$ will be defined. In this proof it is this mapping and its properties that constitute the core of the approach. We make no reference to knowledge of the dual space of $C[a, b]$ nor to the correspondence between bounded sesquilinear functionals on X and $L(X, X)$ anywhere in this presentation, despite the fact that acquaintance with these notions was vital to the other proof (given in Chapter 25).

Using the mapping mentioned above, we shall be able to get a resolution of the identity and to obtain a spectral theorem. The mapping is of interest for its own sake, however. Specifically, the mapping will be an algebraic homomorphism from (the algebra) C_2 into $L(X, X)$—it preserves sums, products, and scalar multiples. One often refers to the homomorphism taking $f \in C_2$ into $f(A) \in L(X, X)$, where $A \in L(X, X)$ is some particular self-adjoint operator, and its ramifications as an *operational calculus for A*. A use for the operational calculus, for example, is that it allows computation of inverse operators in certain cases.

The mapping will be built up in steps: First a rather natural correspondence between polynomials (real coefficients) and operators will be defined; this mapping will then have its domain extended to C_1, the class of nonnegative, upper semicontinuous functions on a certain interval. This extension will be effected by noting that a sequence of *polynomials* exists, for each upper semicontinuous function, converging to it monotonically from above, taking the image of each polynomial in $L(X, X)$, and then taking the strong limit of these images in $L(X, X)$. Of course there are problems: The strong limit must exist and, even more fundamentally, we must be assured that we get the same strong limit even if a different sequence of polynomials is chosen that also converges to the upper semicontinuous function. After doing this, it is then an easy matter to extend the domain of definition a little further to C_2—differences of elements of C_1. Under this correspondence, we find that the functions

$$k_\mu(\lambda) = \begin{cases} 1 & (\lambda \le \mu), \\ 0 & (\lambda > \mu), \end{cases}$$

have as their corresponding images in $L(X, X)$ the family of orthogonal projections $E(\mu)$ which, as it happens, is a resolution of the identity for A.

26.1 A Second Approach to the Spectral Theorem for Bounded, Self-Adjoint Operators

A second proof of the theorem of Sec. 25.3, due to F. Riesz, is given here. The notation of that section will be adhered to throughout: X is a complex Hilbert space, $A \in L(X, X)$ is self-adjoint, $m = \inf_{\|x\| = 1}(Ax, x)$ and $M = \sup_{\|x\| = 1}(Ax, x)$.

The first link in the chain to be established is forged by defining a mapping from P, the class of polynomials with real coefficients, into $L(X, X)$ as follows:

$$\varphi_1 : P \to L(X, X),$$

$$p(\lambda) \to p(A).$$

Noting that we define the product of two polynomials, $p_1(\lambda)$ and $p_2(\lambda)$, $(p_1 p_2)(\lambda)$, as $p_1(\lambda)p_2(\lambda)$, the following facts about the mapping φ_1 are evident:

(1) $\varphi_1(p_1 + p_2) = p_1(A) + p_2(A)$;
(2) $\varphi_1(p_1 p_2) = p_1(A)p_2(A)$;
(3) for real scalars α, $\varphi_1(\alpha p) = \alpha p(A)$.

In addition it also possesses the following property of monotonicity:

(4) If $p(\lambda) \geq 0$ on $[m, M]$, then $p(A) \geq 0$.

Clearly this last requirement implies that, if $p_1(\lambda) \geq p_2(\lambda)$ on $[m, M]$, then $p_1(A) \geq p_2(A)$.

Viewing $L(X, X)$ as a real space (which we can always do), properties (1)–(3) are seen to constitute an algebraic homomorphism between the algebras P and $L(X, X)$. The first three properties all follow directly from the definition of the mapping; the last property, the property of monotonicity, will be demonstrated now.

Denote the real roots of $p(\lambda)$ that are less than or equal to m by $\alpha_1, \ldots, \alpha_j$; denote those that are greater than or equal to M by $\beta_1, \ldots, \beta_l$; last, denote all the rest, real or complex, by $v_1 + i\mu_1, \ldots, v_k + i\mu_k$. Thus if $\mu_i = 0$, we require $v_i \in (m, M)$. The assumption that $p(\lambda) \geq 0$ in $[m, M]$ implies that any real root of $p(\lambda)$ in (m, M) must be of even multiplicity. Since the complex roots must occur in conjugate pairs, the following representation for $p(\lambda)$ is assured:

$$p(\lambda) = a \prod_{i=1}^{j}(\lambda - \alpha_i) \prod_{i=1}^{l}(\beta_i - \lambda) \prod_{i=1}^{k}(\lambda - v_i)^2 + \mu_i^2, \qquad (26.1)$$

where $a \geq 0$.

We wish to show

$$A - \alpha_i \geq 0,$$

that is,

$$((A - \alpha_i)x, x) \geq 0$$

or

$$(Ax, x) \geq (\alpha_i x, x) = \alpha_i(x, x)$$

for all x. Since $\alpha_i \leq m = \inf_{\|x\| = 1}(Ax, x)$, the last requirement is indeed seen to be satisfied; this implies that $A - \alpha_i \geq 0$. In similar fashion,

$$\beta_j - A \geq 0.$$

Since $A - v_k$ is self-adjoint, its square is positive, and since the sum of positive operators is positive,

$$(A - v_k)^2 + \mu_k^2 \geq 0.$$

Thus each factor in (26.1) is positive, and since all the factors commute, it follows by Theorem 22.8 that their product is positive. Property (4) has now been verified.

It is our wish now to greatly expand the domain of definition of φ_1 to a much larger class. Before doing this a new definition is necessary, however.

DEFINITION 26.1. Let S be a subset of the real numbers R and denote its derived set by S'. Consider the function

$$f: S \to R$$

and let $x_0 \in S \cap S'$. f is said to be *upper semicontinuous at* x_0 if

$$\lim_{x \to x_0} \sup f(x) = \overline{\lim_{x \to x_0}} f(x) \leq f(x_0),$$

where, it is recalled,

$$\overline{\lim} f(x_0) = \inf_{\varepsilon > 0} \sup \{f(S_\varepsilon(x_0) \cap S - \{x_0\})\}.$$

It can be shown that upper semicontinuity at $x_0 \in S \cap S'$ is equivalent to being able to find a real number $\delta(\varepsilon) = \delta$, such that, for each $\varepsilon > 0$,

$$f(x) < f(x_0) + \varepsilon$$

for all $x \in S_\delta(x_0) \cap S$. If f is upper semicontinuous at every point of $S \cap S'$, then f is said to be *upper semicontinuous on* S. We denote by C_1 the class of all real-valued, nonnegative upper semicontinuous functions on $[m, M]$, and it is to this class† that we wish to extend the domain of definition of φ_1. To facilitate this, the following fact about upper semicontinuous functions is needed‡:

(A) If $f \in C_1$, there exists a family of continuous functions g_n such that, pointwise, $g_n(\lambda)$ converges monotonically to $f(\lambda)$ from above; that is, for $i > j$, $g_i(\lambda) \geq g_j(\lambda) \geq f(\lambda)$.

The convergence mentioned in (A) will be denoted symbolically by writing $g_n(\lambda) \downarrow f(\lambda)$.

In view of the Weierstrass theorem, since the functions $g_n \in C[m, M]$, we can find polynomials $p_n \in P$ such that

$$d(g_1 + 1, p_1) \leq \tfrac{1}{4}$$

$$d(g_2 + \tfrac{1}{2}, p_2) \leq \tfrac{1}{8}$$

$$\vdots \qquad \vdots$$

the distance function d, here being the *sup* norm distance function of $C[m, M]$. Having taken this care to preserve the monotonicity of the sequence, we have now achieved a sequence of polynomials that converge monotonically to the upper semicontinuous function f from above; symbolically, $p_n(\lambda) \downarrow f(\lambda)$.

Although every continuous function is upper semicontinuous, it is quite easy to show that the converse is not true. For example, the function defined on $[0, 1]$ to be 0 at every point except $x = \tfrac{1}{2}$ and defined to be $\tfrac{1}{2}$ there is upper semicontinuous at $\tfrac{1}{2}$ but not continuous there. In view of this, the convergence of the above-mentioned polynomials to the upper semicontinuous function f cannot be uniform in most cases, for if it were it would imply that the limit function is continuous.

Since $p_n(\lambda) \downarrow f(\lambda) \geq 0$,

$$p_1(A) \geq p_2(A) \geq \cdots \geq p_n(A) \geq p_{n+1}(A) \geq \cdots \geq 0$$

† Note that if $f_1, f_2 \in C_1$ and α is a *nonnegative* scalar, then $f_1 + f_2, f_1 f_2$, and αf_1 are all upper semicontinuous. As an example of an upper semicontinuous function, consider the intersection of the interval $[a, b]$ with any closed set. Denoting this intersection by E, the characteristic function of E is semicontinuous.

‡ A slight rewording of Theorem 10 on p. 152 of Natanson, *Theory of Functions of a Real Variable*, Vol. 2, yields this result.

by (4). Since each $p_n(A)$ is self-adjoint and since, for any j and k, $p_j(A)p_k(A) = p_k(A)p_j(A)$, by Theorem 23.1 the strong limit of $\{p_n(A)\}$ exists. We denote this limit, naturally enough, by $f(A)$; that is,

$$p_n(A) \xrightarrow{s} f(A) \geq 0.$$

Also, by the theorem just mentioned, $f(A)$ is self-adjoint, $f(A) \in L(X, X)$, and any transformation that commutes with A will also commute with $f(A)$. In this way we have achieved a correspondence between members of C_1 and members of $L(X, X)$; whether this correspondence is well-defined or not is something else, and we shall verify this vital consideration now.

Clearly it is possible that two different sequences of polynomials, $\{p_n(\lambda)\}$ and $\{q_n(\lambda)\}$, converge to $f(\lambda)$ monotonically from above. And by the same reasoning as used above, the strong limits of $\{p_n(A)\}$ and $\{q_n(A)\}$ must exist. It will now be demonstrated that these two strong limits are the same. Before doing this the reader is asked to recall Dini's theorem:

If the sequence of continuous functions $\{g_n(\lambda)\}$ defined on a compact set S converges monotonically to the continuous function $g(\lambda)$ on S, then the convergence is *uniform*.

Having recalled this fact, denote the maximum (point-wise) of the polynomials $p_n(\lambda)$ and $q_k(\lambda)$ on $[m, M]$ by $p_n(\lambda) \vee q_k(\lambda)$:

$$\max_{\lambda} (p_n(\lambda), q_k(\lambda)) = p_n(\lambda) \vee q_k(\lambda).$$

Since $p_n(\lambda)$ and $q_n(\lambda)$ are each continuous then, clearly, $p_n(\lambda) \vee q_k(\lambda)$ is continuous. In view of this, the sequence

$$h_k(\lambda) = p_k(\lambda) \vee q_n(\lambda) \qquad (k = 1, 2, \ldots)$$

is a sequence of continuous functions. From the monotone character of the sequence $\{p_n(\lambda)\}$ and $\{q_n(\lambda)\}$, it is evident that

$$h_k(\lambda) \downarrow q_n(\lambda).$$

By Dini's theorem, therefore, the convergence must be uniform, and it follows that for all sufficiently large k and every $\lambda \in [m, M]$,

$$h_k(\lambda) - q_n(\lambda) = p_k(\lambda) \vee q_n(\lambda) - q_n(\lambda) \leq \frac{1}{n}.$$

Since

$$p_k(\lambda) - q_n(\lambda) \leq h_k(\lambda) - q_n(\lambda),$$

it follows that, for all sufficiently large k, and all $\lambda \in [m, M]$,

$$p_k(\lambda) - q_n(\lambda) \leq \frac{1}{n}. \tag{26.2}$$

Similarly, interchanging the roles of the polynomials, we can also show that, for all sufficiently large k,

$$q_k(\lambda) - p_n(\lambda) \leq \frac{1}{n} \qquad \text{(for all } \lambda \in [m, M]\text{)}. \tag{26.3}$$

By the monotonicity mentioned in (4), (26.2) implies that (for the strong limit)

$$\lim_{k} p_k(A) \leq q_n(A) + \frac{1}{n}. \tag{26.4}$$

Since the self-adjoint transformation $q_n(A) + 1/n$ commutes with $\lim_k p_k(A)$ for every n, then, by the remark following Theorem 23.1, it follows that

$$\lim_{k} p_k(A) \leq \lim_{n}\left(q_n(A) + \frac{1}{n}\right) = \lim_{n} q_n(A).$$

In an exactly analogous fashion, Eq. (26.3) implies that

$$\lim_{k} q_k(A) \leq \lim_{n} p_n(A),$$

and we conclude that the two strong limits must be the same. The following extension of φ_1 to φ_2 has now been established:

$$\varphi_2 : C_1 \to L(X, X),$$

$$f(\lambda) \to f(A),$$

with the properties that, for $f_1, f_2 \in C_1$,

(1') $\varphi_2(f_1 + f_2) = f_1(A) + f_2(A)$;
(2') $\varphi_2(f_1 f_2) = f_1(A) f_2(A)$;
(3') for a *nonnegative* scalar α, $\varphi_2(\alpha f) = \alpha f(A)$;
(4') if $f_1(\lambda) \geq f_2(\lambda)$ on $[m, M]$, then $f_1(A) \geq f_2(A)$.

It is now a simple matter to extend the domain of the mapping even a little further to

$$C_2 = \{f - g \,|\, f, g \in C_1\}$$

by taking the image of a typical element of C_2, $h = f - g$, to be

$$h(A) = f(A) - g(A).$$

To show that this mapping is well-defined, suppose that we can write $h \in C_2$ in two different ways as differences of elements of C_1; that is, suppose $f_i, g_i \in C_1$ ($i = 1, 2$) and that

$$h = f_1 - g_1 = f_2 - g_2.$$

In this case $f_1 + g_2 = f_2 + g_1 \in C_1$, so, since φ_2 is well-defined,

$$f_1(A) + g_2(A) = f_2(A) + g_1(A),$$

which implies

$$f_1(A) - g_1(A) = f_2(A) - g_2(A).$$

This extension of φ_2 will be denoted by φ_3,

$$\varphi_3 : C_2 \to L(X, X),$$

$$h = f - g \to h(A) = f(A) - g(A).$$

This mapping clearly preserves sums and products just as φ_2 does. In addition, it also preserves *any* scalar multiple; this being demonstrated by noting that, for $\alpha < 0$,

$$\alpha(f - g) = (-\alpha)g - (-\alpha)f; \quad -\alpha g, \ -\alpha f \in C_1.$$

Thus, by allowing differences of elements of C_1, any scalar multiple is preserved; in addition it is a simple matter to verify that φ_3 also has the property of monotonicity. Now that this mapping has been established, we are in a position to get the spectral theorem. It will be shown that the images of the upper semicontinuous functions

$$k_\mu(\lambda) = \begin{cases} 1 & (\lambda \leq \mu), \\ 0 & (\lambda > \mu), \end{cases}$$

under φ_3, letting $\varphi_3(k_\mu(\lambda)) = k_\mu(A) = E(\mu)$, are a resolution of the identity.

Certain properties that the family $\{E(\mu)\}$ possesses are already clear. Certainly each $E(\mu)$ is self-adjoint; each commutes with A, hence with any polynomial in A, and therefore with each other. To show that each $E(\mu)$ is an orthogonal projection, since we have already noted that each is self-adjoint, it only remains to demonstrate their idempotency. Since, by the definition of the functions $k_\mu(\lambda)$,

$$k_\mu(\lambda)k_\mu(\lambda) = k_\mu(\lambda)$$

for any μ, it follows that

$$E(\mu)^2 = E(\mu)$$

and proves that the family $\{E(\mu)\}$ is a family of orthogonal projections. Another direct consequence of the definition of the functions $k_\mu(\lambda)$ is the monotonicity of the family $\{E(\mu)\}$; for certainly, if $v \geq \mu$, then

$$k_v(\lambda)k_\mu(\lambda) = k_\mu(\lambda)k_v(\lambda) = k_\mu(\lambda),$$

which, since φ_3 preserves products, implies

$$E(v)E(\mu) = E(\mu)E(v) = E(\mu) \qquad (\text{for } \mu \leq v).$$

If $\mu < m$, $k_\mu(\lambda) = 0$ on $[m, M]$ and a sequence of polynomials converging monotonically to $k_\mu(\lambda)$ from above is $0, 0, \ldots, 0, \ldots$; this implies by Theorem 23.1 that

$$E(\mu) = 0 \qquad (\text{for } \mu < m).$$

Similarly, if $\mu \geq M$, $k_\mu(\lambda) = 1$ on $[m, M]$ and the sequence $1, 1, \ldots, 1, \ldots \downarrow k_\mu(\lambda)$ on $[m, M]$.

Therefore,

$$E(\mu) = 1 \qquad (\text{for } \mu \geq M).$$

Next it is shown that the family $\{E(\mu)\}$ is strongly continuous from the right. To this end we first establish the existence of a sequence of polynomials $\{p_n(\lambda)\}$ satisfying simultaneously the conditions

$$p_n(\lambda) \downarrow k_\mu(\lambda) \tag{26.5}$$

and

$$p_n(\lambda) \geq k_{\mu + 1/n}(\lambda). \tag{26.6}$$

For any fixed n, there exists a sequence of polynomials $\{p_j^{(n)}(\lambda)\}$ such that

$$p_j^{(n)}(\lambda) \downarrow k_{\mu+1/n}(\lambda).$$

Thus we have

$$p_1^{(1)}(\lambda), p_2^{(1)}(\lambda),\ldots \qquad \downarrow k_{\mu+1}(\lambda)$$

$$p_1^{(2)}(\lambda), p_2^{(2)}(\lambda), \ldots \qquad \downarrow k_{\mu+\frac{1}{2}}(\lambda)$$

$$p_1^{(3)}(\lambda), p_2^{(3)}(\lambda), \ldots \qquad \downarrow k_{\mu+\frac{1}{3}}(\lambda)$$

$$\vdots \qquad\qquad\qquad \vdots$$

Consider the monotone sequence

$$g_n(\lambda) = \inf\{p_s^{(r)}(\lambda) \mid r, s = 1, 2, \ldots, n\}.$$

It is evident that

$$g_n(\lambda) \geq k_{\mu+1/n}(\lambda). \qquad (26.7)$$

We also have $g_n(\lambda) \downarrow k_\mu(\lambda)$ by virtue of the following: Because of (26.7), we have

$$\lim_n g_n(\lambda) \geq \lim_n k_{\mu+1/n}(\lambda) = k_\mu(\lambda). \qquad (26.8)$$

By the way $g_n(\lambda)$ was defined,

$$g_n(\lambda) \leq p_n^{(r)}(\lambda) \qquad (r = 1, 2, \ldots, n),$$

which implies

$$\lim_n g_n(\lambda) \leq \lim_n p_n^{(r)}(\lambda),$$

for any r. Taking the limit as $r \to \infty$,

$$\lim_r \lim_n p_n^{(r)}(\lambda) = k_\mu(\lambda),$$

we see that

$$\lim_n g_n(\lambda) = k_\mu(\lambda).$$

The functions $g_n(\lambda)$ need not be polynomials, however. To get polynomials satisfying (26.5) and (26.6), we shall "stretch" the continuous functions $g_n(\lambda)$ apart [that is, consider $\{g_n(\lambda) + (1/n)\}$] and then "approximate" each term in the stretched sequence uniformly by a polynomial $p_n(\lambda)$ ($p_n(\lambda)$ is chosen so that $|g_n(\lambda) + (1/n) - p_n(\lambda)| < 1/2^{n+1}$ for all $\lambda \in [m, M]$). The sequence of polynomials so obtained satisfies (26.5) and (26.6).

We now have

$$p_n(A) \to E(\mu)$$

and

$$p_n(A) \geq k_{\mu+1/n}(A) = E\left(\mu + \frac{1}{n}\right) \geq E(\mu). \qquad (26.9)$$

The sequence $\{E[\mu + (1/n)]\}$ must converge strongly to some self-adjoint transformation, F say, by Theorem 23.1. Using this and (26.9), we have, in the limit as $n \to \infty$,

$$(E(\mu)x, x) \geq (Fx, x) \geq (E(\mu)x, x)$$

for any x; hence,

$$(E(\mu)x, x) = (Fx, x)$$

for every x, which implies

$$E(\mu) = F,$$

and we have shown

$$E\left(\mu + \frac{1}{n}\right) \xrightarrow{\ s\ } E(\mu).$$

Since the family $\{E(\mu)\}$ is monotone, this implies that, for any sequence of positive terms $\{\varepsilon_n\}$ converging to zero,

$$E(\mu + \varepsilon_n) \xrightarrow{\ s\ } E(\mu),$$

which, in turn, implies strong continuity from the right.

To complete the proof it only remains to show that we have the desired integral representation of A in terms of the family $\{E(\lambda)\}$. We show now that the integral representation of Theorem 25.4 prevails here.

For $\mu \le v$,

$$\mu(k_v(\lambda) - k_\mu(\lambda)) \le \lambda(k_v(\lambda) - k_\mu(\lambda)) \le v(k_v(\lambda) - k_\mu(\lambda)). \tag{26.10}$$

To verify this, it is noted that, if λ is strictly less than μ or strictly greater than v, everything equals zero. If $\mu \le \lambda \le v$, then $k_v(\lambda) - k_\mu(\lambda) = 1$ and, again, the inequality is verified. The monotonicity of the mapping ϕ_3 then implies

$$\mu(E(v) - E(\mu)) \le A(E(v) - E(\mu)) \le v(E(v) - E(\mu)). \tag{26.11}$$

Choosing real numbers a and b such that $a < m$ and $b \ge M$, we partition $[a, b]$ as follows:

$$a = \lambda_0 < \lambda_1 < \cdots < \lambda_n = b.$$

Using the same reasoning as in showing that (26.10) implies (26.11), we see that, for any $k = 1, 2, \ldots, n$,

$$\lambda_{k-1}(E(\lambda_k) - E(\lambda_{k-1})) \le A(E(\lambda_k) - E(\lambda_{k-1})) \le \lambda_k(E(\lambda_k) - E(\lambda_{k-1})).$$

Summing with respect to k and noting that $\sum_{k=1}^{n}(E(\lambda_k) - E(\lambda_{k-1})) = 1$, we have

$$\sum_{k=1}^{n} \lambda_{k-1}(E(\lambda_k) - E(\lambda_{k-1})) \le A \le \sum_{k=1}^{n} \lambda_k(E(\lambda_k) - E(\lambda_{k-1})). \tag{26.12}$$

Let λ_k' be any point in $[\lambda_{k-1}, \lambda_k]$ and form

$$A - \sum_{k=1}^{n} \lambda_k'(E(\lambda_k) - E(\lambda_{k-1})).$$

By (26.12) and noting that $\lambda_k - \lambda_k' \le \lambda_k - \lambda_{k-1}$, we have

$$A - \sum_{k=1}^{n} \lambda_k'(E(\lambda_k) - E(\lambda_{k-1}) \le \sum_{k=1}^{n} (\lambda_k - \lambda_{k-1})(E(\lambda_k) - E(\lambda_{k-1})).$$

Letting $\varepsilon = \max_k |\lambda_k - \lambda_{k-1}|$, we have

$$A - \sum_{k=1}^{n} \lambda_k'(E(\lambda_k) - E(\lambda_{k-1})) \le \varepsilon \sum_{k=1}^{n} (E(\lambda_k) - E(\lambda_{k-1})) = \varepsilon 1.$$

Similarly, we can show

$$-\varepsilon 1 \leq A - \sum_{k=1}^{n} \lambda_k'(E(\lambda_k) - E(\lambda_{k-1})).$$

Since $A - \sum_{k=1}^{n} \lambda_k' (E(\lambda_k) - E(\lambda_{k-1}))$ is self-adjoint, it follows that†

$$\left\| A - \sum_{k=1}^{n} \lambda_k'(E(\lambda_k) - E(\lambda_{k-1})) \right\| \leq \varepsilon,$$

which implies

$$A = \int_a^b \lambda \, dE(\lambda) = \int_{m^-}^{M} \lambda \, dE(\lambda).$$

Actually, a little more than this is desired. In the previous approach a result similar to the above was obtained for *any polynomial* in A, $p(A)$, with real co-efficients. To this end, consider, using the same notation as above, any positive integer r.

$$\left(\sum_{k=1}^{n} \lambda_k'(E(\lambda_k) - E(\lambda_{k-1})) \right)^r$$

and note that by the monotonicity of the family $\{E(\lambda)\}$, for $j < k$ [which implies $j \leq (k-1)$],

$$(E(\lambda_j) - E(\lambda_{j-1}))(E(\lambda_k) - E(\lambda_{k-1})) = 0.$$

Combining this with the fact that $(E(\lambda_j) - E(\lambda_{j-1}))^r$ is just $(E(\lambda_j) - E(\lambda_{j-1}))$ for any j, we see that

$$\left(\sum_{k=1}^{n} \lambda_k'(E(\lambda_k) - E(\lambda_{k-1})) \right)^r = \sum_{k=1}^{n} (\lambda_k')^r(E(\lambda_k) - E(\lambda_{k-1})).$$

In the limit as $\varepsilon \to 0$, the quantity on the left goes to A^r, whereas the quantity on the right goes to $\int_a^b \lambda^r \, dE(\lambda)$; therefore, for any $n \geq 0$,

$$A^n = \int_a^b \lambda^n \, dE(\lambda),$$

which implies, for any polynomial $p(\lambda)$ with real coefficients,

$$p(A) = \int_a^b p(\lambda) \, dE(\lambda) = \int_{m^-}^{M} p(\lambda) \, dE(\lambda).$$

The proof is now complete.

EXERCISES 26

1. If $f_1, f_2 \in C_1$ and if $f_1 \geq f_2$ on $[m, M]$, show that $f_1(A) \geq f_2(A)$ under the correspondence introduced. Do the same thing for $f_1, f_2 \in C_2$.

---

† If $C \in L(X, X)$ is self-adjoint and $-\varepsilon 1 \leq C \leq \varepsilon 1$, then $-\varepsilon(x, x) \leq (Cx, x) \leq \varepsilon(x, x)$, which implies $\|C\| = \sup_{\|x\|=1} |(Cx, x)| \leq \varepsilon$.

2. Using the same notations as in the spectral theorem, suppose it has been established that $p(A) = \int_a^b p(\lambda) \, dE(\lambda)$ in a uniform sense, as was done in the treatment given in the second approach to the spectral theorem, then prove that $(p(A)x, y) = \int_a^b p(\lambda) \, d(E(\lambda)x, y)$ for all $x, y \in X$.

3. From $p(A) = \int_a^b p(\lambda) \, dE(\lambda)$ for all real polynomials p, prove that $f(A) = \int_a^b f(\lambda) \, dE(\lambda)$ for all real-valued continuous functions f on $[m, M]$, where $f(\lambda) = f(m)$ for $a \le \lambda \le m$, and $f(\lambda) = f(M)$ for $M \le \lambda \le b$.

4. Using the same convention as in the preceding theorem, suppose f and g are real-valued continuous functions defined on $[m, M]$; prove that

$$\int_a^b f(\lambda) \, dE(\lambda) \int_a^b g(\lambda) \, dE(\lambda) = \int_a^b f(\lambda) \, g(\lambda) \, dE(\lambda).$$

5. Finish the proof that the projections $E(\mu)$ are continuous from the right.

6. Let f be as in Exercise 3. Prove that $\|f(A)\| \le \max_{\lambda \in [m, M]} |f(\lambda)|$, and $\|f(A)x\|^2 = \int_b^a |f(\lambda)|^2 \, d\|E(\lambda)x\|^2$.

7. Let $A \in L(X, X)$ be self-adjoint with resolution of the identity $\{E(\lambda)\}$. Show that, if $B \in L(X, X)$, and $AB = BA$, then $BE(\lambda) = E(\lambda)B$.

8. Let $A \in L(X, X)$ be self-adjoint. Define

$$f(\lambda) = \begin{cases} \dfrac{1}{\lambda} & (0 \le c < \lambda \le d), \\[2mm] 0 & \text{(elsewhere).} \end{cases}$$

Show that $f(A)$ exists.

9. Let $A \in L(X, X)$ be self-adjoint with resolution of identity $\{E(\lambda)\}$. Show that if $BE(\lambda) = E(\lambda)B$ for all λ, then B commutes with $f(A)$, where f is as in Exercise 3.

10. Give an example of a real-valued function $f(\lambda)$ for which $f(A)$ is not defined by the definitions of this chapter.

REFERENCES

1. F. Riesz and B. Sz.-Nagy, " Functional Analysis."
2. I. P. Natanson, " Theory of Functions of a Real Variable," Vol. 2.

CHAPTER 27

A Third Approach to the Spectral Theorem for Bounded, Self-Adjoint Operators and Some Consequences

Theorem 25.3, the spectral theorem for bounded, self-adjoint operators on a Hilbert space, will be proved again in this lecture. The approach given here is due to B. Sz.-Nagy. It relies heavily on the results of Sec. 23.1 especially Theorems 23.1 and 23.2; that is, we rely very heavily on the fact that a positive bounded operator possesses a unique positive square root.

The freshness that this proof provides is its geometric appeal. In neither of the first two proofs does one get any ideas as to what the projections $E(\lambda)$ (the resolution of the identity, that is) "look like." In the first approach (Chapter 25), we got no insight at all into their character, their existence being guaranteed only by judicious application of previously obtained representation theorems. In the second (Chapter 26), the projections were obtained as images of certain upper semicontinuous functions and, again, one does not get much feeling for even the subspaces that are the ranges of the orthogonal projections.

In the proof of this chapter a monotone increasing sequence of closed subspaces is obtained first, and the resolution of the identity is achieved by taking the orthogonal projections on these subspaces.

Using the spectral Theorem (proved in any of the three ways), certain facts about the resolvent set and its relation to the resolution of the identity associated with the self-adjoint, bounded transformation are then obtained. In particular, it is shown, denoting the resolution of the identity by $\{E(\lambda)\}$, that, if $\lambda_0 \in \rho(A)$, then $E(\lambda)$ must be constant in a whole interval about λ_0. After getting this result, a convenient representation for $(\lambda_0 - A)^{-1}$ is obtained.

27.1 A Third Approach to the Spectral Theorem for Bounded, Self-Adjoint Operators

We present here a proof of Theorem 25.3 due to Sz.-Nagy and will adhere to the notation of that section; we assume X is a complex Hilbert space, $A \in L(X, X)$ is self-adjoint and a and b are real numbers such that

$$a < m = \inf_{\|x\|=1} (Ax, x),$$

and

$$b \geq M = \sup_{\|x\|=1} (Ax, x).$$

In Exercise 20.2 we noted the existence of self-adjoint, bounded linear transformations,

$$B = \frac{D + D^*}{2} \quad \text{and} \quad C = \frac{D - D^*}{2i},$$

such that, for $D \in L(X, X)$,

$$D = B + iC;$$

thus we have expressed an arbitrary bounded transformation in terms of self-adjoint transformations. In Exercise 23.5 we went a little further and noted that any self-adjoint transformation $A \in L(X, X)$ can be expressed as a difference of positive operators; that is, there exist positive operators, A^+ and A^-, $A^+ = \frac{1}{2}(\sqrt{A^2} + A)$, $A^- = \frac{1}{2}(\sqrt{A^2} - A)$, such that

$$A = A^+ - A^-. \tag{27.1}$$

The operators A^+ and A^- have the additional properties that $A^+A^- = 0$ and each commutes with anything that commutes with A. If it happens that A is positive, then $A = \sqrt{A^2}$ and A^+ and A^- reduce to

$$A^+ = A, \qquad A^- = 0. \tag{27.2}$$

If $A \leq 0$ (that is, $-A \geq 0$), then

$$A^+ = 0 \qquad \text{and} \qquad -A^- = A \tag{27.3}$$

In this notation, then, let N denote the null space of A^+ and let E be the orthogonal projection on this closed subspace. For any $x \in X$, $Ex \in N$ and $A^+Ex = 0$; therefore,

$$A^+E = 0, \tag{27.4}$$

which, taking adjoints, implies

$$EA^+ = 0. \tag{27.5}$$

Since $A^+A^- = 0$, $A^+(A^-x) = 0$ for any x, which implies $A^-x \in N(A^+) = R(E)$; hence $EA^-x = A^-x$ for every x, which implies

$$EA^- = A^- \qquad \text{and (taking adjoints)} \qquad A^-E = A^-. \tag{27.6}$$

Since A^+ and A^- each commute with E, $A^+ + A^-$ commutes with E; therefore, in view of (27.1) and

$$\sqrt{A^2} = A^+ + A^-, \tag{27.7}$$

it follows that

$$A \text{ and } \sqrt{A^2} \text{ commute with } E. \tag{27.8}$$

Thus, using (27.5) and (27.6),

$$AE = EA = E(A^+ - A^-) = EA^+ - A^- = 0 - A^- = -A^-;$$

so

$$AE = EA = -A^-. \tag{27.9}$$

Equation (27.9) in turn implies that

$$A(1 - E) = (1 - E)A = A - EA = A + A^- = A^+;$$

hence,

$$A(1 - E) = (1 - E)A = A^+. \tag{27.10}$$

Since A^+ and A^- are each positive it follows that

$$\sqrt{A^2} = A^+ + A^- \geq A^+ - A^- = A$$

Similarly,

$$\sqrt{A^2} = A^+ + A^- \geq A^- - A^+ = -A;$$

hence,

$$\sqrt{A^2} \geq \pm A. \tag{27.11}$$

Also we clearly have

$$A^+ \geq A^+ - A^- = A \tag{27.12}$$

and similarly,

$$A^- \geq -A. \tag{27.13}$$

We now wish to demonstrate a contention.

CONTENTION 1. If the self-adjoint transformation $B \in L(X, X)$ commutes with A, then B commutes with E.

Proof. If B commutes with A, B certainly commutes with A^2, hence with $\sqrt{A^2}$ and therefore with A^+. This implies that $N(A^+)$ is invariant under B which is equivalent to†

$$BE = EBE.$$

Taking adjoints, we see that we also have

$$EB = EBE,$$

which proves that B commutes with E.

CONTENTION 2. Suppose $B \in L(X, X)$ is self-adjoint; B commutes with A; $B \geq \pm A$ (that is, B has all the properties, essentially, of $\sqrt{A^2}$). Then $B \geq \sqrt{A^2}$ (that is, $\sqrt{A^2}$ is the "smallest" transformation having the above properties).

Proof. By the preceding result, B commutes with E; by hypothesis B commutes with A; therefore the positive operators $(B - A)$ and $(1 - E)$ commute. This implies that

$$(B - A)(1 - E) \geq 0,$$

or, using (27.10),

$$B(1 - E) \geq A(1 - E) = A^+. \tag{27.14}$$

Since $(B + A)E = E(B + A)$, $E \geq 0$, and $(B + A) \geq 0$,

$$BE + AE \geq 0$$

or, using (27.9),

$$BE \geq -AE = A^-. \tag{27.15}$$

Adding (27.14) and (27.15), we get

$$B \geq A^+ + A^- = \sqrt{A^2},$$

which proves Contention 2.

CONTENTION 3. Suppose $B \in L(X, X)$ is positive; $B \geq A$; B commutes with A. Then $B \geq A^+$.

Proof. Clearly Eq. (27.14) also holds here:

$$B - BE \geq A^+.$$

Since B and E are each positive and commute,

$$BE \geq 0.$$

It is now evident that $B \geq A^+$.

In a similar fashion one can also establish another contention.

† See Result (8), Sec. 22.1.

CONTENTION 4. Suppose $B \in L(X, X)$, $B \geq 0$, B commutes with A and $B \geq -A$. Then $B \geq A^-$

We now introduce the following abbreviation:

$$A(\lambda) = A - \lambda.$$

In this notation, if $\lambda_1 \leq \lambda_2$, then, clearly,

$$A(\lambda_1) \geq A(\lambda_2). \tag{27.16}$$

By (27.12)

$$A(\lambda_1)^+ \geq A(\lambda_1) \geq A(\lambda_2).$$

Since $A(\lambda_1)^+ \in L(X, X)$, $A(\lambda_1)^+ \geq 0$ and $A(\lambda_1)^+$ commutes with anything that commutes with $A(\lambda_1)$, in particular $A(\lambda_2)$, then by Contention 3,

$$A(\lambda_1)^+ \geq A(\lambda_2)^+. \tag{27.17}$$

For $\lambda \leq m = \inf_{\|x\|=1}(Ax, x)$ and any x,

$$\lambda(x, x) \leq m(x, x) \leq (Ax, x)$$

which implies

$$((A - \lambda)x, x) = (A(\lambda)x, x) \geq 0 \qquad \text{(for any } x\text{)};$$

consequently,

$$A(\lambda) \geq 0 \qquad \text{(for } \lambda \leq m\text{)}. \tag{27.18}$$

By (27.2) then, for $\lambda \leq m$,

$$A(\lambda)^+ = A(\lambda). \tag{27.19}$$

For $\lambda \geq M$,

$$(Ax, x) \leq M(x, x) \leq \lambda(x, x),$$

which implies

$$(A(\lambda)x, x) \leq 0 \qquad \text{(for } \lambda \geq M\text{)} \tag{27.20}$$

for any x; therefore $A(\lambda) \leq 0$, which [by (27.3)] implies

$$A(\lambda)^+ = 0 \qquad \text{(for } \lambda \geq M\text{)}. \tag{27.21}$$

For $\lambda_1 \leq \lambda_2$, $A(\lambda_1)^+ \geq A(\lambda_2)^+$ by (27.17); since they commute and each is positive,

$$A(\lambda_2)^+(A(\lambda_1)^+ - A(\lambda_2)^+) \geq 0$$

or

$$A(\lambda_2)^+ A(\lambda_1)^+ \geq (A(\lambda_2)^+)^2.$$

So, for any x,

$$(A(\lambda_2)^+ A(\lambda_1)^+ x, x) \geq (A(\lambda_2)^+ x, A(\lambda_2)^+ x) = \|A(\lambda_2)^+ x\|^2. \tag{27.22}$$

If $x \in N(A(\lambda_1)^+)$,

$$A(\lambda_1)^+ x = 0,$$

which, by (27.22) implies

$$A(\lambda_2)^+ x = 0;$$

therefore,

$$N(A(\lambda_1)^+) \subset N(A(\lambda_2)^+) \qquad \text{(for } \lambda_1 \leq \lambda_2\text{)}. \tag{27.23}$$

In set-theoretic language one says that the family of closed subspaces $\{N(A(\lambda)^+)\}$ forms a monotone increasing sequence of sets. This immediately implies that the orthogonal projections defined on this family will be a monotone family. Letting $E(\lambda)$ denote the orthogonal projection on $N(A(\lambda)^+)$, we now show that $E(\lambda) = 0$ for $\lambda < m$ and $E(\lambda) = 1$ for $\lambda \geq M$. For $\lambda < m$, by the definition of m,

$$((A - \lambda)x, x) > 0 \qquad \text{(for } x \neq 0). \tag{27.24}$$

Using (27.19) this implies

$$((A(\lambda)^+)x, x) > 0 \qquad (x \neq 0, \lambda < m). \tag{27.25}$$

Therefore, for $\lambda < m$, $N(A(\lambda)^+) = \{0\}$, which implies

$$E(\lambda) = 0. \tag{27.26}$$

Similarly, for $\lambda \geq M$ we can show
$$A(\lambda)^+ = 0,$$

which implies
$$N(A(\lambda)^+) = X$$

and, consequently,

$$E(\lambda) = 1. \tag{27.27}$$

To complete the proof it only remains to show that $E(\lambda)$ is strongly continuous from the right and that we have an integral representation for A like the one in Theorem 25.4. As for the integral representation, consider $\lambda_1 \leq \lambda_2$ and define

$$F(\lambda_1, \lambda_2) = E(\lambda_2) - E(\lambda_1) \geq 0.$$

In this notation†

$$E(\lambda_2)F(\lambda_1, \lambda_2) = E(\lambda_2) - E(\lambda_1) = F(\lambda_1, \lambda_2) \tag{27.28}$$

and

$$E(\lambda_1)F(\lambda_1, \lambda_2) = 0. \tag{27.29}$$

In view of this we can write

$$(\lambda_2 - A)F(\lambda_1, \lambda_2) = -A(\lambda_2)E(\lambda_2)F(\lambda_1, \lambda_2). \tag{27.30}$$

Using (27.9) and the fact that $A(\lambda_2)^-$ commutes with anything that $A(\lambda_2)$ commutes with, we have that the last term in the preceding equation is just

$$A(\lambda_2)^- F(\lambda_1, \lambda_2) \geq 0. \tag{27.31}$$

Similarly, using (27.10)

$$(A - \lambda_1)F(\lambda_1, \lambda_2) = A(\lambda_1)F(\lambda_1, \lambda_2) = A(\lambda_1)(1 - E(\lambda_1))F(\lambda_1, \lambda_2)$$

$$= A(\lambda_1)^+ F(\lambda_1, \lambda_2) \geq 0. \tag{27.32}$$

The last three results (27.30)–(27.32), imply that

$$\lambda_1 F(\lambda_1, \lambda_2) \leq AF(\lambda_1, \lambda_2) \leq \lambda_2 F(\lambda_1, \lambda_2) \tag{27.33}$$

or

$$\lambda_1(E(\lambda_2) - E(\lambda_1)) \leq A(E(\lambda_2) - E(\lambda_1)) \leq \lambda_2(E(\lambda_2) - E(\lambda_1)). \tag{27.34}$$

† Section 22.3, Theorem 22.4, parts 1, 4 and 5.

In the previous chapter we had arrived at precisely the same condition [Eq. (26.11)]. Once it was seen that the family $\{E(\lambda)\}$ gotten in the previous proof satisfied this condition, it followed that

$$A = \int_a^b \lambda \, dE(\lambda)$$

and also, for any polynomial with real coefficients $p(\lambda)$,

$$p(A) = \int_a^b p(\lambda) \, dE(\lambda).$$

In showing this in the preceding chapter, no use was made from this point on of how the $E(\lambda)$ were obtained; thus exactly the same procedure could be gone through here to verify that we do, indeed, have the desired integral representation. Last, strong continuity from the right will be investigated.

In the same notation as before, consider

$$\lim_{\lambda_2 \to \lambda_1^+} F(\lambda_1, \lambda_2)x = G(\lambda_1)x,$$

it being clear from the monotonicity of the family $\{E(\lambda)\}$ that the strong limit $G(\lambda_1)$ exists; it will be shown that $G(\lambda_1) = 0$. Using (27.33) and letting $\lambda_2 \to \lambda_1^+$, we have

$$(\lambda_1 G(\lambda_1)x, x) \le (AG(\lambda_1)x, x) \le (\lambda_1 G(\lambda_1)x, x),$$

which implies

$$AG(\lambda_1) = \lambda_1 G(\lambda_1)$$

or

$$A(\lambda_1)G(\lambda_1) = 0. \tag{27.35}$$

Now consider, using (27.10) and (27.35),

$$A(\lambda_1)^+ G(\lambda_1) = (1 - E(\lambda_1))A(\lambda_1)G(\lambda_1) = 0;$$

thus, for any x, $G(\lambda_1)x \in N(A(\lambda_1)^+)$. This implies

$$E(\lambda_1)G(\lambda_1)x = G(\lambda_1)x$$

or

$$E(\lambda_1)G(\lambda_1) = G(\lambda_1). \tag{27.36}$$

Since, by (27.29), $E(\lambda_1) F(\lambda_1, \lambda_2) = 0$,

$$(1 - E(\lambda_1))F(\lambda_1, \lambda_2)x = F(\lambda_1, \lambda_2)x, \tag{27.37}$$

for any x. Letting $\lambda_2 \to \lambda_1^+$ in (27.37), we get

$$(1 - E(\lambda_1))G(\lambda_1)x = G(\lambda_1)x,$$

which, since x is arbitrary, implies

$$(1 - E(\lambda_1))G(\lambda_1) = G(\lambda_1),$$

or

$$E(\lambda_1)G(\lambda_1) = 0.$$

In view of (27.36) the proof is seen to be complete.

27.2 Two Consequences of the Spectral Theorem

In this section we shall adhere to the notation of the preceding section.

Under the correspondence used in the proof of the spectral theorem given in Chapter 26, if f is a real-valued, continuous function defined on $[m, M]$, there is an associated transformation $f(A) \in L(X, X)$ obtained by taking the strong limit of a certain sequence of polynomials in A. It is exceedingly simple to verify that we also have†

$$f(A) = \int_a^b f(\lambda) \, dE(\lambda) \tag{27.38}$$

(the reader was asked to verify this in Exercise 26.3); that is, the Stieltjes integral (27.38) exists and is equal to the same quantity as obtained by taking the strong limit of the abovementioned sequence of polynomials.

In Sec. 25.1 the infinite series respresentation of completely continuous, self-adjoint operators was rephrased as an integral. The resolution of the identity used there, $\{E(\lambda)\}$, was a continuous function everywhere except at eigenvalues [or possibly 0 if $0 \in C\sigma(A)$]. $E(\lambda)$ was in fact *constant* at all other points. Using the results of the preceding paragraph, we shall examine the jumps of $E(\lambda)$ when $E(\lambda)$ is the resolution of the identity of a bounded, self-adjoint operator A.

THEOREM 27.1. Let $A \in L(X, X)$ be self-adjoint and let $\{E(\lambda)\}$ be the resolution of the identity associated with A. In this case the point $\lambda_0 \in \rho(A)$, the resolvent set of A, if and only if there exists $\varepsilon > 0$ such that $E(\lambda)$ is constant in the interval $[\lambda_0 - \varepsilon, \lambda_0 + \varepsilon]$.

Proof. Suppose the condition is satisfied and consider the functions:

$$f(\lambda) = \lambda_0 - \lambda \quad (-\infty < \lambda < \infty),$$

$$g(\lambda) = \frac{1}{\lambda_0 - \lambda} \quad (\lambda \notin [\lambda_0 - \varepsilon, \lambda_0 + \varepsilon]).$$

We now wish to define $g(\lambda)$ within the interval $[\lambda_0 - \varepsilon, \lambda_0 + \varepsilon]$ and in such a way as to make it continuous everywhere. One such acceptable extension is obtained by defining

$$g(\lambda) = \frac{-1}{\varepsilon^2}(\lambda - \lambda_0) \quad (\lambda \in [\lambda_0 - \varepsilon, \lambda_0 + \varepsilon])$$

as shown by the dotted line in Fig. 27.1. At any rate, we now assume that $g(\lambda)$ has been defined for every λ as a continuous function. By virtue of Exercises 26.3 and 26.4, we have

$$\int_a^b f(\lambda)g(\lambda) \, dE(\lambda) = \int_a^b f(\lambda) \, dE(\lambda) \int_a^b g(\lambda) \, dE(\lambda) = f(A)g(A).$$

† We take $f(\lambda) = f(m)$, $a \leq \lambda \leq m$, and $f(\lambda) = f(M)$, $M \leq \lambda \leq b$ for this consideration.

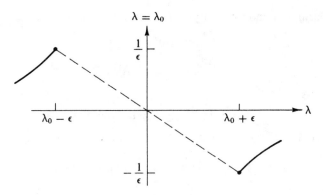

FIG. 27.1. A continuous extension of $g(\lambda)$.

Since $\sigma(A) \subset [m, M]$ if $\lambda_0 \notin [m, M]$, $\lambda_0 \in \rho(A)$; thus we assume $\lambda_0 \in [m, M]$, and note that, by hypothesis,

$$\int_{\lambda_0 - \varepsilon}^{\lambda_0 + \varepsilon} f(\lambda)g(\lambda) \, dE(\lambda) = 0.$$

For $\lambda \notin [\lambda_0 - \varepsilon, \lambda_0 + \varepsilon]$, $f(\lambda)g(\lambda) = 1$; therefore

$$\int_a^b f(\lambda)g(\lambda) \, dE(\lambda) = \int_a^{\lambda_0 - \varepsilon} 1 \, dE(\lambda) + 0 + \int_{\lambda_0 + \varepsilon}^b 1 \, dE(\lambda)$$

$$= E(\lambda_0 - \varepsilon) + E(b) - E(\lambda_0 + \varepsilon).$$

Since $E(\lambda_0 - \varepsilon) = E(\lambda_0 + \varepsilon)$ and $E(b) = 1$, we see that

$$g(A)f(A) = f(A)g(A) = (\lambda_0 - A)g(A) = 1;$$

so $(\lambda_0 - A)$ is $1 : 1$ and onto.

Also, by the way $g(A)$ was obtained, $g(A) \in L(X, X)$, and we have shown that $(\lambda_0 - A)$ has a *bounded* inverse [namely, $g(A)$]. In view of this

$$\lambda_0 \in \rho(A).$$

Conversely, suppose that no ε satisfying the condition of the theorem exists; that is, for every $\varepsilon > 0$, we assume existence of $\mu_1, \mu_2 \in [\lambda_0 - \varepsilon, \lambda_0 + \varepsilon]$ such that

$$E(\mu_1) \neq E(\mu_2).$$

Since the family $\{E(\lambda)\}$ is a monotone family then, assuming $\mu_1 < \mu_2$, we must have

$$E(\mu_1) \leq E(\mu_2).$$

This implies

$$R(E(\mu_1)) \subset R(E(\mu_2)),$$

and this inclusion must be proper. By Theorem 10.6, there must be some nonzero vector $w \in R(E(\mu_2))$ such that $w \perp R(E(\mu_1))$ (that is, $w \in R(E(\mu_1))^\perp = N(E(\mu_1))$; thus

$$E(\mu_1)w = 0 \qquad \text{and} \qquad E(\mu_2)w = w.$$

For $\lambda \leq \mu_1$, using the monotonicity of $E(\lambda)$ again, we have

$$E(\lambda)w = E(\lambda)E(\mu_1)w = 0.$$

Consequently,

$$E(\lambda)w = 0 \qquad (\text{for } \lambda \leq \mu_1),$$

whereas for $\lambda \geq \mu_2$,

$$E(\lambda)w = E(\lambda)E(\mu_2)w = E(\mu_2)w = w.$$

By Exercise 26.6 with $f(\lambda) = \lambda - \lambda_0$ and $x = w$, we then have

$$\|(A - \lambda_0)w\|^2 = \int_a^b |\lambda - \lambda_0|^2 \, d\|E(\lambda)w\|^2$$

$$= \int_{\mu_1}^{\mu_2} |\lambda - \lambda_0|^2 \, d\|E(\lambda)w\|^2$$

$$\leq (2\varepsilon)^2 (\|E(\mu_2)w\|^2 - \|E(\mu_1)w\|^2)$$

$$= (2\varepsilon)^2 \|w\|^2;$$

hence, since ε is arbitrary and such a w must exist for every ε, we see that $\lambda_0 \in \pi(A)$, the approximate spectrum which for the self-adjoint transformation A coincides with $\sigma(A)$ by Theorem 21.6. The theorem is now proved.

For $\lambda \in \rho(A)$ we define the *resolvent operator associated with A* as

$$R_\lambda = (\lambda - A)^{-1}.$$

Most often we shall refer to R_λ as simply the resolvent operator. A convenient representation for R_λ in terms of an integral and the resolution of the identity associated with A is obtained in the next theorem.

THEOREM 27.2. If $\lambda_0 \in \rho(A)$, then $R_{\lambda_0} = \int_a^b [dE(\lambda)/(\lambda_0 - \lambda)]$.

Proof. We must consider two cases: the case when $\lambda_0 \notin [a, b]$ and the case when $\lambda_0 \in [a, b]$. In the first case

$$\frac{1}{\lambda_0 - \lambda} \in C[a, b]$$

and, as in the preceding theorem,

$$\int_a^b \frac{dE(\lambda)}{\lambda_0 - \lambda} \int_a^b (\lambda_0 - \lambda) \, dE(\lambda) = \int_a^b dE(\lambda) = 1.$$

Noting that

$$\frac{1}{\lambda_0 - \lambda} \in C[a, b] \Rightarrow \int_a^b \frac{dE(\lambda)}{\lambda_0 - \lambda} \in L(X, X),$$

we conclude

$$\int_a^b \frac{dE(\lambda)}{\lambda_0 - \lambda} = (\lambda_0 - A)^{-1} = R_{\lambda_0}.$$

In the second case we can assume $\lambda_0 \in (a, b)$. By the preceding result, there exists an $\varepsilon > 0$ such that $E(\lambda)$ is constant on $[\lambda_0 - \varepsilon, \lambda_0 + \varepsilon]$. Now, interpreting the improper integral

$$\int_a^b \frac{dE(\lambda)}{\lambda - \lambda_0}$$

in the customary sense, we get the same result as before, the singularity causing no trouble because $E(\lambda)$ is constant in a whole neighborhood of λ_0.

Our next result shows that eigenvalues of A reflect themselves in the resolution of the identity by causing jumps.

THEOREM 27.3.
(1) The value $\mu \in P\sigma(A)$ if and only if $E(\mu) \neq E(\mu^-)$. The corresponding eigenmanifold is then given by $N(A - \mu) = R(E(\mu) - E(\mu^-))$.
(2) $\mu \in C\sigma(A)$ if and only if $E(\mu) = E(\mu^-)$ and $E(\lambda)$ is not constant in any neighborhood of μ.

REMARK. If the first part of the theorem can be established, the second part follows by applying Theorem 27.1 and using the fact that $R\sigma(A) = \emptyset$.

Proof. From Exercise 25.2 it follows that, for any μ, $E(\mu^-)$ is an orthogonal projection and $E(\mu^-) \leq E(\mu)$; moreover,

$$E(\lambda)E(\mu^-) = E(\lambda) \qquad \text{(for } \lambda < \mu),$$

$$E(\lambda)E(\mu^-) = E(\mu^-) \qquad \text{(for } \lambda \geq \mu).$$

Suppose now that $E(\mu) \neq E(\mu^-)$ or, equivalently, that

$$E(\mu) - E(\mu^-) \neq 0.$$

This implies that there is some $x \neq 0$ such that

$$y = (E(\mu) - E(\mu^-))x \neq 0.$$

Applying $E(\lambda)$ to y, we get

$$E(\lambda)y = (E(\mu) - E(\mu^-))x = y \qquad \text{(if } \lambda \geq \mu),$$

$$E(\lambda)y = (E(\lambda) - E(\lambda))x = 0 \qquad \text{(if } \lambda < \mu).$$

Using Exercise 26.6, we now consider

$$\|(A - \mu)y\|^2 = \int_a^b (\lambda - \mu)^2 \, d\|E(\lambda)y\|^2.$$

We have just noted that $E(\lambda)y$ is continuously constant at all points except $\lambda = \mu$; at that point, however, $\lambda - \mu = 0$. Therefore,

$$\|(A - \mu)y\|^2 = 0,$$

which implies $\mu \in P\sigma(A)$ and since y was any nonzero vector in $R(E(\mu) - E(\mu^-))$,

$$R(E(\mu) - E(\mu^-)) \subset N(A - \mu). \tag{27.39}$$

Conversely if $\mu \in P\sigma(A)$, there is a nonzero vector y such that $(A - \mu)y = 0$:

$$0 = \|(A - \mu)y\|^2 = \int_a^b (\lambda - \mu)^2 \, d\|E(\lambda)y\|^2. \tag{27.40}$$

Also $\mu \in P\sigma(A)$ implies $\mu \in [m, M]$, which assures us that (by choosing b strictly greater than M)

$$a < \mu < b.$$

Hence there must exist a $\delta > 0$ such that

$$a < \mu - \delta \quad \text{and} \quad \mu + \delta < b,$$

which implies that, for $a \le \lambda \le \mu - \delta$,

$$(\lambda - \mu)^2 \ge \delta^2.$$

Since $\|E(\lambda)y\|^2$ is a monotone function and since $(\lambda - \mu)^2 \ge 0$, it follows that (27.40) implies

$$0 = \int_a^{\mu - \delta} (\lambda - \mu)^2 \, d\|E(\lambda)y\|^2 \ge \delta^2(\|E(\mu - \delta)y\|^2 - \|E(a)y\|^2).$$

Since $E(a) = 0$, the above result implies that $E(\mu - \delta)y = 0$. Since δ can be made arbitrarily small, it follows that

$$E(\mu^-)y = 0.$$

In similar fashion we have

$$0 = \int_{\mu + \delta}^b (\lambda - \mu)^2 \, d\|E(\lambda)y\|^2 \ge \delta^2(\|E(b)y\|^2 - \|E(\mu + \delta)y\|^2).$$

Noting that $E(b) = 1$, this implies that

$$\|y\|^2 = \|E(\mu + \delta)y\|^2$$

which, since $E(\mu + \delta)$ is an orthogonal projection, implies [Sec. 22.1, property (6)]

$$y = E(\mu + \delta)y.$$

Letting $\delta \to 0$ and using the continuity from the right of $E(\lambda)$, we have

$$y = E(\mu)y;$$

therefore,

$$E(\mu) \neq E(\mu^-)$$

and

$$y = (E(\mu) - E(\mu^-))y$$

or

$$y \in R(E(\mu) - E(\mu^-)).$$

Combining this with (27.39), the theorem is completely proved.

EXERCISES 27

1. Let $B \in L(X, X)$ and suppose $BA = AB$, where A is a bounded self-adjoint transformation. Using the customary notations, show that if f is continuous on R, then $f(A)B = Bf(A)$.

2. Establish the existence of $\sqrt{A^2}, A^+, A^-$ on the basis of the correspondence set up between C_2 and functions of A.

3. Using the same notation as in the development of the spectral theorem for a completely continuous normal transformation A, show that if A is self-adjoint and $\lambda_0 \in \rho(A)$, then, if $x = R_{\lambda_0} y$,

$$x = \frac{1}{\lambda_0} y + \frac{1}{\lambda_0} \sum_{i=1}^{\infty} \sum_{j=1}^{n_i} \frac{\lambda_i(y, x_{ij})}{\lambda_0 - \lambda_i} x_{ij}.$$

4. Recall that $E(\lambda)$ is the projection on $N(A(\lambda)^+)$, where $\{E(\lambda)\}$ is the resolution of the identity for the self-adjoint operator $A \in L(X, X)$. Let $\lim_{\nu \to \mu^-} E(\nu)x = E(\mu^-)x$. What is the range of $E(\mu^-)$?

5. Give an alternate proof that $P\sigma(A) = \{\mu | E(\mu) \neq E(\mu^-)\}$ and that $N(A - \mu) = R(E(\mu) - E(\mu^-))$ using the fact that $R(E(\lambda)) = N(A(\lambda)^+)$.

6. From the spectral decomposition for a bounded self-adjoint transformation A, obtain the spectral decomposition for a self-adjoint completely continuous transformation as a special case.

7. Although we have seen earlier (Sec. 19.3) that $\sigma(A)$ is a closed set for $A \in L(X, X)$, prove this fact again for A bounded and self-adjoint just on the basis of the results of this chapter.

REFERENCES

1. F. Riesz and B. Sz.-Nagy, "Functional Analysis."
2. L. Liusternik and W. Sobolev, "Elements of Functional Analysis."
3. A. Taylor, "Introduction to Functional Analysis."

CHAPTER **28**

Spectral Theorem
for Bounded, Normal Operators

In this brief chapter we wish to obtain a spectral decomposition theorem for bounded, *normal* transformations on a Hilbert space. The result obtained here follows rather directly from the theorem for bounded, self-adjoint operators, proved previously, after one realizes that any bounded transformation can be expressed as a linear combination of self-adjoint transformations (Exercise 20.2). It is desirable to note any significant changes from the theorem for self-adjoint operators. One recalls that in the case of a self-adjoint transformation, the entire spectrum is real. This is not so for normal transformations. A reflection of this property is found in that in the theorem of the chapter, we can no longer represent the transformation as a single integral—a double integral is necessary; we have been forced to extend our thinking and our result from the line to the plane.

For the reader familiar with measure theory, the result is then modified slightly after the notion of a complex *spectral measure* (a countably additive set function defined on the Borel sets of the plane that assumes orthogonal projections as values) is briefly introduced. Much, indeed most, of the material presented about this function is so analogous to standard results from measure theory that only a sketch is given. (For example, one of the things that is sketched is the process of extending a set function defined only on half-open rectangles to the Borel sets of the plane.) Only very limited use will be made of this highly useful tool, however.†

The special case of a unitary transformation is treated next. In this case it turns out that, once again, the transformation can be represented as a single integral.

28.1 The Spectral Theorem for Bounded, Normal Operators on a Hilbert Space

First the spectral theorem for bounded, normal operators on a Hilbert space will be presented. The theorem itself follows quite readily from the theorem for bounded self-adjoint operators.

THEOREM 28.1. Let X be a complex Hilbert space and let A be a normal transformation belonging to $L(X, X)$. Let $B = \frac{1}{2}(A + A^*)$ and $C = 1/2i(A - A^*)$. Then B and C are self-adjoint operators such that $A = B + iC$ and, denoting the resolutions of the identity associated with B and C, respectively, by $E_B(\lambda)$ and $E_C(\mu)$,

$$A = \int_{-\infty}^{\infty} \int_{-\infty}^{\infty} (\lambda + i\mu)\, dE_B(\lambda)\, dE_C(\mu).$$

Proof. It is obvious that the transformations B and C are self-adjoint and that $A = B + iC$; moreover, each of these transformations is clearly bounded. (In fact $\|B\| \leq \|A\|$ and $\|C\| \leq \|A\|$.) Thus, each is a self-adjoint transformation belonging to $L(X, X)$ and Theorem 25.3, the spectral theorem for bounded self-adjoint operators on Hilbert space, is applicable. We denote the resolutions of the identity associated with B and C by $E_B(\lambda)$ and $E_C(\mu)$ and can write

$$B = \int_{-\infty}^{\infty} \lambda\, dE_B(\lambda) \qquad \text{and} \qquad C = \int_{-\infty}^{\infty} \mu\, dE_C(\mu),$$

the limits on the integrals being written this way just for symmetry. Since A is

† For a further discussion see Halmos, "Introduction to Hilbert Space".

485

assumed to be normal, B and C must commute with each other. Since B and C commute, certainly any polynomials in B and C must also commute. By the proof of the spectral theorem for bounded self-adjoint operators given in Sec. 26.1 for any given fixed λ and μ, there exist monotonic sequences of polynomials in B and C, $\{p_n(B)\}$ and $\{q_n(C)\}$ such that $p_n(B)$ and $q_n(C)$ converge from above to $E_B(\lambda)$ and $E_C(\mu)$ strongly. Since all the terms in the above sequences commute, the strong limits [namely, $E_B(\lambda)$ and $E_C(\mu)$] must commute, too.

Bearing these facts in mind, we now write

$$A = B + iC$$

$$= \int_{-\infty}^{\infty} \lambda\, dE_B(\lambda) + i \int_{-\infty}^{\infty} \mu\, dE_C(\mu).$$

Since $\int_{-\infty}^{\infty} dE_B(\lambda) = \int_{-\infty}^{\infty} dE_C(\mu) = 1$, we can rewrite the preceding equation as

$$A = \int_{-\infty}^{\infty} \lambda\, dE_B(\lambda) \int_{-\infty}^{\infty} dE_C(\mu) + i \int_{-\infty}^{\infty} \mu\, dE_C(\mu) \int_{-\infty}^{\infty} dE_B(\lambda). \qquad (28.1)$$

We now claim that the interated integrals can be merged into a double integral; that is,

$$A = \int_{-\infty}^{\infty} \int_{-\infty}^{\infty} (\lambda + i\mu)\, dE_B(\lambda)\, dE_C(\mu), \qquad (28.2)$$

where the double integral denotes

$$\lim_{\substack{i,j \to \infty \\ \text{area of max } \Delta_{ij} \to 0}} \sum_{i,j} \xi_{ij}\big(E_B(\lambda_i) - E_B(\lambda_{i-1})\big)\big(E_C(\mu_j) - E_C(\mu_{j-1})\big), \qquad (28.3)$$

in the uniform sense on $L(X, X)$, Δ_{ij} is the half-open rectangle shown in Fig. 28.1 and ξ_{ij} is an arbitrary point in Δ_{ij}.

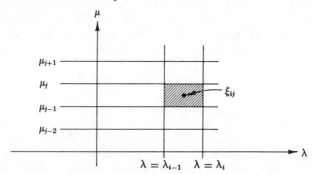

FIG. 28.1. The rectangle Δ_{ij}; $\Delta_{ij} = \{(\lambda, \mu) \mid \lambda_{i-1} \leq \lambda < \lambda_i,\ \mu_{j-1} \leq \mu < \mu_j\}$.

The validity of this last assertion will be verified now.

We write

$$H_i = E_B(\lambda_i) - E_B(\lambda_{i-1}),$$

$$K_j = E_C(\mu_j) - E_C(\mu_{j-1}),$$

$$\xi_{ij} = \lambda_{ij} + i\mu_{ij}.$$

In this notation, a typical term in (28.3) becomes

$$\sum_{i,j} \xi_{ij} H_i K_j = \sum_{i,j} \lambda_{ij} H_i K_j + i \sum_{i,j} \mu_{ij} H_i K_j.$$

Suppose that $|\lambda_i - \lambda_{i-1}| = \varepsilon$. Then, for every j (see Fig. 28.1),

$$|\lambda_{ij} - \lambda_i| < \varepsilon$$

or

$$\lambda_i - \varepsilon < \lambda_{ij} < \lambda_i + \varepsilon.$$

This implies

$$\sum_{i,j} (\lambda_i - \varepsilon) H_i K_j \le \sum_{i,j} \lambda_{ij} H_i K_j \le \sum_{i,j} (\lambda_i + \varepsilon) H_i K_j. \tag{28.4}$$

Since these are finite sums, the mode of summation is irrelevant; in particular, we can write

$$\sum_{i,j} (\lambda_i + \varepsilon) H_i K_j = \sum_i H_i (\lambda_i + \varepsilon) \sum_j K_j.$$

But

$$\sum_j K_j = \sum_j (E_C(\mu_j) - E_C(\mu_{j-1})) = 1. \tag{28.5}$$

[We assume, of course, that the spectrum of C, $\sigma(C)$, is contained in the grid imposed by $\{\mu_j\}$ in the finite sum above.]

A similar result holds for the left inequality in (28.4); therefore, (28.4) becomes

$$\sum_i \lambda_i H_i - \varepsilon 1 \le \sum_{i,j} \lambda_{ij} H_i K_j \le \sum_i \lambda_i H_i + \varepsilon 1. \tag{28.6}$$

According to our previous remarks about commutativity, it follows that for any i and j, H_i commutes with K_j. Since each H_i and K_j is an orthogonal projection, it now follows that the product $H_i K_j$ is an orthogonal projection (Theorem 22.2). The orthogonal projections $H_i K_j$ are clearly an orthogonal family because the families $\{E_B(\lambda)\}$ and $\{E_C(\mu)\}$ are monotonic families that commute with each other. Hence (Theorem 24.4) the middle term in (28.6) is an orthogonal projection; consequently the middle term in (28.6) is certainly self-adjoint. Hence, (28.6) implies that

$$\left\| \sum_{i,j} \lambda_{ij} H_i K_j - \sum_i \lambda_i H_i \right\| \le \varepsilon. \tag{28.7}$$

[The positive definite statement such as (28.6) always implies the norm statement (28.7), for self-adjoint transformations; its validity is verified in the discussion following Eq. (26.12).] In (28.7) we may now let $i, j \to \infty$ and let $\varepsilon \to 0$; since

$$\lim_{i,j} \sum_{i,j} \lambda_{ij} H_i K_j = \int_{-\infty}^{\infty} \int_{-\infty}^{\infty} \lambda \, dE_B(\lambda) \, dE_C(\mu)$$

and

$$\lim_{i,j} \sum_i \lambda_i H_i = \int_{-\infty}^{\infty} \lambda \, dE_B(\lambda),$$

we have shown that

$$\int_{-\infty}^{\infty} \int_{-\infty}^{\infty} \lambda \, dE_B(\lambda) \, dE_C(\mu) = \int_{-\infty}^{\infty} \lambda \, dE_B(\lambda) = \int_{-\infty}^{\infty} \lambda \, dE_B(\lambda) \int_{-\infty}^{\infty} dE_C(\mu).$$

A corresponding statement immediately follows for the imaginary part of (28.2), and we have demonstrated the validity of that representation for A. The theorem is now proved.

Note, however, that integrating over the whole plane is somewhat superfluous; that is, the double integral of the theorem is not really improper, because each of the functions $E_B(\lambda)$ and $E_C(\mu)$ are constant outside the spectra of B and C, respectively. For this reason one need actually integrate only over a sufficiently large rectangle and, thus, one's fears about convergence can be allayed. In the next section we shall show that we have only to integrate over the closed circle of radius $\|A\|$.

28.2   Spectral Measures; Unitary Transformations

In this section we shall adhere precisely to the notation of the preceding section. We wish to introduce the concept of a spectral measure and, briefly, apply it to the theorem of the preceding section. In addition we shall also treat the special case of a unitary transformation; more precisely, we shall get a spectral theorem for unitary transformations.

Consider the family of half-open rectangles Δ_{ij} described in Fig. 28.1 and the set function E defined on these rectangles by taking

$$E(\Delta_{ij}) = (E_B(\lambda_i) - E_B(\lambda_{i-1}))(E_C(\mu_j) - E_C(\mu_{j-1})). \tag{28.8}$$

Let us note some properties of E now†:

(1) $E(\Delta_{ij})$ is an orthogonal projection;

(2) for $k \neq i$ or $m \neq j$, $E(\Delta_{km})E(\Delta_{ij}) = 0$; that is, the family $\{E(\Delta_{ij})\}$ is an orthogonal family of orthogonal projections.

In view of these observations, Theorem 24.4 enables us to assert that $\{E(\Delta_{ij})\}$ is summable to an orthogonal projection whose range is given by

$$\sum_{i,j} \oplus R(E(\Delta_{ij})).$$

Since the number of distinct terms involved in the family $\{E(\Delta_{ij})\}$ is finite, the summation can be carried out in an iterated fashion; doing this we note [as in Eq. (28.5)] that

$$\sum_{i,j} E(\Delta_{ij}) = 1,$$

which implies

$$\sum_{i,j} \oplus R(E(\Delta_{ij})) = X. \tag{28.9}$$

Although the union of two half-open rectangles Δ_1 and Δ_2 is rarely a half-open rectangle, it *is* possible; moreover, it is possible that Δ_1 and Δ_2 could be disjoint and still have as their union a half-open rectangle. If the latter occurs, then

$$E(\Delta_1 \cup \Delta_2) = E(\Delta_1) + E(\Delta_2). \tag{28.10}$$

Adopting the convention that

$$E(\emptyset) = 0, \tag{28.11}$$

---

† These results are discussed in more detail in the preceding section; see especially the discussion following inequality (28.6).

then we can say [see Exercise 28.1] that, for any two half-open rectangles Δ_1 and Δ_2,

$$E(\Delta_1 \cap \Delta_2) = E(\Delta_1)E(\Delta_2). \tag{28.12}$$

Suppose now that δ is the closed rectangle given by

$$\{(\lambda, \mu) \mid m_B \le \lambda \le M_B, m_C \le \mu \le M_C\},$$

$$m_B = \inf_{\|x\|=1} (Bx, x), \qquad m_C = \inf_{\|x\|=1} (Cx, x),$$

$$M_B = \sup_{\|x\|=1} (Bx, x) \quad \text{and} \quad M_C = \sup_{\|x\|=1} (Cx, x).$$

Clearly for any half-open rectangle $\Delta \supset \delta$, we have

$$E(\Delta) = 1, \tag{28.13}$$

whereas if Δ is a half-open rectangle such that $\delta \cap \Delta = \emptyset$,

$$E(\Delta) = 0. \tag{28.14}$$

Note how very similar this entire development has been to the case of developing Lebesgue–Stieltjes measure on the real line. In that case it is customary to define a set function P on the intervals as follows: If F is a monotone nondecreasing function, continuous from the right, we take

$$P(a, b] = F(b) - F(a),$$

$$P[a, b) = F(b^-) - F(a^-),$$

$$P[a, b] = F(b) - F(a^-)$$

$$P(a, b) = F(b^-) - F(a);$$

in the special case when $b = a$, we take

$$P[a, b] = F(a) - F(a^-).$$

If we adopt the convention

$$[a, b) = (a^-, b^-],$$

$$[a, b] = (a^-, b],$$

$$(a, b) = (a, b^-],$$

we need only consider entities of the form $(u, v]$ and take

$$P(u, v] = F(v) - F(u).$$

In the case of the function E we have been working with, we assume an analogous convention has been adopted concerning the representation of half-open rectangles. In this way we now have a function defined on the class I of all rectangles. In the event that $\Delta = \bigcup_{k=1}^{\infty} \Delta_k \in I$, $\Delta_k \in I$ for all k, $\Delta_k \cap \Delta_j = \emptyset$ for $k \ne j$, then

$$E(\Delta) = \sum_{k=1}^{\infty} E(\Delta_k).$$

Now consider the class I_y of all countable, disjoint unions of rectangles (note that this class is the same as all countable unions of rectangles, disjoint or not) and define, for $\bigcup_{k=1}^{\infty} I_k \in I_\Sigma$, $I_k \in I$,

$$E\left(\bigcup_{k=1}^{\infty} I_k\right) = \sum_{k=1}^{\infty} E(I_k).$$

By defining, for an arbitrary collection of orthogonal projection $\{E_\alpha\}$,

$$\inf_\alpha E_\alpha = \text{orthogonal projection on } \bigcap_\alpha R(E_\alpha)$$

and

$$\sup_\alpha E_\alpha = \text{orthogonal projection on } \overline{[\bigcup_\alpha R(E_\alpha)]}$$

we can define the function E^* for any plane set S as follows:

$$E^*(S) = \inf\{E(I) \mid S \subset I \in I_\Sigma\}.$$

And so on, in this fashion, we obtain a set function E defined on all Borel sets in the complex plane C (the smallest σ-algebra containing all rectangles) S such that

(a) $E(C) = 1$;
(b) for a disjoint union of Borel sets, $\bigcup_{n=1}^{\infty} S_n$,

$$E\left(\bigcup_{n=1}^{\infty} S_n\right) = \sum_{n=1}^{\infty} E(S_n).$$

(The notion of summability of $\{E(S_n)\}$ here is intended in the sense of Definition 24.1.)

The function E constitutes what is called a *complex spectral measure*. This is a special instance of a general spectral measure—an orthogonal projection-valued, set function E defined on a σ-algebra of subsets of an abstract set X such that $E(X) = 1$ and E is countably additive. The integral representation obtained for A in the previous section can now be rewritten in terms of the spectral measure we have obtained: Letting $\xi = \lambda + i\mu$, we have

$$A = \int_C \xi \, dE(\xi). \tag{28.15}$$

Equation (28.15) implies that, for any $x, y \in X$,

$$(Ax, y) = \int_C \xi \, d(E(\xi)x, y). \tag{28.16}$$

It is to be noted here that the set function $\mu(S) = (E(S)x, y)$ is a *complex* Borel measure and $(E(S)x, x)$ is actually a Borel measure. We now wish to show that it suffices in (28.15), to integrate just over the set $G = \{\xi \mid |\xi| \leq \|A\|\}$; that is,

$$A = \int_C \xi \, dE(\xi) = \int_G \xi \, dE(\xi). \tag{28.17}$$

[Although we shall not prove it here, one need actually only integrate over the set of points, $\sigma(A)$.]

Since

$$\int_C \xi \, dE(\xi) = \int_G \xi \, dE(\xi) + \int_{CG} \xi \, dE(\xi)$$

(where CG denotes complement of G) if we can show E vanishes on every closed rectangle Δ outside G—hence on every Borel set outside G—we shall have established (28.17). To show this, let $y = E(\Delta)x$, where x is arbitrary; in this case

$$\|A\|^2\|y\|^2 \geq \|Ay\|^2 = (Ay, Ay)$$

$$= (A^*Ay, y)$$

$$= \int_C \xi\bar{\xi}\, d(E(\xi)y, y) \qquad \text{[Similar to Exercise 26.4]}$$

$$= \int_\Delta \xi\bar{\xi}\, d(E(\xi)y, y) + \int_{C\Delta} \xi\bar{\xi}\, d(E(\xi)y, y).$$

$$\geq \int_\Delta \xi\bar{\xi}\, d(E(\xi)y, y). \tag{28.18}$$

We now contend that

$$\int_{C\Delta} \xi\bar{\xi}\, d(E(\xi)y, y) = 0.$$

To prove it, we show that, for any $\tilde{\Delta}$ such that $\tilde{\Delta} \cap \Delta = \emptyset$, $E(\tilde{\Delta})y = 0$. Consider

$$E(\tilde{\Delta})y = E(\tilde{\Delta})E(\Delta)x$$

$$= E(\tilde{\Delta} \cap \Delta)x$$

$$= E(\emptyset)x$$

$$= 0.$$

This establishes the contention.

Since Δ and G are compact sets and $\Delta \cap G = \emptyset$, it follows that

$$\inf\{|z - w|\,|\,z \in \Delta, w \in G\} = \varepsilon > 0.$$

Therefore, continuing (28.18),

$$\|A\|^2\|y\|^2 \geq (\|A\| + \varepsilon)^2 \int_\Delta d(E(\xi)y, y)$$

$$= (\|A\| + \varepsilon)^2 \int_C d(E(\xi)y, y)$$

$$= (\|A\| + \varepsilon)^2 (E(C)y, y)$$

$$= (\|A\| + \varepsilon)^2 \|y\|^2,$$

which implies

$$0 = y = E(\Delta)x,$$

which in turn implies

$$E(\Delta) = 0.$$

This proves the validity of (28.17).

In the above discussion we have shown that $E(\Delta)$ vanishes whenever Δ is outside G and has nothing in common with G. In quite similar fashion one can show that if there exists an $\alpha > 0$ such that

$$\|Ay\| \geq \alpha\|y\|$$

for all y, then $E(\Delta) = 0$ for all rectangles Δ lying wholly within the closed circle of radius α. Consider the special kind of normal transformation now—the unitary transformation. Denoting such a transformation by U, we have $\|Ux\| = \|x\|$ for every $x \in X$. Thus, for unitary transformations, an α such as was mentioned above exists, and it follows that the associated spectral measure will vanish on every closed rectangle—hence on every Borel set—within the closed circle of radius 1. (Note that every unitary transformation has norm one.) Loosely speaking, the entire spectral measure associated with U is concentrated on the rim of the unit circle. In this case we need not integrate over the whole circle but only over the circumference. Let C_φ denote the segment of arc on the unit circle: $0 \le \theta \le \varphi$. Denoting the spectral measure generated by U as E, we now define the function

$$
E_U(\varphi) = \begin{cases} 0 & (\varphi \le 0), \\ E(C_\varphi) = \int_{C_\varphi} dE(\xi) & 0 \le \varphi \le 2\pi), \\ 1 & (\varphi \ge 2\pi). \end{cases}
$$

It will now be verified that $E_U(\varphi)$ is continuous from the right. With n a positive integer, we clearly have

$$
\lim_n C_{\varphi + 1/n} = \bigcap_n C_{\varphi + 1/n} = C_\varphi,
$$

which, using the countable additivity of E, implies [in the strong sense]

$$
\lim_n E_U\left(\varphi + \frac{1}{n}\right) = \lim_n E\left(C_{\varphi + 1/n}\right) = E(C_\varphi).
$$

From the monotonicity of E, it now follows that $E_U(\varphi)$ is continuous from the right. We now claim that

$$
U = \int_{C_{2\pi}} \xi\, dE(\xi) = \int_0^{2\pi} e^{i\varphi}\, dE_U(\varphi).
$$

To prove this, consider a partition of $[0, 2\pi]$,

$$
0 = \varphi_0 < \varphi_1 < \cdots < \varphi_n = 2\pi,
$$

let $\varphi_k' \in [\varphi_{k-1}, \varphi_k]$, let $\varepsilon = \max_k |\varphi_k - \varphi_{k-1}|$, and form the difference

$$
U - \sum_{k=1}^n e^{i\varphi_k'}(E_U(\varphi_k) - E_U(\varphi_{k-1}))
$$
$$
= \sum_{k=1}^n \int_{C_{\varphi_k} - C_{\varphi_{k-1}}} (\xi - e^{i\varphi_k'})\, dE(\xi).
$$

Thus, letting $\Gamma_k = C_{\varphi_k} - C_{\varphi_{k-1}}$, for any x,

$$
\alpha_n(x) = \left(\left[U - \sum_{k=1}^n e^{i\varphi_k'}(E_U(\varphi_k) - E_U(\varphi_{k-1}))\right]x, x\right)
$$
$$
= \sum_{k=1}^n \int_{\Gamma_k} (\xi - e^{i\phi_k'})\, d(E(\xi)x, x),
$$

which implies

$$
|\alpha_n(x)| \le \varepsilon \sum_{k=1}^n \int_{\Gamma_k} d\|E(\xi)x\|^2 = \varepsilon \int_{C_{2\pi}} d\|E(\xi)x\|^2.
$$
$$
= \varepsilon\|x\|^2
$$

Consequently

$$\sup_{n}\{|\alpha_n(x)| \mid \|x\| = 1\} \le \varepsilon. \qquad (28.19)$$

Since

$$U - \sum_{k=1}^{n} e^{i\varphi_k{}'}(E_U(\varphi_k) - E_U(\varphi_{k-1}))$$

is normal because U commutes with all $E_U(\varphi)$ and so does $U^* = U^{-1}$, its norm is given by the left half of (28.19); hence we have shown

$$\left\| U - \sum_{k=1}^{n} e^{i\varphi_k{}'}(E_U(\varphi_k)) - E_U(\varphi_{k-1}) \right\| \le \varepsilon,$$

and it follows that

$$U = \int_0^{2\pi} e^{i\varphi} dE_U(\varphi). \qquad (28.20)$$

EXERCISES 28

1. Prove that the projection-valued function $E(\Delta)$ associated with the bounded normal transformation A is, as a function on rectangles, additive and multiplicative.

2. Let E be a projection (orthogonal) valued function defined on $\mathscr{S}$, all Borel sets of the complex plane C. Show that E is a complex spectral measure if and only if
   (a) $E(C) = 1$;
   (b) for arbitrary $x, y \in X$, $\mu(M) = (E(M)x, y)$, $M \in \mathscr{S}$ is countably additive [that is, for $\{M_i\}$, a disjoint sequence of sets in $\mathscr{S}$, $\mu(\bigcup_{i=1}^{\infty} M_i) = \sum_{i=1}^{\infty} \mu(M_i)$].

3. If E is a complex spectral measure, prove that $E(\emptyset) = 0$.

4. A complex spectral measure is called *regular* if $E(N)$, where N is any Borel set of C, is the projection on $\overline{[U_\alpha R(E(M_\alpha))]}$, where $\{M_\alpha\}_{\alpha \in \Lambda}$ is the set of all compact sets of C contained in N. Prove that every complex spectral measure is regular (see Halmos Ref. 2).

5. Prove that a bounded normal operator A is unitary, self-adjoint, or positive if and only if $\sigma(A)$ is contained in the unit circle, the real axis, or the non-negative real axis.

REFERENCES

1. F. Riesz and B. Sz.-Nagy, "Functional Analysis."
2. P. Halmos, "Introduction to Hilbert Space."

CHAPTER 29

Spectral Theorem for Unbounded, Self-Adjoint Operators

We call attention to the fact that when the concept of adjoint was first introduced (Sec. 20.1), we did not require the transformation to be bounded; in fact we did not even require it to be linear.† Thus, since the notion of adjoint is still sensible for unbounded transformations, we shall now consider a spectral theorem for such operators.

The theorem will be arrived at by (loosely) writing the unbounded (self-adjoint) transformation as a limit of *bounded* self-adjoint operators and judiciously using the spectral theorem for these bounded operators. The form of the result for unbounded operators is very similar to the previous result for bounded operators. More specifically, in this case we also get a monotone family of orthogonal projections $\{E(\lambda)\}$, strongly continuous from the right and yielding an integral representation for the original transformation. In the previous situation, however, the family of orthogonal projections was zero to the left of some point on the real line and equal to the identity transformation to the right of some other point. In this case the situation will not be so clear-cut: The *limiting* values of $E(\lambda)$ will be the zero transformation and the identity transformation as $\lambda \to -\infty$ and $\lambda \to \infty$, respectively. Another departure from the previous result caused by omitting the assumption of boundedness is that we cannot retain finite limits of integration in the integral representation for the operator. The main cause of these differences is that in the former instance the spectrum of the transformation was a bounded set of real numbers; in this case, even though the spectrum of an unbounded, self-adjoint operator is purely real too, it is no longer bounded. Two proofs of this spectral theorem will be given. The first is due to Riesz and Lorch and uses heavily Theorem 20.6, the second, due to von Neumann, relies upon the Cayley transform (cf. Theorem 20.12).

Before carrying out this program, we introduce the notion of *permutativity*, a generalization of the notion of commutativity of transformations. It is remarked that the term "permutativity" as used here is not a standard usage.

29.1 Permutativity

The notion of two transformations A and B *commuting*, which we have used so far, has meant that they have a common domain and that

$$ABx = BAx$$

for all x in the domain. In the notation of extensions (recall that B extends A, written $A < B$ or $B > A$, if and only if the domain of A is contained in the domain of B and for every x in the domain of A, $Ax = Bx$). This means that $BA < AB$ and $AB < BA$. We would like to weaken this requirement slightly at this time. As motivation for why a weakened notion is desirable suppose B is a 1 : 1 mapping of (the set) X into X. In this case $B^{-1}B = 1$, but it does not always make sense to

---

† In spite of this, when the notions of Hermitian, symmetric, and self-adjoint transformations were introduced, it was assumed that the underlying mapping was linear. This was done only so that we would not have to bother writing "*linear*, symmetric mapping," and "*linear*, Hermitian mapping" in all our theorems, the linear case being the case of most interest. (Note that a self-adjoint transformation must always be linear.) Since we did not and do not intend to afford ourselves of any of the greater generality available, nothing has been lost by this policy—the case of the linear transformation still remains the most interesting one.

even consider BB^{-1}, because the domain of B^{-1} will not be all of X unless B is an onto mapping. If that were the case, the consideration of BB^{-1} leads to no problems, $BB^{-1} = 1$, and we can say that B commutes with B^{-1}. What we desire is a generalized notion of commutativity that will allow us to say, for any 1 : 1 mapping, that a mapping always "commutes" with its inverse. Tailoring things to meet our needs, we shall define the new notion only for the case when the underlying space is a Hilbert space, and one of the transformations is bounded and defined on the whole space.† In the following definition and discussion we reserve the symbol B, even if it appears with a subscript, for bounded linear transformations defined on a Hilbert space X, whose range is also in X. A (even with a subscript) will denote a linear transformation such that

$$X \supset D_A \xrightarrow{A} X. \tag{29.1}$$

DEFINITION 29.1. With B and A as above, B is said to *permute with A*, written $B \,!\, A$ (and only in this order) if $BA < AB$. (Note that only the bounded transformation can go on the left.) To show that this *generalizes* our old notion of commutativity, several properties that commuting (this term being reserved for the previous notion) transformations possess will be shown to hold for permuting operators.

(1) If B_1 and B_2 permute with A, then $B_1 + B_2$ and $B_1 B_2$ permute with A.

(2) If B permutes with A_1 and with A_2, then B permutes with $A_1 + A_2$ and $A_1 A_2$.

(3) Assuming A to be a 1 : 1 mapping, its inverse A^{-1} can be considered on the range of A, $R(A)$. In this case, $B \,!\, A$ implies $B \,!\, A^{-1}$.

(4) If $B \,!\, A_n$ for $n = 1, 2, \ldots$, then $B \,!\, A$, where‡ $A_n \xrightarrow{s} A$.

(5) If A is a closed transformation, $B_n \xrightarrow{s} B$ and $B_n \,!\, A$ for all n, then $B \,!\, A$.

(6) Assuming $\bar{D}_A = X$, we have that $B \,!\, A$ implies $B^* \,!\, A^*$.

The spirit of proof of (1) is very similar to that in (2); hence only (2) will be proved.

Proof (2). By hypothesis,

$$BA_1 < A_1 B \qquad \text{and} \qquad BA_2 < A_2 B,$$

which implies

$$BA_1 A_2 < A_1 B A_2 < A_1 A_2 B.$$

Therefore,

$$B \,!\, A_1 A_2 .$$

† If B is a bounded linear transformation such that $X \supset D_B \xrightarrow{B} Y$, where X is a Hilbert space and Y is a Banach space, then, by the results of Sec. 17.2, we can extend B, by continuity, to $\bar{D}_B$. Moreover, the extension will be a bounded linear transformation having the same norm as B. Since $\bar{D}_B$ is a closed subspace of a Hilbert space, we have the orthogonal direct sum decomposition of X, $X = \bar{D}_B \oplus \bar{D}_B^{\perp}$. Hence any $z \in X$ can be written uniquely as $z = x + y$, where $x \in \bar{D}_B$ and $y \in \bar{D}_B^{\perp}$. We now define $\hat{B}z$ as a linear transformation such that $\hat{B}z = Bx$. It is evident that $\hat{B}$ is a bounded linear transformation, extending B to the whole space and such that $\|\hat{B}\| = \|B\|$. (Note that we have denoted the extension of B to $\bar{D}_B$ by B, too.) Therefore, for any transformation such as B, there is no loss in generality in assuming that it is defined on the whole space, and we shall make this assumption throughout.

‡ The notion of $A_n \xrightarrow{s} A$, where the A_n ($n = 1, 2, \ldots$) are arbitrary linear transformations is defined in identically the same way as when the transformations are bounded.

Proof (3). If $x \in D_{A-1}$, then

$$Bx = B(AA^{-1}x) = A(BA^{-1}x),$$

which implies $Bx \in R(A) = D_{A-1}$ and $A^{-1}Bx = BA^{-1}x$. Thus

$$BA^{-1} < A^{-1}B.$$

Proof (4). If $x \in D_A = \{x \mid$ there exists N such that $x \in D_{A_n}$ for all $n > N\}$,

$$\lim_n A_n x = Ax.$$

The continuity of B implies that

$$BAx = B(\lim_n A_n x) = \lim_n BA_n x$$

$$= \lim_n A_n Bx$$

$$= ABx,$$

where we note that $Bx \in D_{A_n}$ for all $n > N$.

Proof (5). By the corollary to Theorem 15.3, we note that the strong limit of a sequence of bounded linear transformations is bounded so $B \in L(X, X)$; for any x, $B_n x \to Bx$. Let $x \in D_A$ and consider

$$AB_n x = B_n Ax \to BAx.$$

Since A is closed, this implies

$$Bx \in D_A \qquad \text{and} \qquad ABx = BAx;$$

hence

$$BA < AB.$$

Proof (6). Let $x \in D_A$ and $y \in D_{A^*}$. Since $B \in L(X, X)$, B^* exists and $D_{B^*} = X$ (Theorem 20.7), and it is sensible to consider

$$(Ax, B^*y) = (BAx, y)$$

$$= (ABx, y)$$

$$= (Bx, A^*y)$$

$$= (x, B^*A^*y).$$

Since this is true for any $x \in D_A$,

$$B^*y \in D_{A^*}$$

and

$$A^*B^*y = B^*A^*y,$$

which implies $B^*A^* < A^*B^*$ and proves (6).

Thus one sees that *permuting* transformations do indeed possess many of the familiar properties that *commuting* transformations have. It will now be shown that, if $B = E$, an orthogonal projection, then $E! A$ implies A reduces $M = E(X)$, that is

$$ED_A \subset D_A,$$

$$A(M \cap D_A) \subset M,$$

$$A(M^\perp \cap D_A) \subset M^\perp$$

for clearly since $EA < AE$, then $ED_A \subset D_A$. Of course

$$AE = EA$$

on D_A so, analogous to Sec. 22.1(9), we see that A reduces M. Thus we have established:

THEOREM 29.1. If E is an orthogonal projection on the Hilbert space X and A is as in (29.1), $E \,!\, A$ implies $EA = AE$.

29.2 The Spectral Theorem for Unbounded, Self-Adjoint Operators

Letting X denote a complex Hilbert space and A a linear transformation such that

$$X \supset D_A \overset{A}{\to} X \tag{29.2}$$

where $\bar{D}_A = X$, we obtain a spectral theorem for the case when A is self-adjoint. Before doing this, two facts which were proved for *bounded*, self-adjoint transformations will be shown to hold for arbitrary self-adjoint transformations. The first such consideration is to show that the spectrum of a self-adjoint transformation is real (proved for the bounded case in Theorem 21.6).

THEOREM 29.2. If A, as in (29.2), is self-adjoint, the spectrum of A, $\sigma(A)$, is real; symbolically, $\sigma(A) \subset R$.

Proof. The following statement is equivalent to the theorem: If $\lambda = \alpha + i\beta$ ($\beta \neq 0$), then $\lambda \in \rho(A)$, the resolvent set of A, and this is what will be proved here. If $\lambda \in P\sigma(A)$, there exists $x \neq 0$ such that $Ax = \lambda x$ and, thus,

$$\lambda(x, x) = (\lambda x, x) = (Ax, x) = (x, Ax)$$

$$= (x, \lambda x) = \bar{\lambda}(x, x),$$

which implies $\lambda = \bar{\lambda}$—an impossibility if $\beta \neq 0$; hence $\lambda \notin P\sigma(A)$. In view of this, the inverse of $(A - \lambda)$ must exist on the range of $A - \lambda$, $R(A - \lambda)$. To complete showing the $\lambda \in \rho(A)$ we must show that $\overline{R(A - \lambda)} = X$ and that $(A - \lambda)^{-1}$ is continuous. With respect to the latter, consider

$$y = (A - \lambda)x;$$

$$\|y\|^2 = ((A - \alpha)x - i\beta x, (A - \alpha)x - i\beta x)$$

$$= \|(A - \alpha)x\|^2 - i\beta(x, (A - \alpha)x) + i\beta((A - \alpha)x, x) + |\beta|^2\|x\|$$

$$= \|(A - \alpha)x\|^2 + |\beta|^2\|x\|^2.$$

This implies $|\beta|^2 \|x\|^2 \le \|y\|^2$ or

$$\|x\| \le \frac{1}{|\beta|} \|y\|$$

or

$$\|(A - \lambda)^{-1} y\| \le \frac{1}{|\beta|} \|y\|$$

for any $y \in R(A - \lambda)$; therefore, $(A - \lambda)^{-1}$ is continuous. With regard to showing $\overline{R(A - \lambda)} = X$, consider, denoting the null space of $A - \lambda$ as usual by $N(A - \lambda)$,

$$\overline{R(A - \lambda)}^{\perp} = N((A - \lambda)^*)$$

(cf. Eq. (20.37), the proof for A unbounded being almost identical to the bounded case). This implies $\overline{R(A - \lambda)} = N((A - \lambda)^*)^{\perp}$

$$= N(A^* - \bar{\lambda})^{\perp}$$

$$= N(A - \bar{\lambda})^{\perp}.$$

Since Im $\bar{\lambda} = -\beta \ne 0$, it follows from first observation that $(A - \bar{\lambda})^{-1}$ exists, which implies

$$N(A - \bar{\lambda}) = \{0\};$$

hence

$$\overline{R(A - \lambda)} = X,$$

and the proof is complete.

For bounded self-adjoint operators, we showed that the spectrum was closed, bounded, and real. For unbounded self-adjoint operators, the spectrum is closed (see Exercise 29.3), and real but not necessarily bounded.

Previously (Theorem 21.7), it was shown that the residual spectrum of a bounded normal transformation was empty, which implied that the residual spectrum of any bounded self-adjoint transformation was empty. (The use of boundedness in that theorem lay in being able to make the assumption that the transformation was defined on the whole space.) We now prove that the residual spectrum of any self-adjoint transformation is empty.

THEOREM 29.3. With A as in (29.2) assumed to be self-adjoint, the residual spectrum of A is empty.

Proof. Suppose

$$\lambda \in R\sigma(A) \subset \sigma(A) \subset R.$$

Since $\lambda = \bar{\lambda}$, and $\lambda \in R\sigma(A)$ implies $\overline{R(A - \lambda)} \ne X$,

$$N(A^* - \bar{\lambda}) = N(A^* - \lambda)$$

$$= N(A - \lambda)$$

$$= \overline{R(A - \lambda)}^{\perp} \ne X^{\perp} = \{0\}.$$

The fact that $N(A - \lambda) \neq \{0\}$ is equivalent to

$$\lambda \in P\sigma(A).$$

Since $R\sigma(A) \cap P\sigma(A) = \emptyset$, the assumption that $\lambda \in R\sigma(A)$ must be false; hence

$$R\sigma(A) = \emptyset.$$

In our next theorem we show that, given an orthogonal family of closed subspaces and a sequence of self-adjoint transformations defined on those subspaces, there exists a self-adjoint transformation that reduces to these transformations on each of the given closed subspaces. In Theorem 29.5 it is shown that the converse holds; actually a little more than the converse holds, for it is shown that, given a self-adjoint transformation, there exists a complete orthogonal family of subspaces and a sequence of *bounded*, self-adjoint transformations defined on each of the subspaces that the original reduces to on those subspaces. Once these two results have been obtained, we obtain the spectral theorem.

THEOREM 29.4. Let $M_1, M_2, \ldots$ be a *complete orthogonal family* of (closed) subspaces of the Hilbert space X; that is,

$$\left[\bigcup_{i=1}^{\infty} M_i \right] = \sum_i \oplus M_i = X.$$

Suppose, in addition, that $A_1, A_2, \ldots$ is a sequence of linear transformations such that the restriction of A_i to M_i, $A_i | M_i$, is a self-adjoint transformation mapping M_i into M_i. Then there exists a unique, self-adjoint transformation A such that

$$A|_{M_i} = A_i|_{M_i}.$$

Proof. Since $X = \sum_i \oplus M_i$, then for any $x \in X$ there exist unique $x_i \in M_i$ such that

$$x = \sum_i x_i$$

In this notation then, it will be shown that the transformation A, whose domain D_A is

$$\left\{ x \left| \sum_i \| A_i x_i \|^2 < \infty \right. \right\}$$

defined by†

$$Ax = \sum_i A_i x_i,$$

satisfies the conditions of the theorem.

Consider the set

$$S = [\cup_i M_i]$$

consisting of all *finite* sums of vectors chosen from the M_i. Since S is clearly contained in D_A,

$$\bar{S} \subset \bar{D}_A.$$

---

† Recall that, in a Hilbert space, $\sum_i y_i$ converges (y_i orthogonal) if and only if $\sum_i \|y_i\|^2 < \infty$; see Theorem 9.5.

But $\bar{S}$ must also contain $\sum_i \oplus M_i = X$, so it follows that

$$\bar{D}_A = X. \tag{29.3}$$

It is clear that A is linear and also, if $x_j \in M_j$, that $Ax_j = A_j x_j$. In view of (29.3), to show that A is a symmetric transformation, it suffices to show that A is Hermitian. To this end let x and y belong to D_A. As noted before there exist x_i and y_i belonging to M_i such that $x = \sum_i x_i$ and $y = \sum_i y_i$. We now have

$$(Ax, y) = (\sum_i A_i x_i, y) = \sum_i (A_i x_i, y)$$

$$= \sum_i \sum_j (A_i x_i, y_j).$$

Since $(A_i x_i, y_j) = 0$ if $i \neq j$ because M_i is invariant under A_i and the family $\{M_i\}$ is orthogonal, the last term can be replaced by

$$\sum_i (A_i x_i, y_i) = \sum_i (x_i, A_i y_i) = (x, Ay),$$

which proves, by using (29.3), that A is symmetric; that is, $A < A^*$. We wish to show that A is self-adjoint, so it remains to prove that $A^* < A$, which means that y in D_{A^*} is also in D_A and $A^* y = Ay$. For this purpose, let $y = \sum_i y_i \in D_{A^*}$ and $x = \sum_i x_i \in D_A$ and consider

$$(Ax, y) = (x, A^* y)$$

or

$$\sum_i (A_i x_i, y) = \sum_i (A_i x_i, y_i) = \sum_i (x_i, A^* y).$$

Writing $A^* y$ as $\sum_i (A^* y)_i$, the last term becomes $\sum_i (x_i, (A^* y)_i)$. Taking $x = x_j$, we get

$$(A_j x_j, y_j) = (x_j, (A^* y)_j),$$

and, since A_j is self-adjoint on M_j,

$$A_j y_j = (A^* y)_j.$$

Now we have

$$A^* y = \sum_j (A^* y)_j = \sum_j A_j y_j = Ay.$$

Note that this tacitly implies $y \in D_A$ because $\sum_j \|A_j y_j\|^2 = \sum_j \|(A^* y)_j\|^2 < \infty$ and we have proved $A^* < A$; hence A is self-adjoint.

It will now be shown that the transformation A is the only such transformation with these properties. Suppose there exists some self-adjoint transformation B such that B is defined on each M_i and that

$$B|_{M_i} = A_i|_{M_i}.$$

We contend that

$$\left\{ x = \sum_i x_i \left| \sum_i \|Bx_i\|^2 < \infty \right. \right\}$$

is contained in the domain of B, D_B.

To prove this contention consider a vector, $x = \sum_i x_i$, from this set. In this case we have

$$x_1 + x_2 + \cdots + x_n \to x$$

and

$$Bx_1 + Bx_2 + \cdots + Bx_n \to y.$$

Since each $x_i \in D_B$ and the domain of B is a subspace, $\sum_{i=1}^n x_i \in D_B$, and the latter convergence can be written

$$B(x_1 + x_2 + \cdots + x_n) \to y.$$

Since $B = B^*$ is closed, we must have

$$x \in D_B \quad \text{and} \quad y = Bx;$$

hence

$$Bx = \sum_i Bx_i,$$

and we have proved the contention. Since, for $x_i \in M_i$, $Bx_i = A_i x_i$,

$$\left\{ x = \sum_i x_i \,\middle|\, \sum_i \|Bx_i\|^2 < \infty \right\} = \left\{ x = \sum_i x_i \,\middle|\, \sum_i \|A_i x_i\|^2 < \infty \right\} = D_A.$$

Consequently $D_A \subset D_B$; moreover, for any $x = \sum_i x_i \in D_A$,

$$Ax = \sum_i A_i x_i = \sum_i Bx_i = Bx,$$

which implies

$$A < B.$$

Taking adjoints, we have $B^* = B < A = A^*$, which proves $A = B$. The theorem is now proved.

In the notation of the above theorem, note that if each A_i is bounded and there is some real number M such that $\|A_i\| \le M$ for every i, then for any $x = \sum_i x_i \in D_A$,

$$\|Ax\|^2 = \sum_i \|A_i x_i\|^2 \le \sum_i \|A_i\|^2 \|x_i\|^2 \le M^2 \|x\|^2,$$

which implies that A is bounded and $\|A\| \le M$. In view of the footnote at the beginning of Sec. 29.1, it follows that we can view A as being defined on the whole space in this case.

We now prove a result slightly stronger than the converse of Theorem 29.4.

THEOREM 29.5. If A, as in (29.2), is self-adjoint, there exists an orthogonal family of closed subspaces $\{M_i\}$ ($i = 1, 2, \ldots$) such that

$$X = \sum_i \oplus M_i \tag{29.4}$$

and a family of linear transformations

$$A_i : M_i \to M_i,$$

where each A_i is a bounded, self-adjoint operator on M_i such that $A|_{M_i} = A_i$.

Proof. It has been shown (Theorem 20.6) that for a self-adjoint operator such as A, the transformations $B = (1 + A^2)^{-1}$ and $C = AB$ exist and are defined on all of X; B is positive (therefore self-adjoint, since it is defined on the whole space), and B and C are continuous with norm less than or equal to one. For a bounded self-adjoint operator such as B,

$$\sigma(B) \subset [m, M],$$

where $m = \inf_{\|x\|=1}(Bx, x)$ and $M = \sup_{\|x\|=1}(Bx, x)$ (Theorem 25.1). Since B is positive and continuous with norm less than or equal to one,

$$m \geq 0 \quad \text{and} \quad M \leq 1.$$

By the spectral theorem for bounded self-adjoint operators (Theorem 25.3), we have a resolution of the identity $\{F(\lambda)\}$ associated with B. From the estimates on m and M above, it follows that

$$F(\lambda) = 0 \quad (\text{for } \lambda < 0) \quad \text{and} \quad F(\lambda) = 1 \quad (\text{for } \lambda \geq 1).$$

Since $\{F(\lambda)\}$ is a resolution of the identity, it must be strongly continuous from the right everywhere, and the only places it will not also be continuous from the left [that is, where $F(\lambda^-) \neq F(\lambda)$] are at points λ belonging to $P\sigma(B)$, the point spectrum of B (Theorem 27.3(1)). Since B^{-1} exists, $0 \notin P\sigma(B)$, and we must have

$$F(0^-) = F(0).$$

We shall use this fact shortly. To prove the theorem we must get an orthogonal family of closed subspaces satisfying (29.4). To this end, consider

$$H_n = F\left(\frac{1}{n}\right) - F\left(\frac{1}{n+1}\right);$$

since $1/n > 1/(n+1)$ and the family $\{F(\lambda)\}$ is a monotone family of orthogonal projections, it follows that H_n is an orthogonal projection. Consider now the closed subspace $R(H_n)$, the range of H_n, and let $M_n = R(H_n)$. We claim that $\{M_n\}$ is the required family of subspaces. Since, using the monotonicity of the family $\{F(\lambda)\}$, it is easily verified that $\{H_n\}$ is an *orthogonal* family of orthogonal projections, it follows that $\{M_n\}$ is an orthogonal family of closed subspaces. This latter fact implies (Theorem 24.4) that $\{H_n\}$ is summable to the orthogonal projection on $\overline{[\cup_n M_n]}$. Letting "lims" denote the *strong* limit of a sequence of bounded linear transformations, knowing that $\sum_n H_n$ exists, we write

$$\sum_n H_n = \sum_n \left[F\left(\frac{1}{n}\right) - F\left(\frac{1}{n+1}\right)\right]$$

$$= \text{lims} \sum_{k=1}^{n} \left[F\left(\frac{1}{k}\right) - F\left(\frac{1}{k+1}\right)\right]$$

$$= \text{lims}_n \left[F(1) - F\left(\frac{1}{n+1}\right)\right]$$

$$= F(1) - F(0^+) = F(1) - F(0).$$

As noted previously $F(0^-) = 0 = F(0)$; therefore,

$$\sum_n H_n = 1.$$

This implies that any $x \in X$ can be written

$$x = \sum_n H_n x.$$

Since $H_n x \in M_n$, then, letting $H_n x = x_n$, we have shown that any $x \in X$ can be written as $x = \sum_n x_n$, where $x_n \in M_n$, and we have thus shown that $\sum_n \oplus M_n = X$.

With this out of the way, we now show that:

(1) M_n is invariant under A for every n;
(2) $A|_{M_n}$ is a bounded, self-adjoint transformation.

Thus, we shall be able to take the transformations A_n mentioned in the statement of the theorem to be $A|_{M_n}$. Furthermore, if the following two conditions can be shown, (1) and (2) will follow† :

(1') $H_n \,!\, A$ for all n;
(2') AH_n is defined on all of X and is bounded.

Using Theorem 29.1, (1') and (2') clearly imply (1). To see that (2') implies (2), note first that $A|_{M_n}$ is trivially self-adjoint on M_n once we have shown (1), because A is self-adjoint; on the other hand, since for $x \in M_n$, $x = H_n x$, $Ax = AH_n x$. Therefore, $A|_{M_n}$ will certainly be bounded on M_n if AH_n is bounded on all of X. We now demonstrate the validity of (1') and (2').

Clearly,

$$A(1 + A^2) = (1 + A^2)A,$$

which implies

$$BA(1 + A^2)B = B(1 + A^2)AB.$$

But

$$(1 + A^2)B = 1 \qquad \text{and} \qquad B(1 + A^2) < 1,$$

so the previous relation implies

$$BA < AB$$

or $B \,!\, A$. Now consider

$$BC = BAB < ABB = CB;$$

hence $B \,!\, C$. Since B and C are each defined on all of X, this implies $BC = CB$, so we also have $C \,!\, B$. Since $B \,!\, C$, any polynomial in B permutes with C and any strong limit of polynomials in B will therefore‡ permute with C. According to the proof of the spectral theorem for bounded self-adjoint operators given in Chapter 26, there exists a sequence of polynomials in B that converge strongly to $F(\lambda)$. By virtue of the preceding remarks, then, $F(\lambda) \,!\, C$, which implies $H_n \,!\, C$.

Consider the function (see Fig. 29.1)

$$\psi_n(\lambda) = \begin{cases} \dfrac{1}{\lambda} & \left(\dfrac{1}{n+1} < \lambda \le \dfrac{1}{n} \right), \\[2mm] 0 & \text{(elsewhere)}. \end{cases}$$

In Chapter 26 a mapping taking $f \in C_2$, differences of upper semicontinuous

---

† The meaning of $H \,!\, A$ is defined and discussed in Sec. 29.1.
‡ Section 29.1, (5); note that C is closed because it is bounded and defined on the whole space.

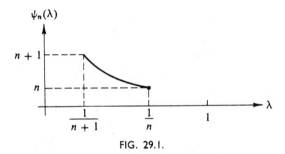

FIG. 29.1.

functions, into $f(\hat{A}) \in L(X, X)$, where $\hat{A} \in L(X, X)$ is a fixed, self-adjoint transformation is defined. Thus if we can show that ψ_n can be written as the difference of upper semicontinuous functions, we can consider the image of ψ_n under the above map with $\hat{A} = B$. Letting

$$w_n(\lambda) = \begin{cases} \dfrac{1}{\lambda} & \left(\dfrac{1}{n+1} \le \lambda \le \dfrac{1}{n}\right), \\ 0 & \text{(elsewhere)}, \end{cases}$$

and

$$\hat{w}_n(\lambda) = \begin{cases} n+1 & \left(\lambda = \dfrac{1}{n+1}\right), \\ 0 & \text{(elsewhere)}, \end{cases}$$

it is clear that w_n and $\hat{w}_n$ are each upper semicontinuous and that $\psi_n = w_n - \hat{w}_n$. We obtain the image of $\psi_n(\lambda)$ under the above map, $\psi_n(B)$, as the strong limit of a certain sequence of polynomials in B. Since this is the case $\psi_n(B) \, ! \, B$ and $\psi_n(B) \, ! \, C$. By direct computation, we have

$$\lambda\psi_n(\lambda) = \psi_n(\lambda)\lambda = \begin{cases} 1 & \left(\dfrac{1}{n+1} < \lambda \le \dfrac{1}{n}\right), \\ 0 & \text{(elsewhere)}. \end{cases}$$

Letting $k_\mu(\lambda)$ denote the same function it did in Chapter 26, we can write

$$\lambda\psi_n(\lambda) = k_{1/n}(\lambda) - k_{1/(n+1)}(\lambda),$$

which implies

$$B\psi_n(B) = \psi_n(B)B = k_{1/n}(B) - k_{1/(n+1)}(B)$$
$$= F\left(\frac{1}{n}\right) - F\left(\frac{1}{n+1}\right)$$
$$= H_n.$$

Hence

$$AH_n = AB\psi_n(B) = C\psi_n(B).$$

Since C and $\psi_n(B)$ are each defined everywhere and are bounded, (2′) is established. To show (1′), we note that

$$H_n A = \psi_n(B)BA < \psi_n(B)AB = \psi_n(B)C,$$

which implies

$$H_n A < \psi_n(B)C = C\psi_n(B) = AB\psi_n(B) = AH_n$$

and proves that $H_n \,!\, A$. The proof is now complete.

By virtue of this result, with any self-adjoint transformation A, as in (29.2), we associate an orthogonal family of closed subspaces $\{M_n\}$ and a sequence of bounded, self-adjoint transformations $A_n : M_n \to M_n$, where $A_n = A|M_n$. Since, for any n, A_n is a *bounded* self-adjoint transformation on M_n, there exists an associated resolution of the identity $\{E_n(\lambda)\}$, by virtue of the spectral theorem for bounded self-adjoint operators. For each fixed value of λ, the sequence $E_1(\lambda)$, $E_2(\lambda)$, ... satisfies the conditions of Theorem 29.4; therefore, by that theorem, there exists a self-adjoint transformation $E(\lambda)$ such that

$$E_n(\lambda) = E(\lambda)|_{M_n},$$

the domain of $E(\lambda)$ being given by

$$D_{E(\lambda)} = \left\{ x = \sum_i x_i \,\Big|\, \sum_i \|E_i(\lambda)x_i\|^2 < \infty \right\};$$

$$E(\lambda)x = \sum_i E_i(\lambda)x_i.$$

Since, for any i and any λ, $\|E_i(\lambda)\| \le 1$, we have for any x

$$\sum_i \|E_i(\lambda)x_i\|^2 \le \sum_i \|x_i\|^2 = \|x\|^2$$

which implies that $D_{E(\lambda)} = X$ and that $E(\lambda)$ is continuous with norm less than or equal to one. For any λ we claim that $E(\lambda)$ is idempotent; this, coupled with the fact that $E(\lambda)$ is self-adjoint, implies that $\{E(\lambda)\}$ is a family of orthogonal projections. We now prove its idempotency: For $x = \sum_i x_i$ consider

$$E(\lambda)E(\lambda)x = E(\lambda)\sum_i E_i(\lambda)x_i.$$

Since $E(\lambda)$ is continuous, and $E_i(\lambda)x_i \in M_i$, this is equal to

$$\sum_i E(\lambda)E_i(\lambda)x_i = \sum_i E_i^2(\lambda)x_i$$

$$= E(\lambda)x$$

and proves the idempotency of $E(\lambda)$. Things are beginning to take shape now in that, associated with A, we have shown the existence of a certain family of orthogonal projections. Three additional properties that the family $\{E(\lambda)\}$ possesses are:

(1) for $\lambda_1 \le \lambda_2$, $E(\lambda_1) \le E(\lambda_2)$ (monotonicity);
(2) for $\varepsilon > 0$, $E(\lambda + \varepsilon) \xrightarrow{s} E(\lambda)$ as $\varepsilon \to 0$ (strong continuity from the right);
(3) $E(\lambda) \xrightarrow{s} 0$ as $\lambda \to -\infty$; $E(\lambda) \xrightarrow{s} 1$ as $\lambda \to \infty$.

Properties (1) and (2) are identical to (1) and (2) of Theorem 25.3, whereas (3) is only similar. It is in (3) that the fact A need not be a *bounded* linear transformation manifests itself.

DEFINITION 29.2. A family of orthogonal projections $\{F(\lambda)\}$ satisfying (1)–(3) is called a *generalized resolution of the identity*.

Proof that $\{E(\lambda)\}$ *satisfies* (1): Suppose $\lambda_1 \leq \lambda_2$ and consider, using the monotonicity of each of the families $\{E_n(\lambda)\}$ and the continuity of the inner product,

$$(E(\lambda_1)x, x) = \sum_i (E_i(\lambda_1)x_i, x_i)$$

$$\leq \sum_i (E_i(\lambda_2)x_i, x_i) = (E(\lambda_2)x, x).$$

Proof that $\{E(\lambda)\}$ *satisfies* (2): The convergence of $\sum_i E_i(\lambda + \varepsilon)x_i$ is equivalent to the convergence of the series of real numbers, $\sum_i \|E_i(\lambda + \varepsilon)x_i\|^2$. Since

$$\|E_i(\lambda + \varepsilon)x_i\|^2 \leq \|x_i\|^2$$

for every i, and $\sum_i \|x_i\|^2 < \infty$, it follows by the Weierstrass M-test that $\sum_i \|E_i(\lambda + \varepsilon)x_i\|^2$ converges uniformly, that is, in a manner independent of ε. Hence $\sum_i E_i(\lambda + \varepsilon)x_i$ converges uniformly too.

For sequences of real-valued functions the following result is valid (Ref. 4, pp. 119–120).

THEOREM. Suppose $f_n \to f$ uniformly on E. Let x be a limit point of E and suppose that

$$\lim_{t \to x} f_n(t) = A_n \qquad (n = 1, 2, \ldots).$$

Then $\{A_n\}$ converges and $\lim_{t \to x} f(t) = \lim_n A_n$. In other words,

$$\lim_{t \to x} \lim_n f_n(t) = \lim_n \lim_{t \to x} f_n(t).$$

By merely replacing absolute values by norms, it is straightforward to extend this result to the case of vector-valued functions—in particular to the sequence of partial sums of $\sum_i E_i(\lambda + \varepsilon)x_i$. For $\varepsilon > 0$, we now have, since each $E_i(\lambda)$ is strongly continuous from the right,

$$\lim_{\varepsilon \to 0} E(\lambda + \varepsilon)x = \lim_{\varepsilon \to 0} \sum_i E_i(\lambda + \varepsilon)x_i$$

$$= \sum_i \lim_{\varepsilon \to 0} E_i(\lambda + \varepsilon)x_i$$

$$= \sum_i E_i(\lambda)x_i = E(\lambda)x.$$

Proof that $\{E(\lambda)\}$ *satisfies* (3): Let $x = \sum_i x_i$ be an arbitrary vector and consider, interchanging limit and summation as above,

$$\lim_{\lambda \to -\infty} E(\lambda)x = \lim_{\lambda \to -\infty} \sum_i E_i(\lambda)x_i$$

$$= \sum_i \lim_{\lambda \to -\infty} E_i(\lambda)x_i = 0.$$

Similar reasoning proves the second statement in (3).

To finish off the spectral theorem for unbounded self-adjoint operators, we would like to show that $E(\lambda)$ permutes (instead of commutes) with A and get an integral representation for A in terms of $E(\lambda)$. Since A need not be bounded, we cannot say that its spectrum is wholly contained in some closed interval on the real axis and, as a result, shall not be able to put finite limits on the integral for A. What we shall do is get the following representation for A:

$$Ax = \int_{-\infty}^{\infty} \lambda \, dE(\lambda)x \qquad D_A = \left\{ x \left| \int_{-\infty}^{\infty} \lambda^2 \, d\|E(\lambda)x\|^2 < \infty \right. \right\};$$

but first we must give meaning to $\int_{-\infty}^{\infty} \lambda \, dE(\lambda)$.

The Meaning of $\int_{-\infty}^{\infty} \lambda \, dE(\lambda)$.

Let $\{E(\lambda)\}$ be an *arbitrary* generalized resolution of the identity, and consider the orthogonal family of orthogonal projections

$$H_n = E(n) - E(n - 1) \qquad (n = 0, \pm 1, \pm 2, \ldots).$$

Let the range of H_n, $R(H_n)$, be denoted by $\hat{M}_n$. Since $\{H_n\}$ is an orthogonal family, $\{\hat{M}_n\}$ is an orthogonal family of closed subspaces. If it can be shown that any $x \in X$ can be written as $x = \sum_i x_i$, where $x_i \in \hat{M}_i$, it will follow that $X = \sum_i \oplus \hat{M}_i$. Since

$$(E(i) - E(i - 1))x \in \hat{M}_i$$

and

$$\lim_n \sum_{i=-n}^{n} (E(i) - E(i - 1))x = \lim_n (E(n)x - E(-n - 1)x)$$

$$= x,$$

we do indeed have this representation for any $x \in X$; therefore $X = \sum_i \oplus \hat{M}_i$. Let f be a real-valued, continuous function, defined on the closed interval $[a, b]$ and let x and y be arbitrary vectors. Using the polarization identity for sesquilinear functionals [Eq. (20.48)] and the monotonicity of $\{E(\lambda)\}$, it is easily verified that the real and imaginary parts of $(E(\lambda)x, y)$ can be written as differences of monotone functions (of λ), which implies that $(E(\lambda)x, y)$ is of bounded variation on $[a, b]$.[†] In view of this there is no problem about the existence of the Riemann–Stieltjes integral

$$L_f\langle x, y \rangle = \int_a^b f(\lambda) \, d(E(\lambda)x, y).$$

Moreover, since $E(\lambda)$ is self-adjoint, $(E(\lambda)x, y)$ is a symmetric sesquilinear functional; hence, since f is real-valued, it can readily be verified that $L_f\langle x, y \rangle$ is a symmetric sesquilinear functional on X.

There must be some real number M such that $|f(t)| \leq M$ for all $t \in [a, b]$, because f is a continuous function defined on a compact set. Hence

$$|L_f\langle x, x \rangle| = \left| \int_a^b f(\lambda) \, d\|E(\lambda)x\|^2 \right| \leq M(\|E(b)x\|^2 - \|E(a)x\|^2)$$

$$\leq M\|x\|^2.$$

[†] An almost identical thing was done in the proof of Theorem 25.3.

Thus the quadratic form associated with $L_f\langle x, y\rangle$ is bounded which implies (Theorem 20.15) that $L_f\langle x, y\rangle$ is a *bounded*, symmetric sesquilinear functional. As such there must be some $\hat{A} \in L(X, X)$ such that (Theorem 21.1)

$$L_f\langle x, y\rangle = (\hat{A}x, y). \tag{29.5}$$

Since $L_f\langle x, y\rangle$ is symmetric, $\hat{A}$ must be self-adjoint (Theorems 20.13, 20.14). Now, exactly as in the proof of Theorem 25.4, it can be shown that (29.5) implies

$$\hat{A} = \int_a^b f(\lambda)\, dE(\lambda).$$

[The only modification that needs to be made is that where, at the end of that proof, we wrote $E(\lambda_n) = 1$, we have here $E(\lambda_n) \le 1$; similarly, instead of having $E(\lambda_0) = 0$ we have $E(\lambda_0) \ge 0$.]

By these remarks we have established the existence of the self-adjoint operator $I_n \in L(X, X)$ defined by

$$I_n = \int_{n-1}^n \lambda\, dE(\lambda).$$

CONTENTION I. $\hat{M}_n$ is invariant under I_n.

Proof. Letting $n - 1 = \lambda_0 < \lambda_1 < \cdots < \lambda_k = n$ be a partition of $[n-1, n]$, it follows from the existence of the integral I_n that

$$I_n = \lim_{\substack{k\to\infty \\ \max_i|\lambda_i - \lambda_{i-1}|\to 0}} \sum_{i=1}^k \lambda_i(E(\lambda_i) - E(\lambda_{i-1}))$$

(the convergence here is meant to be in the uniform sense). From the monotonicity of $\{E(\lambda)\}$ we have that each $[E(\lambda_i) - E(\lambda_{i-1})]$ is an orthogonal projection and

$$E(\lambda_i) - E(\lambda_{i-1}) \le E(n) - E(n - 1),$$

which implies

$$R\big([E(\lambda_i) - E(\lambda_{i-1})]\big) \subset R\big([E(n) - E(n - 1)]\big) = \hat{M}_n.$$

Hence, for every k,

$$\sum_{i=1}^k \lambda_i[E(\lambda_i) - E(\lambda_{i-1})]x \in \hat{M}_n.$$

Recalling that $\hat{M}_n$ is a *closed* subspace, it follows that $I_n x \in \hat{M}_n$, for any x and establishes the contention.

The sequence of transformations $\{I_n\}$, and family of subspaces $\{\hat{M}_n\}$, therefore, satisfies the hypothesis of Theorem 29.4, and it follows, by that theorem, that there exists a self-adjoint transformation I such that

$$I|_{\hat{M}_n} = I_n|_{\hat{M}_n}.$$

It is this transformation I that we shall take as $\int_{-\infty}^\infty \lambda\, dE(\lambda)$; that is, we *define*

$$I = \int_{-\infty}^\infty \lambda\, dE(\lambda).$$

Letting $x_n = [E(n) - E(n-1)]x$ and appealing to Theorem 29.4 again, it follows that the domain of I, D_I, is given by

$$D_I = \left\{ x = \sum_n x_n \,\middle|\, \sum_{n=-\infty}^{\infty} \|I_n x_n\|^2 < \infty \right\}.$$

Recalling that improper Riemann–Stieltjes integrals involving real-valued functions have a customary interpretation, we now state and prove contention 2.

CONTENTION 2. $D_I = \{x \mid \int_{-\infty}^{\infty} \lambda^2 d\|E(\lambda)x\|^2 < \infty\}$.

Proof. In Chapter 26, when A was a bounded, self-adjoint operator, we obtained the relation

$$A = \int_a^b \lambda \, dE(\lambda).$$

Using the monotonicity of the family $\{E(\lambda)\}$ obtained there, it was a simple, indeed trivial, matter to show that the above relation implied the validity of

$$A^k = \int_a^b \lambda^k \, dE(\lambda).$$

Hence, in exactly the same fashion, since $\{E(\lambda)\}$ is a monotone family in this situation too,

$$I_n = \int_{n-1}^n \lambda \, dE(\lambda)$$

implies

$$I_n^{\ k} = \int_{n-1}^n \lambda^k \, dE(\lambda)$$

for any k; in particular, it is true for $k = 2$ and, hence, for any y:

$$(I_n^2 y, y) = \int_{n-1}^n \lambda^2 \, d(E(\lambda)y, y).$$

Letting $x_n = [E(n) - E(n-1)]x$, then, we have

$$\|I_n x_n\|^2 = (I_n x_n, I_n x_n)$$

$$= (I_n^2 x_n, x_n)$$

$$= \int_{n-1}^n \lambda^2 \, d\|E(\lambda)x_n\|^2.$$

Since, for any $\lambda \in [n-1, n]$,

$$E(\lambda)x_n = E(\lambda)[E(n) - E(n-1)]x$$

$$= E(\lambda)x - E(n-1)x, \tag{29.6}$$

it follows that

$$\|E(\lambda)x_n\|^2 = \|E(\lambda)x\|^2 - \|E(n-1)x\|^2$$

(i.e. we can use x instead of x_n); therefore,

$$\|I_n x_n\|^2 = \int_{n-1}^{n} \lambda^2 \, d\|E(\lambda)x\|^2, \tag{29.7}$$

which establishes the contention.

By Theorem 29.4, for $x \in D_I$,

$$Ix = \sum_{n=-\infty}^{\infty} I_n x_n$$

$$= \sum_{n=-\infty}^{\infty} \int_{n-1}^{n} \lambda \, dE(\lambda)x_n.$$

Using (29.6) [noting that $E(n-1)x$ is constant], x_n can be replaced by x to yield

$$Ix = \sum_{n=-\infty}^{\infty} \int_{n-1}^{n} \lambda \, dE(\lambda)x$$

or

$$\int_{-\infty}^{\infty} \lambda \, dE(\lambda)x = \sum_{n=-\infty}^{\infty} \int_{n-1}^{n} \lambda \, dE(\lambda)x. \tag{29.8}$$

CONTENTION 3. If $x \in D_I$, then $E(\mu)x \in D_I$ for any μ.

Proof. To show that $E(\mu)x \in D_I$, we must show that

$$\int_{-\infty}^{\infty} \lambda^2 \, d\|E(\lambda)E(\mu)x\|^2 < \infty.$$

Since $x \in D_I$, it follows that

$$\int_{-\infty}^{\infty} \lambda^2 \, d\|E(\lambda)x\|^2 = \int_{-\infty}^{\mu} \lambda^2 \, d\|E(\lambda)x\|^2 + \int_{\mu}^{\infty} \lambda^2 \, d\|E(\lambda)x\|^2 \tag{29.9}$$

is finite. For all $\lambda \geq \mu, E(\lambda)E(\mu)x = E(\mu)x$ is constant; for all $\lambda \leq \mu, E(\lambda)E(\mu) = E(\lambda)$; hence, using (29.9)

$$\int_{-\infty}^{\infty} \lambda^2 \, d\|E(\lambda)E(\mu)x\|^2 = \int_{-\infty}^{\mu} \lambda^2 \, d\|E(\lambda)x\|^2 < \infty.$$

Contention 3 has now been established.

CONTENTION 4. For any μ, $E(\mu) ! I$. (Note that I need not be defined on the whole space.)

Proof. In the same notation as before, for any x such that $E(\mu)x \in D_I$,

$$IE(\mu)x = \sum_{n=-\infty}^{\infty} I_n[E(\mu)x]_n. \tag{29.10}$$

We claim that

$$[E(\mu)x]_n = E(\mu)x_n, \tag{29.11}$$

where, as usual, $x_n = [E(n) - E(n-1)]x$. This is shown by observing that if

$$x = \sum_i x_i \qquad (x_i \in \hat{M}_i),$$

then, by the continuity of $E(\mu)$,

$$E(\mu)x = \sum_i E(\mu)x_i. \tag{29.12}$$

Since $E(\mu)$ must commute with $E(\lambda)$ for any μ and λ (by the monotonicity of $\{E(\lambda)\}$), it follows that $E(\mu)$ commutes with the orthogonal projection on $\hat{M}_i$, $E(i) - E(i-1)$. Therefore, $\hat{M}_i$ is invariant under $E(\mu)$, and, in the above expression, $E(\mu)x_i \in \hat{M}_i$; that is, in (29.12) we have succeeded in writing $E(\mu)x$ as an infinite series of terms, from $\{\hat{M}_i\}$. Since any $y \in X$ can be written *uniquely* as $y = \sum_i y_i$, where y_i is a (unique) vector in $\hat{M}_i$, it follows that (29.11) is correct.

Since I_n is defined as the uniform limit of a sequence of linear combinations of $[E(\lambda_i) - E(\lambda_{i-1})]$ and $E(\mu)$ commutes with each term in the sequence, it follows that $E(\mu)$ commutes with I_n.

By the preceding contention then, this and Eq. (29.11) in (29.10), imply (for $x \in D_I$)

$$IE(\mu)x = \sum_{n=-\infty}^{\infty} I_n E(\mu)x_n$$

$$= E(\mu) \sum_{n=-\infty}^{\infty} I_n x_n$$

$$= E(\mu)Ix,$$

which implies

$$E(\mu) \,!\, I,$$

and proves Contention 4.

We note that any sequence $\{\lambda_n\}$ $(n = 0, \pm 1, \pm 2, \ldots)$ such that $\lambda_{+n} \to \infty$ and $\lambda_{-n} \to -\infty$ would serve just as well as the sequence $\{n\}$; that is, it would yield the same value for I.

At this point we conclude our discussion of an arbitrary generalized resolution of the identity and return to the case at hand—namely, the generalized resolution of the identity associated with the self-adjoint operator A.

THEOREM 29.6. Consider the linear transformation $A : X \supset D_A \xrightarrow{A} X$, where X is a complex Hilbert space and $\bar{D}_A = X$. If A is self-adjoint, there exists a family of orthogonal projections $\{E(\lambda)\}$ such that:

(1) $\lambda_1 \leq \lambda_2$ implies $E(\lambda_1) \leq E(\lambda_2)$. Note that this is equivalent to $E(\lambda_1)E(\lambda_2) = E(\lambda_2)E(\lambda_1) = E(\lambda_1)$, for $\lambda_1 \leq \lambda_2$;

(2) for $\varepsilon > 0$, $E(\lambda + \varepsilon) \xrightarrow{s} E(\lambda)$ as $\varepsilon \to 0$;

(3) $E(\lambda) \xrightarrow{s} 0$ as $\lambda \to -\infty$; $E(\lambda) \xrightarrow{s} 1$ as $\lambda \to +\infty$;

(4) $E(\lambda) \,!\, A$; if $T \in L(X, X)$ and $T \,!\, A$, then $T \,!\, E(\lambda)$ and $E(\lambda) \,!\, T$;

(5) interpreting the improper integral below in terms of the previous discussion,

$$A = \int_{-\infty}^{\infty} \lambda \, dE(\lambda),$$

and we can write the domain of A as

$$D_A = \left\{ x \,\middle|\, \int_{-\infty}^{\infty} \lambda^2 \, d\|E(\lambda)x\|^2 < \infty \right\};$$

(6) if $\{F(\lambda)\}$ is a generalized resolution of the identity [properties (1)–(3)] such that

$$A = \int_{-\infty}^{\infty} \lambda \, dF(\lambda),$$

then

$$E(\lambda) = F(\lambda) \qquad \text{for all } \lambda.$$

Proof. In the discussion following Theorem 29.5, we exhibited a family of orthogonal projections $\{E(\lambda)\}$ associated with A satisfying conditions (1)–(3). We now show that $\{E(\lambda)\}$ satisfies the last three requirements. To show (4) it must be shown that, if T is any bounded linear transformation that permutes with A, then $T \, ! \, E(\lambda)$. Just as we used Theorem 29.5 and certain facts from its proof to obtain the family $\{E(\lambda)\}$, we return to that setting and notation to prove $T \, ! \, E(\lambda)$. Since $T \, ! \, A$, $T \, ! \, (1 + A^2)^{-1} = B$ (Sec. 29.1, (3)). Since $T \, ! \, B$, T must also permute with any polynomial in B. The transformation $\psi_n(B)$, as introduced in Theorem 29.5, is a strong limit of polynomials in B; hence it follows that† $T \, ! \, \psi_n(B)$; consequently, $T \, ! \, B\psi_n(B) = H_n$. Since H_n is the orthogonal projection on M_n, we have proved that M_n is invariant under T (note, too, that if two *permuting* transformations are defined on the whole space, they *commute*).

Viewing things on the space M_n now, we certainly have that T permutes with the bounded self-adjoint operator $A|_{M_n}$:

$$T \, ! \, A|_{M_n}.$$

Since the resolution of the identity associated with $A|_{M_n}$, $\{E_n(\lambda)\}$, is a strong limit of polynomials in $A|_{M_n}$,

$$T \, ! \, E_n(\lambda).$$

But $E(\lambda)x = \sum_n E_n(\lambda)x_n$, where $x_n = H_n x$ [that is, $E(\lambda) = \sum_n E_n(\lambda)H_n$], so

$$T \, ! \, E(\lambda). \tag{29.13}$$

Since T and $E(\lambda)$ are each bounded and defined on the whole space, this implies that

$$E(\lambda)' \, ! \, T. \tag{29.14}$$

This proves half of (4); the fact that $E(\lambda) \, ! \, A$ will follow once (5) has been established by invoking Contention 4 of the preceding discussion.

In the sense of the discussion preceding this theorem, we consider the self-adjoint transformation

$$I = \int_{-\infty}^{\infty} \lambda \, dE(\lambda).$$

We would now like to show that $I = A$. To do this, it suffices to show that

$$I|_{M_n} = A|_{M_n},$$

because, by Theorem 29.4, there can be only one self-adjoint transformation with this property.

† Sec. 29.1, (4).

Consider a particular subspace M_i and the resolution of the identity $\{E_i(\lambda)\}$ associated with $A|_{M_i}$. There must be real numbers a and b such that $E_i(\lambda) = 0$ for $\lambda < a$ and $E_i(\lambda) = 1$ for $\lambda \geq b$. Recalling that

$$E(\lambda)|_{M_i} = E_i(\lambda),$$

we can now say, for $x \in M_i$

$$E(\lambda)x = \begin{cases} 0 & (\lambda < a), \\ x & (\lambda \geq b). \end{cases}$$

Consequently, for such x,

$$\int_{-\infty}^{\infty} \lambda^2 \, d\|E(\lambda)x\|^2 = \int_a^b \lambda^2 \, d\|E(\lambda)x\|^2 < \infty.$$

Since (Contention 2)

$$D_I = \left\{ x \, \middle| \, \int_{-\infty}^{\infty} \lambda^2 \, d\|E(\lambda)x\|^2 < \infty \right\},$$

this implies that

$$M_i \subset D_I,$$

and, as usual,

$$Ix = \int_{-\infty}^{\infty} \lambda \, dE(\lambda)x$$

$$= \int_{-\infty}^{\infty} \lambda \, dE_i(\lambda)x$$

$$= \int_{a^-}^{b} \lambda \, dE_i(\lambda)x$$

$$= (A|_{M_i})x = Ax.$$

Thus

$$A|_{M_i} = I|_{M_i},$$

and it follows that

$$A = I = \int_{-\infty}^{\infty} \lambda \, dE(\lambda), \tag{29.15}$$

which proves (5).

One notes that the only differences between the family obtained here and the family of orthogonal projections $\{E(\lambda)\}$ obtained in the bounded case is that in the latter case there exist real numbers a and b such that $E(\lambda) = 0$ for $\lambda < a$ and $E(\lambda) = 1$ for $\lambda \geq b$; in the former instance we have only the limiting statements. We show now, before proving uniqueness, that if A is bounded, the family obtained by this theorem also has this property; that is, the generalized resolution of the identity reduces to a resolution of the identity if A is bounded. To this end suppose $A \in L(X, X)$ is self-adjoint and consider

$$m = \inf_{\|x\|=1} (Ax, x) \quad \text{and} \quad M = \sup_{\|x\|=1} (Ax, x).$$

Let x be an arbitrary vector. Consider $a < m$, and let $\{F(\lambda)\}$ be the associated generalized resolution of the identity and consider

$$y = F(a)x.$$

In this case

$$F(\lambda)y = \begin{cases} F(\lambda)x & (\lambda < a), \\ y & (\lambda = a), \\ F(a)x & (\lambda > a), \end{cases}$$

and†

$$(Ay, y) = \int_{-\infty}^{\infty} \lambda \, d\|F(\lambda)y\|^2 \tag{29.16}$$

$$= \int_{-\infty}^{a} \lambda \, d\|F(\lambda)y\|^2$$

$$\leq a\|F(a)y\|^2 = a(y, y).$$

But, since (for $y \neq 0$)

$$(Ay, y) \geq m(y, y) > a(y, y),$$

y must be zero; that is,

$$F(a)x = 0$$

for any x, which implies

$$F(a) = 0.$$

That $F(\lambda) = 1$ for $\lambda \geq M$ follows similarly, and we have shown that nothing new or different is obtained if $A \in L(X, X)$.

Having proved this, we are now in a better position to deal with the uniqueness mentioned in part (6). Once again let A be the transformation of the theorem, let $\{E(\lambda)\}$ be as above, and let $\{F(\lambda)\}$ be as in (6). We first observe that since $\{F(\lambda)\}$ is a generalized resolution of the identity, as we have defined it,

$$I_F = \int_{-\infty}^{\infty} \lambda \, dF(\lambda) = A$$

makes sense; moreover, from the way in which I_F was defined, $F(\lambda)$ must permute with $I_F = A$. Since

$$F(\lambda) \, ! \, A$$

we have, by part (4),

$$F(\lambda) \, ! \, E(\lambda),$$

which implies $\hat{M}_n = R(E(n) - E(n - 1))$ is invariant under $F(\lambda)$ as well as $E(\lambda)$. Define

$$F_n(\lambda) = F(\lambda)|_{\hat{M}_n} \quad \text{and} \quad E_n(\lambda) = E(\lambda)|_{\hat{M}_n}.$$

Since

$$A = \int_{-\infty}^{\infty} \lambda \, dE(\lambda) = \int_{-\infty}^{\infty} \lambda \, dF(\lambda),$$

then certainly

$$A|_{\hat{M}_n} = \int_{-\infty}^{\infty} \lambda \, dE_n(\lambda) = \int_{-\infty}^{\infty} \lambda \, dF_n(\lambda).$$

† Equality (29.16) follows readily from $A = \int_{-\infty}^{\infty} \lambda \, dF(\lambda)$. See also Exercise 26.2.

For $x \in \hat{M}_n$, however,

$$E_n(\lambda)x = E_n(\lambda)[E(n) - E(n-1)]x = \begin{cases} 0 & (\lambda \leq n-1), \\ [E(\lambda) - E(n-1)]x & (n-1 < \lambda \leq n), \\ [E(n) - E(n-1)]x & (\lambda > n), \end{cases}$$

which implies

$$A|_{\hat{M}_n} = \int_{-\infty}^{\infty} \lambda \, dF_n(\lambda) = \int_{n-1}^{n} \lambda \, dE_n(\lambda).$$

In discussing the meaning of $\int_{-\infty}^{\infty} \lambda dE(\lambda)$, terms such as $\int_{n-1}^{n} \lambda \, dE_n(\lambda)$ appeared [Eq. (29.5)ff], and it was noted at that point that they were *bounded* self-adjoint operators [specifically, self-adjoint transformations belonging to $L(\hat{M}_n, \hat{M}_n)$ in this case], it being clear that $\{E_n(\lambda)\}$ is a generalized resolution of the identity. Thus $\{E_n(\lambda)\}$ and $\{F_n(\lambda)\}$ are generalized resolutions of the identity for the *bounded* self-adjoint operator $A|_{\hat{M}_n}$. Since $A|_{\hat{M}_n}$ is bounded then, by what we have just shown, $\{E_n(\lambda)\}$ and $\{F_n(\lambda)\}$ reduce to just resolutions of the identity. By the theorem for bounded self-adjoint operators, however, we know there is only one such associated resolution of the identity; hence

$$E_n(\lambda) = F_n(\lambda) \qquad \text{or} \qquad E(\lambda)|_{\hat{M}n} = F(\lambda)|_{\hat{M}n}.$$

Finally, since $F(\lambda)$ and $E(\lambda)$ are self-adjoint and $\hat{M}_n$ is invariant under each of them, we can apply Theorem 29.4 to assert that

$$E(\lambda) = F(\lambda).$$

This (at last!) completes the proof.

29.3 A Proof of the Spectral Theorem Using the Cayley Transform

Once again we assume that X is a complex Hilbert space, that A is defined on a dense subspace of X and that A is self-adjoint. For such transformations, it has been shown† that $(A \pm i)$ is onto X and that

$$U = (A - i)(A + i)^{-1} \tag{29.17}$$

exists (on all of X) and is unitary; furthermore (on the domain of A),

$$A = i(1 + U)(1 - U)^{-1}. \tag{29.18}$$

The operator U is referred to as the *Cayley transform of A*. Using the already proven spectral theorem for unitary operators‡, we shall give an alternate proof of Theorem 29.6—the spectral theorem for unbounded self-adjoint operators. We adhere to the notation of the spectral theorem for unitary operators throughout (Sec. 28.2). In this notation, then, the unitary operator U can be written as

$$U = \int_0^{2\pi} e^{i\varphi} d\, F(\varphi), \tag{29.19}$$

† Sec. 20.3, Theorem 20.12.
‡ Eq. (28.20).

where, among other things, $\{F(\varphi)\}$ is a monotone family of orthogonal projections, strongly continuous from the right such that

$$F(\varphi) = \begin{cases} 0 & (\varphi \leq 0), \\ 1 & (\varphi \geq 2\pi). \end{cases} \tag{29.20}$$

Since $F(\varphi) = 0$ for $\varphi \leq 0$ and $F(0) = F(0^+)$,

$$F(0^+) = 0 = F(0^-), \tag{29.21}$$

which proves

$$F(\varphi) \text{ is continuous at } 0.$$

We now contend that

$$F(\varphi) \text{ is continuous at } \varphi = 2\pi.$$

Proof. Since $F(2\pi) = F(2\pi^+)$, we need only show $F(2\pi^-) = F(2\pi)$. Suppose

$$F(2\pi) - F(2\pi^-) \neq 0.$$

If so, there must be some x such that

$$[F(2\pi) - F(2\pi^-)]x = y \neq 0.$$

From the monotonicity of $\{F(\varphi)\}$, it follows that

$$F(\varphi)y = \begin{cases} 0 & (\varphi < 2\pi), \\ y & (\varphi \geq 2\pi). \end{cases} \tag{29.22}$$

It is next observed that

$$U - 1 = \int_0^{2\pi} (e^{i\varphi} - 1)\, dF(\varphi)$$

and

$$U^* - 1 = \int_0^{2\pi} (e^{-i\varphi} - 1)\, dF(\varphi).$$

[This latter fact follows from consideration of the limiting sums leading to U in (29.19).] From the monotonicity of $\{F(\varphi)\}$ and resulting pairwise orthogonality of the family $\{F(\varphi_n) - F(\varphi_{n-1})\}$, where $\{\varphi_n\}$ is any partition of $[0, 2\pi]$, it can be shown (Exercise 26.4) that

$$(U^* - 1)(U - 1) = \int_0^{2\pi} (e^{-i\varphi} - 1)(e^{i\varphi} - 1)\, dF(\varphi)$$

$$= \int_0^{2\pi} |e^{i\varphi} - 1|^2 dF(\varphi),$$

which implies,† for any $z \in X$,

$$((U^* - 1)(U - 1)z, z) = \int_0^{2\pi} |e^{i\varphi} - 1|^2\, d\|F(\varphi)z\|^2.$$

† The proof for Exercise 26.2 will work here, too.

Thus, with y as before,

$$\|(U - 1)y\|^2 = ((U - 1)y, (U - 1)y)$$
$$= ((U - 1)^*(U - 1)y, y)$$
$$= \int_0^{2\pi} |e^{i\varphi} - 1|^2 \, d\|F(\varphi)y\|^2$$
$$= 0,$$

the last equality following from (29.22). Since $y \neq 0$, this states that $1 \in P\sigma(U)$, which is contradictory because $(1 - U)^{-1}$ exists. Therefore $F(\varphi)$ is continuous at $\varphi = 2\pi$.

Choose a sequence of real numbers $\varphi_n \in (0, 2\pi)$ $(n = 0, \pm 1, \pm 2, \ldots)$ such that (see Figs. 29.2 and 29.3)

$$-\cot\left(\frac{\varphi_n}{2}\right) = n,$$

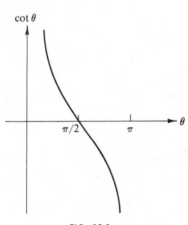

FIG. 29.2.

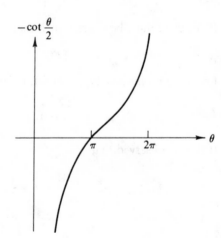

FIG. 29.3.

and let the orthogonal projection $F(\varphi_n) - F(\varphi_{n-1}) = T_n$. Since U commutes with $F(\varphi)$, U commutes with T_n; since $A = i(1 + U)(1 - U)^{-1}$, this implies that $T_n \, ! \, A$. Therefore the range of T_n is invariant under U and under A.

The family of orthogonal projections $\{T_n\}$ is summable to the orthogonal projection on $\sum_n \oplus R(T_n)$; moreover, using the continuity of $F(\varphi)$ at $\varphi = 0$ and $\varphi = 2\pi$,

$$\sum_n T_n = \sum_n (F(\varphi_n) - F(\varphi_{n-1}))$$
$$= \lim_{\varphi \to 2\pi} F(\varphi) - \lim_{\varphi \to 0} F(\varphi)$$
$$= 1 - 0 = 1.$$

Hence,

$$\sum_n \oplus R(T_n) = X.$$

Before continuing the proof per se, a slight digression is necessary. For an arbitrary unitary transformation U to each trigonometric polynomial with complex coefficients

$$p(e^{i\varphi}) = \sum_{k=-n}^{n} c_k e^{ik\varphi},$$

we assign the transformation

$$p(U) = \sum_{k=-n}^{n} c_k U^k.$$

We state without proof that the correspondence so obtained is linear, multiplicative and such that the transformation corresponding to

$$\overline{p(e^{i\varphi})} = \sum_{k=-n}^{n} \bar{c}_k e^{-ik\varphi}$$

is $p(U)^*$, the adjoint of $p(U)$. If $p(e^{i\varphi})$ is real-valued, $p(U)$ is self-adjoint, and if

$$p(e^{i\varphi}) \geq 0 \quad \text{then} \quad p(U) \geq 0.$$

Note how very similar this situation is to the one that prevailed in Chapter 26, where a similar mapping was discussed. A significant difference is that the role played by polynomials $p(\lambda)$ is now played by $p(e^{i\varphi})$, this is significant because, in the previous situation, we were able to extend the mapping to continuous functions by using the fact that the polynomials were dense in the class of continuous functions (dense with respect to the supremum norm, that is). It can also be shown (Ref. 3, p. 111) that the class of trigonometric polynomials is dense (supremum norm) in the class of continuous functions on $[0, 2\pi]$. Following in strict analogy with Chapter 26, we can give meaning to $f(U)$, where f is a continuous function on $[0, 2\pi]$ and prove that (Ref. 1, p. 283)

$$f(U) = \int_0^{2\pi} f(e^{i\varphi}) \, dF(\varphi) \tag{29.23}$$

and the corresponding weak result,

$$(f(U)x, y) = \int_0^{2\pi} f(e^{i\varphi}) \, d(F(\varphi)x, y). \tag{29.24}$$

Returning to the proof again, let $x = T_n x$ belong to the range of T_n. In this case

$$F(\varphi)T_n x = \begin{cases} 0 & (\varphi < \varphi_{n-1}), \\ (F(\varphi) - F(\varphi_{n-1}))x & (\varphi_{n-1} \leq \varphi \leq \varphi_n), \\ (F(\varphi_n) - F(\varphi_{n-1}))x & (\varphi > \varphi_n). \end{cases}$$

Hence,

$$(1 - U)x = \int_0^{2\pi} (1 - e^{i\varphi}) \, dF(\varphi)x$$

$$= \int_{\varphi_{n-1}}^{\varphi_n} (1 - e^{i\varphi}) \, dF(\varphi)x.$$

By Exercise 26.4, then

$$\int_{\varphi_{n-1}}^{\varphi_n} (1 - e^{i\varphi})^{-1} \, dF(\varphi) \int_{\varphi_{n-1}}^{\varphi_n} (1 - e^{i\varphi}) \, dF(\varphi)x$$

$$= \int_{\varphi_{n-1}}^{\varphi_n} (1 - e^{i\varphi})^{-1}(1 - e^{i\varphi}) \, dF(\varphi)x$$

$$= \int_{\varphi_{n-1}}^{\varphi_n} dF(\varphi)x = \int_0^{2\pi} dF(\varphi)x = x.$$

Since $\int_{\varphi_{n-1}}^{\varphi_n}(1 - e^{i\varphi})^{-1} \, dF(\varphi)$ commutes with $\int_{\varphi_{n-1}}^{\varphi_n}(1 - e^{i\varphi}) \, dF(\varphi)$, it follows that

$$\int_{\varphi_{n-1}}^{\varphi_n} (1 - e^{i\varphi})^{-1} \, dF(\varphi) = ((1 - U)|_{R(Tn)})^{-1}$$

$$= (1 - U)^{-1}|_{R(T_n)}.$$

With x as above, then,

$$Ax = \int_{\varphi_{n-1}}^{\varphi_n} i(1 + e^{i\varphi})(1 - e^{i\varphi})^{-1} \, dF(\varphi)x.$$

Letting

$$\lambda = -\cot\frac{\varphi}{2} \quad \text{and} \quad E(\lambda) = F(-2 \cot^{-1} \lambda),$$

we have

$$Ax = \int_{n-1}^{n} \lambda \, dE(\lambda)x, \tag{29.25}$$

and it is simple to verify that $E(\lambda)$ is a generalized resolution of the identity such that

$$R(E(n) - E(n - 1)) = R(F(\varphi_n) - F(\varphi_{n-1})).$$

Since this is so, the discussion of the preceding section [concerning the meaning of $\int_{-\infty}^{\infty} \lambda \, dE(\lambda)$] is applicable the expression in (29.25) being called $I_n x$ in the notation of that section. Therefore, taking the improper integral in the same sense as in the preceding section,

$$Ax = \int_{-\infty}^{\infty} \lambda \, dE(\lambda)x.$$

29.4 A Note on the Spectral Theorem for Unbounded Normal Operators

Let X be a Hilbert space and A a closed linear transformation such that

$$X \supset D \xrightarrow{A} X,$$

where $\bar{D}_A = X$. A is called *normal* if $AA^* = A^*A$. It can be shown that if A is a normal operator, there exists a complex spectral measure† E such that

$$D_A = \left\{ x \,\middle|\, \lim_n \int_{|\xi| \le n} \xi \, dE(\xi)x \text{ exists} \right\}$$

† See Sec. 28.2 for the definition of complex spectral measure.

and, for $x \in D_A$,

$$Ax = \lim_n \int_{|\xi| \le n} \xi \, dE(\xi)x.$$

This can be shown by first proving that an operator A is normal if and only if $A = A_1 + iA_2$, where A_1 and A_2 are self-adjoint with generalized resolutions of the identity $\{E_1(\lambda)\}$ and $\{E_2(\lambda)\}$, where $E_1(S_1)E_2(S_2) = E_2(S_2)E_1(S_1)$ for arbitrary Borel sets of R, where the generalized resolution of identity has been defined on Borel subsets of R analogous to the construction carried out in Sec. 28.1. One then defines a complex spectral measure E such that $E(S) = E_1(S_1)E_2(S_2)$ for all rectangles $S = S_1 \times S_2$, where S_1 and S_2 are Borel sets of R. This can be achieved by defining a finite complex measure μ on the σ-algebra of all finite, disjoint unions of rectangles by taking $\mu(S_1 \times S_2) = (E_1(S_1)E_2(S_2)x, y)$ and extending μ to all Borel sets of C in the usual measure-theoretic fashion. This complex measure is written $\mu(S; x, y)$; it is a symmetric sesquilinear functional to which corresponds a bounded operator $E(S)$. One then shows that E has the desired properties. We leave the details to the reader.

We shall return to spectral considerations in Volume 2 from a different point of view and will there go into the concept of spectral measures in much greater detail. An analogous construction for *bounded* normal operators is carried out in Halmos, Ref. 9. See also Sz.-Nagy, Ref. 8, for some other related considerations.

EXERCISES 29

1. Let X be a Hilbert space and $X \supset D_A \overset{A}{\to} X$ be such that $\bar{D}_A = X$. Show that, if A is self-adjoint, then:
   (a) If $\lambda \in \rho(A)$, then $R(A - \lambda) = X$.
   (b) $C\sigma(A) \subset \{\lambda | R(A - \lambda) \ne \overline{R(A - \lambda)}\}$; this latter set is called the *essential spectrum* of A and is denoted by $E\sigma(A)$. Show that $E\sigma(A) \subset \sigma(A)$.
   (c) $P\sigma(A) \supset \{\lambda | \overline{R(A - \lambda)} \ne X\}$.

2. Under the same hypothesis as 1, the transformation $R_\lambda = (\lambda - A)^{-1}$ is called the *resolvent operator* of A and is defined for all λ for which $(\lambda - A)^{-1}$ exists and for which $\overline{R(A - \lambda)} = X$. Prove, assuming A is self-adjoint, the following:

   (a) $\|R_\lambda x\| \le \dfrac{1}{|\beta|} \|x\|$, where $\beta = \operatorname{Im} \lambda \ne 0$.

   (b) $R_{\lambda_2} - R_{\lambda_1} = (\lambda_1 - \lambda_2) R_{\lambda_1} R_{\lambda_2}$, where $\lambda_1, \lambda_2 \in \rho(A)$.

   (c) $R_\lambda{}^* = R_{\bar\lambda}$.

3. Under the same hypothesis as 1, show that if A is self-adjoint, then $\sigma(A)$ is a closed set.

4. Suppose that X is a Hilbert space, let $X \supset D_A \overset{A}{\rightarrow} X$ and let A be unbounded. A is called *normal* if A is closed, and $\bar{D}_A = X$, and $AA^* = A^*A$. Show that if A is normal, then A^* is normal.

5. Suppose that X is a Hilbert space, let $X \supset D_A \overset{A}{\rightarrow} X$, $\bar{D}_A = X$; also let $U = (A - i)(A + i)^{-1}$ if it exists and call it the *Cayley transform* of A. Show that if U is unitary and A is symmetric, then A is self-adjoint. Show that if A is symmetric, U is isometric.

6. Using the same hypothesis and notation as in 5, let A be a closed symmetric operator. Show that

$$D_U^{\perp} = \{x | A^*x = ix\}$$

and

$$R(U)^{\perp} = \{x | A^*x = -ix\}.$$

These subspaces are called the *deficiency subspaces* of A, and their dimensions, the *deficiency indices* of A. Show that A is self-adjoint if and only if its deficiency indices are both zero.

7. The integral of Theorem 29.6 can be put in a different form using some techniques from measure theory. In particular if, for A as in Theorem 29.6, $\sigma(A) \subset [a,b]$, we can write $A = \int_a^b \lambda \, dE(\lambda)$. Using this, show that if A (self-adjoint) has a bounded spectrum, then A is bounded.

REFERENCES

1. F. Riesz and B. Sz.-Nagy, "Functional Analysis."
2. E. Lorch, "Spectral Theory."
3. I. P. Natanson, "Theory of Functions of a Real Variable," Vol. 1.
4. W. Rudin, "Mathematical Analysis."
5. N. Dunford and J. Schwartz, "Linear Operators. Part 2: Spectral Theory."
6. M. Naimark, "Normed Rings."
7. M. Stone, "Linear Transformations in Hilbert Space."

In these references some alternate approaches to the material of this chapter are available.

8. B. Sz.-Nagy, "Spektraldarstellungen Linearer Transformationen des Hilbertschen Raumes."
9. P. Halmos "Introduction to Hilbert Space."

Bibliography

G. Bachman, "Elements of Abstract Harmonic Analysis," Academic Press, New York, 1964.

G. Bachman, "Introduction to *p*-adic Numbers and Valuation Theory," Academic Press, New York, 1964.

S. Banach, "Opérations Linéaires," Hafner, New York, 1932.

S. Berberian, "Introduction to Hilbert Space," Oxford Univ. Press, London and New York, 1961.

S. Berberian, "The Numerical Range of a Normal Operator," *Duke Math J.*, 31(3), 479-85, September (1964).

R. Boas, "A Primer of Real Functions," Wiley, New York, 1961.

N. Bourbaki, "Topologie Generale," Hermann, Paris, 1951.

R. Buck, "Advanced Calculus," McGraw-Hill, New York, 1956.

D. Bushaw, "Elements of General Topology," Wiley, New York, 1963.

J. Dieudonné, "Foundations of Modern Analysis," Academic Press, New York, 1960.

N. Dunford and J. Schwartz, "Linear Operators. Part I: General Theory," Wiley (Interscience), New York, 1958.

D. Hall and G. Spencer, "Elementary Topology," Wiley, New York, 1955.

P. Halmos, "Finite-Dimensional Vector Spaces," Van Nostrand, Princeton, N.J., 1958.

————, "Introduction to Hilbert Space," Chelsea, New York, 1951.

————, "Measure Theory," Van Nostrand, Princeton, N.J., 1950.

E. Hille and R. Philips, "Functional Analysis and Semigroups," Am. Math. Soc., Providence, R.I., 1957.

K. Hoffman and R. Kunze, "Linear Algebra," Prentice-Hall, Englewood Cliffs, N.J., 1961.

N. Jacobson, "Lectures in Abstract Algebra," Vol. II, Van Nostrand, Princeton, N.J., 1953.

J. Kelley, "General Topology," Van Nostrand, Princeton, N.J., 1955.

K. Knopp, "Theory of Functions," Vol. I, Dover, New York, 1945.

A. Kolmogoroff and S. Fomin, "Elements of the Theory of Functions and Functional Analysis," Vol. I, Graylock Press, Rochester, N.Y., 1957.

L. Liusternik and W. Sobolev, "Elemente Der Funktionalanalysis," Akademie-Verlag, Berlin, 1955.

E. Lorch, "Spectral Theory," Oxford Univ. Press, London and New York, 1962.

M. Naimark, "Normed Rings," Noordhoff, Groningen, 1959.

I. P. Natanson, "Theory of Functions of a Real Variable," Ungar, New York, 1955.

C. Putnam, "On the Spectra of Commutators," *Proc. Am. Math. Soc.*, 5(6), 929-931 (1954).

C. Rickart, "General Theory of Banach Algebras," Van Nostrand, Princeton, N.J., 1960.

F. Riesz and B. Sz.-Nagy, "Functional Analysis," Ungar, New York, 1955.

H. Rubin and J. Rubin, "Equivalents of the Axiom of Choice," North Holland, Amsterdam, 1963.

W. Rudin, "Principles of Mathematical Analysis," McGraw-Hill, New York, 1953.

I. Singer and J. Wermer, "Derivation on Commutative Normed Algebras," *Math. Ann.*, **129**, 260-264 (1955).

M. Stone, "Hausdorff's Theorem Concerning Hermitian Forms," *Bull. Am. Math. Soc.*, 36(4), 259-262 (1930).

M. Stone, "Linear Transformations in Hilbert Space," Am. Math. Soc., Providence, R.I., 1932.

B. v. Sz.-Nagy, "Spektraldarstellungen Linearer Transformationen des Hilbertschen Raumes," Ungar, New York, 1955.

A. Taylor, "Introduction to Functional Analysis," Wiley, New York, 1958.

F. Valentine, "Convex Sets," McGraw-Hill, New York, 1964.

R. Vaidyanathaswamy, "Set Topology," Chelsea, New York, 1960.

D. V. Widder, "The Laplace Transform," Princeton Univ. Press, Princeton, N.J., 1946.

A. Zaanen, "Linear Analysis," Wiley (Interscience), New York, 1952.

————, "An Introduction to the Theory of Integration," Interscience, New York, 1958.

Index of Symbols

Index